Feature Papers in Electronic Materials Section (Volume 2)—15th Anniversary of *Materials*

Feature Papers in Electronic Materials Section (Volume 2)—15th Anniversary of *Materials*

Guest Editor

Fabrizio Roccaforte

MDPI

Basel • Beijing • Wuhan • Barcelona • Belgrade • Novi Sad • Cluj • Manchester

Guest Editor
Fabrizio Roccaforte
Istituto per la Microelettronica
e Microsistemi (IMM)
Consiglio Nazionale delle Ricerche (CNR)
Catania
Italy

Editorial Office
MDPI AG
Grosspeteranlage 5
4052 Basel, Switzerland

This is a reprint of the Special Issue, published open access by the journal *Materials* (ISSN 1996-1944), freely accessible at: www.mdpi.com/journal/materials/special_issues/1OTU796FIK.

For citation purposes, cite each article independently as indicated on the article page online and using the guide below:

Lastname, A.A.; Lastname, B.B. Article Title. *Journal Name* **Year**, *Volume Number*, Page Range.

ISBN 978-3-7258-2798-5 (Hbk)
ISBN 978-3-7258-2797-8 (PDF)
https://doi.org/10.3390/books978-3-7258-2797-8

Contents

About the Editor

Fabrizio Roccaforte

He received his M.Sc. in Physics from the University of Catania (Italy) in 1996 and his PhD from the University of Göttingen (Germany) in 1999. Then, he was a visiting scientist at the University of Göttingen and scientific consultant at STMicroelectronics (Italy). In December 2001, he joined the permanent staff of CNR-IMM in Catania as a researcher, and he became a senior researcher in 2007 and research director in 2020. At CNR-IMM, he is the team leader of the "Power and High-frequency devices group".

His research interests are mainly focused in the field of wide-bandgap (WBG) semiconductors (e.g., SiC, GaN, Ga_2O_3...) and material and device processing for power electronics devices. In particular, he has well-recognized experience in metal/semiconductor and dielectric/semiconductor interfaces on WBG semiconductors.

He is the co-author of more than 370 papers (h-index 45, Scopus) in international journals and proceedings, 10 book chapters, 15 US patents and/or applications, and he has given several invited talks and lectures on SiC and GaN at international conferences.

He has been the chairperson of the conferences HeteroSiC-WASMPE2009, WOCSDICE2011, ICSCRM2015, and WOCSDICE-EXMATEC2023. Currently, he is a member of the Steering Committee of the conferences ICSCRM and EXMATEC and is the section editor-in-chief of the journal *Materials* (MDPI).

He has been responsible for the CNR-IMM research unit in several European and national projects, bilateral collaborations with other European institutions, and industrial research contracts.

Preface

Energy saving, climate neutrality, and sustainability are among the greatest challenges of our society whilst completing the transition to renewable energy sources. In this context, advanced electronic materials play a critical and enabling role in the green and digital transition. Electronic materials include semiconductors for digital power and high-frequency electronics, advanced functional materials for detectors and sensors, conductive and insulating coatings, flexible substrates, etc. All these materials are employed for the electronic components used in computers, wireless systems, solid-state lighting devices, sensors and detectors, wearable electronic devices, telecommunication systems, power devices for energy conversion, etc.

Clearly, the continuous development of electronic device technologies requires significant efforts from the scientific community, which is devoted to the full comprehension of the fundamental properties of these materials and of the related device physics.

This reprint, entitled "*Feature Paper in Electronic Materials Section (Volume 2)*", is a collection of selected regular and review papers on recent advances in electronic materials and devices in different fields (e.g. power and high-frequency electronics, optoelectronic devices, sensors, etc.), published in the MDPI journal *Materials*.

In particular, the first part of the volume collects a series of papers dedicated to the most popular wide-bandgap semiconductors (WBG), i.e., silicon carbide (SiC) and gallium nitride (GaN). Both fundamental and applicative issues of these materials are discussed, such as the mechanisms of the bulk growth of GaN, irradiation effects in SiC, novel processing for SiC devices (e.g. laser annealing), and advanced characterization techniques of materials and interfaces.

The second part of the reprint includes a miscellaneous collection of selected papers in the field of electronic materials such as InGaAsN/GaAs quantum wells and graphene and oxide materials (e.g. Cu_2O thin films or TiO_2 nanoparticles). The latter are discussed focusing on their applications as holes transporting materials for perovskite solar cells or in specific gas sensing technologies.

Fabrizio Roccaforte

Guest Editor

materials MDPI

Review

Ammonothermal Crystal Growth of Functional Nitrides for Semiconductor Devices: Status and Potential

Thomas Wostatek [1], V. Y. M. Rajesh Chirala [1], Nathan Stoddard [2], Ege N. Civas [1], Siddha Pimputkar [2,*] and Saskia Schimmel [1,*]

1 Friedrich-Alexander-Universität Erlangen-Nürnberg, Chair of Electron Devices (LEB), Cauerstraße 6, 91058 Erlangen, Germany
2 Department of Materials Science and Engineering, Lehigh University, 5 E Packer Avenue, Bethlehem, PA 18015, USA
* Correspondence: sip516@lehigh.edu (S.P.); saskia.schimmel@fau.de (S.S.)

Abstract: The state-of-the-art ammonothermal method for the growth of nitrides is reviewed here, with an emphasis on binary and ternary nitrides beyond GaN. A wide range of relevant aspects are covered, from fundamental autoclave technology, to reactivity and solubility of elements, to synthesized crystalline nitride materials and their properties. Initially, the potential of emerging and novel nitrides is discussed, motivating their synthesis in single crystal form. This is followed by a summary of our current understanding of the reactivity/solubility of species and the state-of-the-art single crystal synthesis for GaN, AlN, AlGaN, BN, InN, and, more generally, ternary and higher order nitrides. Investigation of the synthesized materials is presented, with a focus on point defects (impurities, native defects including hydrogenated vacancies) based on GaN and potential pathways for their mitigation or circumvention for achieving a wide range of controllable functional and structural material properties. Lastly, recent developments in autoclave technology are reviewed, based on GaN, with a focus on advances in development of in situ technologies, including in situ temperature measurements, optical absorption via UV/Vis spectroscopy, imaging of the solution and crystals via optical (visible, X-ray), along with use of X-ray computed tomography and diffraction. While time intensive to develop, these technologies are now capable of offering unprecedented insight into the autoclave and, hence, facilitating the rapid exploration of novel nitride synthesis using the ammonothermal method.

Keywords: ammonothermal; synthesis; nitrides; semiconductors; solubility; in situ monitoring; crystal; high-pressure technology; supercritical fluid; GaN

Citation: Wostatek, T.; Chirala, V.Y.M.R.; Stoddard, N.; Civas, E.N.; Pimputkar, S.; Schimmel, S. Ammonothermal Crystal Growth of Functional Nitrides for Semiconductor Devices: Status and Potential. *Materials* **2024**, *17*, 3104. https://doi.org/10.3390/ma17133104

Academic Editor: Fabrizio Roccaforte

Received: 30 April 2024
Revised: 7 June 2024
Accepted: 10 June 2024
Published: 25 June 2024

1. Introduction

Electronic devices play an important role in modern society, and so do processes for the synthesis of semiconductor materials needed for making electronic devices. Several trends further increase the demand for energy- and resource-efficient solutions for harvesting, converting, distributing, and utilizing electric energy. These trends include electrification, digitalization, and the increasing demand on computing power for applications of artificial intelligence, to name just a few. While silicon has been the backbone of electronics and will remain important in cost-driven applications, silicon-based devices are approaching their physical limits.

The global need to reduce CO_2 emissions creates an increasing momentum to focus on the energy-efficiency of devices and systems. Besides environmental aspects, efforts to strengthen energy sovereignty are an additional driving force to aim for an efficient use of electrical energy, as well as to favor the use of earth-abundant elements, preferably with diverse options for supply chains.

To improve energy-efficiency, to widen the device design space by novel physical properties, and to meet the above-outlined additional criteria, new materials are increasingly being researched.

This review delivers an overview of the synthesis and solubility of various nitrides that are accessible via the ammonothermal method (Sections 2–4). In the case of GaN, as the most studied and most reviewed ammonothermally grown material, we give only a compact overview of the state-of-the-art uses and the available literature, except for aspects that have never, or not recently, been reviewed in detail. Accordingly, the sections on point defects (Section 5), reactor technology (Section 6), and in situ monitoring (Section 7) are based on GaN and may contain generalizable information for other nitrides. These sections are almost exclusively based on GaN-specific literature, due to the lack of literature on these aspects for most other ammonothermally synthesized materials.

1.1. Potential of Functional Nitrides for Semiconductor Devices

The III-nitrides have become one of the main and most developed material systems for power electronics based on being (ultra-)wide bandgap semiconductors and their subsequent potential to improve energy-efficiency beyond the capabilities of silicon-based devices [1]. To further widen the material parameter and design space, novel nitrides that are lattice matched to GaN or AlN represent a promising class of materials, as they could be integrated with the maturing material platforms of GaN and, prospectively, AlN. In some cases, their ternary and multinary alloys hold promise to surpass not only the functional properties of competing classes of materials, but also the binary III-nitrides. Examples of areas in which ternary and multinary nitrides could overcome existing limitations include insufficient dielectric strength and electrical conductivity (that is, p- and n-type dopability with sufficient charge carrier mobilities) for power electronic devices, preventing devices with higher energy-efficiency [2,3], insufficient piezoelectric coefficients for piezoacoustic devices [4]; lack of ferroelectrics with high performance and good technological compatibility with a reasonably established semiconductor technology for, e.g., memory and micro/nano-actuator applications [5], too low thermal conductivities and electro-optical coefficients for nonlinear optical materials [6], and too high coercive field strength for energy-efficient data storage and neuromorphic computing devices [7].

Multiple binary and ternary nitride materials that exhibit material properties of interest have emerged, which could be integrated into the GaN or AlN semiconductor platform. These are summarized in Figure 1. The wide material parameter design space facilitates the creation of various heterostructures of semiconducting (e.g., III-N [8], II-IV-N_2 [9,10]), piezo- and ferroelectric (e.g., AlScN [4,5,8]), magnetic (e.g., Mn-IV-N_2 [9]), and superconducting (e.g., NbN [8]) materials.

Nitride semiconductors with enhanced piezoelectric and ferroelectric properties are a particularly recent and exciting emerging topic [11,12]. Important catalysts for this development were the discovery of enhanced piezoelectric properties of AlScN by M. Akiyama in 2009 [4] and S. Fichtner's first report of AlScN being not only piezo- but also ferroelectric in 2019 [5]. Meanwhile, additional AlN-based alloys are being explored, with regard to enhanced piezoelectric properties. Examples are alloys (solid solutions) containing B [11] or certain transition metals [13,14]. Specifically, X.-H. Zha et al. have investigated alloys of the composition $Al_{0.9375}TM_{0.0625}N$ via density functional theory (DFT) calculations (using a $2 \times 2 \times 2$ supercell of 32 atoms) and identified Ca, Cr, Sr, Mo, Ru, and Rh as promising transition metals (TM) for enhancing the piezoelectric modulus d_{33}, in relation to pure AlN, indicating a significant enhancement beyond that achievable by the same concentration of Sc in the cases of Mo, Rh, Ru, Sr, and Cr (with Mo causing the most pronounced effect) [14]. Similarly, J. Startt et al. studied $Al_{1-x}TM_xN$ alloys, likewise by DFT calculations (using a $3 \times 3 \times 2$ supercell of 72 atoms), with the transition metal content x being 0.166, 0.333, and 0.5, respectively, and TM representing different earth-abundant d-block elements [13]. J. Startt et al. concluded that the group 4 metals Ti, Zr, and Hf induce large piezoelectric enhancements comparable to those caused by Sc, while the group

5 elements Nb and Ta also provide enhancements, but are less effective [13]. Very recently, F. Wang et al. also published a related DFT study (using a 2 × 2 × 3 supercells with 48 atoms), investigating the three ternary materials, AlScN, AlHfN, and AlZrN, as well as co-alloying AlN with Sc and either Hf or Zr [15]. Wang et al. reported a piezoelectric coefficient d_{33} as large as 49.18 pC/N and 47.00 pC/N for $(HfSc)_{0.375}Al_{0.625}N$ and $(ZrSc)_{0.375}Al_{0.625}N$, respectively, which is 164% and 156% larger than that of $Sc_{0.375}Al_{0.625}N$, and 926% and 885% larger than that of pure wurtzite AlN [15]. The piezoelectric moduli from X.-H. Zha et al., J. Startt et al., and F. Wang et al. [15] are jointly presented in Figure 2.

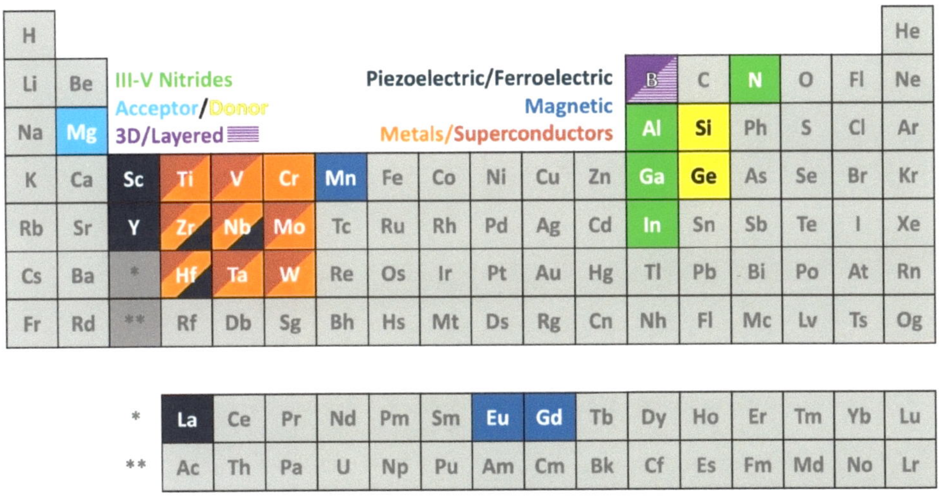

Figure 1. Property design space of "new" nitrides as of 2019 by D. Jena et al. [8]. Periodic table with constituents of "traditional" III-nitride semiconductors and associated dopants highlighted in green and bright blue/yellow, respectively. The other highlighting colors refer to physical properties and the usage of the elements in functional nitride materials. Reprinted from [8] under Creative Commons Attribution 4.0 license.

Together, the aforementioned studies substantiate that alloying of III-nitrides with transition metals other than Sc is promising for enhancing d_{33} further, or for using more accessible elements. However, the theoretical predictions show significant discrepancies and further studies including experimental results and more reliable calculations (larger supercells or more representative selections of inequivalent supercells, see e.g., D. F. Urban et al. [16] for a discussion of this aspect, with the example of AlScN) are necessary to establish confidence regarding the effective choice of alloying elements.

Motivated by the search for enhanced functional properties, novel heterostructures with III-nitrides, and earth-abundant semiconductors, heterovalent ternary nitrides of the stoichiometry II-IV-N_2 [9,10] are increasingly being studied. They typically crystallize in orthorhombic, wurtzite-derived superstructures that feature in-plane lattice constants similar to those of GaN and/or AlN (Figure 3). More recently, additional polymorphs have been discovered for some of these materials (such as wurtzite and rocksalt $MnSnN_2$ [17]), further widening the design space. In addition, this also opens up the possibility of heteroepitaxial integration of different phases of the same material with distinctly different properties, as O. Ambacher recently contemplated for AlScN [18]. Both the properties of ternary nitrides and their epitaxial relationships to different materials are of interest, as will be elaborated in the following paragraphs.

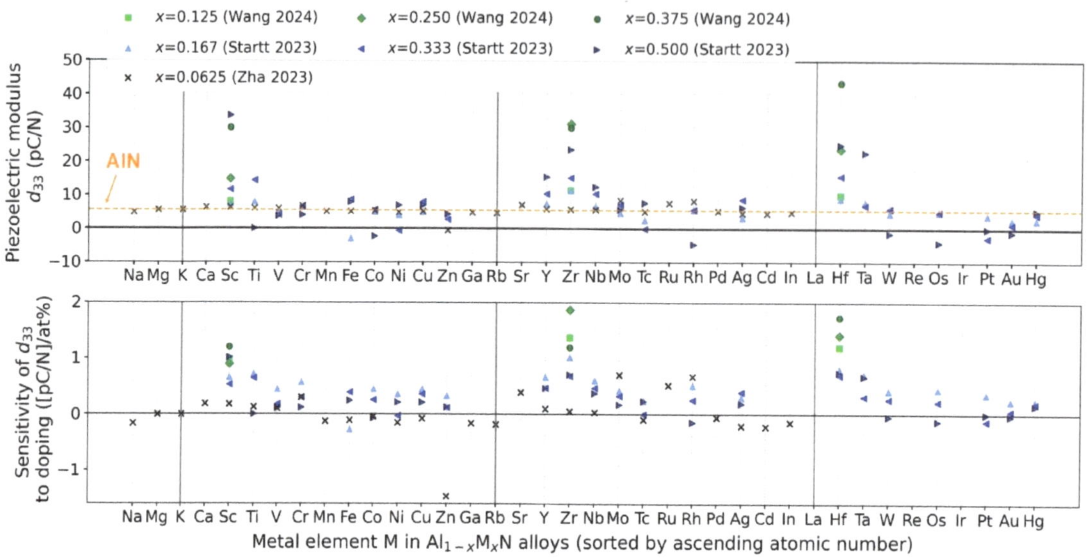

Figure 2. Sensitivity of the piezoelectric modulus d_{33} to doping with transition metal elements according to theoretical studies by X.-H. Zha et al. [14], J. Startt et al. [13], and F. Wang et al. [15]. Note that the original data by X.-H. Zha et al. represent absolute values, whereas those by J. Startt et al. represent changes of d_{33} with respect to pure AlN. This was considered when creating the graphic. The dashed orange line in the upper subplot indicates the piezoelectric modulus of pure wurtzite AlN.

An important obstacle for realizing complementary field-effect devices in nitrides (implementation of CMOS logic directly in the wide bandgap nitride platform) is the low hole mobility in GaN [19]. In this context, II-IV-N_2 compounds provide novel opportunities for strain engineering. For example, epitaxial growth of GaN on $ZnGeN_2$ or $MgSiN_2$ results in an inversion of the heavy hole band and split-off hole band, thereby lowering the effective hole mass in the compression direction and increasing the hole mobility in GaN by 50% and 260%, respectively [20]. In addition, electron and hole effective masses have been calculated for Zn-II-N_2 (II = Si, Ge, Sn), indicating relatively low hole effective masses in the c-direction (Γ-Z, below 0.3 for $ZnGeN_2$ [21], compared to 0.87 for GaN [22], both for the highest sub-band), but also a significant anisotropy with relatively high hole effective masses in the range of 2.5 to 6 in the Γ-X and Γ-Y directions. As an example of a heterostructure with a binary nitride in the wurtzite structure, $ZnSiN_2$/AlN interfaces exhibit a larger conduction-band offset than AlGaN/GaN, in combination with a large polarization charge, and therefore hold promise to yield higher electron sheet densities [23].

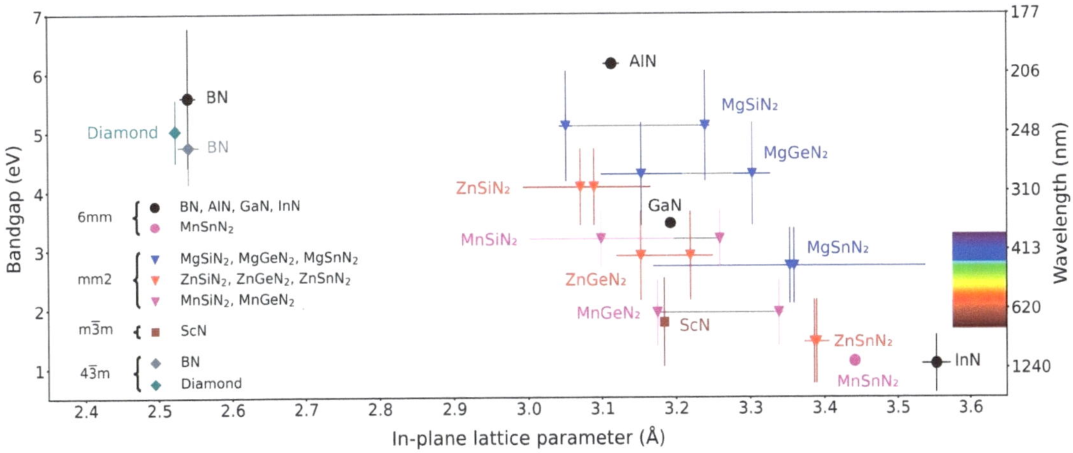

Figure 3. Bandgap and in-plane lattice constants for selected nitrides and lattice-matched materials (labeled with point groups (it is common in the literature to distinguish different polymorphs by an abbreviation of either the crystal system (such as h for hexagonal), or the crystal structure (such as w for wurtzite). We avoid distinguishing the different polymorphs by crystal system in order to avoid confusion, as both wurtzite and graphitic boron nitride would be referred to as h-BN in a consistent labeling by crystal system, while most readers would expect w-BN for BN in the wurtzite structure. The use of point groups allows consistent labeling of all discussed materials with a sufficient level of detail for our purposes.)). The two data points shown for the orthorhombic materials are the wurtzite-equivalent in-plane lattice parameters $a\sqrt{3}$ and $b/2$. Data were obtained from the following references: 6mm-BN [24–26], 6mm-AlN [18,27–29], 6mm-GaN [26–29], 6mm-InN [26–29], 6mm-$MnSnN_2$ [30], mm2-$MgSiN_2$ [31–36], mm2-$MgGeN_2$ [31,33–37], mm2-$MgSnN_2$ [17,33,34,38–40], mm2-$ZnSiN_2$ [21,23,41–44], mm2-$ZnGeN_2$ [21,23,41,42,44], mm2-$ZnSnN_2$ [21,23,40,44–47], mm2-$MnSiN_2$ [36,43,48–50], mm2-$MnGeN_2$ [36,50,51], $\bar{4}3$m-BN [25,52,53], m$\bar{3}$m-ScN [8,18,54], and $\bar{4}3$m-diamond [3,55–57]. The error bars represent the standard deviation of values collected from the different references. Note that the lack of an error bar in the case of 6mm-$MnSnN_2$ does not imply low uncertainty, but an insufficient amount of data for quantifying uncertainty in this way. Note as well that only the experimental bandgap data on $MnSiN_2$ and $MnGeN_2$ from [36] were used (as the spin-specific two bandgaps per material would have made the figure less readable), and that the calculated values for the bandgaps are considerably lower.

According to calculations of elastic constants, both $MgSiN_2$ and $MgGeN_2$ are softer than their III-nitride counterparts AlN and GaN, which eases their heteroepitaxial growth on these binary nitrides [33]. In addition, $MgSiN_2$ and Mg_2PN_3 (which likewise has already been synthesized via the ammonothermal method [58]) have recently been identified as promising multinary wurtzite-type ferroelectrics with comparatively low switching barriers, as needed for next-generation low-power computing [7].

The Mn-II-N_2 alloy system is least explored, however it is known to exhibit spin polarization (resulting in spin-specific bandgaps according to DFT calculations), as Mn introduces a magnetic moment [36]. Both cation-ordered $MnSiN_2$ [49] and $MnGeN_2$ [30,59] maintain anti-ferromagnetic ordering above room temperature, whereas the cation-ordered $MnSnN_2$ shows a magnetic transition at around 10 K [30]. The order or disorder in the cation sublattice appears to play an important role in the formation of polymorphs [30]. The ordering of the heterovalent cations has also been identified as a design parameter for electronic bandgaps, which was derived via a combined experimental and theoretical study on $ZnSnN_2$ and $MgSnN_2$ and should be applicable to a broad group of ternary heterovalent compounds [40].

Investigations on doping of the II-IV-N_2 materials remain limited. Theoretical studies on $ZnGeN_2$ suggest some similarities to the III-nitrides. Oxygen O_N is a shallow donor with a low formation energy and is thus likely to cause n-type conductivity in unintentionally doped (UID) samples, whereas the incorporation of Li on substitutional lattice sites is predicted to yield p-type doping [10].

A current frontier, specifically in energy-efficient power electronics, is ultra-wide bandgap (UWBG) semiconductors, adding to the interest in this subgroup of materials. The motivation for transitioning to materials with an even wider bandgap (E_g) than GaN or SiC is that the critical electric field scales approximately as E_g^2 [3]. The Baliga figure of merit (BFOM) used to assess potential performance for low frequency, for example, unipolar power devices, therefore scales approximately as E_g^6 [3]. Materials for next generation power electronics are evaluated using the BFOM in Figure 4 by comparing the specific on-resistance, R_{on}, with the breakdown voltage, V_{BR}. The bottom right corner is most desired, i.e., lowest R_{on} at the highest V_{BR}. The two contours of constant BFOM shown for $Al_{0.82}Sc_{0.18}N$ [60] differ by dislocation density (thus, different electron mobilities μ_n). This highlights the importance of low dislocation densities for utilizing the full potential of a material. Currently, all UWBG materials are still under development and are actively being researched.

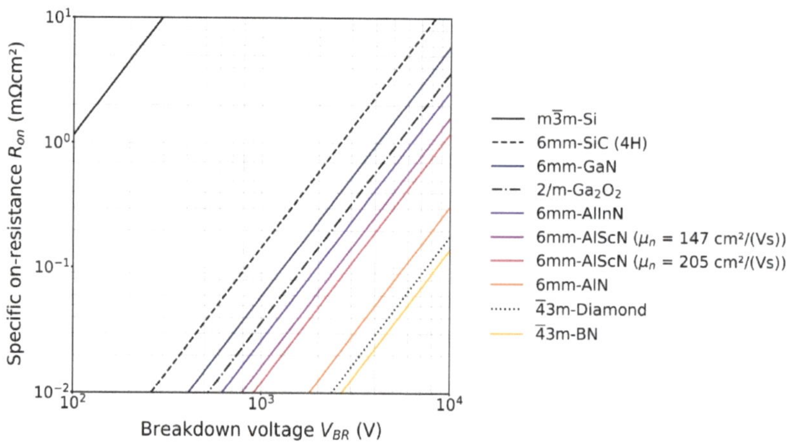

Figure 4. Contours of constant Baliga figure of merit (BFOM) for materials used or considered for power electronic applications. The lower right area represents higher BFOM (lower R_{on} for a given V_{BR}), hence better performance of low-frequency unipolar vertical power switches. Data by Tsao et al. ($\bar{4}$3m-diamond, $\bar{4}$3m-BN) [3] and Tansu et al. (6mm-$Al_{0.82}Sc_{0.18}N$, the composition lattice matched to 6mm-GaN, and $Al_{0.82}In_{0.18}N$) [60], respectively. For 6mm-SiC (specifically, 4H-SiC), 6mm-GaN, 2/m-Ga_2O_3 (β-Ga_2O_3), and 6mm-AlN, average values of both references are given.

For UWBG $Al_xGa_{1-x}N$ alloys, the cohesive energy and subsequent bond strength increase rapidly with increasing Al-content. This complicates doping, due to the formation of deep levels and compensation. As a result, control of the electronic properties is more challenging, causing both n-type and p-type doping to remain active areas of research for Al-contents exceeding 80% [61]. Challenges for n-type doping appear to be related to the formation of DX centers (DX centers were initially thought of as a defect complex involving a substitutional donor and an anion vacancy [62], but are meanwhile thought to be due to a large lattice relaxation and the trapping of two electrons, causing the dopant to transition from a single donor to a single acceptor [63,64]) [65], whereas issues with p-type doping with Mg are caused by a significant deepening of the acceptor level, as well as the decreasing solubility of Mg with increasing Al-content [61]. Recently, p-type doping of AlN using Be at the Al-site has re-gained momentum due to experimental reports [66–68] of p-type conductivity. For both Si doping of Al-rich AlGaN in metal organic chemical vapor deposition (MOCVD) [69] and Be doping of AlN in metal-modulated epitaxy (MME) [66], improvements in doping efficiency appear to be closely linked to low growth temperatures (1050 to 1115 °C in MOCVD [69], 600 to 700 °C in MME [66]). The underlying mechanism is likely related to the dopant-induced strain and appears to favor substitutional incorporation on Al lattice sites (Be_{Al}), as opposed to interstitial sites [66]. An analogous mechanism is likely responsible for the doping efficiency improvement of AlGaN:Si at low growth temperatures [69]. The effect of high temperatures driving light elements into interstitial sites is thought to be universal for light dopants in compound semiconductors [70].

1.2. Brief Overview of Methods for Nitride Synthesis and Crystal Growth

To realize the potential provided by the physical properties of nitride materials in practice, the choice and advancement of suitable methods for their synthesis and crystal growth plays a pivotal role. In particular, the availability of affordable large-area and high-quality native GaN substrates with controlled electrical properties remains a major roadblock [71].

For the synthesis of thin films, vapor-phase approaches, such as (plasma-assisted, ammonia-based, or metal-modulated) molecular beam epitaxy (MBE) [2,66,72] and metal organic vapor phase epitaxy (MOVPE) [69,70] are the primarily used techniques yielding high crystalline quality III-N with controllable layer thicknesses and dopant/alloying control, while reports on MBE [73,74] or MOVPE growth of other nitrides also exist [75,76]. Generally, these techniques are capable of yielding high quality material at temperatures below their decomposition temperature. This is particularly advantageous for nitrides that are prone to phase separation, such as III alloys with high indium content [77–79].

For the growth of thick films or relatively thin bulk crystals, sodium flux [80] and hydride vapor phase epitaxy (HVPE) are the most established techniques, which have been investigated with a strong focus on GaN thus far. Specifically for the bulk growth of GaN via HVPE, sodium flux, and ammonothermal methods, a review of the state-of-the-art aforementioned techniques was provided by R. Kucharski et al. in 2020 [81].

For the growth of thick bulk GaN crystals, the ammonothermal method is thus far the only technique having successfully demonstrated boules with diameters up to nearly 4 inches [82]. In the case of AlN, physical vapor transport (PVT) is, thus far, the main technique explored for bulk growth, that is, a technique based on the sublimation of the material. An important frontier in the PVT growth of AlN has been the expansion of crystal diameters, and remarkable progress has been made in this direction during the last few years [83].

Compared to PVT, the ammonothermal technique is performed at significantly lower temperatures (~600 °C vs. ~>2000 °C [84,85]) and in a different growth environment, offers opportunities to modify the incorporation of species into the growing boule (such as dopants), or leverage kinetic growth processes to modify native defect production or phase selection. One important drawback of the ammonothermal method is the observed lower growth rates (compared to PVT) [85,86]. However, this can be overcome by the simultaneous growth of a large number of crystals in one autoclave (more than 50 has been reported). A more detailed discussion of the ammonothermal method is provided in the following section.

1.3. Potential of the Ammonothermal Method for the Synthesis of Functional Nitrides Materials

The ammonothermal method is a promising yet underexplored synthesis method for nitrides at low to moderate temperatures of about 400 to 900 °C. Its development has primarily been driven by the realization of GaN substrates with superior structural quality [81,87–91] and for exploratory nitride synthesis [92–94]. Despite the associated technical and scientific challenges, the method has seen a remarkable development since the initial reports on ammonothermal GaN growth by R. Dwiliński, H. Jacobs, and coworkers in 1995 [95], as illustrated in Figure 5.

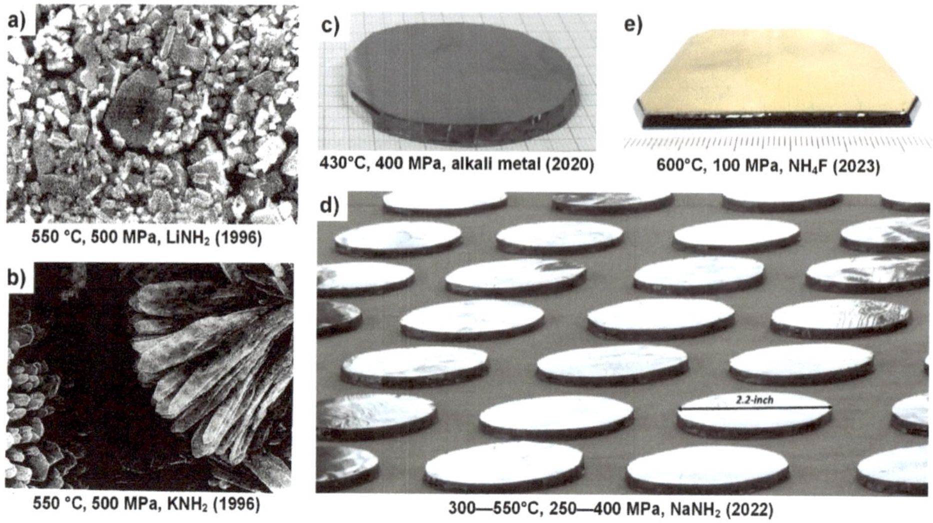

Figure 5. Development of ammonothermal crystal growth of GaN, starting from the initial synthesis experiments by Dwiliński et al. to recent achievements. (**a**) GaN crystallites a few micrometers in size, obtained with $LiNH_2$ mineralizer (SEM image) [96]; (**b**) GaN crystallites of up to 25 μm length, obtained with KNH_2 mineralizer (SEM image) [96]; (**c**) state-of-the-art GaN crystal grown using alkali metal mineralizers [97], (**d**) batch thereof to visualize the method's capability for the simultaneous growth of many crystals [98], and (**e**) state-of-the-art GaN grown with NH_4F mineralizer at comparatively low pressure [84]. All images reproduced from the respective references with permission; (**a**,**b**,**d**) under Creative Commons Attribution Non-Commercial 4.0 International License CC BY-NC 4.0, (**c**) © 2020 Elsevier B.V, (**e**) © 2022 The Japan Society of Applied Physics.

From a crystal properties point of view, the key advantage of the ammonothermal crystal growth method is the exceptionally high structural quality that can be obtained (low amount of threading dislocations, large radius of curvature), whereas the main disadvantage is the difficulty of achieving high purity and low point defect concentrations [99–101]. Importantly, the ammonothermal method as an equilibrium process is suitable for obtaining seeds of excellent structural quality by self-nucleation, in both low-pressure [102,103] and high-pressure [104] variants of the method. Contrary to HVPE as a non-equilibrium process [71], the size of seeds can be increased by consecutive growth runs [102] by taking advantage of lateral growth in non-polar or semi-polar directions [105,106], which allows for simultaneous improvements in structural quality, particularly in the regions that extend laterally beyond the seed [81,86]. This can be exploited by starting from a slender seed obtained by a different growth technique (if available, such as HVPE in case of GaN), which can later be removed from the newly grown ammonothermal crystal, in order to use only the newly grown regions of improved structural quality in subsequent ammonothermal growth runs [81]. It is also possible to combine seeds via so-called tiling technology [107], developed earlier for HVPE [108]. Once a sufficient diameter with appropriate structural quality has been obtained, growth in a polar direction allows for the multiplication of seeds [105], which is important given the comparatively slow growth rates [86], motivating the simultaneous growth of many crystals to achieve cost-competitiveness [88]. The simultaneous growth of large numbers of crystals has proven effective for the well-established yet similar hydrothermal growth of quartz [109]. It is also feasible to use HVPE to speed up the multiplication process [110]. A relevant technological aspect is the importance of surface preparation for avoiding the formation of defect clusters, such as networks of high-density threading dislocations induced by subsurface damage in the seed crystal [107].

While initially developed for bulk GaN, the method is also increasingly being investigated for bulk growth of further binary nitrides with application potential as substrates for electronic devices (such as cubic BN [111]), as well as for the exploratory synthesis of ternary and multinary nitrides [92]. It is also worth noting that a deliberate incorporation of transition metals into nitride crystals during ammonothermal growth has already been achieved, which is of increasing interest given the recent developments in piezo- and ferroelectric nitrides, as described in Section 1.1. The first such report dates back to a study from 2001 which targeted GaN-based dilute magnetic semiconductors (DMS) such as $Ga_{1-x}Mn_xN$ [112].

Besides the ability to use low growth temperatures, the synthesis conditions in the ammonothermal method (mineralizer, temperature, pressure) can be tailored to favor crystallization of the same material in a specific crystal structure, such as wurtzite or zincblende in the case of GaN [113]. While most ammonothermal research has so far focused on wurtzite GaN, both wurtzite and zincblende crystal structures emerge as interesting, from an application point of view, for some of the novel nitrides [18,114].

It should be noted that a synthesis method for self-nucleated bulk material with high structural quality, as demonstrated for GaN [102], is important even for materials that are more likely to be applied as thin films than as substrates. For many of the novel nitrides, small crystals of high quality with low defect concentrations and minimal internal stress are currently lacking, while relevant functional properties tend to be influenced by structural quality. The ferroelectric properties of AlScN are an example of a property of high current interest that was shown to be impacted by structural quality [115]. Even for InN, a bulk material of sufficient quality is not yet available [116]. High quality bulk samples are essential for an accurate experimental determination of fundamental properties, as needed for investigating prospective device applications.

In this review, we focus on the ammonothermal method as a technology for the synthesis and crystal growth of various nitrides with a variety of functional properties, while building on the much further developed example of GaN.

2. Ammonothermal Method

Having motivated the interest of the crystal growth community in the types and attributes of crystals that have been demonstrated using ammonothermal technology, let us take a step back and describe the operating principle and associated equipment of the technique. The tendency of nitrides to decompose into their base metals and triple-bonded nitrogen gas molecules limits the abilities of more traditional melt-growth techniques for bulk crystallization [28,117]. Only in extreme pressure, oxygen-free conditions can the formation of materials like GaN be achieved directly [118]. Therefore, a solution-based growth principle is much more appealing, and the ammonothermal method employs ammonia (NH_3) as a nitrogen analog to water (H_2O) [96,119], aimed at bypassing some of the difficulties of direct nitride formation.

In solution growth methods, the material constituents are typically dissolved using an appropriate solvent. The solution is then brought into a saturated or supersaturated condition by the manipulation of variables like temperature or pressure, or the selective removal of the pure solvent by methods such as evaporation or reverse osmosis [120]. Crystallization may take place spontaneously through nucleation, or may proceed on a seed crystal.

The hydrothermal method, used for making oxide crystals such as quartz, employs subcritical water as the solvent [121,122] and has long served as an enormously successful example of the synthesis of large numbers of bulk single crystals from a solvothermal solution, motivating the development of the ammonothermal method for nitride synthesis in an analogous way [123]. For nitrides and the ammonothermal method, ammonia is the equivalent option. In both hydrothermal and ammonothermal boule growth, temperature-dependent solubility is the key to success, and needs to enable the dissolution of suitable quantities of the material and the control of its transport via the thermal field in the reactor.

In order to attain sufficient solubility of the desired solutes, the methods are operated in the sub- or supercritical fluid regime. Near their critical point, fluids generally have liquid-like dissolving power, in conjunction with transport properties that are in-between those of a gas and a liquid [124]. In supercritical fluids, the relatively high, liquid-like density enhances solubility [125], while temperature and pressure are sufficient to create an extremely mobile (i.e., diffusive [126] and convective [127]) and highly reactive fluid [126]. Most supercritical fluid applications function best in the region of high isothermal compressibility, not too far above the critical point [126]. The ammonothermal method, however, uses a parameter space far above the critical point [128]. As can be seen from the fluid properties and pressure/temperature ranges in Table 1, this differentiates the ammonothermal method from the hydrothermal method, which utilizes even slightly sub-critical conditions ($T/T_c < 1.0$ and $p/p_c < 1.0$). The reasons for this unusual process window have not yet been fully clarified. In this context, the recently increasing interest in the potential of supercritical geothermal resources [129] might lead to potentially transferable results. The conditions in supercritical geothermal reservoirs were estimated to be in the temperature range of 400 to 600 °C (T/T_c = 1.07 to 1.60) and the pressure range of 20 to 30 MPa (p/p_c = 0.91 to 1.36) [129,130], about the same absolute temperature but not pressure range. The interest in supercritical geothermal resources has already led to recent thorough investigations on solubilities of quartz under those conditions [129].

Table 1. Characteristics of the hydrothermal growth of quartz and ammonothermal growth of GaN in comparison, with a focus on pressure and temperature range and the properties of the respective fluid under those conditions. Fluid property data were taken from or calculated based on NIST database [131]. Note that the operation range for the ammonothermal method ranges significantly beyond the available data range, as data are only available up to 451.9 °C. Density data of the upper end of the range were linearly extrapolated, viscosity data were extrapolated using a polynomial fit. The data in the table represent the ends of the ranges obtained from NIST database and extrapolation thereof, based on all combinations of pressure and temperature given.

Fluid Property/Process Characteristic	Hydrothermal (Quartz)	Ammonothermal (GaN)
Solvent (critical parameters) [132]	Water (374.5 °C, 22.1 MPa)	Ammonia (132.1 °C, 11.3 MPa)
Temperature range (°C)	345–360 [122]	400–800 [86]
Reduced temperature range T/T_c	0.92–0.96	3.03–6.06
Pressure range (MPa)	70–150 [122]	100–600 [86]
Reduced pressure range p/p_c	3.17–6.79	8.85–53.10
Solvent density (mol/L) in p-, T-range	40.7–44.2	18.2–36.2
Solvent density (kg/m^3) in p-, T-range	733.8–795.8	310.2–616.6
Dynamic viscosity (µPa·s)	90.3–103.2	49.6–284.3
Kinematic viscosity (m^2/s)	1.230×10^{-7}–1.296×10^{-7}	1.597×10^{-7}–6.228×10^{-7}

However, even the supercritical ammonia atmosphere is often not reactive enough to cause sufficient dissolution of the nitride source material for effective crystal growth. To further increase solubility, so-called "mineralizers" are added to the solution as a co-solvent to form soluble intermediates [133,134]. Mineralizers are termed acidic if they increase the concentration of $[NH_4]^+$ ions (this is the case for ammonium halides [133]). Accordingly, mineralizers that increase the concentration of $[NH_2]^-$ ions are called basic mineralizers (alkali and alkaline earth metals are typical basic mineralizers) [134]. The addition of mineralizers can dramatically increase the solubility through the formation of complex ions, due to the greater availability of $[NH_2]^-$ and $[NH_4]^+$ ions under ammonobasic and ammonoacidic conditions, respectively [119]. Regarding the second ion introduced with the mineralizer, the columns IA, IIA, or VIIA of the periodic table are convenient both for their reactivity with many elements, and for their lack of reactivity with nitrogen to form nitrides [135].

With enough solubility of the starting material achieved in the supercritical ammonia environment, and with mass transport facilitated by buoyancy, the last key element for successful ammonothermal growth is the availability of a solubility gradient as the driving force for crystallization. This solubility gradient, which needs to be established between the region containing the source material and the region containing the seed crystals, is usually created via a temperature gradient. A saturated solution is formed in the zone of the autoclave that contains the source material. When the convective flow transports the saturated fluid to another zone with lower solubility, typically provisioned with seed crystals, the solution becomes supersaturated. This supersaturation acts as the driving force for deposition on available surfaces, moving the solution back towards equilibrium. A schematic of the process and a representative experimental setup are shown in Figure 6, using the example of ammonothermal growth of bulk GaN, as this is the most developed ammonothermal bulk nitride growth process to date.

Since the ammonothermal method relies on the use of ammonia as the solvent, elevated pressures are required to prevent excessive decomposition of ammonia into hydrogen and nitrogen gases [136]. Higher pressures have generally yielded higher crystal growth rates for gallium nitride, at least with several acidic mineralizers [103,137]. Apart from preventing ammonia decomposition, the generally observed increase of solubilities in supercritical fluids with increasing solvent density or pressure [124,129] may also play a role.

A challenge of ammonothermal growth is the currently limited crystal growth rates. In the worst cases, some processes seem incapable of exceeding 30 μm/day on certain stable crystallographic facets. The fastest substantiated steady-state growth rates of GaN single crystal growth advertise 150–200 μm/day [84], although some higher claims exist. The aforementioned challenges can be compensated by growing many crystals in parallel, as demonstrated in industrial-scale hydrothermal processes, as well as ammonothermal processes approaching industrial scales [84,88,138,139].

Active observation of the ammonothermal process is also limited by the thick autoclave walls. In order to gain a more detailed understanding of the inner workings of the process, two approaches have been developed. The first is computational simulations, wherein the temperature distributions and convective flow patterns in the fluid can be predicted. For this approach, a review from 2021 is available [128]. The second approach is direct in situ observation of the hot, pressurized autoclave, which will be reviewed in Section 6.2.

As with any crystal growth process, limitations emerge for the design of systems and crystal grow processes based on the chemical reactivity of species to the nitriding environment. This is no different for the ammonothermal system, and knowledge of what materials can and will go into a solution offers opportunities for nitride crystal growth containing these elements, while limiting opportunities for use of these materials in their soluble forms (elemental form or as part of a specific compound) as functional elements in the system design. The next section discusses the current state of knowledge pertaining to the reactivity and solubility of elements under ammonothermal conditions.

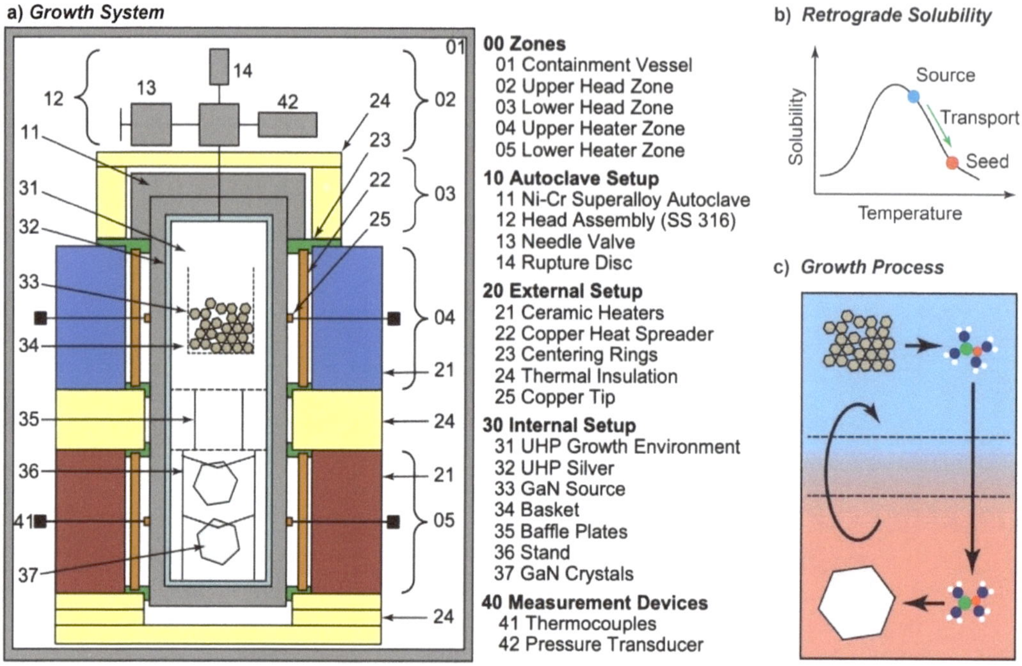

Figure 6. Representative diagram of an ammonothermal growth apparatus and mechanisms. (**a**) The growth system situated inside a containment vessel, wherein the high pressures of the process are monitored and managed safely. Two independent heat zones, 04 and 05, create the conditions necessary for dissolution, transport, and crystal growth. An ammonothermal system with retrograde solubility is depicted accordingly, with the seeds located in the hotter region. (**b**) A graph of solubility illustrating the availability and use of a retrograde temperature dependence, as is available with most basic mineralizers. (**c**) The cycle of the growth process in the supercritical fluid. Applied heat causes the hotter fluid (red) to rise on the perimeter. In the cooler upper zone, source material is dissolved at the higher solubility level. The cooler fluid (blue) then falls through the central region and delivers solute to the seed crystals, where the now supersaturated solution deposits new material. Note that, in the case of normal solubility, as is available with most acidic mineralizers at moderate temperatures, the hot zone remains at the bottom while the seeds are moved to the upper, cooler zone. Reproduced from [140], copyright (2014), with permission from Elsevier (left subfigure) and [87], © 2017 WILEY-VCH Verlag GmbH & Co. KGaA, Weinheim, Germany (right subfigure).

3. Reactivity and Solubility of Elements under Ammonothermal Conditions

As already delineated in general, the solubility of substances is markedly enhanced under ammonothermal conditions, owing to the elevated pressure and temperature coupled with the unique solvent characteristics of supercritical ammonia and the use of mineralizers for solubility enhancement. This environment facilitates the dissolution of a wide range of materials, albeit with a broad variability across different elements and strongly depending on the mineralizer used. Ammonia and the mineralizer are a versatile solvent due to the formation of diverse complex species (so called intermediates) with dissolved elements, thereby modulating their solubility and chemical reactivity within the solution. Precise control of the element solubility is significant for controlling crystal growth rates, composition, and impurity incorporation, as well as for ensuring the longevity of the autoclaves.

Notably, solubility in ammonothermal solutions demonstrates a significant dependency on temperature and pressure, with alterations in these parameters exerting substantial effects on element solubility and chemical equilibria within the solution medium [132].

From the viewpoint of bulk crystal growth, the temperature dependency of solubility is particularly relevant, however, quantitative data are available only for GaN and even for GaN, those data are incomplete and partially contradictory [128]. Many studies from the past and recent times dealt with either crystal growth [88,102,105,140–142], synthesis of compounds [9,92,143] via ammonothermal method and their intermediates [119,133,144], dopant incorporation [87,145,146], corrosion resistance [147,148], or dissolution kinetics and solubility studies of GaN [149–152] or BN [111]. The references mentioned in this paragraph are not intended as a complete list, but rather as examples of the different types of information that are available in the literature, while a full list of references used in this section can be found in the Sections A–C.

Before going further in this section, we describe our approach to data visualization using a periodic table, which will then be used to indicate whether a given element was used in any ammonothermal or closely related environment, and which qualitative information on the reactivity and solubility of the element can be derived from the available literature.

The available types of information were sorted into categories, with negative values representing the availability of data that discourage reactivity and solubility (high corrosion resistance), zero in case of no information being available, and positive values for literature indicating some level of reactivity or solubility. This numerical value is assigned for plotting purposes. It was defined so that it increases with increasing confidence in solubility and will be termed the reactivity and solubility value (R and S value).

Considering that the same element can have rather different reactivity (and, as a consequence, solubility) depending on its chemical environment in a solid, we distinguish between data obtained using elemental samples and those obtained using a compound containing the element. For this differentiation, we define subcategories for data on elemental and compound samples, respectively. The following five distinct categories will be used:

- Category C1 denotes elements or compounds exhibiting exceptional resistance to corrosion within the ammonothermal environment, either in their elemental or compound forms. Negative R and S values are assigned, as the available information suggests a lack of reactivity and solubility;
- Category C2 represents elements or compounds for which no data regarding their reactivity or solubility under ammonothermal conditions are available;
- Category C3 represents compounds or elementals that do undergo reactions under ammonothermal conditions, but no clear indication of significant transport was found. Typical examples are the corrosion of elements or compounds, or the formation of compounds without indications of transport, as is often the case for the synthesis of compounds in nano- or microcrystalline form;
- Category C4 indicates the formation of compounds that are soluble in liquid ammonia. Typical examples are the compounds that were synthesized under ammonothermal conditions and were reported to likely represent soluble intermediates during ammonothermal synthesis and crystal growth, but at times room temperature or even lower temperatures, but on the whole, ranging from 197–873 K [119];
- Category C5 represents elements for which a clear indication of transport under ammonothermal conditions was found. Examples are mineralizer cation in the case of ammonobasic growth, mineralizer anion in the case of ammonoacidic growth, impurities incorporated in grown crystals (mostly GaN, as there are hardly any investigations on impurities in other materials grown via solute transport), elements for which dedicated solubility studies are available, as well as nitrogen and hydrogen as ubiquitous elements in ammonothermal fluids and ammonothermally grown crystals.

For a clear visual representation of the data, a color-coded legend will be utilized. In the periodic tables, each subcategory will be represented by a color, which is assigned according to the legend shown in Figure 7.

Category	Subcategory description	Color	R&S value
C1: High corrosion resistance	Compound		**−2**
	Element		**−1**
C2: No information available			**0**
C3: Reacts but no transport	Corrosion of compound		**1**
	Corrosion of element		**2**
	Formation of compounds		**3**
C4: Forms compounds soluble in liquid ammonia (intermediates)			**4**
C5: Transport under ammonothermal conditions	Akali metal or halogen elements of mineralizers		**5**
	Small quantities (e.g., impurities)		**6**
	Temperature depependent solubility or transport via thermal gradient (crystal growth, solubility studies)		**7**
	Nitrogen and hydrogen: N and H as constituents of mineralizers; N as constituent in nitrides, H as common impurity in nitrides		**8**

Figure 7. Data visualization via color-coding legend describing the qualitative reactivity and solubility information on the periodic table. The first column describes the main categories, while the second column defines subcategories where necessary. The third column indicates the color that will be assigned in the periodic table, while the last column lists the reactivity and solubility (R and S) values assigned for plotting and listing purposes.

Color-coded periodic tables were devised to visually represent the reactivity and solubility of known elements within the ammonothermal environment in general (Figure 8), as well as under ammonoacidic (Figure 9), ammonobasic (Figure 10), and ammononeutral (Figure 11) conditions, respectively. In case of multiple values per element (commonly due to the availability of data for the use of different ammonoacidic/ammonobasic mineralizers, data for both elemental and compound samples, or deviating results), the highest value is displayed for that element. The complete set of data and references can be found in Appendix A (ammonoacidic), Appendix B (ammonobasic), and Appendix C (ammononeutral).

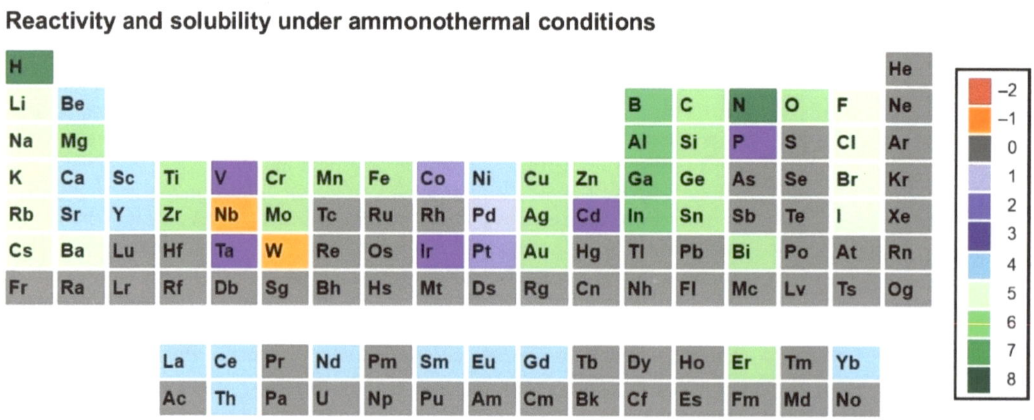

Figure 8. Color-coded periodic table illustrating the reactivity and solubility of elements split into different categories based on their interaction, regardless of the type of ammonothermal environment. For a definition of the categories listed in the legend, see Figure 7.

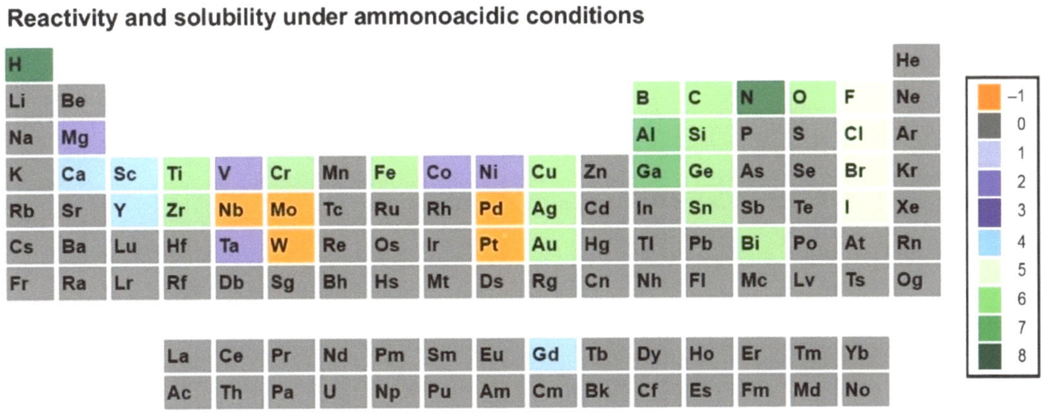

Figure 9. Color-coded periodic table illustrating the reactivity and solubility of known elements split into different categories based on their interaction specifically in ammonoacidic conditions. For a definition of the categories listed in the legend, see Figure 7.

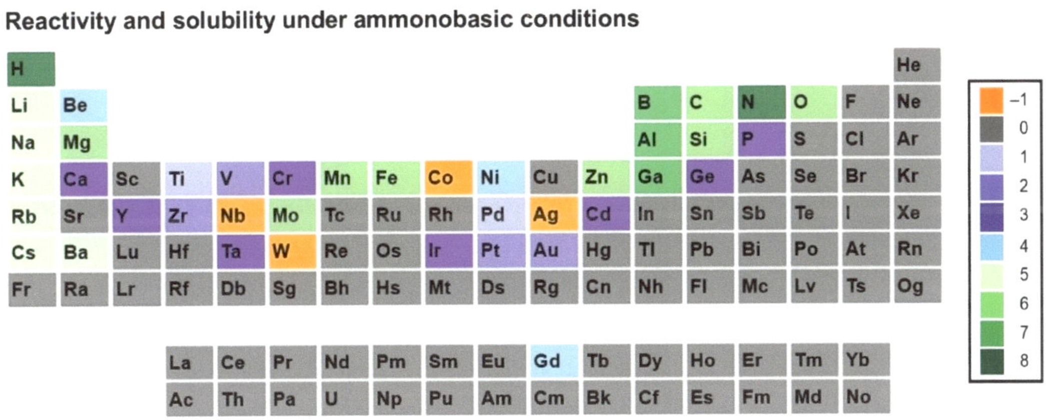

Figure 10. Color-coded periodic table illustrating the reactivity and solubility of known elements split into different categories based on their interaction specifically in ammonobasic conditions. For a definition of the categories listed in the legend, see Figure 7.

This systematic approach provides a consistent overview of both data availability and the feasibility of achieving reactions and/or solubility under the respective ammonothermal conditions. To indicate clearly for which elements there is a complete lack of data, the first periodic table (Figure 8) summarizes the data from acidic, basic, and neutral milieus.

Quite a few elements belonging to actinide and lanthanide series, and also to the last periods of the table, were not employed in the ammonothermal environment (no data available, grey). Considering the recent studies on the enhancement of the piezoelectric modulus of AlN by alloying with certain transition metals, knowledge on further transition metals, particularly Zr and Hf, would be of interest (based on Figure 2 and the references therein). In addition, for elements with data only on reactivity, but not on transport, knowledge of solubility in the sense of proven transport via an ammonothermal solution would be valuable, particularly for those for which applications are on the horizon. This applies to Be, Mg, and Si as dopants for III-nitrides, to the transition metals Sc, Ti, Y, Zr, Nb, Hf, and Ta for piezoelectric $Al_{1-x}TM_xN$ (TM = transition metal), to Mg, Si, Mn, Zn as constituents of II-IV-N_2 semiconductors (based on Figure 3 and the references therein), and to Ti, V, Zr, Nb, Mo, Hf, Ta, and W, as they are constituents of nitride superconductors (based on Figure 1 and the references therein). It should be noted that reactivity and solubility vary greatly with the used temperature and pressure conditions and mineralizers. Therefore, the observation of corrosion resistance under a limited set of ammonothermal conditions does not necessarily imply that reactivity and solubility cannot be achieved under different ammonothermal conditions and used mineralizers.

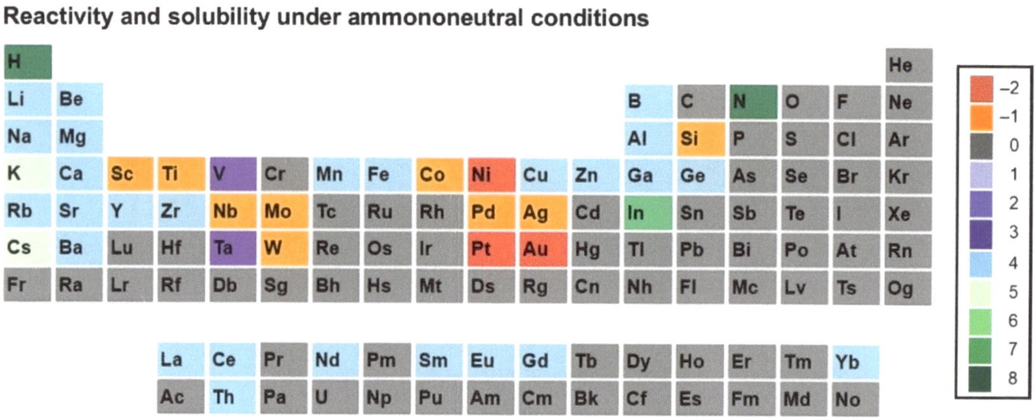

Figure 11. Color-coded periodic table illustrating the reactivity and solubility of known elements split into different categories based on their interaction specifically in ammononeutral conditions. For a definition of the categories listed in the legend, see Figure 7.

In the following, environment-specific information will be given on the basis of the following types of environments:

- Ammonoacidic conditions: acidic mineralizers such as the ammonium halides NH_4X (X = F, Cl, Br, I) are present, enhancing the concentration of NH_4^+ ions;
- Ammonobasic conditions: basic mineralizers such as alkali and alkaline earth metals or amides or azides thereof are present, enhancing the concentration of NH_2^- ions;
- Ammononeutral conditions: the concentration of NH_4^+ ions and NH_2^- ions is balanced, either in supercritical ammonia without mineralizers, or due to the balanced presence of different compounds increasing the concentrations of the NH_4^+ ions, as well as that of the NH_2^- ions, resulting in an ammononeutral milieu.

3.1. *Ammonoacidic Conditions*

As ammonoacidic conditions mean having mineralizers with halogens, these elements, alongside nitrogen and hydrogen, get transported as they are constituents of the solvent i.e., supercritical NH_3 [103,153–155].

Ag showed varied responses in ammonoacidic conditions based on the mineralizer used. It is corrosion resistant with NH_4F [156], whereas it gets corroded with NH_4Cl [147]. It is reported to be transported in small quantities with NH_4Br [157]. Ca was reported to form soluble intermediates in the ammonoacidic system with NH_4F as the mineralizer [158].

Al gets dissolved and transported and crystallizes as AlN in the case of NH_4Cl and NH_4I [159,160], however, soluble intermediates were formed only when NH_4F [161] and NH_4I [162] were present as educts. Other group III elements like Ga form crystals in the presence of all mineralizers [150,154] and B was transported in the presence of NH_4F, NH_4Cl and NH_4Br [163], at a near-ammonoacidic process. These nitrides will be discussed in the further sections. The group III-nitrides will be discussed in more detail in Section 4.1. Fe [87,164] and Cr [164] were reported to be transported as impurities into the system with NH_4Cl as the mineralizer.

Au was seen to be resistant against corrosion with NH_4Cl [147], but gets transported in the presence of NH_4Br [157]. Bi and Cu both get transported as impurities when reacted in the presence of NH_4Br [157]. Co was reported to be corroded in its elemental form, but remained stable as an alloy, $Co_{80}W_{10}Al_{9.4}$ with NH_4Cl [147]. C was found to be transported as an impurity with NH_4Cl as the mineralizer [87,164], similar to that of O [87,103] with all the possible acidic mineralizers.

Gd was observed to be forming soluble intermediates, whilst Ge got transported in little quantities [157]. Mg seemed to be corroded in compound/elemental form [147], but also reacts to form compounds [119].

Mo, Nb, Pd, and Pt all were reported to be corrosion resistant (with Mo and Pt even in alloy forms) when NH_4Cl was the mineralizer [147], with Nb losing its mechanical stability; on the other hand, Ni was observed to be corroded without getting transported [165,166], while maintaining its chemical passivity as an alloy [147].

For both Sc [167] and Y [168], soluble intermediates with NH_4I were reported, while Y also gets corroded like Ta in the presence of NH_4Cl [147]. Ti and V were seen to be corroded superficially, forming respective nitrides flaking off the surface [147]. Si, Zr, and W (in alloy form also) were described as showing high corrosion resistance with NH_4Cl, but in similar conditions, Zr was observed to be corroded [147]. Si was also observed to be transported in the ammonoacidic system during experimentation with all the acidic mineralizers [103]. Sn was reported to be transported in smaller quantities with NH_4Br as the mineralizer [157]. Also, Ti and Zr both were found as impurities when NH_4Cl was used as the mineralizer [164].

3.2. Ammonobasic Conditions

For ammonobasic conditions, elements like Li [169], Na [102], K [170], Rb, and Cs, and [171] also occasionally Ba [172], apart from N and H, get transported into the ammonothermal crystal.

Ag was found to be showing corrosion resistance with $NaNH_2$, while the Au sample was seen to be corroded, turning black [147]. Al was reportedly dissolved and transported and formed an AlN crystal with pure Li and K, $LiNH_2$, KNH_2, and KN_3 [169,173,174]. B also gets transported via the system to form BN [111].

Be was reported form soluble intermediates [175] with NaN_3 as the mineralizer, while C was known to be transported as an impurity with both $NaNH_2$ [145] and KNH_2 [176]. Ca formed compounds which were not reported to be transported in cases of NaN_3 [177] and $LiNH_2$ [178], whereas Cd [119] and Cr react similarly [179] in a KNH_2 environment. Co shows higher corrosion resistance in a $NaNH_2$ environment in both elemental and alloy forms [147]. Mg and Si were both observed to be transported through the system when reacted with metallic Na [87,140].

Ga formed bulk ammonothermal crystals when synthesized in ammonobasic conditions with $NaNH_2$ [149] and KNH_2 [109] serving as mineralizers. Gd was reported to form soluble intermediates [167] with $NaNH_2$, whereas Ge formed compounds with no traceable transport with $LiNH_2$, $NaNH_2$, and KNH_2 [42], apart from being corroded [147]. Fe was also observed as an impurity in the ammonobasic environment [87,140].

Ir was also reported to be corroded [147] in the presence of $NaNH_2$. Mg was not only found to be forming soluble intermediates in the solution with $NaNH_2$ [180], NaN_3, and KNH_2 [181], but corrosion studies were also performed in ammonobasic conditions with both compound and elemental forms [147].

Mn was reported to form soluble intermediates in supercritical ammonia with $NaNH_2$ [182], apart from being incorporated into the crystal [183]. Mo (also as an alloy) showed strong corrosion resistance in a similar environment [147], but was found to be transported into the system upon reacting with metallic Na [87,140]. Ni and Nb were both observed to be strong against corrosion, with the latter losing its mechanical stability in the process, besides being inert as an alloy [147]. Ni was also reported to form a nitride in the presence of $NaNH_2$ [184]. O was found to be transported as an impurity in both cases involving $NaNH_2$ and KNH_2 [145,176]. P was reported to form compounds in the presence of NaN_3, without being transported across the system [185].

Pd and Pt were found to become corroded in the ammonobasic system, with the former in elemental form and the latter as an alloy [147]. Si was observed to form soluble compounds in the presence of KNH_2 [119], while being vulnerable to corrosion with $NaNH_2$ [147]. Ta was found to form compounds with no known transport process with Na, KN_3, and $RbNH_2$ [186].

Ti and V were both seen to be ineffective against corrosion, with a crack found in the V sample [147]. W was observed as having resistance to corrosion, even in its alloy form, while Y and Zr became heavily corroded [147]. Y was also reported to form compounds in the presence of KNH_2 [187], while Zn was reported to form only soluble intermediates [188], but it is also as used as a dopant [189].

3.3. *Ammononeutral Conditions*

Al [190], Be [191], and Ce [192] were all reported to form soluble intermediates in ammononeutral conditions (liq. NH_3), whereas Au and Co were chemically stable in their alloy forms, with the latter being stable in elemental form, too [147]. Cu forms similar compounds, but with CuCl and KNH_2 in a 1:3 ratio, making it an ammonobasic milieu [193]. Ba was found to form soluble compounds with the presence of a secondary cation like Cs [194], K, and Rb [195].

Similarly, Ca also forms similar compounds with Na [196], K [197], Rb [198], and Cs [199] as secondary cations. Cs and K both act as mineralizers in the case of InN crystal growth, with $InCl_3$ and KNH_2 in a 1:3 ratio [141] or InI_3 and $CsNH_2$ in a 1:3 ratio [200], respectively.

Eu [201], Fe [202], Ga [133], Ge [147], and Gd [203] all form soluble intermediates with supercritical ammonia without any mineralizer present. La [204], Mg [205], and Nd [192] were found to form compounds which were soluble in ammononeutral conditions. Elements like Mo, Ni, Pd, Pt, Si, and W showed inertness to the ammononeutral environment in their respective alloy forms (with Mo, Pd, Si, and W being passive even in elemental form, too) [147]. Li forms soluble intermediates with K [206], whereas Mg forms with Cs [207], K, and Rb [208] as secondary cations in the ammononeutral environment.

Nb and Sc are corrosion resistant [147] in the ammononeutral environment, whereas elements like Sm [209], Th [181], Y [168], Yb [210], Zn [211], and Zr [212] formed soluble compounds. Sr was reported to form soluble intermediates with Rb [195], Cs [199], Na, and K [196] as secondary cations in respective compounds. Ta and V were found to form their respective nitrides, with no information being available regarding their transport [147].

In summary, many elements have been identified to react with the ammonothermal environment, leading to their dissolution and compound formation in basic and acidic environments. Of the two, the acidic environment is more reactive, offering more opportunities for crystal growth at the expense of added complexity to the growth system. The basic environment is less reactive, facilitating the design of these growth systems, though it may in turn limit the type of materials that can be grown from these solutions. Knowledge of the solubility of species in solution and the resulting dependence on temperature (and pressure) are important, as this enables the targeted pursuit of crystal growth using these elements as constituents of the growing nitride crystal. Depending on the degree of solubility, it is possible to dope a material at low concentrations to modify its electrical conductivity, or to form alloys/compounds at higher concentrations. The following section explores the current state of ammonothermal synthesis and crystal growth for various compounds.

4. Synthesis and Crystal Growth of Nitrides

In the following sections, the application of the ammonothermal method for the growth of nitride crystals will be reviewed. The first subsection will focus on the binary and ternary III-nitrides, whereas the second subsection is dedicated to other nitrides obtained by ammonothermal crystallization until now.

4.1. III-Nitrides

This section covers the binary III-nitrides BN, AlN, GaN, and InN, as well as the ternary alloy AlGaN. All subsections give a review of the current state of ammonothermal growth of these nitrides, except for the GaN section. The ammonothermal field of crystal growth is mainly investigated for GaN, contrary to other nitrides. Since there are already several review articles on ammonothermal GaN, this review provides a very compact overview and guides the reader towards the other reviews. For the same reason, we start with GaN before moving on to materials for which ammonothermal growth is less well-investigated.

4.1.1. GaN (Overview of Available Literature)

Many reviews and perspectives have targeted the ammonothermal growth of bulk GaN single crystals [81,86,87,98–100,105,128,142,213–216]. Therefore, this section provides an overview of these reviews and perspectives from 2015–2023. However, a very brief summary of the state of the art will be given beforehand.

State-of-the-art ammonothermal GaN crystals (boules) have reached the following characteristic values (reported from 2018–2023): a boule thickness of 1–4 mm [86], a bowing radius of 20–1460 m [86], a threading dislocation density of $10^{3-5} \times 10^4$ cm^2 [86], a rocking curve width of up to 11.9–19.2 arcsec (reflection: (0002)/slit size: 2 mm $\times$ 1 mm) [84], and a wafer size of up to 100 mm [86]. Information on the process conditions of ammonothermal GaN growth is referred to in Section 2.

The previously published reviews and perspective articles on the ammonothermal growth of bulk GaN are listed in Table 2, alongside their focus areas. In Table 2, the term "ammonothermal scope of the article" shows the number of pages in the publication (reference pages are excluded) that cover the ammonothermal growth of GaN. This value serves as an indicator for the depth of discussion. The topics in Table 2 are grouped into four blocks: miscellaneous, crystal properties, growth conditions, and growth technology. All of the articles with a lower ammonothermal scope value (less than eight pages) have their focus on the general description of the process, or the general part of the growth technology. These articles help to quickly get an overview, whereas the others provide more detailed information.

It is necessary to point out that the focus of the publications did not change for the blocks "crystal properties" and "miscellaneous" from 2015–2023. On the contrary, the focus changed for the blocks "growth conditions" and "growth technology". Over the last eight years, the focus of "growth technology" increased, and the focus of "growth conditions" decreased. This change indicates the shift of the development progress from basic research towards a more commercially usable process, where more focus is put on growth technologies. This observation fits to the assessment of the near-equilibrium ammonothermal (NEAT) method of SiX-Point to a technology readiness level (TRL) of six, which stands for a technology which is near to industrial deployment [217]. Although the process shows a TRL of six, recent ongoing research shows a lot of unknowns in the material system and basic research is still needed, for example, the micro and macro defects in ammonothermal GaN newly identified by L. Kirste et al. [107,218].

Besides that, in 2021 a book was published by R. Niewa and M. Meißner which covers the ammonothermal growth of GaN and further nitride materials, autoclave technology for crystal growth and in-situ monitoring, and the chemistry of ammonothermal growth of GaN and related materials. So far, it is the only book that exclusively looks at the ammonothermal process [219].

Table 2. Overview of identified review and perspective papers and book chapters from 2015–2023 [81,86,87,98–100,105,128,142,213–216] about the ammonothermal growth, focusing exclusively on GaN. Legend/Abbreviations: a: acidic, b: basic, X: topic which was mentioned in the publication at least in some sentences, **F**: the publication focuses on this topic, R: review paper, P: perspective paper, BC: book subchapter, SCoRA: Scalable Compact Rapid Ammonothermal, NEAT: near-equilibrium ammonothermal, SCAAT: Super Critical Acidic Ammonothermal, LPAAT: low-pressure acidic ammonothermal.

	Year	2023	2023	2022	2021	2021	2020	2020	2020	2018	2018	2017	2015	2015
	Reference	[213]	[86]	[98]	[128]	[214]	[81]	[215]	[99]	[105]	[100]	[87]	[142]	[216]
	Publication type	R	R	R	R	BC	P	BC	R	R	BC	R	BC	R
	Ammonothermal scope of the article (pages)	4	14	2	25	7	4	5	3	11	13	15	19	3
Miscellaneous	Process	X	X	**F**	X	**F**	X	**F**	X	X	X	X		**F**
	Historical perspective	**F**	X		X					X			**F**	X
	Research group listing		X	X			X	X		X			X	
	Perspective view			X					X					
	Process		X		X		X	X			X	X	X	
Crystal properties	General	X		X	X	X	X			X	X	X	X	X
	Defects	X	**F**			X	X			X	X	**F**	X	
	Purity/doping		X				X		X	**F**	X	**F**	X	
	Device Applications		**F**		X					**F**		X		
Growth conditions	Environment	b & a	b & a	b	b & a	b	b & a	b & a	b	b	b & a	b	b & a	b & a
	Growth rates		X		X					X	X	X	X	
	Morphology	X						X		X	X		X	
	Chemistry	X										X	X	
	Solubility				X						**F**		X	
Growth technology	General	**F**	**F**				**F**		**F**					
	Autoclave	X	X		X						X	X	X	
	SCoRA	X											X	X
	SCAAT/LPAAT	X	X				X							
	NEAT		X											
	2-stage growth		X			X	X	X	X	**F**				
	Tiled seeds		X				X							
	Circular-cross section		X						X					
	Annealing									X				
	Simulations				**F**						X			
	In situ measurements		**F**		**F**									

4.1.2. AlN

In comparison to the ammonothermal growth of GaN, the ammonothermal growth of AlN has not been heavily reported. The primary method for growing bulk AlN single crystals is the PVT method, which has demonstrated to yield high-quality, free-standing AlN wafers [85]. This section reviews the ammonothermal method for AlN growth by offering a historical view and a detailed discussion of the field.

The first ammonothermal synthesis of AlN was reported in 1988 by D. Peters [38]. The last scientific publication about ammonothermal synthesis of AlN was reported by B. T. Adekore et al. in 2006 [174]. To evaluate to which extent the ammonothermal growth of AlN has been explored, all authors, publications, countries, or institutes that published experimental results on ammonothermal AlN growth (original research papers) are listed chronologically:

- D. Peters, 1988 [160]/D. Peters, 1990 [173], Germany, FuE Chemikalien, Hoechst AG, Werk Knapsack, Hürth;
- R. Dwiliński et al. 1998 [169]/R. Dwiliński et al. 1998 [220], Poland, Warsaw University, Institute of Experimental Physics;
- Y. C. Lan et al. 1999 [159]/Y. G. Cao et al. 2000 [221], Y. G. Cao et. al. 2001 [222], China, Institute of Physics, Centre for Condensed Matter Physics, Chinese Academy of Sciences, Beijing;

- A. I. Motchanyy et al. 2005 [223], Russia, Russian Research Institute for the Synthesis of Materials (VNIISIMS);
- B. T. Adekore et al. 2006 [174], USA, NCSU, Department of Materials Science and Engineering.

In addition to these, there exists a patent from Dwiliński et al. from 2008 [224]. There is one perspective paper which briefly touches on the ammonothermal synthesis of AlN [225] and one review paper that focuses on intermediates in the ammonothermal system in general and touches AlN slightly [119].

The list indicates that a global interest in ammonothermal growth of AlN existed from 1988 until 2008. However, nothing has been published within the last 16 years. Obviously, the global interest in growing ammonothermal AlN has leveled down, likely due to the faster success of PVT growth in the case of AlN, including recent breakthroughs in diameter enlargement [83].

All publications about the ammonothermal synthesis/crystal growth of AlN primarily report on the solubility of Al and characterization of synthesized AlN crystalline powder/grains, which will be discussed below. For the following discussion, the ratio [X:X:X] always stands for the ratio [nutrient:mineralizer:ammonia] and all mole ratios of D. Peters [173] were numerically solved from his experimental data.

Solubility Results Will Be Discussed in the Following

There are two types of nutrients mentioned in the literature, Al and AlN.

Al: D. Peters [173] demonstrated that it is possible to dissolve Al in pure ammonia, using Al-powder as the nutrient. An Al to AlN conversion of 32% could be reached, using Al-powder, the ratio [1:0:7.9], at 500 °C and 200 MPa, though it is likely insufficient to pursue crystal growth. As such, a mineralizer is required to dissolve a significant amount of Al, similarly to Ga or GaN solubility. A summary of conditions under which Al is dissolved in ammonothermal environments is provided in Table 3, together with a description of the crystalline products obtained. Unlike GaN [70], irrespective of growth chemistry, only retrograde solubility has thus far been observed for AlN [160,173,174].

D. Peters [173] investigated the conversion of Al to AlN (Al–AlN conversion) in ammonia + mineralizer solutions, in relation to ammonia decomposition, ammonia to mineralizer ratio, and temperature.

He observed an ammonia decomposition dependent on the Al–AlN conversion. At an ammonia decomposition $\geq$ 40%, a complete Al–AlN conversion was possible, using an Al:mineralizer ratio of 1:0.09–0.13 at 600 °C and 200 MPa.

The ammonia decomposition depends mainly on the pressure/amount of ammonia. The higher the pressure/amount of ammonia, the lower the decomposition rate. However, the ammonia decomposition also depends on the deployed amount of mineralizer (KNH_2). According to the reaction of KNH_2 + $3NH_3$ + Al to $KAl(NH_2)_4$ + $1.5H_2$, a higher amount of KNH_2 leads to a higher decomposition of ammonia. D. Peters [173] based his statements on the following amount of ammonia and mineralizer dependent experiments. Firstly, he increased the Al to ammonia ratio from 1:1.85 to 1:4.4 (at a constant Al to mineralizer ratio of 1:0.09 at 600 °C), which resulted in an increase of the Al–AlN conversion from 40% to 100%. The ammonia filling levels caused a pressure increase from ca. 1 MPa to 2 MPa. Secondly, he increased the Al to mineralizer ratio from 1:0.09 to 1:0.1 (at a constant Al to ammonia ratio of 1:7.9 at 600 °C), which resulted in a decrease of the Al–AlN conversion from 100% to 94%. Accordingly, a peak Al–AlN conversion dependent on the mole ratio of mineralizer (KNH_2) was observed by D. Peters [173], contrary to Ga or GaN. The change in the Al–AlN conversion can be explained by an increase in the solubility, but also by a change in the crystal growth (transport, formation, or dissolution of the intermediate and growth kinetics).

In addition, D. Peters [173] measured an increase in the Al–AlN conversion with a temperature increase from 500 °C to 600 °C, at 200 MPa.

Table 3. Process conditions under which Al dissolved and AlN synthesis was demonstrated. Ascending order according to the layer thickness/grainsize. The mole ratios of D. Peters [173] and B. T.Adekore [174] were numerically solved/calculated from their experimental data.

Temperature Nutrient-Zone (°C)	Temperature Growth-Zone (°C)	Pressure (MPa)	Mineralizer	Solubility	Mol Ratio Al:mineralizer:NH_3	Growth Time (days)	Product	Layer Thickness-/ Grainsize	Reference
500	600	200	KNH_2	retrograde	1:0.09:4.5	2.7–4.6	Dense crystalline layers	Some mm	[173]
525	550	246–286	KN_3	retrograde	1:[1.9–2.6]:[23.5–24.5]	21	Crystalline layers	0.1–1.5 mm	[174]
500	600	100–200	KNH_2	retrograde	1:0.1:[>4]	n/a	Dense crystalline layers	~1 mm	[160]
<500		400–500	Li, K and $LiNH_2$	n/a	n/a	n/a	Crystalline compact grains	≤25 µm	[169]
<550		500	KNH_2 and $LiNH_2$	n/a	1:[n/a]:10	14	Crystalline compact grains as well as small and big needles	Few µm	[220]
400	600	50–200	NH_4I	retrograde	1:0.05:[>2]	n/a	Crystalline powder	100 nm	[160]
450		ca. 197	NH_4Cl	n/a	n/a	n/a	Crystalline powder	Ø 32 nm	[159]
350, 400, 450, 500, 550		80–120	NH_4Cl	n/a	1:1:[n/a]	5–7	Crystalline powder	20–30 nm	[223]

AlN: The solubility of AlN is less investigated than that of Al. B. T. Adekore et al. [174] tried unsuccessfully to dissolve AlN source material, using KN_3 as a mineralizer with the ratio [1:n/a:19.6] at 550 °C and 246–286 MPa. Y. G. Cao et al. [222] successfully dissolved AlN source material using NH_4Cl as mineralizer at 580 °C and probably at above 253 MPa (no AlN:mineralizer:ammonia ratios were given). This could point towards a solubility difference between the acidic and basic system, but more data are needed.

Crystallization Results Will Be Discussed in the Following

Every publication in which ammonothermal dissolution of Al or AlN was reported also reported the crystallization of AlN from the respective solution. The crystalline products obtained are listed in Table 3. Within all the process conditions listed in Table 3, only B. T. Adekore et al. [174] approached a seeded crystal growth. The other authors reported the crystallization of layers, grains, or powders. In the following, we discuss the crystallization of AlN with regards to the obtained largest crystalline products, growth environment, fluid flow, and the influence of the solubility on the obtained crystalline products.

Largest crystalline products: To evaluate the current state of crystalline ammonothermal AlN, a comparison of the crystallization results from D. Peters [173] and B. T. Adekore et al. [174] is useful, as both crystallized a layer in the mm-range and reported it in detail. B. T. Adekore et al. [174] demonstrated an ammonothermal growth of a 100-µm-thick, dense, contiguous polycrystalline AlN-layer on the Ga-polar side of a GaN seed (HVPE) and a less dense, 1.5 mm thick, polycrystalline AlN-layer on the N-polar side, grown within 21 days. D. Peters [173] achieved an AlN-layer up to some millimeters in thickness (probably on the inner autoclave wall), which consisted of closely packed, completely parallel columns grown in the <0001> direction (Al-polar side), within a duration of app. 2.7–4.6 days.

Both D. Peters [173] and B. T. Adekore et al. [174] observed the growth of a lower structural quality layer prior to the establishment of stable growth conditions. This initially grown layer could explain why the seeded growth attempt of B. T. Adekore et al. did not lead to a single crystalline layer.

Growth environment: The growth zone temperatures that yielded crystallization in the basic and acidic system are stated in Table 3.

In the acidic system, Y. C. Lan et al. [159] produced their powder at 450 °C with an NH_4Cl mineralizer. A. I. Motchanyy et al. [223] reported that the nucleation results below 450 °C were the worst with the NH_4Cl mineralizer. Correspondingly, 450 °C is the lowest temperature at which noticeable nucleation of AlN has been observed in the acidic environment. The highest reported crystallization temperatures in the acidic environment were reported by D. Peters [160] at 600 °C with a NH_4I mineralizer. Accordingly, the temperature range where crystallization occurred in the acidic system can be delineated between 450 °C and 600 °C.

In the basic system, D. Peters [173] quoted that the temperature in the growth region has to exceed 400 °C with KNH_2 as mineralizer. R. Dwiliński et al. [220] produced their powder at T < 500 °C with Li, K, and $LiNH_2$ as mineralizers. Correspondingly, the T-range 400–500 °C is the lower T-range in which the nucleation of AlN can be expected. The highest reported temperature for the precipitation in the basic environment is 600 °C with KNH_2 as mineralizer [160,173]. Accordingly, the temperature range where crystallization can be expected in the basic system can be delineated between 400 °C and 600 °C.

The crystallization temperature range in the acidic and basic environment does not seem to differ significantly. The crystallization temperature range of ammonothermal AlN seems comparable to the crystallization temperature range of GaN, though according to Y. G. Cao et al. [222], the growth temperature for AlN is 80 °C higher than GaN.

Fluid flow: Besides temperature, pressure, and mineralizer, the flow velocities of the supercritical fluid are also thought to be of importance. D. Peters [173] suggested and B. T. Adekore et al. [174] simulated that lower flow velocities of the supercritical fluid would be beneficial for crystal growth. This is in accordance with recent investigations on the effects of fluid flow velocity on the quality of ammonothermally grown GaN bulk crystals [138].

Influence of the solubility to the obtained crystalline products: Subsequent studies after that of D. Peters [173] had lower crystallization rates in their growth experiments. A possible explanation for this, is that all groups completely dissolved their Al nutrient. Accordingly, these groups used too low Al nutrient amounts, or too high ammonia or mineralizer amounts. From the results of D. Peters [173] a border ratio can be derived at which the Al to AlN conversion starts to decrease below 100% at 600 °C. This ratio is 1:0.09:4.4. Below this ratio, aluminum does not completely react to AlN. With this growth condition, the growth of ammonothermal AlN single crystals should be more feasible. However, there is a tradeoff, as a decreasing amount of mineralizer also increases the reaction time and decreases the crystallization rate [173].

Accordingly, it seems that ammonothermal AlN growth requires a lower amount of mineralizer dissolved in NH_3, in relation to ammonothermal GaN growth. In comparison, the Na mineralizer: NH_3 ratio which is used for GaN growth is 0.05 at 575 °C [140].

4.1.3. AlGaN

The ammonothermal growth of bulk $Al_xGa_{1-x}N$ is less explored, with only two literature reports [222,226]. Currently, no primary method for growing bulk $Al_xGa_{1-x}N$ exists, though approaches to grow bulk $Al_xGa_{1-x}N$ with other methods like HVPE exist [227].

Via the acidic ammonothermal method, Y. G. Cao et al. [222] grew polycrystalline $Al_xGa_{1-x}N$ fragments with an average thickness of 600 μm. The effective growth time was 5 days, using NH_4Cl as mineralizer, holding at a pressure of above 152–253 MPa and a temperature of 500-580°C. In their experiments, Y.G. Cao et al. [222] observed that the solubility of $Al_xGa_{1-x}N$ depends both on the mineralizer concentration and the Al content. Additionally, Y.G. Cao et al. identified the need to use an Al_xGa_{1-x} alloy as feedstock (as opposed to pure Ga and Al metals) to successfully synthesize polycrystalline $Al_xGa_{1-x}N$ over the entire Al content range (Al: 0%/15%/30%/50%/60%/70%/80%/90%/100%). Regardless of the process conditions used, Y.G. Cao et al. did not succeed in growing $Al_xGa_{1-x}N$ if Ga and Al were used as feedstock.

Via the basic ammonothermal method, Dwiliński et al. [226] grew a 10 μm thick single crystalline layer of $Al_{0.2}Ga_{0.8}N$ on single crystalline GaN. The effective growth time was 2 days, using NaN_3 as a mineralizer, holding at a pressure of 360 MPa and a temperature of 550 °C. Contrary to Y.G. Cao et al. [222], R. Dwiliński et al. [226] used Ga and AlN as feedstock materials.

Overall, these results clearly demonstrate that ammonothermal growth of crystalline $Al_xGa_{1-x}N$ is feasible for the entire range of stoichiometries. However, the feasibility of bulk single crystal growth of $Al_xGa_{1-x}N$ via the ammonothermal method is yet to be proven.

4.1.4. InN

Synthesis of indium nitride (InN) in an ammonothermal environment was first reported in 2018 by J. Hertrampf et al. [141]. The material is of interest as a fast switching, relatively narrow bandgap semiconductor. Given its low decomposition temperature, synthesis has been a major challenge with any method. One notable aspect of the experimental procedure is the use of both acidic and basic mineralizers to produce a compensated ammonia environment closer to neutral in pH. A mixture of $InCl_3$ and KNH_2 at ratios between 0.4:1 and 0.25:1 (or between 1.2:1 and 1:0.75 in terms of Cl:K) was used at temperatures ranging from 663–723 K. Liners of either silicon nitride or boron nitride were also included inside the autoclave to help protect the walls from infiltration by indium. Pressures of operation went as high as 300 MPa, while reaction times ranged from 10 to 120 h. At 773 K, aggregates of faceted ball-shaped crystallites were found at a ~400 nm length scale after 24 h in a BN liner, while rods up to 5 microns long were produced from a Si_3N_4 liner within the same time. At 663K, well-formed platelets reached over 2 microns in length after 10 h in a silicon nitride (Si_3N_4) liner, while chunky cylinder agglomerates of about the same size were observed after 100 h. Longer periods of time only resulted in decomposition of the InN into metallic indium spheres. As a result of these demonstrations, a path to larger crystals based on these conditions is not evident, however the first demonstration of growth was a milestone. It was found that excess quantities of either mineralizer species resulted in at least some metallic indium being formed.

A follow-up contribution from the Stuttgart group [144] investigated the intermediate molecules involved in dissolution, transport, and crystal growth of ammonothermal InN. With a combination of powder diffraction and vibrational spectroscopy, they describe the properties of $[In(NH_3)_5Cl]Cl_2$ and $InF_2(NH_2)$. Both reaction products were found in the bottom of the autoclave (the hottest part), from which the authors infer a retrograde solubility in the system when using operating temperatures of 663–773 K.

In 2022, P. Becker and R. Niewa continue to focus on the ammononeutral regime but move to a different combination of mineralizers: InI_3 and $CsNH_2$ [200]. The researchers also transition to a different pressure and temperature regime with 34 MPa and only 423 K, presumably to avoid decomposition of newly grown InN. Cs is referenced to be about twice as soluble in liquid ammonia, and heavier halides are also generally observed to be more soluble in ammonia [228]. The products of the ammonothermal reaction were InN and CsI, but the particle sizes were smaller and less faceted than those obtained in the syntheses at higher temperatures of base and acid ions, but move to a different combination of mineralizers: InI_3 and $CsNH_2$ [200]. Although the morphology is disappointing, the ability to synthesize materials ammonothermally at such low temperatures and pressures is encouraging for the growth of new nitrides.

The second half of this paper looks to investigate the separate contributions of KNH_2 and ICl_3, returning to the pressure and temperature space of the original paper (250 MPa, 663 K). By spatially separating the two mineralizer compounds, placing the one in the crucible and the other below the crucible at the bottom of the autoclave, they were able to demonstrate that the KNH_2 is completely dissolved in the ammonia while synthesis happens close to the ICl_3. This remains true when the locations of the two materials are swapped. When the indium chloride was at the bottom of the autoclave, InN and KCl were found, but when it was in the crucible, only the intermediate $[In(NH_3)_5Cl]Cl_2$ was found. It would seem, then, that the mobility of $InCl_3$ and the intermediate in the autoclave is relatively small.

4.1.5. BN

Both cubic and hexagonal boron nitride have become materials of great interest in the device research community [229]. Cubic boron nitride (in zincblende structure, termed $\bar{4}3$m-BN in this review for consistency but often termed c-BN) has been demonstrated as mm-scale single crystals, derived from the high pressure high temperature (HPHT) synthesis method at temperatures from 1200–1800 K and pressures ranging from 2–7 GPa [230]. It is an UWBG semiconductor with attractive predicted properties for high power switching [231]. Hexagonal boron nitride (6/mmm-BN, usually termed h-BN) is formed of 2D graphene-structured layers held together by van der Waals forces [232]. While both graphene and 6/mmm-BN have high thermal conductivity, unlike graphene, 6/mmm-BN is electrically insulating, making it a useful complement for potential integrated 2D devices [233]. Neither material can currently be synthesized as single crystals in the bulk at centimeter or greater length scales. Because of the success of ammonothermal growth for other nitrides, BN is thought to be a candidate for ammonothermal synthesis, although relatively little published work is available in this direction.

The earliest reported successful attempt is a 2018 paper by Maruyama et al. [234]. Using both Ba_3N_2 and MgB_2, along with hexagonal boron nitride as educts, the authors demonstrate synthesis of both 6/mmm-BN and rhombohedral boron nitride ($\bar{3}$-BN, known as r-BN). With an autoclave filled with supercritical ammonia at 750 °C and 120 MPa, two synthesis conditions were tested, one with the Mg- and Ba-containing additives mixed with 6/mmm-BN and one with 6/mmm-BN alone. In the absence of the alkaline earth additives, no change to the 6/mmm-BN morphology was observed. With the additives, however, triangular features on the order of 100 nm were found and X-ray diffraction (XRD) was able to detect peaks unique to $\bar{3}$-BN. No estimate was made for the solubility of the materials in ammonia, nor was any mass transport necessarily involved, since products and reactants were in the same location.

In a 2023 paper, J. Dooley et al. report on the solubility of boron nitride in ammonia using sodium as a mineralizer [111]. The experiments involved $\bar{4}3$m-BN feedstock placed in a crucible inside an autoclave. The reactor conditions varied from 450–600 °C and 150–190 MPa with 0.2 mol of $\bar{4}3$m-BN, 0.00 to 0.53 mol/L of sodium, and 17.7 to 24.6 mol/L of ammonia. Holding times varied from 24 to 96 h. The data presented support a normal solubility behavior relative to temperature peaking at 0.05 mol_{BN}/mol_{NH3} at a temperature of 600 °C. Full saturation was apparently achieved in less than 24 h, and solubility as a function of sodium content was essentially flat for sodium density higher than 0.08 mol_{Na}/L. The solubility was calculated by comparing $\bar{4}3$m-BN mass in the crucible before and after the ammonothermal experiment, so it is evident that the dissolved BN species was mobile enough to move through the mesh cap of the crucible and deposit itself elsewhere in the autoclave. With both an exploitable solubility curve and nutrient species mobility established, some kind of BN growth is expected to be feasible. To date, however, the nature of such growth has yet to be reported.

Besides that, there exists a paper from 1979, which reports on a process near to the ammonoacidic process, the HPHT synthesis [163]. T. Kobayashi [163] converted 6/mmm-BN powder (1–5 μm in size) to $\bar{4}$3m-BN powder (0.2–1 μm in size) at 5.6 GPa, 1500 °C, in 30 min (process time), with four different mineralizers as solvents: NH_4F, NH_4Cl, NH_4Br, and NH_4I. But, contrary to the ammonothermal process, no ammonia was filled in. He observed that the gained conversion rate was dependent on the mineralizer used. The gained conversation rate decreased with an increase of the melting point of the mineralizer. Accordingly, the highest conversation rate has been reached with NH_4F. That points towards ammonothermal comparable conditions and a transport of B in the fluid phase, because the conversion rate was dependent on the mineralizer.

4.2. Novel Nitrides and Related Materials beyond Group III-Nitrides

A variety of ternary and multinary nitrides have been obtained via variants of ammonothermal synthesis. Those materials are listed in Table 4, together with the main synthesis parameters used. From a crystal growth point of view, it is important to note that in most cases, the synthesis and transport mechanisms are not well understood, and the synthesis of microcrystalline powders does not necessarily imply an appreciable solubility of all constituents in supercritical ammonia with the mineralizer and process conditions applied. In the case of $ZnGeN_2$, the presence of X-ray absorbing solutes likely representing Zn- and Ge-transporting intermediates has been observed via in situ X-ray imaging [42].

Table 4. Nitride materials beyond the III-nitrides that have been synthesized via an ammonothermal synthesis process so far.

Material	Space Group	*p*/MPa	*T*/°C	Mineralizer	Year	Reference
$NaTaN_2$	R$\bar{3}$m	100	600	$NaNH_2$	1988	[235]
$KTaN_2$	Pbca	100	600	KNH_2	1988	[235]
$RbTaN_2$	Pbca	100	600	$RbNH_2$	1988	[235]
$CsTaN_2$	Fd3m	100	600	$CsNH_2$	1988	[235]
$Li_2Ta_3N_5$	C2/m	600	550	$Li/Li_3N/LiNH_2$	1991	[236]
$NaSi_2N_3$	$Cmc2_1$	600	575	$NaNH_2$	1993	[237]
$K_3P_6N_{11}$	$P4_132$	600	500	KNH_2	1997	[238]
$SrAlSiN_3$	$Cmc2_1$	100	500	$NaNH_2$	2012	[94]
$MgSiN_2$	$Pna2_1$	170	797	KN_3	2017	[36]
$MgGeN_2$	$Pna2_1$	230	597	NaN_3	2017	[36]
$MnSiN_2$	$Pna2_1$	170	797	KN_3	2017	[36]
$MnGeN_2$	$Pna2_1$	170	597	NaN_3	2017	[36]
$LiSi_2N_3$	$Cmc2_1$	170	697	LiN_3	2017	[36]
$LiGe_2N_3$	$Cmc2_1$	230	627	LiN_3	2017	[36]
$ZnSiN_2$	$Pna2_1$	230	797	KN_3	2017	[42]
$ZnGeN_2$	$Pna2_1$	230	797	KN_3	2017	[42]
Mg_2PN_3	$Cmc2_1$	140	797	NaN_3	2018	[58]
Zn_2PN_3	$Cmc2_1$	200	527	KN_3	2018	[58]
$Mg_{0.5}Mn_{0.5}SiN_2$	$Pna2_1$	150	797	KN_3	2019	[9]
$Mg_{0.5}Zn_{0.5}SiN_2$	$Pna2_1$	150	797	KN_3	2019	[9]
$Mn_{0.5}Zn_{0.5}SiN_2$	$Pna2_1$	150	797	KN_3	2019	[9]
$Mg_{0.5}Mn_{0.5}GeN_2$	$Pna2_1$	200	597	NaN_3	2019	[9]
$Mg_{0.5}Zn_{0.5}GeN_2$	$Pna2_1$	200	597	NaN_3	2019	[9]

Table 4. *Cont.*

Material	Space Group	p/MPa	T/°C	Mineralizer	Year	Reference
$Mn_{0.5}Zn_{0.5}GeN_2$	$Pna2_1$	200	597	NaN_3	2019	[9]
$NaTaN_2$	$R\bar{3}m$	170	800	$NaN_3/KN_3/RbN_3/CsN_3$	2019	[186]
$KTaN_2$	Pbca	170	800	$NaN_3/KN_3/RbN_3/CsN_3$	2019	[186]
$RbTaN_2$	Pbca	170	627	$NaN_3/KN_3/RbN_3/CsN_3$	2019	[186]
$CsTaN_2$	$Fd\bar{3}m$	170	350	$NaN_3/KN_3/RbN_3/CsN_3$	2019	[186]
$Sr_3P_3N_7$	P2/c	140	797	NaN_3	2020	[239]

In summary, the crystal growth of a few binary nitrides, including GaN, AlN, InN, and BN and a few ternary nitrides has been demonstrated. Of these, GaN is most matured, while AlN, InN, and BN and all of the ternary nitrides are in their early stages of investigation and development. Given the current state of the art, opportunities remain for the exploration of novel nitride compounds of binary, ternary, or higher order. Additionally, looking beyond crystal growth in itself, the control over the functional properties (such as electrical conductivity, optical absorption, magnetic properties, etc.) is a fruitful area to obtain functional materials for use in a larger fabrication chain leading to functional devices (i.e., as bulk materials or substrates). Understanding and controlling defects that contribute to changes in material properties is important and will be discussed in the following section.

5. Generation of Point Defects and Their Impact on Crystal Properties

This section focuses on the effects of point defects (both native and impurity-related) on the properties of the ammonothermally grown crystals. More specifically, Sections 5.1 and 5.2 refer to unintentional point defects/dopants. Section 5.3 refers to intentional point defects/dopants and their interaction with unintentional point defects/dopants, which in turn influence semiconductor properties.

Currently, it is possible to crystallize both n- (O as a dopant) and p-type (Mg as a dopant) conductive, as well as semi-insulating (SI) (Mg or Mn as a dopant) ammonothermal GaN. The mentioned dopant elements represent examples of the most commonly used ones for ammonothermal GaN.

The discussion in the respective sections is based exclusively on the data from ammonothermally grown GaN crystals, as the available information is insufficient for all other materials grown or synthesized ammonothermally thus far. Nevertheless, considering the unique characteristics of the ammonothermal growth environment, it can be assumed that some of the effects observed for GaN will also occur for some other nitrides grown with this method. These characteristics are the supercritical ammonia solution containing mineralizers, the growth temperature up to 600 °C, and pressure up to 450 MPa.

5.1. Native Point Defects and Related Defect Complexes

The most predominant native point defects in GaN are Ga–vacancies and their related complexes [105,240–242]. For GaN-growth using the basic ammonothermal environment, Ga–vacancies are found to be in the range of 10^{18}–10^{19} cm^{-3} [240,241]. Recently, in 2024, M. Zając et al. [243] showed for the first time that N–vacancies form in basic ammonothermal GaN as well (specifically, in Mg-doped GaN), albeit only in small amounts in the order of 10^{17} cm^{-3}. For GaN growth using the acidic ammonothermal environment, mainly Ga–vacancies (with few N–vacancies) are stated to be in the range of 3×10^{16}–9×10^{16} cm^{-3}, according to the recent publication by K. Shima et al. [242]. These Ga–vacancies and their related complexes are of interest for several reasons; they act as electron acceptors and hence compensate n-type doping and limit the carrier mobility, they contribute to sub-bandgap optical absorption via formation of near-bandedge states, and they even contribute to device degradation [105,240].

In the ammonothermal system, the Ga–vacancy complexes with H atoms (V_{Ga}–H complexes) are the predominant complexes [240,244]. Hence, we focus on these defects in this section, which is primarily based on the reviews by S. Suihkonen et al. [87] and M. Zając et al. [105], but includes more recent advances as well. For considerations on the formation of defects, note that the ammonothermal system represents N-rich conditions. The charges state of the point defects will be stated using the "modified" Kröger–Vink notation [245].

Of all bulk crystal growth techniques for GaN, the ammonothermal technique yields GaN with the highest Ga–vacancy concentration. This is speculated to be due to the relatively low growth temperature [240]. According to F. Tuomisto [240], the formation mechanism of Ga–vacancies and their related complexes depends strongly on temperature, pressure, and chemistry of the different crystal growth processes. However, the amount of oxygen impurities has been identified as one of the main variables controlling Ga–vacancy formation; for all bulk GaN crystal growth techniques, the concentration of Ga–vacancies increases with increasing oxygen concentration [240].

Removal of a Ga atom from a Ga site in the GaN crystal lattice takes away three valence electrons and leaves behind four nitrogen dangling bonds, comprised mainly of the 2p orbitals of nitrogen [246]. This vacancy can have four different charge states from V_{Ga}^{-3} till V_{Ga}^{+1} [246]. All these charge states have very high formation energies and are unlikely to be formed (V_{Ga}^{-3} has the lowest formation energy at the conduction band minimum (CBM) near 2 eV at N-rich conditions [247]). With a donor impurity complex, the formation energy of a Ga–vacancy can be considerably lowered [246]. For such a Ga–vacancy complex generation to happen, however, a donor impurity must be present in a high amount. In ammonothermal growth, hydrogen is such an impurity which can act as a donor, and it likely occurs in high quantities due to the ubiquity of hydrogen in the solvent ammonia. The lowered formation energy in combination with the hydrogen oversupply results in the mentioned formation of the predominant Ga–vacancy complexes with multiple H atoms [246,248]. It is hypothesized that hydrogenated Ga–vacancy complexes form by an incomplete dehydrogenation of NH_3, NH_2^{-1}, or, NH_4^{+1} molecules [87]. This assumption is preceded by the assumption that N from NH_3, NH_2^{-1}, or, NH_4^{+1} is the nitrogen source of GaN [87].

According to the four nitrogen dangling bonds, four V_{Ga}–H complexes are: ($V_{Ga}H_1$), ($V_{Ga}H_2$), ($V_{Ga}H_3$), and ($V_{Ga}H_4$). These V_{Ga}–H complexes can have the following charge states: $(V_{Ga}H_1)^{0,-1,-2}$, $(V_{Ga}H_2)^{0,-1}$, $(V_{Ga}H_3)^0$, and $(V_{Ga}H_4)^{+1}$. According to first-principles calculations, $V_{Ga}H_{1-2}$ forms a deep acceptor about 0.8 eV above the valence band maximum (VBM) [87], and $V_{Ga}H_3$ is neutral and forms a shallow defect-state close to the VBM [87]. $V_{Ga}H_4$ would be the only complex that forms a donor defect-state in the valence band [87], but this was not detected so far.

Which of the four V_{Ga}–H complexes will be formed depends on the likelihood of multiple H atoms being available and the formation energy at the time of incorporation [248]. The formation energy depends on the Fermi energy level (FEL). Figure 12 shows the dependence of formation energy on the FEL for wurtzite–GaN at Ga-rich conditions, calculated from A. F. Wright [249]. The difference between Ga-rich and N-rich conditions is typically in the range of <1 eV, accordingly, the order of the formation energies of the V_{Ga}–H complexes is not influenced. The order is: $V_{Ga}H_1 > V_{Ga}H_2 > V_{Ga}H_3 > V_{Ga}H_4$. But it is necessary to mention that according to the used functional and framework approximations, the formation energy order can change. C. E. Dreyer et al. [250] calculated the formation energies of $V_{Ga}H_1$, $V_{Ga}H_2$, $V_{Ga}H_3$ dependent on the FEL, also via DFT at N-rich conditions in the wurtzite structure, but they used other framework approximation. They have calculated for $V_{Ga}H_1$, $V_{Ga}H_2$, $V_{Ga}H_3$ the same formation energy order for an FEL near the VBM, but not near the CBM. For an FEL near the CBM, they have the following order: $V_{Ga}H_1 > V_{Ga}H_3 > V_{Ga}H_2$. M. A. Reshchikov et al. [251] calculated the formation energies of $V_{Ga}H_1$, $V_{Ga}H_2$, $V_{Ga}H_3$, $V_{Ga}H_4$ via the Heyd–Scuseria–Ernzerhof hybrid density functional (HSE) at N-rich conditions, probably in the wurtzite structure and for an FEL near the CBM. Under these conditions, they have obtained the following order of formation energies: $V_{Ga}H_4 > V_{Ga}H_1 > V_{Ga}H_3 > V_{Ga}H_2$. How many H atoms (1–4) are available at the time of incorporation is difficult to determine, but if the complex formation takes place due to an incomplete dehydrogenation of NH_3 (like that mentioned above), ($V_{Ga}H_1$), ($V_{Ga}H_2$) and ($V_{Ga}H_3$) can be expected to be formed.

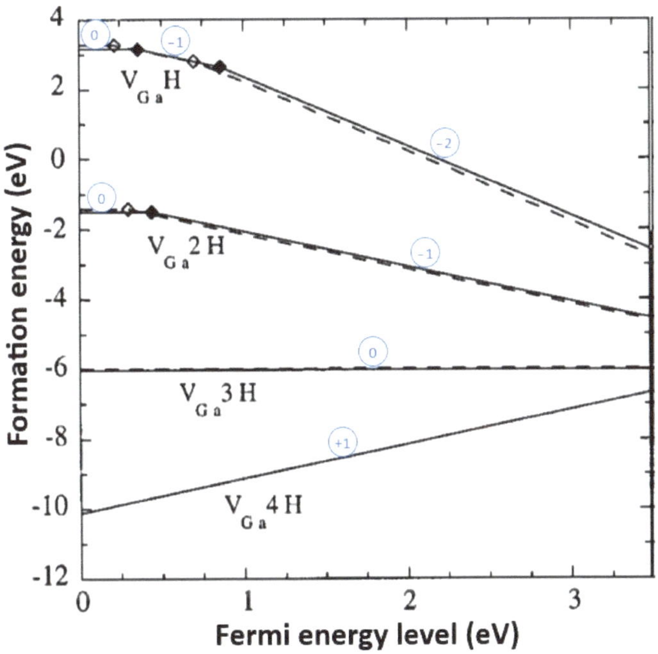

Figure 12. Formation energy as a function of FEL for hydrogenated Ga–vacancies for Ga-rich conditions of wurtzite–GaN, calculated via DFT. The FEL is defined to be 0 eV at the VBM and 3.48 eV at the CBM. Legend: blue circles the oxidation states of the complexes, diamond symbols: transitions of the oxidation stages, solid line: H atoms have a bonding to the N atom located along the c-axis, dashed line: H atoms have no bonding to the N atom located along the c-axis. Reprinted and modified from A. F. Wright 2001 [249] with permission, Copyright (2001) by the American Physical Society.

To detect the type and quantity of complexes in GaN, the following two measurement techniques are used: Fourier transform infrared (FTIR) spectroscopy [241,252] and positron annihilation spectroscopy (PAS) [183,253–255]. Photoluminescence (PL) can also be used, but the technique needs more background information for the interpretation of the spectra [242–244,256]. According to the FTIR data compiled by M. Zając et al. [105] for the basic environment, the ($V_{Ga}H_3$) complex was always predominantly present, given an oxygen content ranging from 1×10^{18} cm^{-3} till 1×10^{19} cm^{-3}. In addition to the ($V_{Ga}H_3$) complex, the ($V_{Ga}H_{1,2}$) were also present at oxygen concentrations > 10^{18} cm^{-3} ($V_{Ga}H_{1,2}$). The amount of ($V_{Ga}H_{1,2}$) increases with the oxygen amount in the GaN crystal [105]. Further, the ($V_{Ga}H_4$) complex was not observed and instead a complex containing O–H bonds was detected. The amount of this O–H bond complex also increased with an increasing amount of oxygen. Using FTIR, W. Jiang et al. [252] investigated GaN samples grown via an ammonoacidic process with varying oxygen content ranging from 1.6×10^{18} cm^{-3} to 5.1×10^{18} cm^{-3}. In contrast to the basic environment, the ($V_{Ga}H_{1,2}$) complex was predominant and not the ($V_{Ga}H_3$) complex (interpretation of W. Jiang et al. [252] spectra with M. Zając et al. [105] Ga–vacancy absorption edges). Further, the amount of ($V_{Ga}H_{1,2}$) increased with an increasing oxygen concentration, like in the ammonobasic environment. Using PAS data from ammonoacidic GaN, S. F. Chichibu et al. [244] exclude the presence of V_{Ga}–H complexes with multiple H atoms. According to the measured positron diffusion length, the ($V_{Ga}H_1$) complex appears to be the only possible V_{Ga}–H complex.

These FTIR and PAS observations of the V_{Ga}–H complexes in both the acidic and the basic environments has the following implications. The detection of the complexes ($V_{Ga}H_1$), ($V_{Ga}H_2$), and ($V_{Ga}H_3$), but not ($V_{Ga}H_4$), supports the hypothesis that Ga–vacancies are formed due to an incomplete dehydrogenation of NH_3. In the ammonoacidic environment, the ($V_{Ga}H_{1,2}$) complexes are the dominant complexes. In the ammonobasic environment, the ($V_{Ga}H_3$) complex with more H atoms is the dominant complex. According to Figure 12, ($V_{Ga}H_3$) is more likely to be formed than ($V_{Ga}H_{1,2}$), provided that enough H atoms are available. Accordingly, in the ammonoacidic system, a lower amount of hydrogen is available at the time of incorporation, contrary to the ammonobasic system, assuming the process takes place near the thermodynamic equilibrium.

From the literature, it is not evident whether crystals grown in the acidic or the basic ammonothermal environment differ systematically in their Ga–vacancy concentrations. According to the following consideration, though, it can be speculated that acidic growth leads to a lower concentration of Ga–vacancies. According to F. Tuomisto [240], higher crystal growth temperatures lead to a lower number of Ga–vacancies, like it is typical for the acidic, in comparison to the basic, ammonothermal growth of GaN. Furthermore, higher oxygen concentrations lead to higher Ga–vacancy concentrations [240]. The mean oxygen concentration in the ammonoacidic system is lower than in the ammonobasic system, according to S. Suihkonen et al. [87]. In addition to that, recent publications indicate a similar trend [240–242] (see also the first paragraph of this chapter).

The concentration of V_{Ga}–H complexes can be reduced with different effectiveness by post-growth annealing. Hydrogen is mobile in high-resistivity/p-type GaN at an annealing temperature > 500 °C [251]. The mobility depends on the charge state of hydrogen, H_i^- and H_i^0 were found to be less mobile than H_i^+ [251,257,258]. This difference explains why the conditions that are effective for the removal of hydrogen differ by conductivity type as follows: 900–1000 °C at 1 atm under N_2 atmosphere for p-type GaN, and at 1400 °C at 1 GPa under N_2 atmosphere for n-type, respectively [251].

Before bonded hydrogen in a complex can diffuse, it must break free from the complex. 2016, in n-type (oxygen-doped) GaN, A. Uedono et al. [255] could not destroy Ga–vacancy complexes containing O or H by annealing at 1000 °C under N_2 atmosphere, at an annealing time of 24 h. 2022 M. A. Reshchikov et al. [251] were able to reduce the amount of ($V_{Ga}H_3$) at an annealing temperature > 1000 °C, at a nitrogen pressure of 1 GPa and an annealing time of 1 h. Moreover, they observed that higher annealing temperatures result in a more efficient reduction of the ($V_{Ga}H_3$) concentration.

Ga–vacancies are known to be mobile at >227–327 °C [259]. K. Horibuchi et al. [260] observed a "helical" deformation of a threading dislocation and the generation of voids in the GaN bulk material via heat treatment (1100 °C and atmospheric pressure). On the GaN substrate, a GaN layer was epitaxially deposited via MOCVD. This layer underwent the same thermal treatment and showed no "helical dislocations" and voids. They explained the generation of voids by the aggregation of vacancies, which deform the threading dislocation into a helical shape. F. Tuomisto et al. [261] showed that ammonothermal GaN has a higher number of Ga–vacancies than MOCVD GaN. This result supports the assumption of K. Horibuchi et al. [260] that the observed voids and the deformation of the threading dislocations are related to vacancies.

The current state of annealing of ammonothermal GaN is summarized in a paper by M. A. Reshchikov et al. [251].

Doping GaN using Mg has been found to strongly suppress the formation of Ga–vacancies [183]. It is possible to eliminate all the V_{Ga}–H complexes in an Mg-doped oxygen-containing GaN crystal, due to an annealing process at 1100 °C under N_2 atmosphere [105,183]. This effect is only detectable at Mg concentrations above 1×10^{18} cm^{-3} [105,183].

5.2. Impurities

The most cited measurement technique for impurities for ammonothermal GaN is SIMS (secondary ion mass spectrometry) [86,87,100,105].

One of the most successful measures implemented thus far for the reduction of impurity concentrations in ammonothermal GaN is the usage of noble corrosion-resistant metal (Ag/Pt) liners, which decreased the amount of most transition metals in the crystals to below the detection limit. For more details on this advancement, see Section 6.2. In this section, we focus mostly on discussing impurities that are incorporated in measurable quantities even with the use of a liner in the basic and/or acidic environment:

- H, O (major impurities) [87];
- C, Na, Mg, Al, Si, Mn, Fe, Zn (minor impurities) [87,105].

Zn is an exception, this element was not found in GaN crystals grown in an autoclave with a liner, but it was found in ammonobasic GaN crystals grown without the use of a liner [105].

This section discusses their concentration, contamination source, potential countermeasures, defect behavior, and their likelihood to form. Additionally, we investigate the effect of the crystallographic orientation of growing surfaces (Facet) and the growth time-dependent incorporation of impurities. Table 5 shows all previously mentioned impurities with their growth environment, concentration, and potential sources.

Table 5. Current state of impurities in GaN grown via the ammonothermal method with a liner (Zn without a liner). All data which are not individually referenced are from [87,100,105] and were measured by SIMS in bulk GaN.

Impurity	Growth Environment	Impurity Type	Concentration Min.	Concentration Max.	Potential Impurity Sources	
Si, Al, Mg	Basic	Minor	High 10^{16}	Low 10^{17}	GaN nutrient (impurities) Mineralizer (impurities) Autoclave	[100] [100] [146]
Mn, Fe, Zn	Basic	Minor	High 10^{16}	Low 10^{17}	Autoclave	[140,146]
C	Basic	Minor	High 10^{16}	Low 10^{17}	Surface contamination	[87,100]
Na	Basic	Minor	low 10^{16}	High 10^{18}	Mineralizer (solvent)	[87,100]
H	Basic Acidic	Major Major	Low 10^{18} Mid 10^{17}	Mid 10^{20} High 10^{19}	Ammonia (solvent) Mineralizer (solvent)	[87,100] [103,262,263]
O	Basic Acidic	Major Major	Low 10^{18} High 10^{17}	Mid 10^{20} High 10^{19}	Surface contamination Mineralizer (impurities) GaN nutrient (impurities)	[87,100] [103,262,263] [100,140,252,262]

Note that the ammonothermal system represents N-rich conditions. Impurities that are incorporated at the Ga site are stabilized at N-rich conditions [264], whereas impurities which go on the N site are more stable at Ga-rich conditions. Accordingly, ammonothermal conditions favor the incorporation of those impurities that are incorporated on the Ga site. The lower the formation energy of an impurity, the more stable and likely it is to be incorporated in GaN. Note, the formation energies are calculated and mainly given in the publications in diagrams like Figure 12. The electronic state of impurities will be given according to the "modified" Kröger–Vink notation [245].

Silicon: Independently of the FEL, Si is stable on the Ga site at the charge state Si_{Ga}^{+} [65,247] and forms a shallow donor 12–27 meV below the CBM [265]. All Ga–vacancy complexes have a higher formation energy than Si on the Ga site at the charge state Si_{Ga}^{+} [247]. Only the neutral charge state of the Si_{Ga}–C_N complex has a lower formation energy near the CBM [266].

Aluminum: Al forms a solid solution with Ga in $Al_xGa_{1-x}N$ (Section 4.1.3). Accordingly, Al has a good solubility in GaN and goes on the Ga site at the neutral charge state. J. H. Lee [267] grew Al-doped GaN via MOCVD and ascribed the reduced concentrations of electron scattering centers and non-radiative recombination centers to the Al doping. That also decreased the compensating acceptors, Ga–vacancies, and their related complexes [267].

Magnesium: Mg is stable on the Ga site and forms an acceptor level at 200 meV above the VBM [246,268]. According to J. L. Lyons et al. [268], Mg is an "accidentally" shallow acceptor, with a $Mg^0_{Ga} + e^-$ to Mg^-_{Ga} transition for an FEL close to the VBM. At the neutral charge state, the hole is highly localized, contrary to a "normal" shallow acceptor where the hole is delocalized and loosely bonded [268]. In the immediate vicinity of the FEL to the VBM, the charge state Mg^+_{Ga} is more stable than the neutral charge state at a formation energy of ca. 2 eV [268].

There are three theoretical studies which deal with different Mg complexes. The first study from J. Neugebauer and C. G. Van de Walle, published in 1995 [269], calculated the formation energy for Mg_{Ga}–H complexes which had a very low formation energy and a neutral charge sate at< -3 eV. In the publication, there is no information whether they calculated it for Ga- or N-rich conditions. The second study published by J. L. Lyons and C. G. Van de Walle in 2017 [270] calculated the formation energy for Mg_{Ga}–V_N complex. The complex had the lowest formation energy at the VBM near 0 eV and the charge state $(Mg_{Ga}–V_N)^{+2}$ under N-rich conditions. D. Lee et al. [271] calculated the formation energy for M_{Ga}–H_i–V_N complexes, which had the lowest formation energy at -2 eV near the VBM and the charge state $(Mg_{Ga}–H_i–V_N)^{+3}$. In the publication, there is no information whether they calculated it for Ga- or N-rich conditions. Due to the oversupply of hydrogen in the ammonothermal process and the lowest formation energy of the complex Mg_{Ga}–H, this complex will be the predominant form. However, every mentioned Mg complex introducing a state for an FEL near the VBM has a lower formation energy than Mg_{Ga}. The Mg_{Ga}–H complex is the dominant defect formed in GaN when growing in a hydrogen-containing environment. This defect is neutral with a deep level at 1.02 eV above the VBM [268]. As is now well known, this Mg_{Ga}–H complex can be separated into H_i and Mg_{Ga} by annealing (Section 5.1).

Si, Al, Mg group: This group presents all metal impurities which are not transition metals and occur in GaN crystals despite the use of a liner. All of them, in their pure form, do not act as a deep level carrier trap [246,265,268]. Prospective methods for the reduction of their concentration will be discussed in the paragraph on oxygen (for contamination via polycrystalline GaN or mineralizer) and in the paragraph on the Mn, Fe, Zn group (for contamination via the inner autoclave wall), respectively.

Iron: Fe is stable on the Ga site at the charge states Fe^0_{Ga}, Fe^+_{Ga}, or Fe^-_{Ga} and forms a deep acceptor 0.5 eV below the CBM [246]. The charge state Fe^0_{Ga} is stable for an FEL in the middle of the bandgap [272]. For an FEL near the CBM, the charge state Fe^-_{Ga} gets more stable and near the VBM, the charge state Fe^+_{Ga} gets more stable [272]. With about 3 eV under N-rich conditions, the formation energy is relatively high [272]. For an FEL near the VBM, the complexes Fe_{Ga}–V_N, Fe_{Ga}–H_i, and Fe_{Ga}–O_N are more stable than Fe_{Ga} [272]. Near the CBM, only Fe_{Ga}–O_N is more stable [272]. All mentioned complexes act as deep defects in the bandgap [272]. Due to the high presence of all the complex partners H_i and O_N in the ammonothermal system, the formation of the complexes Fe_{Ga}–H_i and Fe_{Ga}–O_N is likely.

Zinc: Zn is stable on the Ga site at the charge states Zn^0_{Ga}, Zn^+_{Ga}, or Zn^-_{Ga}, and forms a deep acceptor 0.40–0.46 eV above the VBM [273,274]. The formation energy for an FEL near the VBM is relatively high at ca. 2 eV and decreases to -1 eV near the CBM under N-rich conditions [274].

Manganese: Mn is stable on the Ga site at the charge states Mn^0_{Ga}, Mn^+_{Ga} or Mn^-_{Ga} and forms a deep acceptor 1.5–1.8 eV above the VBM [105]. The charge state behavior is similar to Fe; the charge state Mn^0_{Ga} is stable for an FEL, in the middle of the bandgap. For an FEL near the CBM, the charge state Mn^-_{Ga} is more stable, whereas the charge state Mn^+_{Ga} is more stable for an FEL near the VBM [275]. The formation energies are relatively high, around 2 eV at N-rich conditions [275].

Mn, Fe, Zn group: This group represents all transition metal impurities in GaN crystals that can be reduced in concentration by using a liner. Typically, electron-rich transition metals form deep level carrier traps, so they are of the greatest concern [86]. The solubility limit of Mn in GaN crystals is <1%, 0.03% for Fe, and > 1% for Zn [205], in the basic growth atmosphere. The incorporation risk of these impurities in GaN decreases with a decreasing solubility and an increasing formation energy. Accordingly, the contamination risk by Fe should be lower than by Zn (at least for an FEL near the CBM). The contamination of Mn, Fe, and Zn has its origin from the inner autoclave wall. A significant improvement should be reached via the usage of a hermetically sealed liner instead of a pressure-balanced liner. The effect of the liner will be discussed in more detail in the Section 6.1.

Carbon: C goes predominately to the N site [246]. Experiments showed that the Ga site becomes only relevant at high C contamination concentrations [246]. The neutral complex C_N–H_i has a lower formation energy for an FEL near the VBM than C_N [266] at N-rich and Ga-rich conditions. This C_N–H_i complex will likely form in the ammonothermal system at an FEL near the VBM, where the formation energy is lower than the formation energy of C_N. At the N site, C acts as a deep acceptor at 1 eV above the VBM [246]. The formation energy of the C_N state decreases for an FEL near the CBM. For an FEL, near the VBM, the neutral charge state, and nearest to the VBM, a donor charge state gets more stable at 0.4 eV above the VBM [246]. A further cleaning alongside a refined pre-growth etching process is expected to lead to the complete elimination of the C impurity [87].

Sodium: Na goes predominantly on the interstitial site as a donor, with a lower likelihood on the Ga site as an acceptor [264]. Both sites are stable, so they are self-compensating. Other mineralizers like Li and Ka can potentially show the same behavior, however, the formation energy of K at the Ga site is dramatically high, causing it to act predominantly as a donor [264]. The incorporation of sodium in GaN is non-uniform from run-to-run [87], and it differs according to the growth facet [105]. Other mineralizers in the basic (alkali metals) and acidic (halogenides) environment have been reported to behave similarly [87]. So far, a mineralizer is needed for the ammonothermal growth of GaN and cannot be omitted. However, the growth conditions can be changed to increase the formation energy of sodium on the interstitial-site in GaN. Another way is the usage of other mineralizers that are not as easily incorporated, for example potassium in GaN.

Hydrogen: H incorporates in GaN as an interstitial or forms complexes (for more details, see Section 5.1).

Oxygen: Independently of the FEL, O is stable on the N site at the O_N^+-state [65] and forms a shallow donor at 4–29 meV below the CBM [276].For an FEL near the CBM, the complexes V_{Ga}–O_N–2H, V_{Ga}–O_N–H, and V_{Ga}–O_N are predicted to form more stable states as O_N at Ga-rich conditions [250,266]. The formation energy of O_N varies by only about 0.5 eV between Ga-rich and N-rich conditions [266]. Accordingly, the formation energy relationship between the complexes and O_N should not change much. The mentioned relationship should also be valid for N-rich conditions. Due to the high amount of H and V_{Ga} in the ammonothermal system, it is likely that V_{Ga}–O_N–2H, V_{Ga}–O_N–H, and V_{Ga}–O_N will be formed. Experiments showed that the mentioned complexes are formed, but are not dominant. Instead, the V_{Ga}–H_x complexes are predominant in ammonothermal GaN (see Section 5.1). The complexes V_{Ga}–O_N–2H, V_{Ga}–O_N–H form deep acceptor states [250]. V_{Ga}–O_N–2H has the lowest formation energy and an activation energy of ca. 1 eV above the VBM [277]. As mentioned in Table 5, there are three known oxygen sources.

Reduction of oxygen impurities originating from adsorbates on surfaces is likely to be achieved via improved cleaning, dehydration, and pre-growth etching process refinements [87].

For the basic ammonothermal system, a reduction in oxygen contamination from the mineralizer can be achieved via selection of precursors with lower oxygen and moisture sensitivity. For example, in sodium, it decreases in the following order: sodium amide, sodium azide, and metallic sodium [263]. The experiments showed a clear tendency that the growth rate and crystal quality increases with a decreasing sensitivity to oxygen and moisture of the Na mineralizers, which implies a lower oxygen/contamination content of the GaN crystal [263]. This observed behavior for a Na mineralizer should be transferable to other basic mineralizers.

For the acidic ammonothermal system, the purity can be improved via two methods. The first one is to use NH_4F as the mineralizer and perform crystal growth in the observed retrograde solubility regime, instead of NH_4Cl, NH_4Br, and NH_4I in the normal solubility regime. That has been observed to lead to a reduction of oxygen from 1–1.5 $\times$ 10^{20} cm^{-3} to 8 $\times$ 10^{18} cm^{-3} [103]. This improvement can be attributed to the resulting higher growth temperature and/or the different chemical characteristics of NH_4F (the smallest radius/charge ratio) [103]. The second method is the sequential introduction of purified NH_3 and HCl gasses instead of highly hygroscopic NH_4Cl powder, which enables the reduction of the oxygen concentration from 1–1.5 $\times$ 10^{20} cm^{-3} till 3–5 $\times$ 10^{19} cm^{-3} [262]. According to Tomida et al. [262], that technique should also be transferable to other halides.

All these mineralizer purification steps lead to a reduction of the concentrations of the elements from the Si, Al, Mg group as well.

The last contamination source is the GaN nutrient, which is usually polycrystalline GaN. To improve the purity of the polycrystalline GaN source material, S. Suihkonen et al. [87] point out that the purification processes of starting materials need to be developed further. No publications were found that further pursued this path.

Impurity getters: If introduction of oxygen into the autoclave is unavoidable, the use of getters is a remaining means of reducing the oxygen concentrations in the crystals. Getters have the primary focus of chemically interacting with the impurity element in the fluid phase and binding them in an insoluble compound, thereby effectively removing them from the growth environment.

Some earlier reports on impurity getters were published in 1997 by R. Dwiliński et al. [278] and in 2007 by B. T. Adekore et al. [170]. R. Dwiliński et al. [278] used erbium (Er) as the getter material with a molar Er to GaN ratio of 1:10. Grown GaN crystals exhibited a decreased intensity of the yellow band and a sharper PL-emission peak for GaN in the PL spectrum. That indicates a lower concentration of free electrons [278], likely due to reduced presence of oxygen as a donor. B. T. Adekore et al. [170] also used Er as an oxygen getter with an Er to GaN weight ratio of 1:3–4. With these conditions, Er became unintentionally doped into the bulk GaN single crystal [170]. They mentioned oxygen was gettered, as they measured an O concentration of 1 $\times$ 10^{19} cm^3 and 7 $\times$ 10^{19} cm^3 at the Ga-polar and N-polar sites, respectively (measured via SIMS). However, Er was unintentionally incorporated into the bulk GaN single crystals [170], with a concentration of 10^{18} cm^3 to 10^{17} cm^3. In 2018, D. Tomida et al. [158] found that Al works well as an oxygen getter material in the acidic environment. They explained their results with the low standard Gibbs' formation energies of Al_2O_3 at 600 °C.

Facet-dependency: Incorporation mechanisms of impurities depend on the atomical scale morphology and the crystallographic orientation. That the incorporation of impurities depends on the growth crystallographic orientation is well known and is found in other growth techniques, like the Czochralski method by Si [279]. In the ammonothermal growth method, six growth faces are typically observed, though only four of these are stable [100,105,145,280]:

- (0001)/Ga-face/+c-plane;
- (000-1)/N-face/−c-plane;
- {10-10}/m-plane;
- {11-20}/a-plane;
- {10-1-1};
- {11-22}.

The stable growth faces in the basic system are the +c-plane, −c-plane, and m-plane [100]. In the acidic system the stable growth faces are the same, but the {10-1-1}-plane may terminate the habitus of the crystal instead of the −c-plane [100].

S. Sintonen et al. [145] observed in basic ammonothermal GaN a higher oxygen incorporation in wing regions (which is mainly formed by a-plane growth) and a lower incorporation in the c-direction, measured via SIMS. For MOCVD, the same behavior was explained with the nitrogen surface saturation of a-plane, which defines the oxygen incorporation efficiency [145]. Likewise, for the ammonobasic environment, S. Pimputkar et al. [140] observed the highest oxygen incorporation in the a-plane, measured via SIMS. The oxygen concentration they measured decreased in the following order, depending on the growth direction: a-plane, −c-plane, +c-plane. M. Zając et al. [105] measured a lower oxygen concentration at the m-plane and a higher oxygen concentration at the −c-plane, via SIMS. However, S. Sintonen et al. [145] mentioned that the wing regions (such as the m-plane) show an inhomogeneous incorporation of impurities. The differences between the values for the +c-plane and the m-plane are, however, not high enough to explain the differences by these inhomogeneities. L. Wenhao et al. [280] measured, via SIMS, a higher oxygen incorporation in the {11-22}-plane and a lower oxygen concentration in the a-plane in ammonobasic GaN crystals.

Accordingly, for the ammonobasic crystal growth, the following oxygen impurity incorporation tendency can be given: {11-22}-plane > a-plane >> −c-plane and m-plane > +c-plane.

Considering that the growth rates in the acidic ammonothermal system are higher on the m-plane and +c-plane in relation to the basic ammonothermal system [103,105], the facet-dependent dopant incorporation behavior will likely be different for the two environments. K. Shima et al. [242] measured the oxygen concentration in ammonoacidic GaN, via SIMS, for GaN grown on different crystallographic planes; their measurements show that in the −c-plane, oxygen incorporation was enhanced by a factor of 10 or more, in relation to the +c-plane. For the ammonoacidic crystal growth, the following oxygen impurity incorporation tendency can therefore be stated: −c-plane >> +c-plane.

Growth time-dependency: S. Sintonen et al. [145] measured no significant time-dependent impurity changes of an ammonobasic grown GaN crystal.

A different report by S. Sintonen et al. 2016 [146] on a GaN crystal grown from a different group found that the impurity incorporation increases at the beginning of the growth and slowly decreases after reaching a maximum. This behavior was clearly observable by laser ablation inductively coupled plasma mass spectrometry (LA-ICP-MS) measurements at Mn and Mg (Figure 13). In their experiments, S. Sintonen et al. [146] grew one crystal in five growth steps and in three different autoclaves, which results in five layers/lamellae in one crystal. They explained this behavior with the effect of a decreasing Ga concentration in the supercritical NH_3 via the reduction of the feedstock amount during the growth. This reduced Ga concentration in the solution leads to a lowering of the viscosity of the solution and hence, to a reduced convective transport of impurity species from the autoclave wall to the sample.

The results of B. Wang et al. [281] showed crystal growth time-dependent impurity changes in the ammonobasic system, as did S. Sintonen et al. [146]. B. Wang et al. [281] measured a decay of the yellow and blue luminescence band (measured via cathodoluminescence) along the growth direction of this crystal. The measurements indicated a time-dependent decay of the impurity concentration and/or the defect concentration along the growth direction [281].

It could be that the GaN feedstock of S. Sintonen et al. [145] was so large that they could not measure significant changes. The enlargement of the growth period with a lower incorporation rate could be suitable measure for a further reduction of the impurity incorporation rate in ammonothermal GaN crystals.

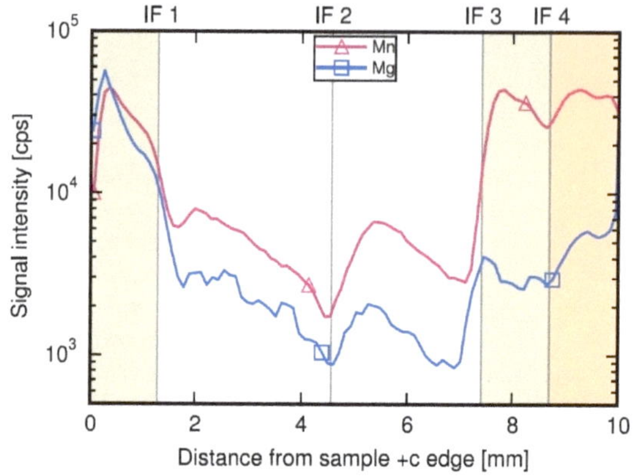

Figure 13. LA-ICP-MS spectrogram for Mn and Mg as function of the distance from the seed in growth direction (+c-edge of the sample). IF1-IF4 indicating the start of a new lamellae, the colored regions in the graph give the visual impression of the crystal. Reprinted from S. Sintonen et al. [146], copyright (2016), with permission from Elsevier.

5.3. *Effect of Point Defects on Electrical, Oprical and Structural Properties*

Intentional and unintentional doping significantly influence crystal properties. The better the intentional dopability of a crystal, the higher the range of crystal properties that can be reached (in a controlled way).

In this section, we discuss the effect of all point defects and their effects on the electrical, optical, and structural properties. The formation of unintentional point defects, both native defects and impurities, is discussed in Sections 5.1 and 5.2.

5.3.1. Electrical Properties

In this first section, we concentrate on dopability, because doping influences not only electrical properties, but also the other crystal properties, which will be discussed in the subsequent sections.

Currently, is it possible to crystallize n- and p-type conductive and semi-insulating ammonothermal GaN. These conductivity types are dependent on further development of the reduction of various impurities and native point defects, above all, oxygen. Oxygen as an impurity enormously hinders the use of other dopants in GaN and hinders the oxygen dopability in a low carrier concentration/high carrier mobility range. Currently, it is not possible to dope ammonothermal GaN without oxygen as a counter-dopant (compensating donor) or co-dopant. Accordingly, oxygen is the mostly used dopant in GaN [86]. To achieve lower doping levels, better cleaning procedures (Section 5.2) or oxygen getters are seen as prospective measures to reduce oxygen concentrations, which cannot currently be removed in advance. H and Fe (for example) can be additionally present in ammonothermal GaN, with these elements leading to a degradation of electrical devices. H can be mobile at elevated temperatures or high electron fluxes during device operation, leading to unintended or uncontrolled passivation of p-type GaN layers [87]. Fe can form effective SRH centers, leading to device degradation for optoelectronic emitters or current collapse in AlGaN/GaN high electron mobility transistors (HEMTs) [87].

Table 6 shows the currently used dopant elements yielding conductive or SI ammonothermal GaN, along with their growth environment and electrical properties. We discuss the dopability in the sense of doping efficiency: Which fraction of the incorporated atoms acts as a dopant. The doping efficiency values in Table 6 were calculated by dividing the charge carrier concentration by the dopant concentration. The doping efficiency can be max. 100%, like for oxygen doping in the HVPE method [252]. Only in the presence of co-dopants, the value can be higher than 100%.

Table 6. Current state of electrical properties of conductive n- and p-type doped conductive and semi-insulating (SI) ammonothermal GaN. The addition (+an.) means that the GaN crystal was annealed under N_2 atmosphere at 1100 °C for 4 h. The resistivity, charge carrier concentration, and mobility were measured at room temperature.

Process Variant	Dopant	Doping Concentration (cm^{-3})	Conductivity Type	Carrier Concentration (cm^{-3})	Doping Efficiency (%)	Resistivity (Ωcm)	Carrier Mobility ($cm^2/(V{\cdot}s)$)	Ref.
basic	O	50–100 × 10^{18}	n	6–10^{18}	12–10	n/a	68	[282]
basic	O	3.5–10 × 10^{18}	n	2.8–7.5 × 10^{18}	80–75	1–10 × 10^{-3}	ca. 250–200	[105]
acidic	O	1.5–27 × 10^{18}	n	0.66–15 × 10^{18}	44–56	n/a	565–155	[252]
acidic	O	1–10 × 10^{18}	n	0.9–20 × 10^{18}	90–50	2–20 × 10^{-3}	300–100	[283]
basic	Mg	7–8 × 10^{18}	p	n/a	--	10^6	n/a	[105]
basic (+an.)	Mg	7–8 × 10^{18}	p	3.8 × 10^{16}	0.54–0.47	30	6	[105]
basic	Mg	1–2 × 10^{18}	SI	<10^{12}–10^{13}	--	≥10^{10}	n/a	[105]
basic	Mn	10 × 10^{19}	SI	n/a	--	>10^{12}	n/a	[105]

In an ammonothermally grown n-type oxygen-doped GaN crystal, the doping efficiency is 10% to 88%, compared to 100% by GaN crystals from the HVPE process. The doping efficiency of p-type magnesium-doped GaN is two orders of magnitude lower than for oxygen-doped ammonothermal GaN.

In the following, we discuss Table 6 according to the conductivity types.

p-type: The low doping efficiency for p-type Mg-doped GaN can be explained by three contributions. The first one is unintentional compensating impurities. In the ammonothermal method, specifically, high oxygen concentrations, are currently unavoidable. The oxygen acts as a compensating donor that reduces the concentration of free holes. For example, the oxygen concentration of the Mg-doped GaN listed in Table 6 is 1.5 × 10^{18} cm^{-3} [105]. The second contribution is the relatively large activation energy of the acceptor Mg_{Ga} at 200 meV above the VBM [105]. The third contribution is the high likelihood of Mg_{Ga} to form the neutral Mg_{Ga}–H complex (see Section 5.2) [105]. The third contribution can be addressed by the removal of H from the Mg–H complex via annealing (Table 6), thereby increasing the charge carrier concentration and decreasing the resistivity from 10^6 to 30 Ωcm [105].

n-type: The doping efficiency < 1 of oxygen in ammonothermal GaN crystals can be explained by the formation of V_{Ga}–H complexes, which can act as acceptors (Section 5.2). W. Jiang et al. [252] compared their Hall mobility measurements of oxygen-doped ammonoacidic GaN with HVPE GaN in the same carrier concentration range. The charge carrier mobility values were similar in their charge carrier concentration ranges of 1.5–27 × 10^{18} cm^{-3} and 0.66–15 × 10^{18} cm^{-3}, respectively. Y. Mikawa et al. [284] observed, with their Hall mobility measurements of oxygen-doped ammonoacidic GaN, the same similarity of the carrier mobility from ammonoacidic GaN and HVPE GaN.

The doping efficiency of the oxygen-doped basic ammonothermal GaN boules reported by S. Pimputkar et al. [282] is the lowest listed in Table 6. Their crystals also had the highest doping level and lowest charge carrier mobility, as can be seen in Table 6. At such high oxygen concentrations, with an FEL close to the CBM, the previously mentioned behavior does not seem to be valid. Given the high concentration of oxygen, it is possible that a large number of hydrogenated Ga–vacancies formed and acted as acceptors, due to an FEL closer to the CBM. This, in turn, would lead to reduced doping efficiency and a high concentration of additional point defects, further reducing mobility.

Accordingly, via the growth of oxygen-doped basic ammonothermal GaN, two doping efficiency ranges exist. Below an oxygen doping level of 27 × 10^{18} cm^{-3}, there are no predominant acceptors and the doping efficiency is high. Above an oxygen doping level of 50×10^{18} cm^{-3}, predominant acceptors exist, and the doping efficiency is low.

SI: Mg and Mn are used as p-type dopants to compensate the process-related unavoidable oxygen n-type doping. The oxygen concentration of the SI Mg-doped GaN listed in Table 6 is 1–2 $\times 10^{18}$ cm^{-3}, whereas the oxygen concentration of the SI Mn-doped GaN is 5×10^{18} [105].

As can be seen from Table 6, for Mg, the concentrations of Mg-doped SI GaN are lower than those of Mg-doped conductive p-type GaN, to realize a lower resistivity. Contrary to the conductive Mg-doped GaN, the annealing (N_2 atmosphere/1100 °C/4 h) of the SI Mg-doped GaN leads not to a resistivity reduction [105]. Conductivity of the SI Mg-doped GaN could only be measured above 400 °C. At this temperature, the GaN is n-type conductive with a carrier concentration of 10^{12}–10^{13} cm^{-3}, due to an activation energy of 1.46 eV above the VBM [105]. FTIR analysis shows no evidence of Mg–H bonds before and after the annealing [105]. Accordingly, no Mg–H bonds were broken, which would result in a lower resistivity.

SI Mn-doped GaN shows the highest resistivity, as shown in Table 6. This results from two circumstances: firstly, the charge carrier compensation effect and secondly, the characteristic of transition metals to form deep levels that trap charge carriers. A high-temperature-annealing (N_2 atmosphere/1100 °C/4 h) does not yield a resistivity reduction, like for SI Mg-doped GaN [105]. The energy level of Mn in this SI Mn-doped GaN was determined to be 1.5 eV above the VBM [105].

5.3.2. Optical Properties

A first assessment of the presence of optically active point defects can be made via a visual inspection by the naked eye. The ammonoacidic and ammonobasic systems have different amounts of specific point defects, which influence the coloration [100]. Accordingly, crystals have a growth environment dependent coloring:

- Ammonobasic: grey, yellow, green, or orange-brown to red [100,105,280];
- Ammonoacidic: grey, yellow, or green [100,285]

Independently of the growth environment, crystal greying depends on n-type doping [100]. As discussed in Section 5.3.1., the n-type doping is currently only represented by oxygen doping. Accordingly, the greying of the crystal gives a first indication of the oxygen doping level of the crystal. In the ammonobasic and -acidic system, the yellow coloring is connected to the presence of deep level defects. These defects are primarily Ga–vacancies and their related complexes, according to their predominant amount (Section 5.1) [105,242]. The green coloring of ammonoacidic GaN and p-type Mg-doped GaN is predicted to be linked to N-vacancies [244,256]. The color of SI Mn-doped GaN crystals ranges from orange-brown to red [105,280].

The visually observable greying of the crystals can be described via the increase of free charge carriers. In the ammonobasic and ammonoacidic systems, an increase of the absorption coefficient can be observed with increasing amounts of free electron/oxygen concentration, likely due to phonon-assisted free carrier absorption [252,282,286]. Figure 14 shows the absorption spectra for six crystals grown in an ammonoacidic environment, with an oxygen concentration from 1.6×10^{18} to 2.7×10^{19} [252]. According to W. Jiang et al. [252], the optical absorption at wavelengths > 600 nm (<2 eV) is similar to that of HVPE GaN. In the region between 500 nm (2.5 eV) and 400 nm (3.1 eV), closer to the absorption band edge, the absorption coefficient of ammonothermal GaN increases in comparison to the HVPE GaN. S. Pimputkar et al. [282] observed a comparable behavior in GaN grown in an ammonobasic environment. They observed this sharp increase of the absorption coefficient for ammonothermal GaN at 2.9 eV (427 nm) and for HVPE GaN at 3.2 eV (387 nm). They measured their ammonothermal and HVPE samples via FTIR and could detect hydrogenated Ga–vacancies in the ammonothermal GaN, but not in HVPE GaN. Additionally, they measured the refractive indexes at the same samples, the indexes suggest defect states in the ammonothermal GaN close to the VBM, but not in HVPE GaN.

Accordingly, it can be suggested that hydrogenated Ga–vacancies are the primary cause of the increase of the absorption in the region of 2.8–3.3 eV (442–375 nm). In the FTIR spectra of ammonoacidic GaN reported by W. Jiang et al. [252], absorption peaks that correspond to V_{Ga}–H complexes were likewise detected. In ammonobasic GaN, E. Letts et al. [286] also verified an oxygen concentration-dependent increase of the absorption coefficient. In conclusion, the absorption behavior of the ammonoacidic and ammonobasic GaN seems comparable.

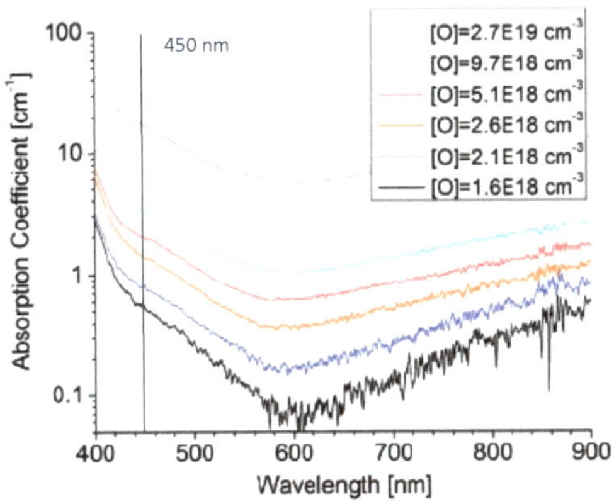

Figure 14. Dependency of the optical absorption coefficient (α) on the wavelength for six oxygen-doped ammonoacidic GaN crystals, with a marking of blue LEDs relevant 450 nm wavelength (in blue). Reprinted and modified from [252] with permission, © 2017 The Japan Society of Applied Physics.

According to Section 5.3.1, the oxygen doping efficiency in ammonothermal GaN ranges from 10% to 90%. Due to that, it can be assumed that the reduction of the phonon-assisted free carrier absorption ranges in ammonothermal GaN from 90% to 10% in relation to the case if every incorporated atom would act as a dopant.

The yellow luminescence band (YL) measured via PL can be observed in ammonothermal GaN, as well as in HVPE GaN, and is linked to deep level defects [273,287], like the visually observable yellowing of GaN crystals. M. Reshchikov [273] detected an ammonothermal GaN-specific yellow luminescence band with a maximum at 2.2–2.3 eV [244,273], which he attributed to the $V_{Ga}H_3$ complex. In addition to the YL, the green luminescence band (GL) at 2.33–2.4 eV (maximum), red luminescence band (RL) at 1.72–1.82 eV (maximum), and blue luminescence band (BL) 2.7–3.0 eV (maximum) can be observed in ammonothermal GaN [244,273]. The GL is mainly linked to N-vacancies, the RL is mainly linked to N–vacancy complexes, and the BL is mainly linked to impurities [244,273].

W. Jiang et al. [288] reported the growth of a highly transparent yellow-tinted GaN crystal in an ammonoacidic environment. That crystal had an absorption coefficient of 0.75 cm^{-1} at the emission wavelength of blue LEDs (450 nm), with a carrier concentration around 1–3 $\times$ 10^{18} cm^{-3}. This absorption coefficient is one of the lowest published absorption coefficients in that oxygen concentration level. Point defects decrease the active region efficiency of LASER diodes (LDs) [216,289] and are linked to increase the threshold current of LDs [290], so they are of great concern and thus have to be reduced.

Further, L. Wenhao [280] showed the possibility to shift the longitudinal optical mode E1(LO) of ammonobasic GaN towards a lower frequency via Mn doping. That opens the possibility to use Mn-doped GaN for infrared emission devices.

5.3.3. Structural Properties

To avoid stress generation during the crystal growth of bulk GaN [110], or during the epitaxial deposition of crystalline layers [291], the lattice parameters have to be as similar as possible. Too high a lattice mismatch, in combination with layer thicknesses over 30–40 μm, lead to large bending and an increase of dislocation density [292]. Even the use of HVPE seeds for the ammonothermal growth of GaN, which results in a relatively small lattice mismatch, leads to multiple negative effects [110,293], including (1) a reduction of the radius of curvature (as a measure of the bending/plastic deformation), (2) an increase of the rocking curve width (as a measure for the mosaicity), (3) an increase of the etch pit density (as a measure of the amount of threading dislocations), and (4) the generation of cracks. E. Letts et al. [293] observed cracking of their crystals grown on HVPE seeds only above a crystal layer thickness of 1 mm. They explained this behavior by a flipping of the sign of the radius of curvature at a critical length where the radius of curvature becomes infinite. They assumed that the differences of the radius of curvature originate from the differences of the lattice spacing due to the different level of contamination of the two growth methods.

Four main factors influence the lattice parameters of GaN: free electron concentration, dopant concentration, native defects, and lattice mismatch to the substrate [291]. The first three points are linked to each other and can be summarized as two effects of point defects on the GaN lattice [294]. Firstly, the lattice deformation, due to a size effect, related to the difference in size of the impurity atom and the substituted lattice atom. Secondly, the lattice deformation due to the electronic effect related to a change of the conduction or valence bands by the excitation of free carriers into the bands and the associated volume change.

Free carriers (holes and electrons) expand the GaN lattice [292,294]. According to C. Van de Walle's [294] calculations, electrons expand the lattice three times more than holes. The size effect can expand or contract the lattice, depending on the size of the impurity atom in relation to the substituted lattice atom [294]. A general weighting relationship between the size effect and the electrical effect does not exist. For example, the size effect-related lattice deformation is 33% lower for oxygen at the N-site and 76% higher for Si at the Ga site compared to the electrical effect for electrons in GaN. The size effect can mainly be treated as isotropic, and the electrical effect is likewise assumed to be isotropic [294]. The element in which charge state occupies which lattice site is described in Sections 5.2 and 5.3.

Calculations of the lattice parameter of GaN show a linear dependency on the dopant concentrations [291,292,294]. M. Leszynski et al. [291] showed experimentally the increase of the lattice parameters a and c with the free electron concentration of GaN for homo- and heteroepitaxial, as well as bulk crystal growth.

A difference of the lattice parameters of 0.02% can already lead to technical implications [292]. Depending on the theory for the calculation of the lattice parameters, the values differ, for example by 3% [270]. At the moment, it is not possible to grow GaN without any lattice deformation. For example, the lattice-deforming oxygen impurity is currently unavoidable in ammonothermal GaN. Accordingly, the determination of precise undistorted lattice parameters of GaN is currently not possible. According to C. Van de Walle [294], an oxygen concentration of 10^{20} cm^{-3} leads to the lattice parameters a = 3.19798 Å and c = 5.18293 Å, assuming every O atom occupies a N site and leads to a free carrier, and the lattice parameters for pure GaN are a = 3.19 Å and c = 5.17 Å [270] (modeled lattice parameters). For the lattice parameter a and c, that means an increase of 0.25%. For an ammonothermal GaN crystal, R. Kucharski et al. [110] measured two values for the lattice parameter a: 3.1881(5) Å and 3.1887(5) Å, and one value for the lattice parameter c: 5.1854(5) Å. The crystal was grown on an HVPE seed (radius of curvature of the seed ~100 m) with an oxygen concentration of 5×10^{17} cm^{-3}. Due to the uncertainty of undisturbed lattice parameter values of GaN, no quantitative interpretation of any measured lattice parameter of ammonothermal GaN is possible, regarding the influence of a point defect on the lattice parameter. However, the statement can be made that the oxygen concentration in ammonothermal GaN crystals plays a huge role in the lattice parameter.

Besides their influence on the lattice parameters, it should be noted that point defects are needed for the dislocation climbing process and thus, they influence the propagation of dislocations, as reported for V-shaped dislocations in epitaxial GaN layers grown by MOVPE [295].

In summary, a wide range of point defects and associated complexes have been identified to exist in GaN and likely, in similarly grown ammonothermal crystals. Control over these defects is critical to maintaining desired crystal properties, though the persevering oxygen contamination of the system can make the pursuit of certain material properties (such as p-type doped semiconductors) a challenge that requires further targeted research efforts. Nonetheless, significant demonstrations have been made, yielding controlled changes to electron conductivities in materials along with changes to optical transparency, both of which are critical to control for substrates, especially for optoelectronic devices.

Significant efforts were made to explore these material properties and how they relate to the growth environment and growth processes. These experiments are generally considered tedious and time consuming, as numerous in situ technologies developed for lower pressure system cannot readily be applied to these high-pressure systems. Combining knowledge of the reactivity and solubility of species with the evolving needs for demonstrating high quality crystal growth has led to continuous development of the growth systems themselves. The following section discusses the various aspects associated with development of reactor technologies suitable for ammonothermal environments, including systems with improved materials leading to higher purity, and/or higher temperature systems, or via integration of windows or feedthroughs to permit the development of in situ technologies.

6. Advances in Reactor Technology

The reactor in the ammonothermal process has to primarily meet two mechanical requirements: high strength and creep resistance permitting high pressure retention (up to ~500 MPa [96]) at elevated temperatures (until~600 °C [84], or beyond), and high corrosion resistance under basic or acidic ammonothermal growth environments.

For the load-bearing components that are exposed to high temperatures, nickel–base superalloys such as Inconel 718 or Rene41 are commonly used, though a molybdenum–base alloy has also been employed [164].

Initially, ammonothermal reactors were developed for crystal growth or the synthesis of new materials, resulting in a long tubular geometry that is both comparatively simple to manufacture and permits the use of relatively thin autoclave walls. The moderate wall thickness prevents excessive heat conduction via the autoclave walls from one temperature zone to another. Thereby, it facilitates the implementation of thermal gradients inside the autoclave, which is an important requirement of crystal growth exploiting the temperature dependency of solubility [296]. An example of such a tubular autoclave geometry can be seen in Figure 15a. The inner and outer diameters of this autoclave are 21 mm and 50 mm, respectively.

More recently, specialized types have been developed for the purpose of implementing different in situ monitoring techniques, starting with an optical cell for video-optical and UV-Vis measurements developed by N. Alt and E. Schlücker [297,298]. Optical cells are defined as autoclaves that are designed to incorporate viewing windows. To achieve this, a material transparent to the radiation in question (such as optical or UV-light, or X-rays) is utilized. In addition to sufficient transparency, the window material needs to possess sufficient mechanical strength and chemical stability in the targeted ammonothermal environment [299]. For 2D imaging using optical light or X-rays, as well as for spectroscopic measurements, the same design of optical cells can be used. These optical cells can flexibly be equipped with the aforementioned window materials [300].

Examples of state-of-the-art optical cells are shown in Figure 15b–d. A defining characteristic of uniaxial optical cells (Figure 15b) is the single, straight beam path with windows at both ends. A biaxial optical cell Figure 15c, on the other hand, comprises a cuboid-shaped viewing cell with two beam paths running perpendicular to each other. Moreover, a specialized autoclave with windows set at the diffraction angle was developed specifically for in situ X-ray diffraction studies (Figure 15d). Unlike the other optical cells, the optical cell in Figure 15d is designed to allow for crystal growth via a thermal gradient and fluid convection.

The simplest (uniaxial) optical cells of the construction depicted in Figure 15b–d have an outer diameter of 108 mm. The thicker walls of the autoclave can impact the thermal field within the system significantly. Firstly, the greater thermal inertia of the thicker walls can lead to slower responses to changes in heating or cooling conditions, affecting the overall thermal responses of the system. Secondly, the enhanced heat conduction along the autoclave walls causes a very uniform temperature distribution inside the autoclave [299]. Such a uniform temperature distribution is intended in the case of studying solubility or dissolution kinetics for a distinct temperature condition, however, measures to mitigate this effect need to be taken if combining optical cells with crystal growth experiments via a thermal gradient [299]. Such a design, with a thinned autoclave wall in-between nutrient and crystal growth zone, is depicted in Figure 15d.

Another important design aspect is the orientation of the tubular inner volume in relation to gravity and, if a thermal gradient is imposed, in relation to the thermal field. The horizontal orientation of the reaction chamber in uniaxial and biaxial optical cells serves multiple functions. Firstly, it enables the convenient alignment of optical components along a single axis or multiple axes, depending on the configuration of the cell. This alignment is critical for accurately measuring light interactions with samples and ensuring precise spectroscopic analysis. Importantly, the horizontal orientation with regard to gravity suppresses fluid convection (optical cells in Figure 15b,c), whereas a vertical orientation with respect to gravity can be applied to enhance fluid convection (optical cell designed for crystal growth, Figure 15d [299].

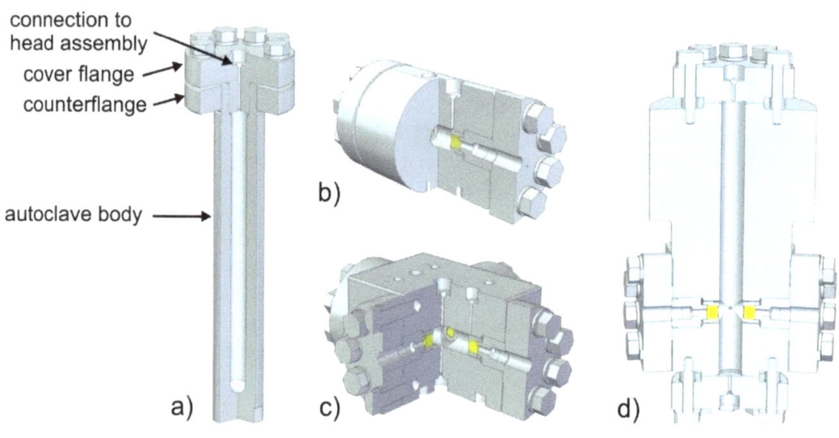

Figure 15. Visualization of the geometries of different types of autoclaves: (**a**) autoclave as used for crystal growth using a temperature gradient, as well as for exploratory synthesis or studies on intermediates using ex situ characterization methods, (**b**) uniaxial optical cell, (**c**) biaxial optical cell, (**d**) optical cell designed for in situ diffraction. Reproduced from [299], with minor adjustments to collect only the autoclave types relevant to this review in one figure.

Current topics in ammonothermal reactor technology are driven by the following factors: general design considerations and the availability of alloys with high mechanical strength at high temperatures, improving both reactor lifetime and crystal purity by mitigating corrosion issues, and advancing the fundamental understanding by gaining experimental access to the physical and chemical processes during the growth. The following sections are dedicated to these aspects.

6.1. General Design Considerations

The most important challenges for the load-bearing parts of ammonothermal autoclaves will be delineated in the following. We aim to shine a light on specific aspects that are and have been important to past and future development of the field, rather than providing comprehensive information. Readers who require more comprehensive information are referred to books covering the design of high-pressure vessels in general, and for ammonothermal experimentation in particular [301,302].

Given the need to simultaneously withstand high temperature and high mechanical stress due to the inner pressure over extended periods of time, the alloy is the main limiting factor preventing experiments at simultaneously higher temperatures and higher pressures [166]. An important aspect is that yield strength $R_{p0.2}$ generally decreases with increasing temperature, as is evident from the data shown in Figure 16. Nickel–base alloys are the state-of-the-art materials, primarily due to their excellent mechanical strength at elevated temperature, achieved via precipitation hardening [166]. An added layer of complexity originates from the limited chemical stability of most prospective load-bearing materials in ammonothermal solutions, which depends strongly on the mineralizer and conditions used [147,166]. While nickel–base alloys generally exhibit good corrosion resistance, as well as oxidation resistance at high temperature, this is not necessarily the case for ammonothermal conditions, and stress corrosion cracking is a major concern in contact with ammonoacidic solutions [165,166,303]. More recently, cobalt-based alloys have been considered [166,304], as the system Co-Al-W provides the opportunity to harden the alloys by the precipitation of the ordered phase (γ') in the Co solid solution matrix phase (γ), and the higher melting point of Co results in high liquidus and solidus temperatures [166]. Aiming for higher temperatures at the expense of maximum operating pressure, the molybdenum-based alloy titanium–zirconium–molybdenum (TZM) [164] and the nickel-base alloy Haynes 282 [177,186,301,305,306] have been used for ammonothermal autoclaves. For comparison, an iron-based alloy with high mechanical strength at elevated temperatures is also shown in Figure 16. The high temperature strength of iron-based alloys (stainless steels) is insufficient for most ammonothermal applications, as their mechanical properties are insufficient at temperatures above 500 °C [301]. In this context, the significantly different temperature range of 345–360 °C for hydrothermal growth of quartz in comparison to 400–800 °C ammonothermal growth of GaN (see Table 1) should be noted.

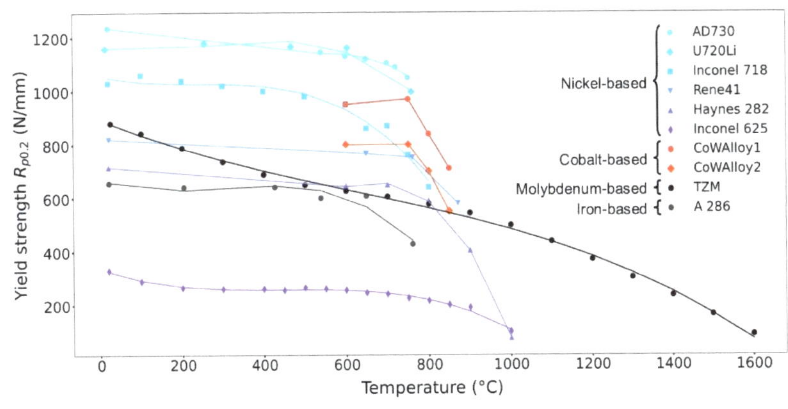

Figure 16. Decline of yield strength with increasing temperatures for several alloys used or considered for use as load-bearing material for hydrothermal and ammonothermal reactors, respectively. Data taken from the same references as in [304], while the data for stainless steel (A 286) were taken from [307]. The fit lines model the yield strength as a third degree polynomial and serve as a guide to the eye.

Another important requirement is sufficient creep resistance [166,301], which is due to the combination of high temperatures (in relation to the melting points of the alloys) and the long exposure to these elevated temperatures (often several weeks or even months per experiment). To predict the creep life of components, parametric models such as the Larson–Miller relation [308] are commonly used [166,309]. In the high temperature range (at temperatures above half the melting temperature), the creep rate is proportional to the diffusion coefficient of atoms and consequently to temperature, as well as to the applied stress [310,311]. Emerging superalloys appear to be promising also in terms of better creep resistance: Specifically, the cobalt–base alloys CoWAlloy1 and CoWAlloy2 were reported to have significantly higher creep resistance than the nickel-based alloys 718 and 282 [166].

Ductility (often evaluated by fracture toughness) is also an important consideration, especially from a safety point of view. This is a particular concern in the case of the molybdenum-based alloy TZM, for which the brittle-to-ductile transition occurs above room temperature, specifically at 100 to 150 °C [164] or even 400 °C depending on the strain rate [312].

The permittable internal pressure p depends on the yield strength $R_{p0.2}$, the inner diameter d_{in}, the outer diameter d_{out} as follows [302]:

$$p = R_{p0.2} \cdot \frac{1}{\sqrt{3}} \cdot ln\left(\frac{d_{out}}{d_{in}}\right) \quad (1)$$

To increase the size and number of crystals grown in one run, the inner diameter of the autoclaves needs to be increased. The difference between the stress at the inside and the outside increases with the square of the diameter ratio, with the stress being substantially higher at the inside [302]. Consequently, if the ratio of outer and inner diameter decreases, the permittable internal pressure decreases. A proportional increase of both, however, increases both the outer diameter and the wall thickness. Depending on the alloy, the former can be a challenge in terms of availability and the latter can likely be a challenge in terms of maintaining a suitable thermal gradient between the two zones. With the hydrothermal growth of quartz, a similar method for the simultaneous growth of large numbers of single crystals has been long demonstrated to have impressive scalability [122]. To enable the use of a steel in place of the commonly employed nickel–base superalloys as the pressure-bearing material, M. D'Evelyn and coworkers from Soraa Inc. (now KYOCERA SLD Laser, Inc., Goleta, CA, USA) developed a new design for an ammonothermal reactor using internal heating [90]. By placing the heater inside the pressure vessel (protected against corrosive attack by a hermetically sealed capsule) and by using a ceramic shell as thermal insulation between the heater and the pressure-bearing reactor walls, they drastically reduced the temperature at which the pressure-bearing parts are held [90,313]. For a heater temperature of 750 °C, the temperature drops to 200 °C within the ceramic shell [90]. This allows this autoclave design to achieve pressures as high as 600 MPa, which has demonstrated benefits such as higher growth rates [90]. Even without an internal heating system, a considerable amount of scaling has been achieved, amongst others, for a pressure variant working at much lower pressures around 100 MPa and likewise, with acidic mineralizers, as reported by K. Kurimoto et al. [84], but also for ammonobasic growth [98].

6.2. Reactor Lifetime and Crystal Purity

The reactor lifetime and the purity of the crystals grown in autoclaves, depends strongly on the chemical stability of the material which the inner surface of the autoclave is made of. Many materials which are stable in the acidic system do not maintain their stability in the basic system, like noble metals and Ni, Co, and Ni–Co alloys vice versa [147,314]. But Ag and Mo are stable in the Na-basic and NH_4F-acidic environments [147,314]. S. Pimputkar et al. [147] and A.-C. L. Kimmel and E. Schlücker [314] published a good overview of the material solubility according to the used ammonia solution.

In general, three approaches for corrosion-resistant reactor designs can be distinguished: manufacturing the entire autoclave from a corrosion-resistant material, protecting the inner wall with a hermetically sealed liner (or coating), or using a pressure-balanced liner (capsule) [314]. The first approach faces the challenge of finding a corrosion-resistant material that is simultaneously able to fulfil all other requirements for a load-bearing pressure vessel material (outlined in Section 6.1 and the references therein). The second approach requires careful design and manufacturing considerations to realize a hermetic seal and avoid a variety of potential issues that originate from the liner being exposed to the internal pressure and the dimensional changes that the load-bearing parts undergo during operation, but hardly any detailed information is mentioned in the literature. The pressure-balanced liner or capsule concept as the third approach avoids these complications by balancing the pressure on the inside and outside of the liner. As a consequence, it is easier to implement and comes with the advantage of easy exchangeability of the liner, at the cost of less effective protection due to the possibility of fluid exchange between the interior of the liner and the gap between liner and inner autoclave wall. Consequently, pressure-balanced liners are often used if the liner material is ceramic (usually resulting in a large mismatch in thermal expansion coefficients between the load-bearing material and the liner, which also is a limitation for the applicability of coatings) [315,316], and if easy exchangeability is desired (as for exploratory synthesis of new materials or prospective intermediates). In spite of its limitations, the pressure-balanced concept is sufficient to prevent, for example, the rapid corrosion of nickel–base alloys by indium [200].

The further discussions in this section are based exclusively on the data from ammonothermal crystal growth of GaN, because for the other nitrides, no or little information is available. Besides the mineralizer, the source material used has an influence on the corrosive behavior of the solution as well [314]. However, as a first approximation, it can be assumed that the mineralizer (alkali metals and halides) has a stronger impact on the corrosive behavior of the solution than the source material. Accordingly, the following insights about the reactor technology should be transferable to other nitrides as well.

To improve on the corrosion resistance of the autoclave, some autoclaves are lined with an inert liner (commonly used in acidic systems) to prevent corrosion of the load-bearing material. Additionally, liners have been used to reduce contamination by elements released from the autoclave wall [314]. Extensive investigations have been performed on the stability of materials in ammonothermal environments, aiding in the development of liners and autoclave technology.

Generally, nickel–chromium alloys are used due to their high strength at high temperature, along with their excellent toughness properties. Most of these alloys are resistant to corrosion by ammonia under basic but not under acidic conditions [101,147,314]. Figure 17 offers an overview of the published autoclave technologies over time, based on four selected groups. The chronological development of the basic ammonothermal field is based on the publications of the group from Poland (IHPP/NL-3 [95,98,101]) and USA (SixPoint/UCSB [293,317]). The chronological development of the acidic ammonothermal field in Figure 17 is based on the publications of the group from Japan (Tohoku/MCC [84,89,102,103,113,153,156,158,242,318,319]) and the first publication on the acidic ammonothermal system from the USA (NRL [320]). We use these groups to show the chronological autoclave material and liner technology development, because:

- They more likely mentioned in their publications which materials they used, contrary to other groups;
- Their publication record is wide, from the first publication on the topic for the basic environment and nearly on the first publication from the acidic environment, until now (contrary to other groups).

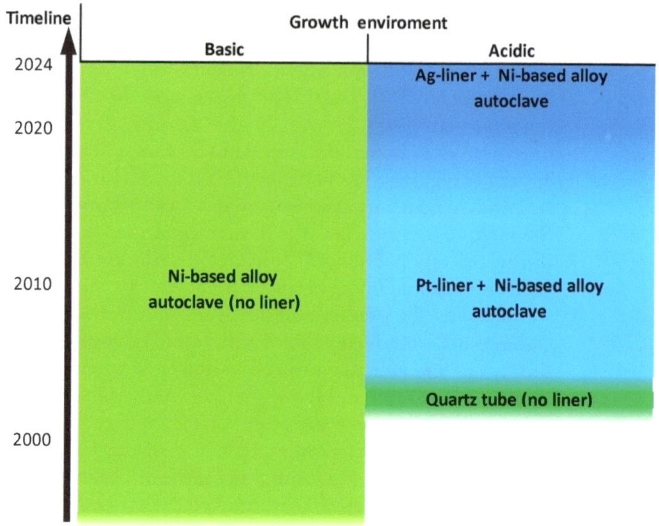

Figure 17. Chronological development of the autoclave material and liner technology, since the first publication of the basic and acidic growth environment by four selected groups that contributed to the development of ammonothermal growth over an extended period of time. For the ammonobasic growth environment, the groups IHPP/NL-3 and SixPoint/UCSB were selected and for the ammonoacidic growth environment, the groups Tohoku/MCC and NRL were selected. Every color is a chronological step. The greenish fields are chronological steps without a liner, the bluish with a liner: Ni-based alloy autoclave (no liner) [95,98,101,293,317], quartz tube (no liner) [320], Pt liner + Ni-based alloy autoclave [89,102,103,113,153,318,319], Ag liner + Ni-based alloy autoclave [84,156,242].

Autoclaves for ammonoacidic experiments: The development of adapted autoclave materials for the ammonoacidic GaN crystal growth started in 2004 with the covering of the inner autoclave wall with Pt [153]. A. Yoshikawa et al. [153] mentioned the important role of the Pt liner for the prevention of possible contamination (Figure 17). The autoclave was likely a Ni-based alloy, according to the group's later publications [153,319].

In 2014, B. Hertweck et al. demonstrated Ag as a coating for acidic ammonothermal conditions [321]. In their studies, Ag showed good corrosion resistance in the NH_4F environment [321]. They pointed out the potential cost reduction with the usage of an Ag liner instead of a Pt liner [321]. Additionally, they also proposed that an Ag coating on a Ni-based alloy has good elastic properties and relaxation behavior, which results in a low stress condition between the two materials [321]. As mentioned by Bao et al. [103], the usage of NH_4F instead of the other halides is beneficial for growth characteristics like purity and growth speed. According to A.-C. L. Kimmel and E. Schlücker [314], Pt is not stable in NH_4F, at least not as stable as Ag. The use of an Ag liner for a Ni-based autoclave in combination with the mineralizer NH_4F was first reported by D. Tomida et al. [156]. In their latest publication on ammonothermal GaN growth from early 2024, K. Shima et al. from the Tohoku/MCC group [242] still used the same combination of mineralizer and reactor materials. The combination of the beneficial mineralizer NH_4F with the cost-reductive and soluble stable Ag liner in a Ni-based autoclave shows the advantage and reasonableness of this latest development step of the Tohoku/MCC group.

Parallel to the liner development of the Tohoku/MCC group, T. F. Malkowski et al. published two articles in 2016 and 2018 [164,322] regarding the usage of an unlined titanium–zirconium–molybdenum alloy (TZM) in the ammonoacidic environment. The advantages of Mo-based alloy autoclaves are the price reduction in comparison to the usage of a Pt-lined Ni-based autoclave [156], the technological simplification (due to the absence of a liner), and the potential of higher growth temperatures (>600 °C) [322]. The main disadvantage of these materials is their brittleness at low temperatures, which poses an increased safety risk [164]. Additionally, these materials have high yield strength so, accordingly, a higher load must be attended for sealing, contrary to Ni-based autoclaves [164].

In addition, the unique approach by Soraa Inc. (now KYOCERA SLD Laser, Inc.) [90] to autoclave technology, which deviates from the traditional sealed cylindrical tube geometry, should be mentioned. Their approach is based on the utilization of internal heaters located in-between the load-bearing autoclave walls and a sealed volume containing the ammonothermal solution. This approach is discussed in greater detail in Section 6.1.

Autoclaves for ammonobasic experiments: The development of the autoclave material for the ammonobasic GaN crystal growth is not clearly mentioned in the literature for the IHPP/NL-3 group. Usually, they only state that it is a Ni-based autoclave [95,98,101]. According to the publication by S. Sintonen et al. [146], the IHPP/NL-3 group developed a so called "next generation autoclave". In their experiments, S. Sintonen et al. [146] grew one crystal in five growth steps and in three different autoclaves, which results in five layers/lamellae in one crystal (Figure 13). They performed the first growth run in a small older autoclave, grew the next two layers in a larger "next generation autoclave", and finally, grew the last two layers in a third autoclave of the older generation [146]. For the two growth runs using the next generation autoclave, they observed an impurity concentration reduction in the order of one magnitude for impurities originating from the inner autoclave wall, contrary to their observations with an autoclave of the older generation [146].

Because Ni alloy autoclaves are more stable in basic ammonothermal fluids, compared to acidic ammonothermal fluids, fewer incentives exist to further refine the technology and include liners. In 2014 and 2016, a notable advance in the reduction of impurities originating from the inner autoclave wall by the usage of a Ag liner in an ammonobasic experiment was reported by S. Pimputkar et al. [140,323]. Nevertheless, even in the crystals grown in the ammonobasic system with a liner, some impurities originating from the autoclave wall were detected, specifically Mn, Fe, and Zn. This can be explained by the lack of a hermetic seal of the capsule. The use of a hermetically sealed capsule would likely have further reduced these concentrations [89,140,323].

6.3. In Situ Monitoring

Autoclave designs made for in situ monitoring purposes require modifications. These modifications include feedthroughs for thermocouples, window assemblies for incorporating optically transparent or X-ray transparent windows (in different locations as indicated earlier in Figure 15b–d), an autoclave designed as a rolling ball viscosimeter, and a rotatable feedthrough. Those technical developments will be described in the following paragraphs.

Feedthroughs for thermocouples: Thermocouples have been introduced either individually [127], or as a group of up to two [127,324]. For sealing, a blind made of a nickel–base alloy with a tight fit bore was used [127]. The space in-between the thermocouple sheath (likewise made from a nickel–base superalloy) and the bore wall was filled either by soldering or by welding, simultaneously providing a firm connection between the thermocouple sheath and the blind [127].

Window assemblies: The autoclave windows represent optical windows of the Poulter type, distinguished by enhanced contact force under internal pressure and relatively low susceptibility to stress distribution irregularities caused by tolerances [297]. The seal between the window and the window rest is achieved by plastic deformation of a gold layer. Since most measurement methods require two optical windows, these autoclaves must include a cross hole for connecting peripheral devices (such as a hand valve, pressure transducer, thermocouple, and burst disc). Both the cross hole itself and the mechanical stresses induced in the autoclave body during peripheral screwing necessitate an increase in wall thickness [299]. To mount the window in the optical cell, specialized components made from Inconel 718 are employed, instead of a conventional cover flange, as illustrated in Figure 18. The process of assembling the window involves applying a compressive force of 30 kN, which remains constant during the phase of contact pressure. This sustained pressure enables the gold layer to deform plastically, effectively filling any surface irregularities and ensuring a secure seal [299].

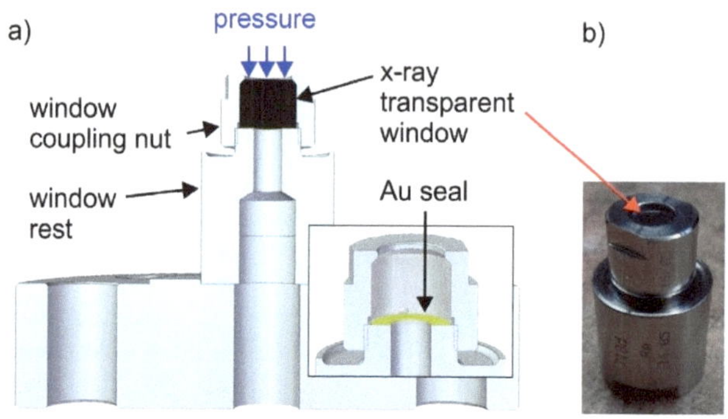

Figure 18. Window assembly with the flange for mounting it to an autoclave, as used for experimental access using optical light or X-rays of moderate photon energy (typically in the 40 to 100 keV range). (**a**) Computer-aided design (CAD) model in cross-section view, the inset shows a close-up of the sealing region and the exchangeable ceramic window displayed as transparent. (**b**) Photograph of window assembly equipped with a boron carbide window. Reproduced from [299].

While considering suitable materials for the windows of autoclaves employed in ammonothermal processes, the primary choices can be narrowed down to single-crystalline sapphire and ceramic boron carbide (B_4C).

Sapphire, particularly when oriented along the a-axis, is favored for its mechanical robustness, making it less susceptible to structural failures under high-pressure conditions. Moreover, sapphire exhibits supreme optical transparency, facilitating the transmission of light across a wide spectral range, which is advantageous for optical measurements and spectroscopic analyses. However, the mechanical strength of sapphire windows can depend on crystal orientation and temperature. Sapphire windows typically undergo extensive factory polishing procedures to attain the requisite surface smoothness and parallelism necessary for optimal functionality in high-pressure environments.

Although B_4C does not exhibit the same degree of optical transparency as sapphire, B_4C showcases exceptional transparency to X-rays, rendering it well-suited for applications involving X-ray analysis or imaging. Furthermore, B_4C represents a polycrystalline ceramic material, potentially conferring advantages in terms of cost-effectiveness and manufacturability, compared to single-crystalline sapphire. Nonetheless, the polycrystalline nature of B_4C may lead to marginally lower mechanical strength and optical quality in comparison to sapphire windows.

Additionally, the chemical stability of both materials in diverse ammonothermal environments needs careful examination. While sapphire demonstrates remarkable resistance to chemical corrosion and degradation across various conditions, the chemical stability of B_4C may exhibit variability with respect to the specific chemical composition, and the pH value of the ammonothermal solution. Hence, thorough evaluation of the intended application and environmental parameters is required when selecting the appropriate window material for optical cells used in ammonothermal processes [299].

Avoiding notches, particularly in the central portion of the window, is crucial to prevent the notch effect, which can lead to window breakage upon insertion into the sealing gold layer, or during operation. Additionally, maintaining low surface roughness in the sealing area is vital to ensure that the plastically flowing gold layer adequately fills any remaining irregularities during sealing. While the precise impact of shape and position tolerances of the upper bevel of the viewing windows on window longevity or assembly success is not fully understood, visible deviations from the target geometry and the occurrence of cracks, often originating from the contact area between the window holder and the bevel, suggest that these tolerances may contribute to localized stress concentrations. Thus, attention to detail in the manufacturing process and adherence to specified tolerances are critical to ensure the integrity and longevity of the window assembly [299].

Rolling ball viscosimeter: A rolling ball viscometer, shown in Figure 19, was fabricated from a nickel-based superalloy, Inconel 718 [325]. It was designed and rigorously tested for a maximum temperature and pressure of 600 °C and 300 MPa, respectively. Design features specific to viscometry were the aspect ratio of the inner volume (realized by connecting two autoclaves), and a special conical shape at both ends to ensure that the ball only starts to roll at a certain angle of inclination [325]. A specific technical challenge lies in the ball itself, which should be as light as possible to maximize rolling time, but should also be resistant to corrosion and be available in spherical geometry [325].

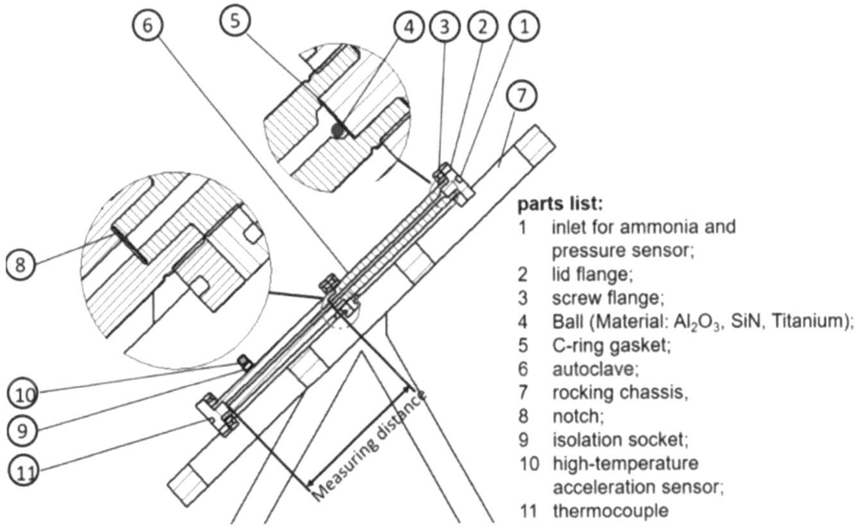

Figure 19. The illustration of the rolling ball viscometer engineered for accuracy in measuring viscosity across different fluid conditions. Reproduced from T.G. Steigerwald et al. [325] with permission, © Springer Nature Switzerland AG 2021.

Rotatable feedthrough: For realizing X-ray diffraction measurements (with monochromatic radiation) on GaN crystal during ammonothermal growth, a rotatable feedthrough for ammonothermal process conditions was also developed [299,326], which was used in conjunction with the autoclave shown in Figure 15d. A CAD model of the respective autoclave, with windows implemented at the diffraction angle to be used, is shown in Figure 20a, while a photograph of the part holding and rotating the crystal is depicted in Figure 20b. A first challenge lies in the high demands on the gasket to withstand the pressure of 300 MPa, while allowing for the movement (grinding seal). To avoid the simultaneous requirement of withstanding temperatures of up to 600 °C, the grinding seal was positioned outside the actively heated region so that PTFE could be used for the sealing package (see Figure 20c). A second challenge lies in transferring the torque from the shaft to the remotely located GaN crystal, which required a shrink fit connection to join the separately manufactured shaft to a long thin nickel–base alloy tube, and a second shrink fit connection between the tube and the crystal mount. The overall assembly, except for the crystal mount, is depicted in Figure 20d.

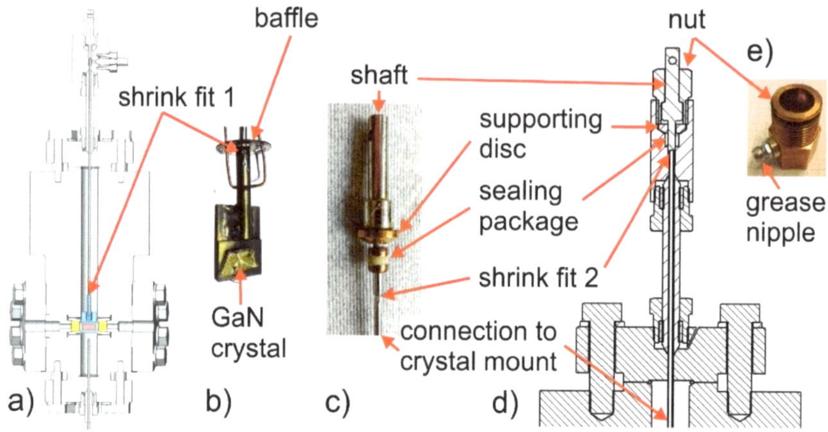

Figure 20. Rotatable feedthrough for ammonothermal autoclaves: (**a**) implementation in an autoclave designed for in situ diffraction experiments (same type of autoclave as in Figure 15d), (**b**) photo of a GaN crystal fixed to a crystal mount, which in turn is connected to the rotatable feedthrough; (**c**) photo of the shaft with supporting disc, sealing package, and shrink fit connection to the drive shaft passing on the torque to the crystal mount, (**d**) drawing of the feedthrough mounted to the cover flange, (**e**) photo of nut with grease nipple to reduce friction to a manageable level. Reproduced from [299].

Windowless autoclaves with improved X-ray transparency: In addition to the above developments, which have been tested experimentally under ammonothermal process conditions, the use of adapted reactor designs in conjunction with computed tomography with high photon energies has been estimated to be a feasible means of enabling in situ monitoring of crystal growth, even in windowless autoclaves [304]. An encouraging factor beyond that published feasibility study is that other application fields of superalloys would likewise benefit from superalloys with lower density [327], suggesting that the availability of suitable alloys with improved X-ray transparency may improve further in the future.

7. Emerging Technologies for In Situ Monitoring

For several reasons, ammonothermal reactors are challenging systems to obtain information from during operation. Firstly, the high pressures necessitate thick metal reactor walls, which lack optical transparency. Secondly, the corrosiveness of most ammonothermal reaction media towards most construction materials [147,148,315] strongly limits the choice of materials for windows with improved transparency for either optical light or X-rays [148]. Consequently, insights via in situ monitoring or numerical simulations are highly desirable.

Despite the above-outlined challenges, in situ monitoring techniques have been developed and have proved insightful for gathering information on the physical properties of the fluid and crystal growth/dissolution characteristics [328]. Understanding the progression of the reaction is vital for enhancing crystal quality in a targeted way. Real-time observation of the reaction offers invaluable insights into the system, surpassing post-reaction analyses. For instance, it can be difficult to unambiguously link distinguishable layers to process steps post-run [329].

In the field of ammonothermal crystal growth, numerical models have only been developed to a limited extent, and discrepancies often arise between simulation outcomes and actual experimental findings. One significant challenge lies in the difficulty of directly validating simulation results against experimental data. This difficulty stems from the intricate nature of ammonothermal experiments, which makes it challenging to measure crucial parameters. For instance, internal temperatures can only be measured for a limited number of measurement locations per reactor, and flow velocities have so far only been observed indirectly via their impact on fluid temperatures [127]. Unlike hydrothermal crystal growth, which benefits from well-established connections with geological research, the ammonothermal growth of nitrides lacks such a robust foundation [128]. Consequently, there remains a considerable gap in our understanding of the underlying thermodynamics, kinetics, and fluid properties specific to ammonothermal systems. As a result, the assumptions made in simulations carry a heightened risk of inadequacy due to the lack of validation and lack of accurate thermophysical data [128].

For instance, fluid properties play a pivotal role in ammonothermal processes, particularly considering the use of supercritical ammonia as a solvent. Data on the properties of pure ammonia are available [131], but the introduction of decomposition products and solutes can significantly alter these properties [330]. Experimental investigations focusing on thermophysical properties of ammonothermal fluids would provide invaluable insights into the intricacies of ammonothermal systems [128].

Viscosity, for example, is a critical factor in processes like ammonothermal synthesis. Its prediction is complicated by variables such as pressure, temperature, and dissolved substances altering fluid properties. While familiar fluids like water and oils typically exhibit Newtonian behavior, more complex ones like polymer melts behave non-Newtonian. In the case of supercritical ammonia used in ammonothermal processes, initial assumptions lean towards linear viscous flow resembling gases according to kinetic gas theory. However, this simplification overlooks the departure from ideal gas behavior under high pressures. To estimate viscosity in such conditions, Sutherland's semi-empirical formula is employed, although it has limitations, especially concerning high pressures and mixtures of substances [331,332]. Understanding viscosity in solutions adds another layer of complexity. The widely used Jones–Dole equation, applied to describe viscosity in aqueous solutions, considers factors such as concentration and ion interactions [333]. However, predicting behavior becomes more intricate with solutions containing multiple components. For example, the impact of ammonium chloride in water can vary depending on factors like the presence of other substances, such as urea or ethanol [334]. Such complexities underscore the difficulty in making broad predictions for solution viscosity. Consequently, in the realm of ammonothermal synthesis, assessing viscosity in situ becomes crucial due to the intricate interplay of factors affecting fluid behavior [335].

Furthermore, the decomposition of ammonia during crystal growth [136] introduces additional complexities, such as variations in ammonia mole fractions and the influence of catalysts on decomposition kinetics. Understanding these factors is essential for modeling the ammonothermal growth process with sufficient confidence in model accuracy. Moreover, uncertainties persist regarding the optical characteristics of the solution and the possible presence of nanoparticles, both of which can affect heat transfer and convection within the system [128]. Additionally, the impact of solutes on heat capacity remains unclear, as does the accurate determination of solubility data under varying conditions [128,150–152]. The establishment of comprehensive databases containing solubility information across different parameters would greatly enhance the reliability of numerical simulations, while also supporting experimentalists in designing their experiments. Furthermore, while some data exist on kinetics, particularly dissolution [152] and growth kinetics [324], they remain limited. Addressing these detailed uncertainties, together with improved model validation, is crucial for advancing the accuracy and reliability of simulation outcomes [128]. Thus far, mainly X-ray imaging [150], optical methods like UV/Vis or Raman spectroscopy [336], and to some extent, ultrasonic viscometry [337], have been applied for in situ monitoring of chemical or physical processes occurring in ammonothermal reactors under conditions of ammonothermal growth processes. In addition, the feasibility of X-ray computed tomography has been evaluated [304]. These techniques are listed in Table 7, alongside the types of information that they are able to provide.

Table 7. In situ measurement techniques for the ammonothermal method and the derivable information from each of them.

Applicable Measurement Techniques	Required Information
Thermometry	Continuous temperature measurement
X-ray absorption imaging	Crystal dissolution kinetics, solubility, solute concentrations and solvent density changes, potentially crystal growth rates
Computed tomography	Time-dependent etch-back and crystal growth rates, crystal geometry, parasitic deposition, nutrient morphology changes (tracking of mass transport in 3D)
Optical spectroscopy (UV/Vis, Raman)	Monitoring reaction kinetics via composition of the formed compounds Prospectively contribute to the identification of solutes such as intermediates
Viscometry	Viscosity

Given that each technique has different strengths and limitations, a synergistic application of several techniques is often beneficial for an unambiguous interpretation of the results. In the following, key aspects of the respective techniques in the context of their application to the ammonothermal method will be delineated, and an overview of the state of the art will be given for each technique.

7.1. Internal Temperature Measurements

Thus far, thermocouples have been the device of choice for measuring temperatures inside ammonothermal reactors during operation, due to their comparatively convenient integration into ammonothermal autoclaves (see Section 6.3). Initial experiments concerned with investigating the effects of fluid temperatures on crystal growth kinetics in supercritical ammonia–sodium (NH_3–Na) solutions were conducted by S. Griffiths and coworkers [324], who evaluated the mass change of source material and seeds in combination with the fluid temperatures to better understand the crystal growth process. The primary focus of the experiment was to delve into the intricate interplay between internal fluid temperatures and crystal growth rate in supercritical NH_3–Na solutions. By measuring temperature changes in both the dissolution and growth zones, the team aimed to understand the mechanisms behind crystal growth and find the best conditions for maximizing growth rates. Inconel-sheathed type K thermocouple probes were placed within the autoclave. One of them was positioned above the polycrystalline GaN source material in the dissolution zone and the other near the top-most seed crystal in the growth zone. The setup was designed to accommodate thermocouples without disrupting fluid flow, using smaller diameter probes, and including open areas in the baffled region. Furthermore, a comprehensive redesign of the loading procedure and hardware setup was undertaken to enable the lower thermocouple probe to reach the growth environment. This thermocouple was carefully positioned near the top-most seed crystal, just below the bottom-most baffle. The other thermocouple probe was situated in the dissolution zone, directly above the polycrystalline GaN source material. For both growth and dissolution zone, they evaluated not only the internal temperatures, but also the mass changes of GaN normalized by surface area and time (termed source loss flux and seed mass flux, respectively). In addition, they analyzed the fluid density difference between the dissolution and growth zone [324].

The aforementioned quantities are presented in Figure 21. From comparison of subfigures (a) and (b), one can see that there is a strong positive correlation between fluid density difference and source loss flux, particularly for low fluid density differences or low dissolution zone temperatures. S. Griffiths et al. determined a critical ammonia density difference of about 1.2 mol/L, above which transport from dissolution to growth zone, but also parasitic deposition occur [324]. Above this critical ammonia density difference, source mass flux increased with decreasing dissolution zone temperature, confirming retrograde solubility of GaN in NH_3–Na solutions [324]. Considering the temperatures and pressure, the overall ammonia density can be estimated to be about 20 mol/L [136], thus, the determined critical density difference corresponds to about 6%.

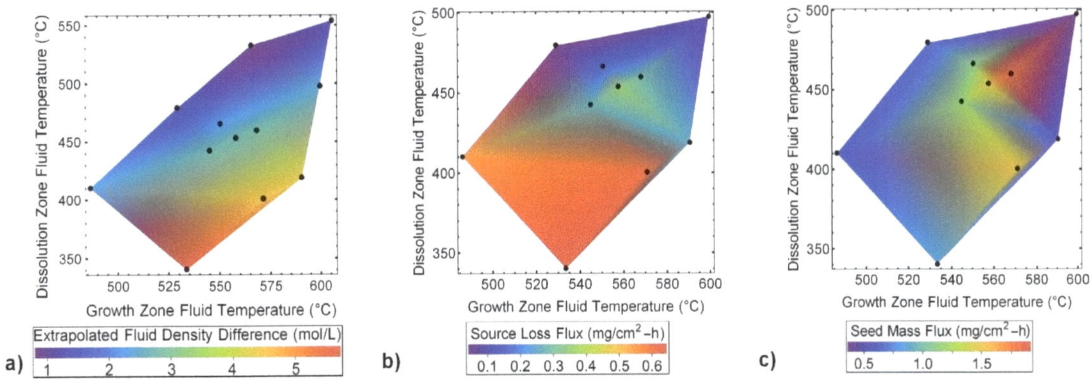

Figure 21. Quantities determined as a function of growth zone and dissolution zone temperatures by internal temperature measurements for investigating the role of surface kinetics and mass transport for crystal growth rates of GaN using Na mineralizer. (**a**) Extrapolated fluid density difference, (**b**) source loss flux, and (**c**) seed mass flux. Reprinted with permission, © 2018 Elsevier B.V. [324].

A comparison of Figure 21b,c reveals that peak source loss flux does not correspond to peak seed mass flux, indicating that for most conditions, seeded GaN crystal growth is limited by surface reaction kinetics, rather than by the transport from dissolution to growth zone [324]. Accordingly, S. Griffiths et al. determined endothermic activation energies of 145, 136, and 191 kJ/mol for the (0001), (000-1), and {1 0-10} surfaces, respectively, for growth zone temperatures below 570 °C. Above 570 °C, a reduction in seed mass flux with increasing growth zone fluid temperature was observed, which S. Griffiths et al. speculate to be related to changes of the fluid composition at high temperatures [324].

Besides the absolute fluid temperatures, S. Griffith et al. also mentioned fluctuations of fluid temperatures, which were in the range of 1 to 10 K in the growth zone and in the range of 5 to 15 K in the dissolution zone [324]. Considering that they mention that these fluctuations primarily depended on the heater setpoints (for the growth zone at the bottom of the autoclave) and on the insulation of the autoclave head (for the dissolution zone in the upper part of the autoclave) [324], these fluctuations are very likely related to fluid flow, which is also strongly influenced by both of these factors [338].

Following comparable observations, Schimmel et al. explored the potential of in situ fluid temperature measurements as indicators of flow dynamics [127]. The investigation involved carefully monitoring temperature fluctuations within the autoclave, particularly focusing on how these fluctuations can be applied as indicators of flow velocity changes. A schematic of one of the experimental setups is depicted in Figure 22a. Both experimental setups, which differed in the length of the autoclave, were equipped with one thermocouple in the upper and one in the lower zone. By analyzing temperature data collected under different experimental conditions, such as the presence of baffles and variations in autoclave geometry, they aimed to decipher the underlying convective heat transfer mechanisms. In Figure 22b, several types of information that can be obtained are denoted, using the first part of an experiment in which the temperature was increased from room temperature to elevated temperatures, with the higher temperatures being reached in the bottom part of the autoclave to facilitate natural convection. The fluid temperature measurements (T_{DZ} for the dissolution zone in the upper part of the autoclave and T_{CZ} for the growth zone in the lower part of the autoclave) gradually developed the before-mentioned short-term fluctuations (broadening of the respective lines in the graph). Furthermore, changes in heat transfer, due to the fluid reaching the thermocouple and subsequently the top inner wall, could be detected. Likewise, the decomposition of the mineralizer precursor sodium azide was detectable, both as a sudden temporary jump in fluid temperatures and as a pressure jump [127].

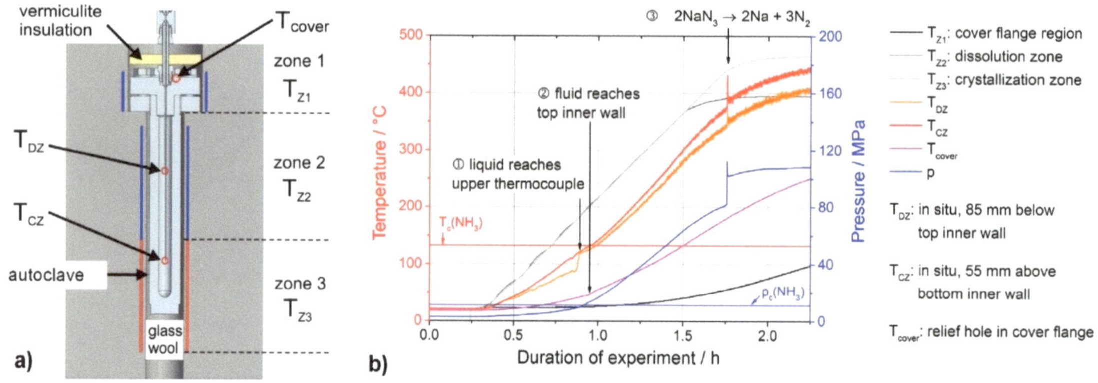

Figure 22. (**a**) Experimental setup consisting of a tube-shaped windowless autoclave equipped with two internal thermocouples for measurements in the dissolution zone (DZ) and the crystallization zone (CZ); (**b**) temperature and pressure measurements recorded in the process of temperature increase from room temperature to growth conditions (the critical temperature T_c and the critical pressure p_C of ammonia are also indicated, and events that can be observed via the internal temperatures are labeled). Reprinted from [127] under open access Creative Commons CC BY 4.0 license.

Numerical simulations were employed to complement experimental findings, shedding light on the response times of thermocouples and their impact on temperature measurements. Considering that flow velocities in the proximity of the growing crystals appear to have significant effects on crystal quality [138], this holistic approach promises not only to enhance our understanding of fluid flow dynamics in ammonothermal crystal growth, but also to provide valuable insights for optimizing growth conditions and ensuring consistent crystal quality [127]. From a methodological standpoint, a crucial consideration is how deeply an internal thermocouple must penetrate the fluid to accurately measure its temperature, without being unduly influenced by wall temperatures due to heat conduction within the thermocouple itself. Based on simulation results indicating a deviation of 15 K or less for a thermocouple inserted a few millimeters into the fluid, it becomes evident that significant thermal gradients between the autoclave wall and the fluid can affect temperature measurements [127].

In previous studies utilizing an optical cell for various measurements, including optical and X-ray absorption studies, thermocouples positioned near the inner wall of the autoclave did not typically exhibit temperature fluctuations, particularly during the main phases of the experiment [148,149,336]. This absence of fluctuations is primarily attributed to the designed isothermal conditions of the optical cell, minimizing convective heat transfer effects. However, it was possible to induce temperature fluctuations in the optical cell by removing the heating sleeve during cool-down, confirming that a distance of approximately 2 mm between the thermocouple tip and the autoclave wall is sufficient for detecting convection. This insight underscores the suitability of internal temperature measurements as probes for convective flow dynamics in experimental setups. Practicality states that the maximum deviation of 15 K is unlikely to manifest, given the typically small axial thermal gradients within the autoclave wall, which mitigate significant thermal differentials between the wall and the fluid. Therefore, the positioning of thermocouples within the fluid, even near the autoclave wall, remains a reliable method for monitoring convective heat transfer [127].

7.2. Imaging Techniques

In the area of imaging techniques, both the use of optical light and X-rays have been tested. As will be elaborated in the respective sections, these techniques differ in the range of ammonothermal process conditions in which they can be used, their requirements for autoclave window materials, and the information that can be obtained.

7.2.1. Video Optical Measurements

With sapphire as the optically transparent material being a usual choice for pressure vessel windows [339–341], video optical measurements were among the first imaging techniques investigated for application to ammonothermal processes. The construction of the first optical cell developed for ammonothermal conditions [298] can be seen in Figure 23. This optical cell was made from Inconel 718 alloy (a nickel–base superalloy) with sapphire windows at both ends to provide a straight optical path through the cell. This optical cell was designed for pressures as high as 200 MPa and temperatures up to 650 °C.

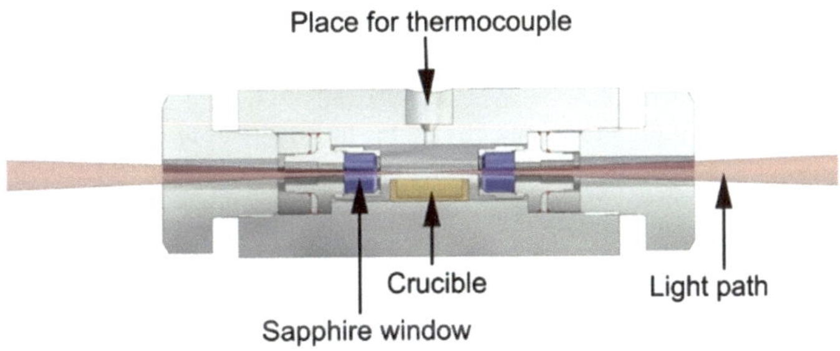

Figure 23. Sectional view of optical cell setup with positions for thermocouple and passage for light. Reproduced from [342] with permission, © 2011 Elsevier B.V.

This optical cell was equipped with a type K thermocouple that extended from the side wall into the reaction chamber, so that fluid temperatures could simultaneously be monitored [298]. For the video optical measurements, a halogen light source was used in conjunction with a standard video camera for recording, and the thermocouple tip served as an object to focus on inside the reaction chamber [298,342]. Ammonia, ammonium chloride, and polycrystalline gallium nitride powder were used as starting materials. It is worth noting that the experimental setup used provides independent control of pressure and temperature, which is achieved utilizing two high-pressure pumps for fill level adjustments during operation [298]. This enables operational flexibility and adjustment of reaction parameters beyond the thermodynamic stability field [298]. Examples of obtained video-optical measurements are depicted in Figure 24. For temperatures up to 350 °C, the video-optical measurements allowed monitoring of the movement of GaN particles, including the velocity and direction of the fluid flow [298]. An axial circulatory flow was clearly detectable, with flow speeds in the order of millimeters per second [298]. As is expectable for the optical cell as a near-isothermal system, the observed order of velocity magnitude corresponds to that found in numerical simulations of ammonothermal autoclaves for crystal growth, if the simulations consider conditions of low driving force for convection during the etch-back process prior to growth (dissolution stage) [329]. At 370 °C, the fluid turned yellow and then, with increasing temperature, became more opaque, leading to barely any light passing through the system at 420 °C [342]. This demands the use of other in situ monitoring techniques for ammonothermal systems under conditions of crystal growth, particularly if high solute concentrations are present in the location of measurement [342]. In spite of this limitation, it is remarkable that video-optical measurements remain the only technique that has been reported to yield quantitative information of fluid flow velocities in ammonothermal autoclaves to date.

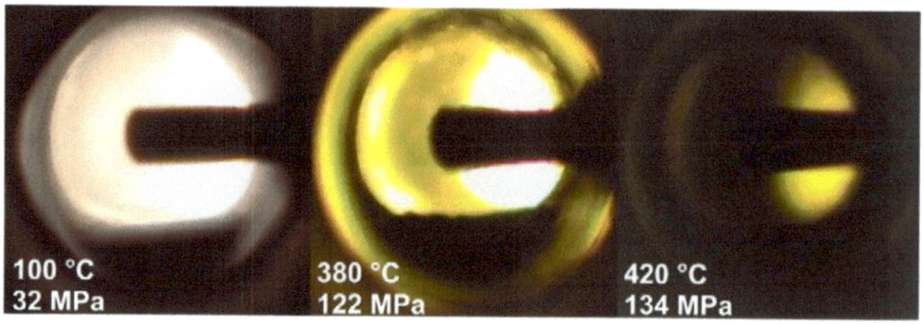

Figure 24. Video optical measurements of ammonoacidic GaN system at increasing temperatures showing decrease in visibility of the internal system [342]. The part extending into the area of view is the tip of the thermocouple. Reproduced from [342] with permission, © 2011 Elsevier B.V.

7.2.2. X-ray Imaging and Computed Tomography

Contrary to optical light, X-ray absorption is not as heavily affected by the presence of solutes that are formed upon the dissolution of GaN. The remaining X-ray intensity after penetration of an object I depends on the initial intensity I_0, the length t of the radiographed path through the object, the linear absorption coefficient μ of the radiographed material, and the density ρ of the radiographed material (or alternatively the concentration, if the absorbing element is a solute). This is evident from Lambert–Beer's law (Equation (2)) [343]:

$$I = I_0 \cdot e^{-\frac{\mu}{\rho} \cdot \rho \cdot t} \tag{2}$$

For objects to be distinguishable in an X-ray image, the objects need to differ sufficiently in their X-ray attenuation to yield a visible contrast in the image. As a measure of contrast, the Michelson contrast C_m is simple and sufficient for our purposes [344], though it overestimates the contrast of dark pairs of points compared with bright pairs with the same level of gray [345]. The Michelson contrast is defined as follows (assuming that I_{max} represents the maximum X-ray intensity reaching the detector and I_{min} represents the minimum X-ray intensity reaching the detector [300]):

$$C_m = \frac{I_{max} - I_{min}}{I_{max} + I_{min}} \tag{3}$$

Consequently, for a crystal to be distinguishable from the reaction medium, their linear attenuation coefficient or radiographed path must be sufficiently different, for which the linear attenuation coefficient usually plays a dominating role. In Figure 25, the linear attenuation coefficients of different materials relevant to ammonothermal experiments are shown as a function of photon energy. Note that the linear attenuation coefficient μ is the product of the density ρ and the mass attenuation coefficient μ/ρ, and therefore already accounts for the element- and density-specific contributions to X-ray absorption. The general trend is that the linear attenuation coefficients decrease with increasing X-ray energy, motivating the use of intermediate to high X-ray energies if the intensity that reaches the detector is a main limitation. An extreme example of this is the prospective application of X-ray computed tomography with acceleration voltages of several hundred volts to penetrate windowless autoclaves designed for crystal growth [304].

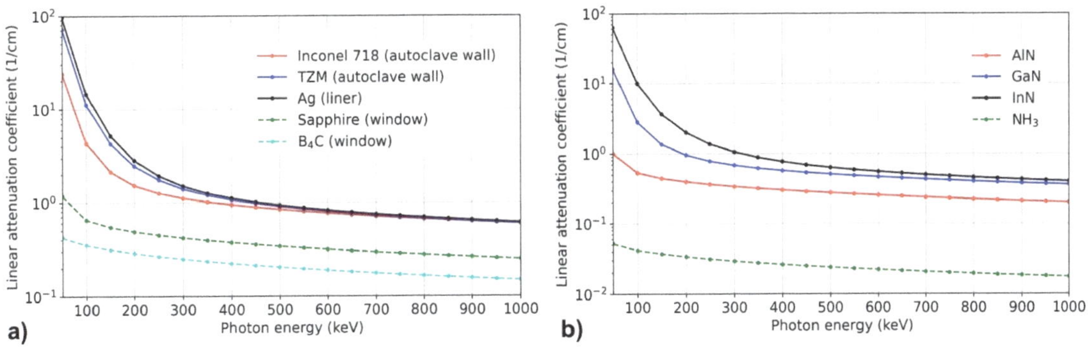

Figure 25. Linear attenuation coefficients plotted against photon energy for (**a**) construction materials of the autoclave and (**b**) educts involved in the ammonothermal synthesis. Mass attenuation coefficients for calculating the linear attenuation coefficients were taken from the Photon Cross Sections Database XCOM of the National Institute of Standards and Technology (NIST) [346]. For ammonia, a density of 0.23395 g/cm^3 was considered.

In the left subfigure of Figure 25, different construction materials typically used for ammonothermal reactors are compared with regard to their linear attenuation coefficients. It can be seen that regardless of the photon energy, the use of sapphire or boron carbide represents a major improvement in linear attenuation coefficient, compared to the nickel–base superalloy Inconel 718, the molybdenum–base alloy TZM, or the liner material silver. Moreover, B_4C yields an even lower linear attenuation coefficient than sapphire. More generally, materials that consist of light elements and have low density are the best choices for window materials if X-ray absorption is to be minimized, including materials like vitreous carbon, diamond, or silicon carbide [148]. However, the additional requirements of chemical stability under ammonothermal conditions and availability of a sufficient thickness to withstand the pressure need to be considered as well. Vitreous carbon exhibited significant damage in solutions containing sodium azide, though only exhibited superficial damage when tried for ammonoacidic conditions. Silicon carbide remained unaltered under ammonobasic conditions, but noticeable corrosion damage occurred under ammonoacidic conditions. From a chemical stability point of view, diamond is the ideal window material, as it withstands various ammonothermal environments, however, it has not yet been applied as a load-bearing window material due to the very limited availability and high cost of thick diamond materials [148].

In the right subfigure of Figure 25, linear attenuation coefficients of different binary nitrides, as well as pure supercritical ammonia, are provided. It is evident that the linear attenuation coefficient of ammonia is much lower than that of any of the binary nitrides, particularly for InN and GaN. This explains the good contrast between ammonia and GaN that is generally observed. An exemplary series of X-ray images obtained during an ammonothermal dissolution experiment with GaN is depicted in Figure 26.

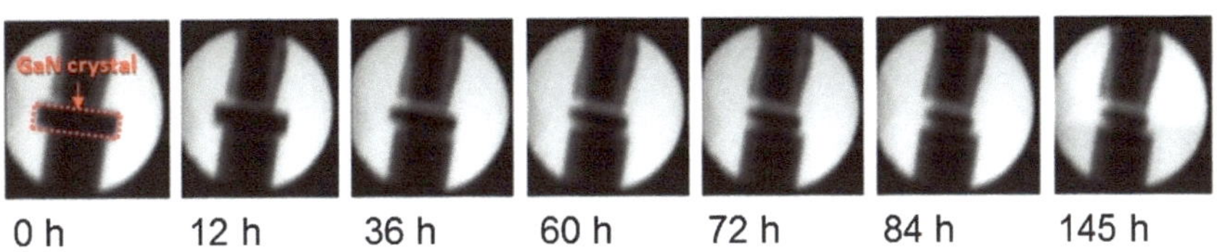

Figure 26. X-ray images showing the dissolution of a GaN crystal during an ammonothermal experiment conducted using an optical cell. Below and above the GaN sample, the Inconel 718 pins of the crystal mount are visible. Reproduced from [149] with permission, © 2017 Elsevier B.V.

For a better understanding of the images shown in Figure 26, imagery depicting a uniaxial optical cell, as used for obtaining the X-ray images, is shown in Figure 27. One can see that the area of view is somewhat small in relation to the inner diameter of the autoclave. Sample holders (Figure 27b–d) are needed to properly position the sample in the volume so that it can be observed. To mitigate corrosion, a gold coating was used for ammonoacidic (Figure 27b), whereas uncoated Inconel 718 for ammonobasic conditions (Figure 27c,d) shows a mount made of SiC, which yields good X-ray contrast to GaN, as can be seen by comparing the top and bottom images shown in Figure 27e [299]. This experimental setup was used for the majority of investigations on solubility and dissolution kinetics of nitrides via in situ X-ray imaging thus far.

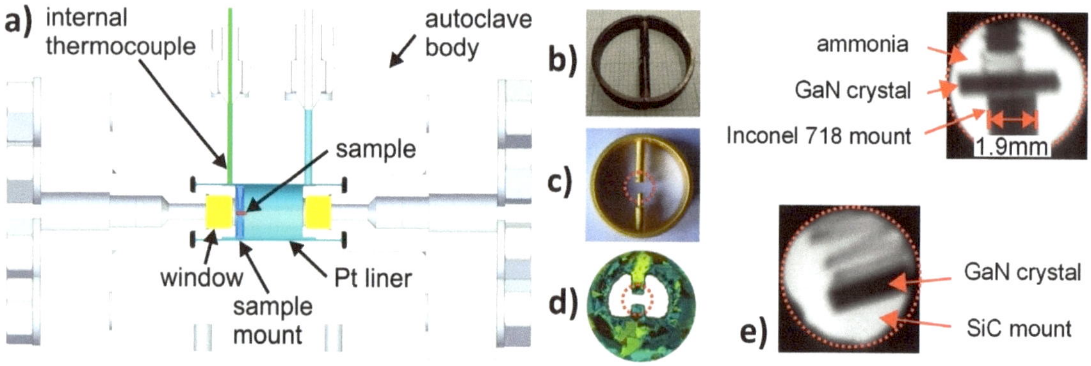

Figure 27. (**a**) Optical cell fitted with transparent windows for X-rays, (**b**) sample mount from Inconel 718 (uncoated), (**c**) sample mount from Inconel 718 (gold-coated), (**d**) sample mount from polycrystalline SiC grown via physical vapor transport, and (**e**) X-ray images obtained with crystal mounts made of Inconel 718 (**top**) and SiC (**bottom**). Regardless of the mount material, an Inconel 718 spring is used in addition, to hold the sample in place (visible above the samples in subfigure (**e**)). The dashed red circles in subfigures (**c**–**e**) indicate the area of view that can be monitored by X-ray imaging. Adapted from [299].

By equipping optical cells with sapphire or boron carbide windows, 2D X-ray imaging becomes feasible with photon energies in the range of 40–100 eV [149,150,330]. Note that beam hardening plays a nonnegligible role for the mean energy of photons that reach the detector and therefore contribute to the image: the white X-ray spectrum (consisting of characteristic radiation and bremsstrahlung) undergoes a shift to higher photon energies as the lower energy photons more likely to be absorbed by the setup. For an acceleration voltage of 100 keV and two sapphire windows of 10 mm thickness each, the mean effective photon energy has been estimated to be about 50 keV [330]. Using as low photon energies as possible is particularly advantageous if contrast is a critical limitation, as lower X-ray energies provide better contrast between different materials. This is due to the fact that the linear attenuation coefficients of materials differ more significantly for low photon energies, which can be seen in both subfigures of Figure 25.

Given the high linear attenuation coefficients of the alloys from which the autoclaves are made (examples shown in Figure 25 are the nickel–base superalloy Inconel 718 and the molybdenum–base alloy TZM), much higher X-ray energies are required if the use of autoclave windows shall be avoided (for instance for thermal field and thermal inertia engineering or to overcome their limitations, such as the small area of view). To penetrate the walls of an Inconel 718 autoclave designed for up to 300 MPa operating pressure at up to 600 °C, high X-ray energies above about 300 keV are required [304,344], and even higher X-ray energies were found to further improve quality of the X-ray images [304].

In Figure 28, exemplary images obtained using X-ray energies in the range of 300 to 600 keV are depicted. Note that these are computed tomography (CT) images, and these were obtained by capturing X-ray projections from multiple angles around the autoclave, in contrast to 2D X-ray imaging, which relies on a single projection. The use of many projections and projections obtained by radiographing the object in various directions yields significantly better image quality (after reconstruction) [344]. Consequently, S. Schimmel et al. have confirmed the feasibility of applying high-energy computed tomography for tracking the ammonothermal growth process of GaN and other materials of comparable X-ray absorption [304].

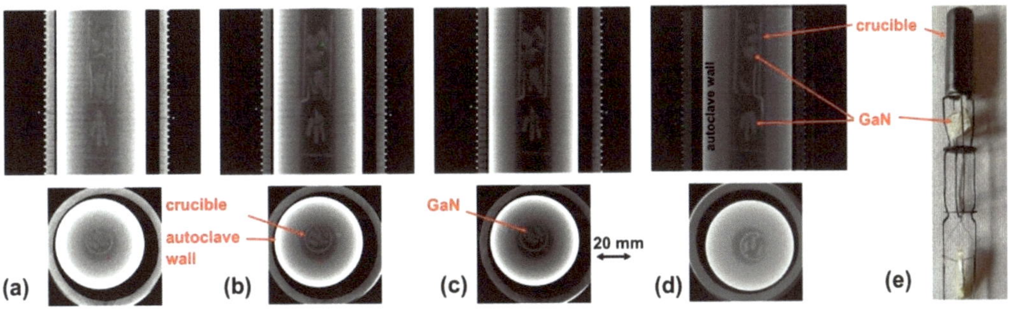

Figure 28. Computed tomography images of an autoclave interior. The (**a**–**c**) images are Inconel 718 autoclaves while (**d**) is Haynes alloy. X-ray energies are (**a**) 300 kV, (**b**) 550 kV, (**c**) 600 kV, (**d**) 590 kV. The photograph in image (**e**) shows the internal furniture including the GaN crystals. The internal furniture is made of Inconel. Reproduced from [304] under open access Creative Commons CC BY 4.0 license.

Up to now, 2D in situ X-ray imaging has primarily been used for studying both kinetic and thermodynamic aspects of GaN dissolution. Dimensional changes in a series of images like the one depicted in Figure 26 directly reveal the dissolution velocities for the crystallographic directions aligned perpendicular to the path of rays. Consequently, information on dissolution kinetics is straight forward to obtain for two crystallographic directions at a time [149,150]. From the left subfigure of Figure 29, it can be seen that the dissolution velocity (slope in the decrease of crystal width over time) depends on the crystallographic direction or, more specifically, the dissolution velocity in the c-direction is different from that of the a-direction. From the right subfigure of Figure 29, one can see that the dissolution velocity differences between the two non-polar directions (m and a) are negligible. The near-identical dissolution velocities in two crystallographic directions open a simple and therefore robust means of estimating the dimension in the third direction, which is not observable as a contour in 2D imaging. Provided that the third dimension represents one of the non-polar directions, it can be assumed to behave in the same way as in the directly observable direction [149,150]. While the dimensional change of the crystal in the direction of the path of rays can, in theory, be obtained from the changes in grayscale values in the respective region of the image, the outlined approach avoids errors that can occur, amongst others, due to changes in fluid absorption before or behind the crystal.

Using the outlined methodology, information on the sample volume as a function of time can be obtained (an exemplary such dataset is shown in Figure 30), which is necessary for investigating solubility. A second prerequisite is to observe the saturation of the solution. As can be seen from the remaining crystal volume in Figure 30, the dissolution slows down and eventually no further dissolution is observed (unless unintended factors like thermal gradients lead to deposition of GaN elsewhere in the reactor, therefore the geometry of the uniaxial and biaxial cells is ideal for the purpose of solubility measurements, see Section 6.3).

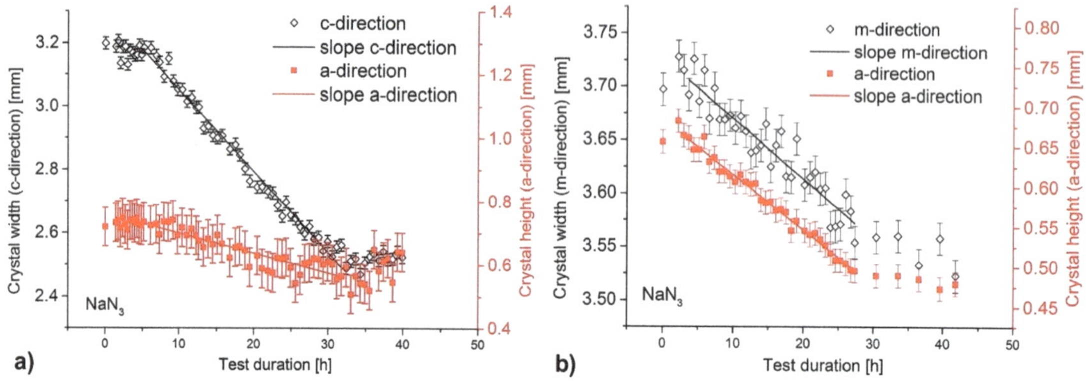

Figure 29. GaN crystal–dissolution kinetics with sodium azide mineralizer for (**a**) polar and non-polar orientations (c- and a-directions), (**b**) non-polar orientations (m- and a-directions). Reproduced from [149] with permission, © 2017 Elsevier B.V.

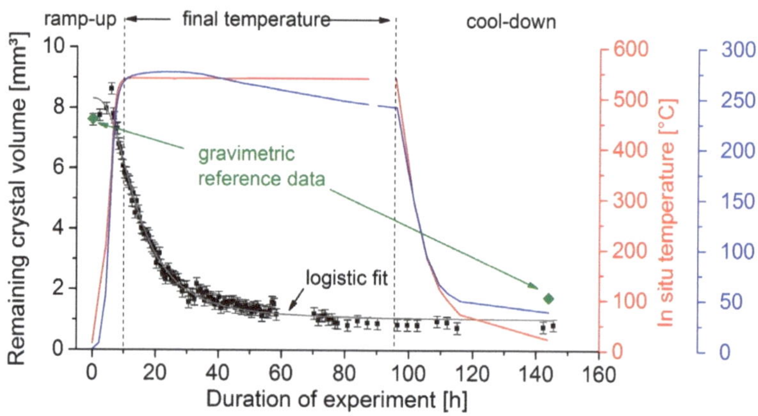

Figure 30. Change of GaN crystal volume over time shown alongside internal temperature and pressure. This ammonobasic experiment was conducted with 3 mol% NaN_3 mineralizer, corresponding to a molar concentration of 0.72 mmol/mL. Reproduced from [149] with the permission, © 2017 Elsevier B.V.

Such studies of dissolution kinetics and solubility were carried out for selected ammonobasic (NaN_3) and ammonoacidic conditions (NH_4F, NH_4Cl), with similar observations conditions [149,150,330]. Solubility data for GaN under different ammonothermal conditions, obtained by different methods of measurement, remain controversial and incomplete. Data have been obtained for some mineralizers, but temperature-dependent data for constant ammonia density are partially lacking, and the absolute magnitudes of solubilities remain inconclusive. A summary of all published data, as of 2021, can be found in [128].

The spatial resolution (both lateral and in the third, axial dimension), as well as the time resolution of 2D X-ray imaging of GaN in optical cells, has also been evaluated [299]. Since several subsequently taken X-ray measurements were commonly averaged for noise reduction and the air-cooled X-ray source allows only for a limited number of exposures shortly after each other [150], the achievable temporal resolution depends on the requirements for the quality of the images. Common time intervals between different series of measurements are in the range of 20 to 60 min [299,330]. While absolute positions are difficult to measure, changes of positions and dimensions can be determined rather accurately, with a lateral resolution of about 15 μm [299]. Regarding the axial resolution, changes in linear attenuation coefficient as small as about 0.12 cm^{-1} are detectable, which corresponds to Ga concentration changes of 0.13 mmol/mL [330].

Besides dimensional changes of GaN crystals, changes in X-ray absorption of the fluid can also yield valuable information. An observed increase in absorption of the solution revealed an increase in the concentration of dissolved gallium-containing species, and a local inhomogeneity in concentrations was observed [330]. A sphere-like region appeared around the crystal that slowly grew and diminished over time, suggesting slow dissolution and diffusional transport of Ga away from the crystal via formation and diffusion of intermediates [330]. The observed slow diffusion hinted towards an unexpectedly low diffusion constant of dissolved Ga complexes, which was at least partially explained by to the diffusion of larger $[Ga_xF_y]^{3x-y}$ aggregates, on the grounds of molecular dynamics simulations [330].

Several binary and ternary nitride materials have linear attenuation coefficients comparable to GaN, as is evident from the right subfigure in Figure 25 of this work (linear attenuation coefficients of additional nitrides can be found in Figure 7 in [304]). Consequently, 2D in situ X-ray imaging has been applied to nitride materials other than GaN, specifically to $ZnGeN_2$ [42], which has a linear attenuation coefficient very similar to that of GaN [304] and Zn_3N_2 [299]. A specific challenge for investigations on such novel nitrides can be a lack of available bulk material. To some extent, this can be circumvented by using pressed powder samples (microcrystalline) [42]. X-ray images of the dissolution and disintegration of a $ZnGeN_2$ pressed powder sample in an ammonobasic experiment with NaN_3 mineralizer are shown in the upper part of Figure 31. For quantitative analysis of the X-ray images, plotting the grayscale values along a line of interest is helpful. Such profile lines are shown in the bottom part of Figure 31. The Figure 31a,b arbitrarily divide the experiment into two parts for better readability of the profile line graphs. The gas phase, liquid phase, and supercritical phase of ammonia can be distinguished, as indicated in Figure 31a. Importantly, a decrease in transmitted intensity was observed in the fluid regions, in conjunction with the dissolution and disintegration of the pressed powder sample. This indicates that while a mechanical disintegration, with powder particles falling out of the area of view, may also have occurred, part of the increasing X-ray transmission in the region of the sample can confidently be ascribed to at least one element with high X-ray absorption going into the solution from there [42]. Considering that prospective intermediates for an ammonothermal solution of Zn have been reported for several ammonobasic mineralizers (amongst others) [188,347,348], it is expected that Zn at least did go into solution. A remaining limitation of X-ray imaging as a stand-alone method is that contributions of different elements in the solution (such as Zn and Ge in the described case) cannot be distinguished. Combining X-ray imaging with simultaneously conducted spectroscopic techniques (see Section 7.3) is considered to be a promising approach to improving the interpretability of fluid absorption changes in dissolution experiments with ternary and multinary nitrides.

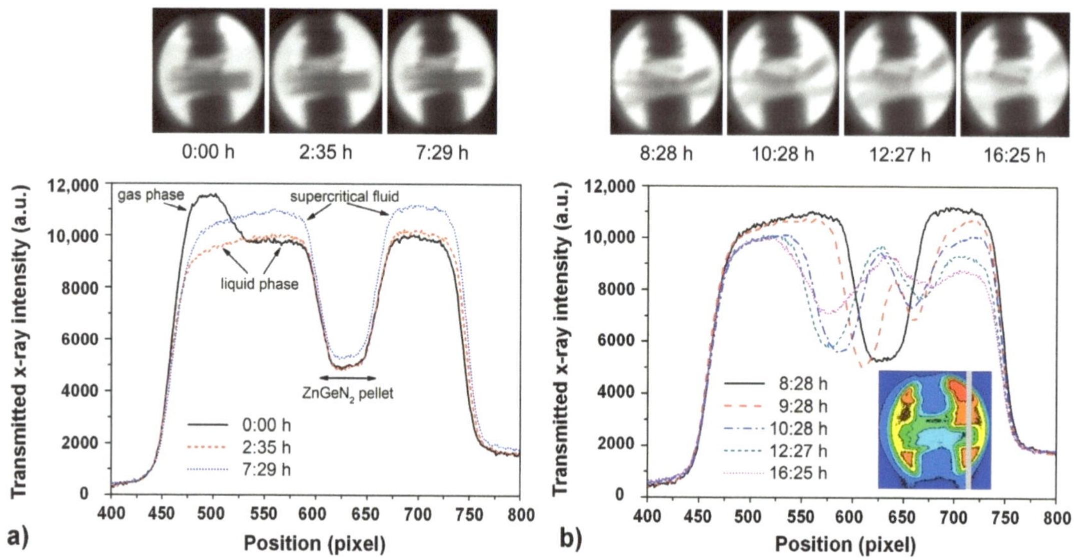

Figure 31. Dissolution of $ZnGeN_2$ pellet (pressed powder) over the time, as reported by Häusler et al., analyzed via X-ray absorption imaging [42]. (**a**) First part of the experiment (top: X-ray images, bottom: profile lines through the X-ray image along the line indicated in the inset of subfigure (**b**)), (**b**) first part of the experiment (top: X-ray images, bottom: profile lines through the X-ray image along the line indicated in the inset of subfigure (**b**)). In both subfigures, the X-ray images are reprinted and slightly modified with permission from [38], © 2017 Wiley-VCHVerlag GmbH&Co. KGaA, Weinheim, Germany, and the profile line graphs are reprinted from [299]. The colors in the inset of subfigure (**b**) represent different grayscale values in the X-ray image.

7.3. *Spectroscopic Techniques*

A broad range of spectroscopic techniques has been applied to high pressure systems [349] and more specifically, supercritical fluids [341]. However, only UV/Vis and Raman spectroscopy have been applied to ammonothermal solutions thus far, and we will therefore focus solely on spectroscopic techniques. The fundamental advantage of spectroscopic techniques is that the contributions of different photon energies (wavelengths) to the measurement signal can be distinguished. Unlike the imaging techniques, they can therefore provide insight into chemical bonding and hereby help to identify, for instance, intermediates present in the solution.

7.3.1. UV/Vis Spectroscopy

The feasibility of applying UV/Vis spectroscopy for investigating ammonothermal solutions under process conditions of crystal growth has been demonstrated using a transmission geometry, with a deuterium lamp as the light source to provide sufficient emission in the UV range [298,328]. Changes in the UV/Vis transmission spectra (as well as the reduced transmission observed by the video optical measurements) are thought to be mostly due to changes in fluid absorption, with UV/Vis being the more sensitive technique for detecting these changes [328]. Exemplary UV/Vis transmission spectra are shown in Figure 32, which were obtained by N. Alt and colleagues during their experimentation with GaN in an ammonoacidic environment with an NH_4Cl mineralizer [342].

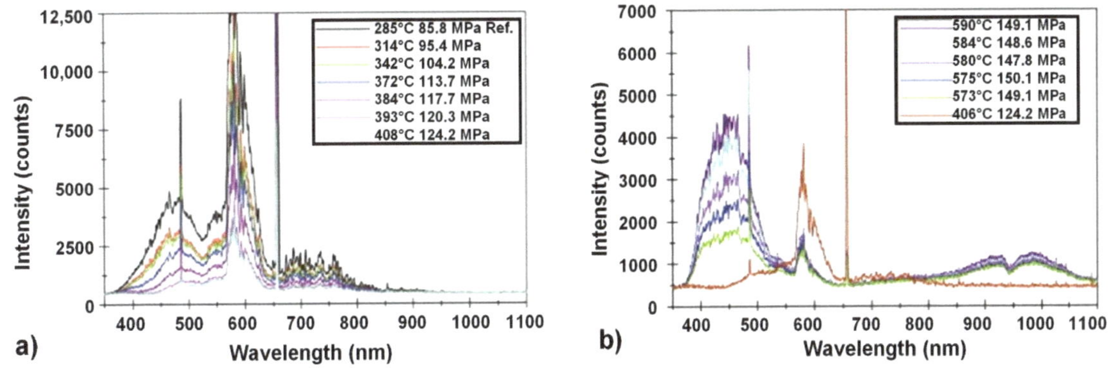

Figure 32. UV/Vis transmission spectra of the $GaN/NH_4Cl/NH_3$ ammonothermal system were examined across a wide range of temperatures, revealing changes in the transmitted intensity across various wavelength regions. Subfigure (**a**) shows the lower, whereas subfigure (**b**) shows the higher temperature range. Reprinted and slightly modified/corrected from [342] with the permission, © 2011 Elsevier B.V.

Notably, temperatures exceeding 300 °C prompted a consistent decrease in overall light intensity, particularly prominent in the shorter wavelength range. This trend closely paralleled observations from video optical analyses, indicating a concurrent reduction in total light transmission up to approximately 540 °C, beyond which minimal transmission was observed. The observed decline in transmission corresponds well with the documented solubility behavior of GaN in supercritical ammonia solutions containing NH_4Cl, implying a rapid increase in solubility beyond 300 °C. Moreover, intriguing alterations in the transmitted UV/Vis spectrum were discerned beyond 560 °C, characterized by notable intensity increases within the 380–550 nm range and reductions within the 550 nm to 800 nm range. These changes suggest the potential formation of intermediary compounds arising from reactions between the fluid and GaN powder, hinting at complex chemical transformations occurring at higher temperatures. The changes in different wavelength ranges point towards more than one chemical species contributing to the changes in fluid absorption [342].

However, an unambiguous assignment of specific absorption changes to specific solutes is yet to be demonstrated. A complication for interpretation lies in the possibility of further elements being present in the solution, with corrosion of reactor materials being a main point of concern. N. Alt and colleagues have therefore employed energy-dispersive X-ray spectroscopy and found that indeed several elements such as iron, chromium, or nickel were present in reaction products investigated after the ammonothermal experiment [342]. While exact information about the intermediate species formed has not yet been obtained, UV-Vis spectroscopy has already proven useful in monitoring solubility by spectroscopic means. Care must be taken when interpreting signals in the infrared range, as such signals could potentially be caused by an emission of a chemical compound formed during the experiment, but infrared emission can also be due to the thermal emission of reactor materials. Thus far, infrared emission has only been observed due to thermal emission of reactor components, given that the observed infrared emission did not show any dependency on the amounts of reactants introduced [342].

UV/Vis spectroscopy was also used in investigating the decomposition of ammonia, both with and without additives, experimentally simulating conventional ammonothermal syntheses performed in standard autoclaves [336]. A range of media was utilized, encompassing pure ammonia (NH_3), ammonia with a ruthenium catalyst, ammonia with sodium azide (NaN_3), and ammonia with sodium azide and gallium nitride (GaN). A type K thermocouple was fitted for internal temperature monitoring, allowing direct observation of fluid temperature during experimentation. An additional thermocouple was affixed externally to one of the windows. The experimental setup further incorporated heating jackets, high-pressure valves, rupture discs, and connection pipes, enabling meticulous control and monitoring of experimental conditions. Optical measurements were completed using a xenon lamp and spectrometer, also shedding light on the transmittance properties of sapphire windows utilized in the setup [336].

7.3.2. Raman Spectroscopy

Fundamentally, Raman spectroscopy relies on the inelastic scattering of light by matter, that is, the energy of the scattered photon is different from the energy of the incident photon (Raman effect) [350,351]. Due to its sensitivity with respect to crystalline, amorphous, and molecular species, Raman spectroscopy allows the monitoring of phase transitions and chemical reactions [341], and can be used for retrieving information on molecular structures [352]. It has been widely applied to a plethora of subject areas and a variety of variants have been developed to address specific challenges, such as the weak signal of spontaneous Raman scattering [350]. While a number of studies at high temperatures up to 1000 K have been reported, studies at high pressures, or under supercritical conditions, are less common [341]. Nevertheless, some applications to supercritical fluids have been reported, besides those on ammonothermal systems, including studies of phase equilibria in fluid mixtures [353] and contributions to revealing structural organization in binary mixtures, with notable effects on diffusion processes [352].

In the context of ammonia decomposition by T.G. Steigerwald and coworkers [336], Raman spectroscopy emerged as a robust tool for investigating high-pressure systems, thanks to the dense molecular environment involved, as elucidated in Braeuer's book on spectroscopic techniques for in situ monitoring at high pressure [349]. The adoption of spectroscopic techniques, particularly Raman spectroscopy, holds significant potential for deepening our understanding of ammonothermal synthesis [336,354].

T.G. Steigerwald et al. [336] applied Raman spectroscopy to ammonothermal experiments, utilizing a continuous wave laser emitting green light with a wavelength of 532 nm and adjustable power up to 2 W. The experimental setup involved backscattering, where the laser beam was directed onto the sample in the optical cell, and the backscattered light was collected at a 0° angle using the same optics. To enhance signal clarity, a dichroic mirror was used to separate the redshifted Raman signal from laser reflections and Rayleigh scattered light, while a long-pass filter further refined the sample signal before reaching the spectrometer. Despite these measures, interference from elastically scattered light was observed below a Raman shift of 100 cm^{-1} [336].

As part of their study, they employed Raman spectroscopy to analyze pure ammonia, in order to better understand ammonia decomposition kinetics. All the spectra pertaining to this experiment are shown in Figure 33. With a low density of 8.69 mol/L, the experiment aimed to explore how pressure influences the equilibrium of NH_3 decomposition. Spectra were normalized to the ammonia signal to facilitate comparison. At 563 °C and 70.0 MPa, characteristic peaks for ammonia, nitrogen, and hydrogen were observed, albeit with initially low signal-to-noise ratios for hydrogen and nitrogen peaks due to short integration times. Subsequent measurements showed increased peak intensities over time, reaching their maximum after 48 h [336].

Another focus of the study was to investigate whether the decomposition of NaN_3 and the formation of $NaNH_2$ occur concurrently or independently. Raman spectroscopy was utilized to analyze changes in the electron shell polarizability of nitrogen and hydrogen. Spectra were collected over a duration of 21 h at a consistent temperature of 325 °C, revealing the emergence of N_2 signals after 3.5 h, coinciding with a progressive increase in intensity alongside a corresponding elevation in pressure from 719 bar to 843 bar [336].

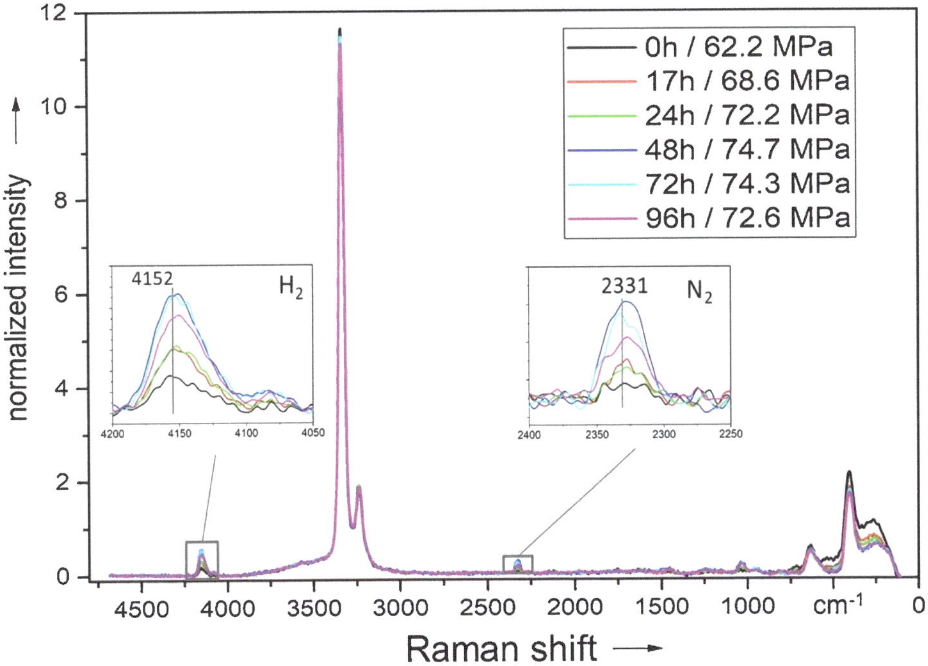

Figure 33. Raman spectra of ammonia under specific experimental conditions; a filling density of ammonia at 8.69 mol/L, a maximum temperature of 563 °C, and a maximum pressure of 74.7 MPa, over an integration time of 100 ms, and a single measurement. Reproduced from [336] with the permission, © 2017 Elsevier B.V.

Further experimentation across varying temperatures provided deeper insights into the kinetics of ammonia (NH_3) decomposition and the catalytic influence of ruthenium. Raman spectroscopy played a crucial role in observing clear signals for ammonia, nitrogen, and hydrogen, facilitating the tracking of their changes over time, and enabling the assessment of the ruthenium catalyst's influence. Through comparisons of different experiments involving various mineralizers and catalysts, the team identified their respective effects on the speed of ammonia breakdown and the production of nitrogen and hydrogen. These investigations unveiled the complexity of ammonia decomposition and highlighted the significant impact of ruthenium, challenging established notions in autoclave-based ammonothermal synthesis. While Raman spectroscopy provided valuable insights into reaction kinetics and catalyst-mediated processes, there is still a need to unravel the underlying mechanisms governing ammonia decomposition and optimize synthesis protocols, especially concerning gallium nitride production [336].

7.4. Ultrasonic Velocity Measurements

In general, ultrasonic techniques can provide information on various properties of materials and local changes of these in an investigated sample, and different ultrasonic parameters such as attenuation, velocity, and acoustic non-linearity can be analyzed and are applied, amongst others, in non-destructive testing [355]. In the field of the ammonothermal method, the application of ultrasonic techniques for in situ monitoring has so far focused on ultrasonic velocity measurements, which were applied for determining the ammonia fill level prior to the ammonothermal experiment [356] and measurements of sodium azide solubility [337]. In addition, a concept for using ultrasonic velocity measurements for determining the viscosity of ammonothermal fluids has been elaborated and is considered to be a particularly promising concept for viscosity measurements under ammonothermal conditions. Both standard tube-shaped autoclaves (fill level, sodium azide solubility) and optical cells (viscometry) have been used for ultrasonic velocity measurements on ammonothermal systems, depending on the measurement goals [325].

Ultrasonic viscometers provide real-time measurement of fluid density and viscosity using acoustic- and shear-impedance analysis, with practical implementations often employing normal incidence configurations for ease of use [357–359], which is also the case in the concept that was elaborated for its application to ammonothermal optical cells.

Ultrasonic viscometers utilize either the pulse-echo or through-beam method, both sharing similar construction. The pulse-echo method, which is widely employed, enables easy determination of medium viscosity and density by emitting and capturing sound waves' reflections [337]. It is adaptable for different mediums and temperatures, exemplified by the Cohen-Tenoudji high-temperature viscometer. However, its reliance on medium and frequency can complicate viscosity determination [360]. Meanwhile, the through-beam method employs two sensors to emit and receive ultrasonic waves, ensuring accurate measurements, even near surfaces. While the pulse-echo method has been preferred for ammonothermal synthesis, through-beam offers potential advantages, particularly considering the need for high-temperature adaptability and application in non-Newtonian fluids [361–363].

Additionally, advancements in measurement accuracy and temperature range, up to 1500 °C with cooling measures, highlight the suitability of ultrasonic viscometers for diverse applications, including ammonothermal synthesis, offering insights into fluid behavior at extreme temperatures [364–366].

The fundamental structure of the ultrasonic viscometer, as well as its signal processing methodology, draws upon the groundwork laid by S. H. Sheen et al. [363]. Notably, in the foundational work, only transducers for longitudinal wave measurement had been utilized, limiting accessibility to density information alone [337]. However, the latest design, as depicted in Figure 34, incorporates adjustments to address this limitation. Specifically, two extension bases are affixed to the optical cell, enabling direct opposite positioning. While the simplest configuration involves bases positioned directly opposite to each other, there is consideration for potential angled autoclaves, pending initial measurements to evaluate the added manufacturing complexity against the benefits in terms of accuracy. Additionally, the exact geometries of these bases remain subject to future research, with potential designs exploring staggered surfaces to ensure the requisite reference reflection for accurate measurements. Furthermore, to concurrently determine viscosity, a sensor for shear waves must be integrated, with an emphasis on optimizing space utilization by installing transducers in pulse-echo mode opposite each other [325].

Incorporating optical cells into the viscometer setup, as illustrated in Figure 34, offers several additional advantages. Optical cells can enhance the precision and versatility of viscosity measurements by providing real-time visual feedback on fluid behavior. By leveraging optical techniques such as laser doppler velocimetry or particle image velocimetry, researchers can prospectively gain insights into flow patterns, turbulence, and shear rates, complementing the data obtained from ultrasonic measurements. This integration of optical cells expands the capabilities of the viscometer, allowing for a more comprehensive understanding of fluid dynamics in ammonothermal systems, especially at elevated pressures and temperatures [325].

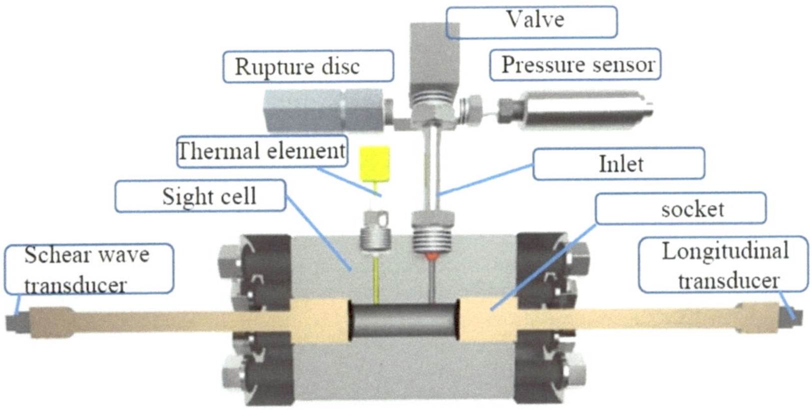

Figure 34. A high-pressure optical cell featuring schematically depicted elements for implementing the ultrasonic pulse-echo measurement technique. Reproduced from [325] with permission, © Springer Nature Switzerland AG 2021.

7.5. Rolling Ball Viscosimeter

Rolling ball viscometers, known for their simplicity and robustness, can be highly precise with proper calibration and fabrication, and have been used by L. T. Carmichael and B. H. Sage to measure the viscosity of liquid ammonia across a wide temperature range [367]. Rolling ball viscometers have been adapted for high-pressure and high-temperature applications [368,369], K. Funakoshi and A. Nozawa achieved viscosity measurements even under extreme conditions using X-ray detection [370]. Adjustments to the standard working equation are necessary, as viscosity changes with temperature and is influenced by factors like diameter ratio and the Reynolds number. Despite the challenges, such as returning the ball to its starting position and determining its velocity, rolling ball viscometers meet the requirements for ammonothermal processes. To address these challenges, a high-temperature borne-noise sensor is proposed, utilizing acoustic signals from the rolling ball for viscosity measurement without significantly affecting the temperature profile within the autoclave [371].

Experimental measurements were conducted by T.G. Steigerwald and coworkers [325,371] using the rolling ball viscometer depicted in Figure 19, within a temperature range spanning from room temperature to 588 °C, coupled with pressures reaching up to 100 MPa, validating the operational integrity of the viscometer. Heating was uniformly distributed along the viscometer via three heating sleeves. Diverse materials were employed for the balls to facilitate density determinations, and calibration ensued by employing fluids with precisely known properties. The setup encompassed high-pressure pumps and sensors, alongside a high-temperature borne-noise sensor meticulously crafted for acoustic logging. To refine accuracy, the reproducibility of the viscometer and meticulous calibration with known fluids were ensured, with more detailed description of the setup found in [325,371]. The rolling ball viscometer evinced accuracy and consistency in viscosity measurement up to 588 °C. The inquiry underscores the imperative of compiling fluid properties as a requisite for the ammonothermal process, with the rolling ball viscometer proffering a precise methodology for viscosity and density determination. The feasibility study underscored the promise of a novel high-temperature borne-noise sensor, giving insights into the autoclave, even at higher temperatures. Subsequent investigations are needed to explore variations in ball density to ascertain the viscosity and density of the fluid system within the ammonothermal process, potentially extending the temperature threshold beyond 588 °C [371].

7.6. X-ray Diffraction

In recent years, a few researchers have developed technologies for an in situ characterization of crystals during their ammonothermal growth. S. Schimmel and coworkers have developed a setup capable of observing the spatial distribution of intensity within the (0002) reflection of GaN (similarly to a rocking curve) [299,372].

8. Summary and Outlook

The ammonothermal method is a promising technique for the synthesis of novel nitrides in crystalline form, along with the development of large scale, single-crystal boules of technologically relevant materials, such as GaN. The method has seen significant advances since its early incarnation for the preparation of amides and imides in the 1960s, to the more rigorous scaling of GaN boules in the 1990s. To date, the method has been adopted worldwide by a few research labs and companies. As we look to the next phase of this method, the synthesis of emerging nitrides, or the experimental realization of theoretically predicted nitrides is becoming a greater focus. There are numerous interesting nitrides that have been identified and actively investigated for single crystal synthesis, including II–IV–N_2, Mn–IV–N_2, $Al_{1-x}TM_xN$ (TM = transition metal), and III–N. These materials have the promise of exhibiting interesting material properties, including strong ferroelectricity, piezoelectricity, non-linear optical properties, spin polarization, and strong magnetization capabilities.

The primary driver for the larger scale adoption and advancement of the technology was the desire to synthesize large boules of GaN for use as substrates in wide bandgap III-N devices. This feat has successfully demonstrated the growth of GaN boules/wafers of up to nearly 4 inches in diameter, with autoclaves capable of growing more than 50 wafers simultaneously, and wafers exhibiting > 1 km wafer curvature, though typically containing relatively high concentrations of oxygen/impurities (typically < 10^{19} cm^{-3}) and hydrogenated gallium vacancies (ammonobasic 10^{18}–10^{19} cm^{-3}/ammonoacidic up to 3×10^{16}–9×10^{16} cm^{-3}).

An important feat for enabling the growth of novel nitrides, or advancing growth conditions, is the development of a rigorous understanding of the solubility of species in the solution. Ammonobasic systems heavily rely on alkali (primarily) and alkaline earth (secondarily) metals, while acidic counterparts use halogens. Both systems have successfully demonstrated dissolution of a range of metal cations, with some of them having been investigated to also demonstrate a temperature-dependent solubility, as required for crystal growth. Some important examples include the solubility of Be, Mg, and Si as dopants for III-nitrides, to the transition metals Sc, Ti, Y, Zr, Nb, Hf, and Ta for piezoelectric $Al_{1-x}TM_xN$, to Mg, Si, Mn, and Zn as constituents of II–IV–N_2 semiconductors, and to Ti, V, Zr, Nb, Mo, Hf, Ta, and W as constituents of nitride superconductors. Only for some of these, and for a limited set of ammonothermal conditions, is qualitative information on reactivity and/or solubility available to date.

Acidic ammonothermal systems generally afford higher solubilities and more aggressive dissolution, though this comes at the cost of the autoclave potentially suffering severe corrosive attack. The use of expensive liners (particularly platinum) enabled the use of the ammonoacidic method, although this added a barrier to entry, particularly for applications that require high purity. More recently, less expensive alternatives have emerged, including liner-free autoclaves based on Mo alloy autoclaves and the use of silver as opposed to platinum.

In the realm of III–N materials, AlN and AlGaN are promising materials to explore using this method. Few demonstrations on their growth exist to date, though solubility and deposition have been demonstrated, albeit at a small scale (~few mm in polycrystalline form). The viability of this method to demonstrate growth of the full AlGaN solid solution range has been successfully performed, though further refinements of the method are needed to develop single crystal boules which may one day be useful for applications involving UV light emitters or power electronics.

Most recently, some efforts have started to explore this method for the growth of BN. Demonstrations of its solubility and synthesis, at least for hexagonal and rhombohedral BN, have already been made.

InN is another intriguing nitride that, thus far, has only been synthesized in bulk, single-crystal form using the ammonothermal method. Initial demonstrations suggest a different growth mechanism at play, requiring the use of ammononeutral conditions, through the presence of both an acidic and basic element. Further refinement of this approach has the potential to yield InN substrates, which could permit development of high frequency/speed electronics based on InN, along with infrared emitters/absorbers based on an all-nitride platform.

With regard to emerging materials, ternary nitrides are of great interest. A relatively small set of materials has been demonstrated thus far (including a range of ternary tantalum, silicon, and germanium nitrides and phosphonitrides), partly due to the lack of large-scale investigations. These materials are of current interest, as they can offer an avenue to earth-abundant materials with desired electronic, optical, magnetic, or structural properties.

The ammonothermal method has clearly demonstrated the ability to grow materials with a range of impurities—some desired, such as dopants (O, Si, Mg, Mn), and some less desired, such as metals (Fe, Zn, Na, Al) or atmospheric contaminants (H, O, C), along with the ability to form hydrogenated point defects, such as the observed hydrogenated gallium vacancy (V_{Ga}–H_x). These defects inevitably cause a change in material properties, ranging from changes in p- and n-type doping, along with introduction of defect levels within the band gap, or formation of point defects with intriguing properties. While advantageous at times, this does pose a challenge for the synthesis of a high purity material using this method. Some demonstrations of the reduction of transition metal incorporation and reduction of oxygen (in part via the use of getters, e.g., Er and Al) exist, although these typically come at the price of more complexity, cost, time, or a combination thereof. No uniformly applicable solution to this challenge is on the horizon.

In accordance with impurity incorporation, a wide range of electronic properties have been demonstrated, at least in the case of GaN, ranging for high free electron concentrations ($>2 \times 10^{19}$ cm^{-3}) and relatively low resistivity (10^{-3} Ω cm). Due to the large number of dopants and defects, mobility in these samples was relatively low. Higher purity material has been demonstrated, yielding higher mobilities up to ~565 cm^2/(V·s). By doping with higher amounts of Mg in GaN crystals than the background oxygen levels, p-type materials have been demonstrated. Additionally, insulating materials have been demonstrated via the addition of compensating defects, introducing defect levels in the middle of the bandgap. Given the demonstrated breadth of ranges of the incorporation of impurities, desired or otherwise, this suggests the ammonothermal method has sufficient flexibility to control a similar range of impurity incorporation for novel nitrides.

Beyond electronic properties, optical characteristics of impurity containing GaN are detailed. The commonly observed colorations of grey, yellow, green, or orange-brown/red are typically due to free electron concentrations, transition metals, hydrogenated gallium vacancies, or other currently unidentified impurities.

Structural properties of grown nitrides can be directly or indirectly influenced via the variation in free electron concentration, generation, or inclusion of impurities and other point defects, along with heteroepitaxial growth on a non-lattice matched substrate.

Autoclave technologies in general are critical to the successful growth of nitrides from a solution, given the requirement to obtain temperature gradients at high pressure. For the ammonothermal method, tube-shaped autoclaves are generally used for bulk crystal growth, as well as for most other purposes, such as exploratory syntheses. The ammonothermal method has demonstrated multiple co-loaded, multi-inch GaN boules. Refinement of these autoclaves has offered the possibility of new material investigations, and the development of techniques to aid in our understanding of the process. A key advancement is the enhanced corrosion resistance via improved material selection (such as Mo alloy or prospectively Co alloy-based autoclaves), or integration of liners (such as noble metals). The integration of in situ technologies is considered critical to better understanding the system, though it is challenging for these systems given the pressures and corrosive chemicals involved. Nonetheless, notable advances have been made.

Early-stage demonstrations were presented in the integration of a thermocouple into the environment allowing for determination of the actual fluid temperature—a value that was found to be significantly lower than the external wall temperature in smaller scale systems. This approach was then refined to include multiple thermocouple junctions and hence temperature gradients, permitting insights into the growth kinetics, e.g. of GaN, and a better understanding of the heat transfer and fluid flow dynamics in these systems. Fluid temperature measurements also offer valuable insights into convective heat transfer and fluid flow dynamics, complemented by observations of chemical reactions with enthalpy changes. Convective heat transfer significantly impacts temperatures at the outer autoclave wall, prompting questions about suitable boundary conditions for simulations. Clarifying fluid properties such as viscosity, possibly through the validation of simulations with experimental data or dedicated viscosity measurements, is crucial for refining simulations and accurately predicting internal temperature distribution and fluid flow dynamics in ammonothermal growth reactors [127].

Tedious efforts were put forth to identify suitable materials that can be used as windows for optical wavelengths up to high energy X-ray beams. Integration of these windows, including sapphire and B_4C, has been successfully demonstrated on small scale systems, allowing for imaging of crystals or solutions in situ during various stages of the process, critically at synthesis pressures and temperatures (up to 300 MPa and 600 °C). These systems have successfully provided insights into the physical and chemical properties of the solution and the kinetics of crystal growth and dissolution. Additionally, the integration of X-rays into the system has allowed for the demonstration of X-ray diffraction of a growing GaN crystal, giving unprecedented insight into the growing material from a structural perspective. Furthermore, sealed, yet rotatable feedthroughs have also been demonstrated allowing for sample rotation.

An exciting advancement is on the horizon in the application of X-ray computed tomography during the growth of nitrides in ammonothermal systems. Demonstrations have already yielded novel insight into the growth environment, along with the ability to clearly distinguish and measure materials placed within the autoclave. While these proofs of concepts were made using GaN, they can apply to any material with a sufficiently high X-ray absorption, at energies to which the autoclave is sufficiently transparent.

UV/Vis spectroscopy has successfully measured changes in the chemical composition of the fluid, and was able to link the dissolution of GaN to the associated changes in optical properties of the fluid. These methods can be further refined to enable more rapid solubility measurements of the systems. They could also potentially be applied to growth systems permitting observations during the run and possibly even provide a growth feedback mechanism.

Lastly, ultrasonic velocity measurements have been demonstrated on ammonothermal autoclaves (from room temperature up to 200 °C), offering assessment of liquid fill levels and measurements of sodium azide solubility, while the prospect of its application on heated systems remains viable.

Overall, while there are technological challenges to the adoption of the ammonothermal method, the field has reached a stage of development with robust autoclaves for both ammonoacidic and ammonobasic environments. The method has evolved and is ready to be used for the exploration of nitrides beyond GaN, which has been the primary driver enabling all these advances.

Author Contributions: S.S. and S.P.: Conceptualization; T.W., V.Y.M.R.C., S.S., S.P. and N.S.: methodology; V.Y.M.R.C., E.N.C., S.S. and T.W.: formal analysis; T.W., V.Y.M.R.C., S.S., N.S. and E.N.C.: investigation; S.S. and S.P.: resources; V.Y.M.R.C., E.N.C. and S.S.: data curation; T.W., V.Y.M.R.C., N.S., S.P. and S.S.: writing—original draft preparation; S.S., S.P., T.W., V.Y.M.R.C., N.S. and E.N.C.: writing—review and editing; E.N.C., V.Y.M.R.C., S.S. and T.W.: visualization; S.S. and S.P.: supervision; S.S. and S.P.: project administration; S.S. and S.P.: funding acquisition. All authors have read and agreed to the published version of the manuscript.

Funding: This research was funded by the Deutsche Forschungsgemeinschaft (DFG, German Research Foundation)—512083685 and Lehigh University New Faculty Start-up Funds.

Institutional Review Board Statement: Not applicable.

Informed Consent Statement: Not applicable.

Data Availability Statement: No new data were created in this study. So far as new plots of data have been created (involving the specified calculations of average values, standard deviations, and conversions to the in-plane lattice constants), data are available upon reasonable request from the corresponding author S.S.

Acknowledgments: S. Schimmel would like to thank Jörg Schulze (FAU Erlangen-Nürnberg, Germany) for serving as the host for her research group, as well as Oliver Ambacher (Albert-Ludwigs-Universität Freiburg, Germany) for scientific discussions on piezo- and ferroelectric ternary nitrides.

Conflicts of Interest: The authors declare no conflicts of interest. The funders had no role in the design of the study; in the collection, analyses, or interpretation of data; in the writing of the manuscript, or in the decision to publish the results.

Appendix A. Ammonoacidic Conditions

Table A1. Available information on reactivity and solubility of elements and their compounds under ammonoacidic conditions. For explanations and interpretation, see the introduction of Sections 3 and 3.1.

Element Symbol	IUPAC Group	CAS	Reactivity and Solubility Value	Mineralizer	Reference
Ag	11	IB	−1	NH_4F	[156]
Ag	11	IB	2	NH_4Cl	[147]
Ag	11	IB	6	NH_4Br	[157]
Al	13	IIIA	4	NH_4I	[162]
Al	13	IIIA	4	NH_4F	[161]
Al	13	IIIA	4	NH_4Cl	[159]
Al	13	IIIA	7	NH_4Cl	[159,223]
Au	11	IB	−1	NH_4Cl	[147]
Au	11	IB	6	NH_4Br	[157]
B	13	IIIA	6	NH_4F	[163]
Bi	15	VA	6	NH_4Br	[157]
Br	17	VIIA	5	NH_4Br	[152,154,373]
C	14	IVA	6	NH_4Cl	[164]
Ca	2	IIA	4	NH_4F	[158]
Cl	17	VIIA	5	NH_4Cl	[103,285,374]
Co	9	VIIIB	2	NH_4Cl	[147]
Co	9	VIIIB	-2	NH_4Cl	[147]
Cr	6	VIB	6	NH_4C	[164]
Cu	11	IB	6	NH_4Br	[157]
F	17	VIIA	5	NH_4F	[84,102,103]
Fe	8	VIIIB	6	NH_4Cl	[87,164]

Table A1. *Cont.*

Element Symbol	IUPAC Group	CAS	Reactivity and Solubility Value	Mineralizer	Reference
Ga	13	IIIA	4	NH_4F	[133]
Ga	13	IIIA	7	NH_4F	[84,102,103,149,150,330]
Ga	13	IIIA	7	NH_4Cl	[103,150,285,319,374,375]
Ga	13	IIIA	7	NH_4Br	[154]
Ga	13	IIIA	7	NH_4I	[154]
Ga	13	IIIA	4	NH_4I	[285,376]
Gd	3	IIIB	4	NH_4I	[167]
Ge	14	IVA	6	NH_4Br	[157]
H	1	IA	8	NH_4F	[84,102,103]
H	1	IA	8	NH_4Cl	[103,285,374]
H	1	IA	8	NH_4Br	[152,154,373]
H	1	IA	8	NH_4I	[103,155,377]
I	17	VIIA	5	NH_4I	[103,155,377]
Mg	2	IIA	2	NH_4Cl	[119,147]
Mg	2	IIA	1	NH_4Cl	[147]
Mo	6	VIB	−1	NH_4Cl	[147]
Mo	6	VIB	−2	NH_4Cl	[147]
Nb	5	VB	−1	NH_4Cl	[147]
Ni	10	VIIIB	2	NH_4Cl	[147,165,166]
N	15	VA	8	NH_4F	[84,102,103]
N	15	VA	8	NH_4Cl	[103,285,374]
N	15	VA	8	NH_4Br	[152,154,373]
N	15	VA	8	NH_4I	[103,155,377]
O	16	VIA	6	NH_4F	[87,103]
O	16	VIA	6	NH_4Cl	[87,103]
O	16	VIA	6	NH_4Br	[87,103]
O	16	VIA	6	NH_4I	[87,103]
Pd	10	VIIIB	−1	NH_4Cl	[147]
Pt	10	VIIIB	−1	NH_4Cl	[147]
Pt	10	VIIIB	−2	NH_4Cl	[147]
Sc	3	IIIB	4	NH_4I	[167]
Si	14	IVA	−1	NH_4Cl	[147]
Si	14	IVA	6	NH_4F	[103]
Si	14	IVA	6	NH_4Cl	[103]
Si	14	IVA	6	NH_4Br	[103]
Si	14	IVA	6	NH_4I	[103]
Sn	14	IVA	6	NH_4Br	[157]
Ta	5	VB	2	NH_4Cl	[147]
Ti	4	IVB	2	NH_4Cl	[147]
Ti	4	IVB	6	NH_4Cl	[164]

Table A1. *Cont.*

Element Symbol	IUPAC Group	CAS	Reactivity and Solubility Value	Mineralizer	Reference
V	5	VB	2	NH_4Cl	[147]
W	6	VIB	−1	NH_4Cl	[147]
W	6	VIB	−2	NH_4Cl	[147]
Y	3	IIB	2	NH_4Cl	[147]
Y	3	IIIB	4	NH_4I	[168,378]
Zr	4	IVB	2	NH_4Cl	[147]
Zr	4	IVB	6	NH_4Cl	[164]

Appendix B. Ammonobasic Conditions

Table A2. Available information on reactivity and solubility of elements and their compounds under ammonobasic conditions. For explanations and interpretation, see the introduction of Sections 3 and 3.2.

Element Symbol	IUPAC Group	CAS	Reactivity and Solubility Value	Mineralizer	Reference
Ag	11	IB	−1	$NaNH_2$	[147]
Al	13	IIIA	7	$NaNH_2$	[147]
Al	13	IIIA	7	KNH_2	[173,220]
Al	13	IIIA	7	KN_3	[174]
Al	13	IIIA	7	$LiNH_2$	[169,220]
Au	11	IB	2	$NaNH_2$	[147]
B	13	IIIA	7	$NaNH_2$	[111]
Ba	2	IIA	5	$Ba(NH_2)_2$	[172]
Be	2	IIA	4	NaN_3	[175]
C	14	IVA	6	$NaNH_2$	[145]
C	14	IVA	6	KNH_2	[176]
Ca	2	IIA	3	NaN_3	[143,177]
Ca	2	IIA	3	$LiNH_2$	[178]
Cd	12	IIB	3	KNH_2	[119]
Co	9	VIIIB	−1	$NaNH_2$	[147]
Co	9	VIIIB	−2	$NaNH_2$	[147]
Cr	6	VIB	3	KNH_2	[179]
Cs	1	IA	5	$CsNH_2$	[171,186,348]
Fe	8	VIIIB	6	Na	[87,140]
Ga	13	IIIA	4	$Ba(NH_2)_2$	[172,379]
Ga	13	IIIA	4	$LiNH_2$	[96,134,205]
Ga	13	IIIA	4	$NaNH_2$	[134]
Ga	13	IIIA	6	$NaNH_2$	[95,324,380]
Ga	13	IIIA	6	$LiNH_2$	[95]
Ga	13	IIIA	6	KNH_2	[170,381]
Ga	13	IIIA	7	$NaNH_2$	[149,374,382]

Table A2. *Cont.*

Element Symbol	IUPAC Group	CAS	Reactivity and Solubility Value	Mineralizer	Reference
Ga	13	IIIA	7	KNH_2	[109,382]
Ga	13	IIIA	4	KNH_2	[205,381]
Gd	3	IIIB	4	$NaNH_2$	[167,192]
Ge	14	IVA	3	$LiNH_2$	[42]
Ge	14	IVA	3	$NaNH_2$	[42]
Ge	14	IVA	3	KNH_2	[42]
Ge	14	IVA	2	$NaNH_2$	[147]
H	1	IA	8	$NaNH_2$	[84,102,103]
H	1	IA	8	KNH_2	[170,173,381]
H	1	IA	8	$RbNH_2$	[383]
H	1	IA	8	$CsNH_2$	[171,186,348]
Ir	9	VIIIB	3	$NaNH_2$	[147]
K	1	IA	5	KNH_2	[170,173,381]
Li	1	IA	5	$LiNH_2$	[169,220]
Mg	2	IIA	2	$NaNH_2$	[119,147]
Mg	2	IIA	2	KNH_2	[119,147]
Mg	2	IIA	1	$NaNH_2$	[147]
Mg	2	IIA	4	$NaNH_2$	[180,191]
Mg	2	IIA	4	NaN_3	[181,191]
Mg	2	IIA	6	Na	[87,140]
Mn	7	VIIB	4	$NaNH_2$	[182,384]
Mn	7	VIIB	6	$NaNH_2$	[183]
Mo	6	VIB	−1	$NaNH_2$	[147]
Mo	6	VIB	−2	$NaNH_2$	[147]
Mo	6	VIB	6	Na	[87,140]
N	15	VA	8	$NaNH_2$	[84,102,103]
N	15	VA	8	KNH_2	[170,173,381]
N	15	VA	8	$CsNH_2$	[171,186,348]
N	15	VA	8	$RbNH_2$	[383]
Na	1	IA	5	$NaNH_2$	[84,102,103]
Nb	5	VB	−1	$NaNH_2$	[147]
Ni	10	VIIIB	−1	$NaNH_2$	[147,165]
Ni	10	VIIIB	−2	$NaNH_2$	[147]
Ni	10	VIIIB	4	$NaNH_2$	[184]
O	16	VIA	6	$NaNH_2$	[145,146]
O	16	VIA	6	KNH_2	[176]
P	15	VA	3	NaN_3	[185,305]
Pd	10	VIIIB	1	$NaNH_2$	[147]
Pt	10	VIIIB	2	$NaNH_2$	[147]

Table A2. *Cont.*

Element Symbol	IUPAC Group	CAS	Reactivity and Solubility Value	Mineralizer	Reference
Pt	10	VIIIB	−2	$NaNH_2$	[147]
Rb	1	IA	5	$RbNH_2$	[383]
Si	14	IVA	4	KNH_2	[119,385]
Si	14	IVA	1	$NaNH_2$	[147]
Si	14	IVA	3	KNH_2	[36,42]
Si	14	IVA	6	Na	[87,140]
Ta	5	VB	2	$NaNH_2$	[147]
Ta	5	VB	3	NaN_3	[186]
Ta	5	VB	3	KN_3	[186]
Ta	5	VB	3	$RbNH_2$	[186]
Ti	4	IVB	1	$NaNH_2$	[147]
V	5	VB	2	$NaNH_2$	[147]
W	6	VIB	−1	$NaNH_2$	[147]
W	6	VIB	−2	$NaNH_2$	[147]
Y	3	IIIB	2	$NaNH_2$	[147]
Y	3	IIIB	3	KNH_2	[187]
Zn	12	IIB	4	KNH_2	[188]
Zn	12	IIB	6	KNH_2	[189]
Zr	4	IVB	2	$NaNH_2$	[147]

Appendix C. Ammononeutral Conditions

Table A3. Available information on reactivity and solubility of elements and their compounds under ammononeutral conditions. For explanations and interpretation, see the introduction of Sections 3 and 3.3.

Element Symbol	IUPAC Group	CAS	R and S Value	Mineralizer	Reference
Al	13	IIIA	4	None	[190,386,387]
Al	13	IIIA	4	None	[173]
Au	11	IB	−2	None	[147]
Ba	2	IIA	4	None	[181,194,195,388]
Be	2	IIA	4	None	[191]
Ca	2	IIA	4	None	[195–199,389]
Ce	3	IIIB	4	None	[192,390,391]
Co	9	VIIIB	−1	None	[147]
Co	9	VIIIB	−2	None	[147]
Cs	1	IA	4	None	[378,392]
Cs	1	IA	5	$n(InI_3):n(CsNH_2) = 1:3$	[200]
Cu	11	IB	4	None	[193,393]
Eu	3	IIIB	4	None	[187,195,199,201,210]

Table A3. *Cont.*

Element Symbol	IUPAC Group	CAS	R and S Value	Mineralizer	Reference
Fe	8	VIIIB	4	None	[202,394,395]
Ga	13	IIIA	4	None	[133,134,396]
Ge	14	IVA	4	None	[147]
Gd	3	IIIB	4	None	[192,203]
H	1	IA	8	None	[111,141,182,200,348,397]
H	1	IA	6	None	[241,254]
In	13	IIIA	6	$n(InF_3):n(KNH_2) = 1:3.12$	[144]
In	13	IIIA	7	$n(InCl_3):n(KNH_2) = 1:3$	[141,144,200]
In	13	IIIA	7	$n(InI_3):n(CsNH_2) = 1:3$	[200]
K	1	IA	5	$n(InCl_3):n(KNH_2) = 1:3$	[141,200]
K	1	IA	5	None	[193,397]
La	3	IIIB	4	None	[192,398–401]
La	3	IIIB	2	None	[147]
Li	1	IA	4	None	[206,402]
Mg	2	IIA	4	None	[180,207,208]
Mn	7	VIIB	4	None	[182,205,211,384,394]
Mo	6	VIB	−1	None	[147]
Mo	6	VIB	−2	None	[147]
N	15	VA	8	Neutral with InX_3 and KNH_2	[200,397]
N	15	VA	8	None	[111,182,348]
N	15	VA	6	None	[403,404]
Na	1	IA	4	None	[195,196]
Nb	5	VB	−1	None	[147]
Nd	3	IIIB	4	None	[192,391,400]
Ni	10	VIIIB	−2	None	[147]
Pd	10	VIIIB	−1	None	[147]
Pt	10	VIIIB	−2	None	[147]
Rb	1	IA	4	None	[195,198,208]
Sc	3	IIIB	−1	None	[147]
Si	14	IVA	−1	None	[147]
Si	14	IVA	−2	None	[147]
Sm	3	IIIB	4	None	[192,209,391]
Sr	2	IIA	4	None	[195,196,199]
Ta	5	VB	3	None	[147]
Th	3	IIIB	4	None	[181,405]
V	5	VB	3	None	[147]
W	6	VIB	−1	None	[147]
W	6	VIB	−2	None	[147]
Y	3	IIIB	4	None	[168,187,203,400,406]

Table A3. *Cont.*

Element Symbol	IUPAC Group	CAS	R and S Value	Mineralizer	Reference
Yb	3	IIIB	4	None	[168,187,201,203,210,400]
Zn	12	IIB	4	None	[211]
Zr	4	IVB	−1	None	[147]
Zr	4	IVB	4	None	[212,407]

References

1. Roccaforte, F.; Fiorenza, P.; Greco, G.; Lo Nigro, R.; Giannazzo, F.; Iucolano, F.; Saggio, M. Emerging trends in wide band gap semiconductors (SiC and GaN) technology for power devices. *Microelectron. Eng.* **2018**, *187–188*, 66–77. [CrossRef]
2. Doolittle, W.A.; Matthews, C.M.; Ahmad, H.; Motoki, K.; Lee, S.; Ghosh, A.; Marshall, E.N.; Tang, A.L.; Manocha, P.; Yoder, P.D. Prospectives for AlN electronics and optoelectronics and the important role of alternative synthesis. *Appl. Phys. Lett.* **2023**, *123*, 070501. [CrossRef]
3. Tsao, J.Y.; Chowdhury, S.; Hollis, M.A.; Jena, D.; Johnson, N.M.; Jones, K.A.; Kaplar, R.J.; Rajan, S.; van de Walle, C.G.; Bellotti, E.; et al. Ultrawide-Bandgap Semiconductors: Research Opportunities and Challenges. *Adv. Electron. Mater.* **2018**, *4*, 1600501. [CrossRef]
4. Akiyama, M.; Kamohara, T.; Kano, K.; Teshigahara, A.; Takeuchi, Y.; Kawahara, N. Enhancement of Piezoelectric Response in Scandium Aluminum Nitride Alloy Thin Films Prepared by Dual Reactive Cosputtering. *Adv. Mater.* **2009**, *21*, 593–596. [CrossRef] [PubMed]
5. Fichtner, S.; Wolff, N.; Lofink, F.; Kienle, L.; Wagner, B. AlScN: A III-V semiconductor based ferroelectric. *J. Appl. Phys.* **2019**, *125*, 114103. [CrossRef]
6. Chu, D.; Huang, Y.; Xie, C.; Tikhonov, E.; Kruglov, I.; Li, G.; Pan, S.; Yang, Z. Unbiased Screening of Novel Infrared Nonlinear Optical Materials with High Thermal Conductivity: Long-neglected Nitrides and Popular Chalcogenides. *Angew. Chem. Int. Ed.* **2023**, *62*, e202300581. [CrossRef]
7. Lee, C.-W.; Ud Din, N.; Yazawa, K.; Brennecka, G.L.; Zakutayev, A.; Gorai, P. Emerging Materials and Design Principles for Wurtzite-Type Ferroelectrics. *Matter* **2024**, *7*, 1644–1659. [CrossRef]
8. Jena, D.; Page, R.; Casamento, J.; Dang, P.; Singhal, J.; Zhang, Z.; Wright, J.; Khalsa, G.; Cho, Y.; Xing, H.G. The new nitrides: Layered, ferroelectric, magnetic, metallic and superconducting nitrides to boost the GaN photonics and electronics eco-system. *Jpn. J. Appl. Phys.* **2019**, *58*, SC0801. [CrossRef]
9. Mallmann, M.; Niklaus, R.; Rackl, T.; Benz, M.; Chau, T.G.; Johrendt, D.; Minár, J.; Schnick, W. Solid Solutions of Grimm-Sommerfeld Analogous Nitride Semiconductors II-IV-N2 (II=Mg, Mn, Zn; IV=Si, Ge): Ammonothermal Synthesis and DFT Calculations. *Chemistry* **2019**, *25*, 15887–15895. [CrossRef]
10. Lyu, S.; Skachkov, D.; Kash, K.; Blanton, E.W.; Lambrecht, W.R.L. Band Gaps, Band-Offsets, Disorder, Stability Region, and Point Defects in II-IV-N 2 Semiconductors. *Phys. Status Solidi (a)* **2019**, *216*, 1800875. [CrossRef]
11. Wang, P.; Wang, D.; Mondal, S.; Hu, M.; Liu, J.; Mi, Z. Dawn of nitride ferroelectric semiconductors: From materials to devices. *Semicond. Sci. Technol.* **2023**, *38*, 43002. [CrossRef]
12. Wang, D.; Yang, S.; Liu, J.; Wang, D.; Mi, Z. Perspectives on nitride ferroelectric semiconductors: Challenges and opportunities. *Appl. Phys. Lett.* **2024**, *124*, 150501. [CrossRef]
13. Startt, J.; Quazi, M.; Sharma, P.; Vazquez, I.; Poudyal, A.; Jackson, N.; Dingreville, R. Unlocking AlN Piezoelectric Performance with Earth-Abundant Dopants. *Adv. Electron. Mater.* **2023**, *9*, 2201187. [CrossRef]
14. Zha, X.-H.; Ma, X.; Luo, J.-T.; Fu, C. Enhanced piezoelectric response of AlN via alloying of transitional metals, and influence of type and distribution of transition metals. *Nano Energy* **2023**, *111*, 108390. [CrossRef]
15. Wang, F.; Ye, Q.; He, X.; Luo, K.; Ran, X.; Zheng, X.; Liao, C.; Li, R. Piezoelectric response enhancement of w-AlN by Hf (or Zr) and Sc co-alloying: A first principles study. *Phys. B Condens. Matter* **2024**, *673*, 415470. [CrossRef]
16. Urban, D.F.; Ambacher, O.; Elsässer, C. First-principles calculation of electroacoustic properties of wurtzite (Al,Sc)N. *Phys. Rev. B* **2021**, *103*, 115204. [CrossRef]
17. Greenaway, A.L.; Loutris, A.L.; Heinselman, K.N.; Melamed, C.L.; Schnepf, R.R.; Tellekamp, M.B.; Woods-Robinson, R.; Sherbondy, R.; Bardgett, D.; Bauers, S.; et al. Combinatorial Synthesis of Magnesium Tin Nitride Semiconductors. *J. Am. Chem. Soc.* **2020**, *142*, 8421–8430. [CrossRef]
18. Ambacher, O.; Mihalic, S.; Yassine, M.; Yassine, A.; Afshar, N.; Christian, B. Review: Structural, elastic, and thermodynamic properties of cubic and hexagonal $Sc_xAl_{1-x}N$ crystals. *J. Appl. Phys.* **2023**, *134*, 160702. [CrossRef]
19. Bader, S.J.; Lee, H.; Chaudhuri, R.; Huang, S.; Hickman, A.; Molnar, A.; Xing, H.G.; Jena, D.; Then, H.W.; Chowdhury, N.; et al. Prospects for Wide Bandgap and Ultrawide Bandgap CMOS Devices. *IEEE Trans. Electron Devices* **2020**, *67*, 4010–4020. [CrossRef]
20. Leveillee, J.; Poncé, S.; Adamski, N.L.; van de Walle, C.G.; Giustino, F. Anisotropic-strain-enhanced hole mobility in GaN by lattice matching to $ZnGeN_2$ and $MgSiN_2$. *Appl. Phys. Lett.* **2022**, *120*, 202106. [CrossRef]

21. Ogura, M.; Han, D.; Pointner, M.M.; Junkers, L.S.; Rudel, S.S.; Schnick, W.; Ebert, H. Electronic properties of semiconducting Zn(Si,Ge,Sn)N_2 alloys. *Phys. Rev. Mater.* **2021**, *5*, 024601. [CrossRef]
22. Santic, B. On the hole effective mass and the free hole statistics in wurtzite GaN. *Semicond. Sci. Technol.* **2003**, *18*, 219–224. [CrossRef]
23. Adamski, N.L.; Wickramaratne, D.; van de Walle, C.G. Band alignments and polarization properties of the Zn-IV-nitrides. *J. Mater. Chem. C* **2020**, *8*, 7890–7898. [CrossRef]
24. Kudrawiec, R.; Hommel, D. Bandgap engineering in III-nitrides with boron and group V elements: Toward applications in ultraviolet emitters. *Appl. Phys. Rev.* **2020**, *7*, 041314. [CrossRef]
25. Ahmed, R.; Fazal-e-Aleem; Hashemifar, S.J.; Akbarzadeh, H. First principles study of structural and electronic properties of different phases of boron nitride. *Phys. B Condens. Matter* **2007**, *400*, 297–306. [CrossRef]
26. Ziembicki, J.; Scharoch, P.; Polak, M.P.; Wiśniewski, M.; Kudrawiec, R. Band parameters of group III–V semiconductors in wurtzite structure. *J. Appl. Phys.* **2022**, *132*, 225701. [CrossRef]
27. Vurgaftman, I.; Meyer, J.R. Band parameters for nitrogen-containing semiconductors. *J. Appl. Phys.* **2003**, *94*, 3675–3696. [CrossRef]
28. Ambacher, O. Growth and applications of Group III-nitrides. *J. Phys. D Appl. Phys.* **1998**, *31*, 2653–2710. [CrossRef]
29. Dreyer, C.E.; Janotti, A.; van de Walle, C.G.; Vanderbilt, D. Correct Implementation of Polarization Constants in Wurtzite Materials and Impact on III-Nitrides. *Phys. Rev. X* **2016**, *6*, 021038. [CrossRef]
30. Rom, C.L.; Smaha, R.W.; Melamed, C.L.; Schnepf, R.R.; Heinselman, K.N.; Mangum, J.S.; Lee, S.-J.; Lany, S.; Schelhas, L.T.; Greenaway, A.L.; et al. Combinatorial Synthesis of Cation-Disordered Manganese Tin Nitride $MnSnN_2$ Thin Films with Magnetic and Semiconducting Properties. *Chem. Mater.* **2023**, *35*, 2936–2946. [CrossRef]
31. Råsander, M.; Quirk, J.B.; Wang, T.; Mathew, S.; Davies, R.; Palgrave, R.G.; Moram, M.A. Structure and lattice dynamics of the wide band gap semiconductors $MgSiN_2$ and $MgGeN_2$. *J. Appl. Phys.* **2017**, *122*, 085705. [CrossRef]
32. Quirk, J.B.; Råsander, M.; McGilvery, C.M.; Palgrave, R.; Moram, M.A. Band gap and electronic structure of $MgSiN_2$. *Appl. Phys. Lett.* **2014**, *105*, 112108. [CrossRef]
33. Råsander, M.; Moram, M.A. Elastic constants of the II–IV nitride semiconductors $MgSiN_2$, $MgGeN_2$ and $MgSnN_2$. *J. Phys. D Appl. Phys.* **2018**, *51*, 375101. [CrossRef]
34. Jaroenjittichai, A.P.; Lambrecht, W.R.L. Electronic band structure of Mg−IV−N_2 compounds in the quasiparticle-self-consistent GW approximation. *Phys. Rev. B* **2016**, *94*, 125201. [CrossRef]
35. Huang, J.Y.; Tang, L.-C.; Lee, M.H. Ab initio study of the structural and optical properties of orthorhombic ternary nitride crystals. *J. Phys. Condens. Matter* **2001**, *13*, 10417–10431. [CrossRef]
36. Häusler, J.; Niklaus, R.; Minár, J.; Schnick, W. Ammonothermal Synthesis and Optical Properties of Ternary Nitride Semiconductors Mg-IV-N_2, Mn-IV-N_2 and Li-IV_2-N_3 (IV=Si, Ge). *Chem.–A Eur. J.* **2018**, *24*, 1686–1693. [CrossRef]
37. Kaewmeechai, C.; Laosiritaworn, Y.; Jaroenjittichai, A.P. HSE hybrid functional calculation of band gap deformation potential in $MgGeN_2$. *J. Phys. Conf. Ser.* **2018**, *1144*, 12045. [CrossRef]
38. Dumre, B.B.; Gall, D.; Khare, S.V. Stability, and electronic and optical properties of ternary nitride phases of $MgSnN_2$: A first-principles study. *J. Phys. Chem. Solids* **2021**, *153*, 110011. [CrossRef]
39. Chinnakutti, K.K.; Kirubaharan, A.M.K.; Patra, L.; Pandey, R.; Theerthagiri, J.; Vengatesh, P.; Salammal, S.T.; Paramasivam, N.; Sambandam, A.; Kasemchainan, J.; et al. Modulating the Combinatorial Target Power of $MgSnN_2$ via RF Magnetron Sputtering for Enhanced Optoelectronic Performance: Mechanistic Insights from DFT Studies. *ACS Appl. Mater. Interfaces* **2023**, *15*, 14546–14556. [CrossRef]
40. Makin, R.A.; York, K.; Durbin, S.M.; Senabulya, N.; Mathis, J.; Clarke, R.; Feldberg, N.; Miska, P.; Jones, C.M.; Deng, Z.; et al. Alloy-Free Band Gap Tuning across the Visible Spectrum. *Phys. Rev. Lett.* **2019**, *122*, 256403. [CrossRef]
41. Punya, A.; Lambrecht, W.R.L.; van Schilfgaarde, M. Quasiparticle band structure of Zn-IV-N_2 compounds. *Phys. Rev. B* **2011**, *84*, 165204. [CrossRef]
42. Häusler, J.; Schimmel, S.; Wellmann, P.; Schnick, W. Ammonothermal Synthesis of Earth-Abundant Nitride Semiconductors $ZnSiN_2$ and $ZnGeN_2$ and Dissolution Monitoring by In Situ X-ray Imaging. *Chemistry* **2017**, *23*, 12275–12282. [CrossRef] [PubMed]
43. Zeman, O.E.O.; von Rohr, F.O.; Neudert, L.; Schnick, W. Facile One-step Synthesis of $Zn_{1-x}Mn_xSiN_2$ Nitride Semiconductor Solid Solutions via Solid-state Metathesis Reaction. *Z. Anorg. Allg. Chem.* **2020**, *646*, 228–233. [CrossRef]
44. Bai, Y.; Luo, G.; Meng, L.; Zhang, Q.; Xu, N.; Zhang, H.; Wu, X.; Kong, F.; Wang, B. Single-layer ZnMN2 (M = Si, Ge, Sn) zinc nitrides as promising photocatalysts. *Phys. Chem. Chem. Phys.* **2018**, *20*, 14619–14626. [CrossRef] [PubMed]
45. Paudel, T.R.; Lambrecht, W.R.L. First-principles study of phonons and related ground-state properties and spectra in Zn-IV-N2 compounds. *Phys. Rev. B* **2008**, *78*, 115204. [CrossRef]
46. Senabulya, N.; Feldberg, N.; Makin, R.A.; Yang, Y.; Shi, G.; Jones, C.M.; Kioupakis, E.; Mathis, J.; Clarke, R.; Durbin, S.M. Stabilization of orthorhombic phase in single-crystal $ZnSnN_2$ films. *AIP Adv.* **2016**, *6*, 075019. [CrossRef]
47. Kawamura, F.; Yamada, N.; Imai, M.; Taniguchi, T. Synthesis of $ZnSnN_2$ crystals via a high-pressure metathesis reaction. *Cryst. Res. Technol.* **2016**, *51*, 220–224. [CrossRef]
48. Kautzsch, L.; Georgescu, A.B.; Puggioni, D.; Kent, G.; Taddei, K.M.; Reilly, A.; Seshadri, R.; Rondinelli, J.M.; Wilson, S.D. Canted antiferromagnetism in polar $MnSiN_2$ with high Néel temperature. *Phys. Rev. Mater.* **2023**, *7*, 104406. [CrossRef]
49. Esmaeilzadeh, S.; Hålenius, U.; Valldor, M. Crystal Growth, Magnetic, and Optical Properties of the Ternary Nitride $MnSiN_2$. *Chem. Mater.* **2006**, *18*, 2713–2718. [CrossRef]

50. Koltsov, V.B.; Mikhailova, M.S. Effect of the 3d Electron Self-Interaction Correction on the Electronic and Magnetic Properties of $CoGeN_2$, $CrGeN_2$, $MnSiN_2$, and $MnGeN_2$. *Russ. J. Inorg. Chem.* **2021**, *66*, 1868–1872. [CrossRef]
51. Lazarov, V.K.; Li, L.; Weinert, M.; Gajdardziska-Josifovska, M. Structure Determination of a Magnetic Semiconductor: $MnGeN_2$. *Microsc. Microanal.* **2004**, *10*, 516–517. [CrossRef]
52. Riane, R.; Boussahla, Z.; Matar, S.F.; Zaoui, A. Structural and Electronic Properties of Zinc Blende-type Nitrides $BxAl_{1-x}N$. *Z. Naturforschung B* **2008**, *63*, 1069–1076. [CrossRef]
53. Dreyer, C.E.; Lyons, J.L.; Janotti, A.; van de Walle, C.G. Corrigendum: "Band alignments and polarization properties of BN polymorphs" [Appl. Phys. Express 7 031001 (2014)]. *Appl. Phys. Express* **2020**, *13*, 19301. [CrossRef]
54. Al-Atabi, H.A.; Zhang, X.; He, S.; Chen, C.; Chen, Y.; Rotenberg, E.; Edgar, J.H. Lattice and electronic structure of ScN observed by angle-resolved photoemission spectroscopy measurements. *Appl. Phys. Lett.* **2022**, *121*, 182102. [CrossRef]
55. Zhang, X.W. Doping and electrical properties of cubic boron nitride thin films: A critical review. *Thin Solid Films* **2013**, *544*, 2–12. [CrossRef]
56. Wellmann, P.J. Power Electronic Semiconductor Materials for Automotive and Energy Saving Applications—SiC, GaN, Ga_2O_3, and Diamond. *Z. Anorg. Allg. Chem.* **2017**, *643*, 1312–1322. [CrossRef] [PubMed]
57. Bilgin, A.; Hammock, I.N.; Estes, J.; Jin, Y.; Bernien, H.; High, A.A.; Galli, G. Donor-acceptor pairs in wide-bandgap semiconductors for quantum technology applications. *NPJ Comput. Mater.* **2024**, *10*, 7. [CrossRef]
58. Mallmann, M.; Maak, C.; Niklaus, R.; Schnick, W. Ammonothermal Synthesis, Optical Properties, and DFT Calculations of Mg_2PN_3 and Zn_2PN_3. *Chemistry* **2018**, *24*, 13963–13970. [CrossRef]
59. Wintenberger, M.; Guyader, J.; Maunaye, M. Etude cristallographique et magnetique de $MnGeN_2$ par diffraction neutronique. *Solid State Commun.* **1972**, *11*, 1485–1488. [CrossRef]
60. Fu, H.; Goodrich, J.C.; Ogidi-Ekoko, O.; Tansu, N. Power electronics figure-of-merit of ScAlN. *Appl. Phys. Lett.* **2021**, *119*, 072101. [CrossRef]
61. Kneissl, M.; Seong, T.-Y.; Han, J.; Amano, H. The emergence and prospects of deep-ultraviolet light-emitting diode technologies. *Nat. Photonics* **2019**, *13*, 233–244. [CrossRef]
62. Lang, D.V.; Logan, R.A.; Jaros, M. Trapping characteristics and a donor-complex (DX) model for the persistent-photoconductivity trapping center in Te-doped $Al_xGa_{1-x}As$. *Phys. Rev. B* **1979**, *19*, 1015–1030. [CrossRef]
63. Chadi, D.J.; Chang, K.J. Theory of the atomic and electronic structure of DX centers in GaAs and $Al_xGa_{1-x}As$ alloys. *Phys. Rev. Lett.* **1988**, *61*, 873–876. [CrossRef]
64. Lyons, J.L.; Wickramaratne, D.; Janotti, A. Dopants and defects in ultra-wide bandgap semiconductors. *Curr. Opin. Solid State Mater. Sci.* **2024**, *30*, 101148. [CrossRef]
65. Gordon, L.; Lyons, J.L.; Janotti, A.; van de Walle, C.G. Hybrid functional calculations of D X centers in AlN and GaN. *Phys. Rev. B* **2014**, *89*, 85204. [CrossRef]
66. Ahmad, H.; Lindemuth, J.; Engel, Z.; Matthews, C.M.; McCrone, T.M.; Doolittle, W.A. Substantial P-Type Conductivity of AlN Achieved via Beryllium Doping. *Adv. Mater.* **2021**, *33*, e2104497. [CrossRef] [PubMed]
67. Ahmad, H.; Engel, Z.; Matthews, C.M.; Doolittle, W.A. p-type AlN based heteroepitaxial diodes with Schottky, Pin, and junction barrier Schottky character achieving significant breakdown performance. *J. Appl. Phys.* **2021**, *130*, 195702. [CrossRef]
68. Ahmad, H.; Engel, Z.; Matthews, C.M.; Lee, S.; Doolittle, W.A. Realization of homojunction PN AlN diodes. *J. Appl. Phys.* **2022**, *131*, 175701. [CrossRef]
69. Almogbel, A.S.; Zollner, C.J.; Saifaddin, B.K.; Iza, M.; Wang, J.; Yao, Y.; Wang, M.; Foronda, H.; Prozheev, I.; Tuomisto, F.; et al. Growth of highly conductive Al-rich AlGaN:Si with low group-III vacancy concentration. *AIP Adv.* **2021**, *11*, 095119. [CrossRef]
70. Tuomisto, F.; Prozheeva, V.; Makkonen, I.; Myers, T.H.; Bockowski, M.; Teisseyre, H. Amphoteric Be in GaN: Experimental Evidence for Switching between Substitutional and Interstitial Lattice Sites. *Phys. Rev. Lett.* **2017**, *119*, 196404. [CrossRef]
71. Freitas, J.A.; Culbertson, J.C.; Glaser, E.R. Characterization of Defects in GaN: Optical and Magnetic Resonance Techniques. *Crystals* **2022**, *12*, 1294. [CrossRef]
72. Nanishi, Y.; Yamaguchi, T. Plasma-excited MBE—Proposal and achievements through R&D of compound semiconductor materials and devices. *Jpn. J. Appl. Phys.* **2022**, *61*, SA0810. [CrossRef]
73. Hatate, A.; Horiuchi, T.; Ishiyama, T.; Toko, K.; Amemiya, K.; Suemasu, T. X-ray magnetic circular dichroism of $Mn_{4-x}Ga_xN$ epitaxial thin films confirming ferrimagnetic-ferromagnetic phase transition by nonmagnetic Ga doping. *J. Magn. Magn. Mater.* **2024**, *597*, 171973. [CrossRef]
74. Hardy, M.T.; Downey, B.P.; Nepal, N.; Storm, D.F.; Katzer, D.S.; Meyer, D.J. Epitaxial ScAlN grown by molecular beam epitaxy on GaN and SiC substrates. *Appl. Phys. Lett.* **2017**, *110*, 162104. [CrossRef]
75. Leone, S.; Streicher, I.; Prescher, M.; Straňák, P.; Kirste, L. Metal-Organic Chemical Vapor Deposition of Aluminum Yttrium Nitride. *Phys. Rapid Res. Lett.* **2023**, *17*, 2300091. [CrossRef]
76. Leone, S.; Ligl, J.; Manz, C.; Kirste, L.; Fuchs, T.; Menner, H.; Prescher, M.; Wiegert, J.; Žukauskaitė, A.; Quay, R.; et al. Metal-Organic Chemical Vapor Deposition of Aluminum Scandium Nitride. *Phys. Rapid Res. Lett.* **2020**, *14*, 1900535. [CrossRef]
77. Izyumskaya, N.; Avrutin, V.; Ding, K.; Özgür, Ü.; Morkoç, H.; Fujioka, H. Emergence of high quality sputtered III-nitride semiconductors and devices. *Semicond. Sci. Technol.* **2019**, *34*, 93003. [CrossRef]
78. Matthews, C.M.; Engel, Z.; Motoki, K.; Doolittle, W.A. Kinetic Model for Ternary III-Nitride Epitaxy: The Role of Vertical Segregation on Phase Separation. *Cryst. Growth Des.* **2023**, *23*, 8856–8864. [CrossRef]

79. Doppalapudi, D.; Basu, S.N.; Ludwig, K.F.; Moustakas, T.D. Phase separation and ordering in InGaN alloys grown by molecular beam epitaxy. *J. Appl. Phys.* **1998**, *84*, 1389–1395. [CrossRef]
80. Mori, Y.; Imanishi, M.; Murakami, K.; Yoshimura, M. Recent progress of Na-flux method for GaN crystal growth. *Jpn. J. Appl. Phys.* **2019**, *58*, SC0803. [CrossRef]
81. Kucharski, R.; Sochacki, T.; Lucznik, B.; Bockowski, M. Growth of bulk GaN crystals. *J. Appl. Phys.* **2020**, *128*, 050902. [CrossRef]
82. Mikawa, Y.; Ishinabe, T.; Kagamitani, Y.; Mochizuki, T.; Ikeda, H.; Iso, K.; Takahashi, T.; Kubota, K.; Enatsu, Y.; Tsukada, Y.; et al. Recent progress of large size and low dislocation bulk GaN growth. In *Gallium Nitride Materials and Devices XV, Proceedings of the Volume 11280, Gallium Nitride Materials and Devices XV, San Francisco, CA, USA, 1–6 February 2020*; Morkoç, H., Fujioka, H., Schwarz, U.T., Eds.; SPIE: Bellingham, WA, USA, 2020; p. 1, ISBN 9781510633230. [CrossRef]
83. Hartmann, C.; Kabukcuoglu, M.P.; Richter, C.; Klump, A.; Schulz, D.; Juda, U.; Bickermann, M.; Hänschke, D.; Schröder, T.; Straubinger, T. Efficient diameter enlargement of bulk AlN single crystals with high structural quality. *Appl. Phys. Express* **2023**, *16*, 75502. [CrossRef]
84. Kurimoto, K.; Bao, Q.; Mikawa, Y.; Shima, K.; Ishiguro, T.; Chichibu, S.F. Low-pressure acidic ammonothermal growth of 2-inch-diameter nearly bowing-free bulk GaN crystals. *Appl. Phys. Express* **2022**, *15*, 55504. [CrossRef]
85. Ehrentraut, D.; Sitar, Z. Advances in Bulk Crystal Growth of AlN and GaN. *MRS Bull.* **2009**, *34*, 259–265. [CrossRef]
86. Stoddard, N.; Pimputkar, S. Progress in Ammonothermal Crystal Growth of Gallium Nitride from 2017–2023: Process, Defects and Devices. *Crystals* **2023**, *13*, 1004. [CrossRef]
87. Suihkonen, S.; Pimputkar, S.; Sintonen, S.; Tuomisto, F. Defects in single crystalline ammonothermal gallium nitride. *Adv. Electron. Mater.* **2017**, *3*, 1600496. [CrossRef]
88. Hashimoto, T.; Letts, E.R.; Key, D. Progress in Near-Equilibrium Ammonothermal (NEAT) Growth of GaN Substrates for GaN-on-GaN Semiconductor Devices. *Crystals* **2022**, *12*, 1085. [CrossRef]
89. Ehrentraut, D.; Kagamitani, Y.; Fukuda, T.; Orito, F.; Kawabata, S.; Katano, K.; Terada, S. Reviewing recent developments in the acid ammonothermal crystal growth of gallium nitride. *J. Cryst. Growth* **2008**, *310*, 3902–3906. [CrossRef]
90. Ehrentraut, D.; Pakalapati, R.T.; Kamber, D.S.; Jiang, W.; Pocius, D.W.; Downey, B.C.; McLaurin, M.; D'Evelyn, M.P. High Quality, Low Cost Ammonothermal Bulk GaN Substrates. *Jpn. J. Appl. Phys.* **2013**, *52*, 08JA01. [CrossRef]
91. Tomida, D.; Saito, M.; Bao, Q.; Ishiguro, T.; Chichibu, S.F. Innovative Techniques for Fast Growth and Fabrication of High Purity GaN Single Crystals. In *Ammonothermal Synthesis and Crystal Growth of Nitrides*; Meissner, E., Niewa, R., Eds.; Springer International Publishing: Cham, Switzerland, 2021; pp. 65–76, ISBN 978-3-030-56304-2.
92. Häusler, J.; Schnick, W. Ammonothermal Synthesis of Nitrides: Recent Developments and Future Perspectives. *Chemistry* **2018**, *24*, 11864–11879. [CrossRef] [PubMed]
93. Mallmann, M.; Cordes, N.; Schnick, W. Explorative Synthesis of Novel Nitride Compounds by Ammonothermal Synthesis. In *Ammonothermal Synthesis and Crystal Growth of Nitrides*; Meissner, E., Niewa, R., Eds.; Springer International Publishing: Cham, Switzerland, 2021; pp. 205–225, ISBN 978-3-030-56304-2.
94. Watanabe, T.; Nonaka, K.; Li, J.; Kishida, K.; Yoshimura, M. Low temperature ammonothermal synthesis of europium-doped $SrAlSiN_3$ for a nitride red phosphor. *J. Ceram. Soc. Jpn.* **2012**, *120*, 500–502. [CrossRef]
95. Dwilinski, R.; Wysmolek, A.; Baranowski, J.; Kaminska, M.; Doradzinski, R.; Garczyński, J.; Sierzputowski, L.; Jacobs, H. GaN synthesis by ammonothermal method. *Acta Phys. Pol. A* **1995**, *88*, 833–836. [CrossRef]
96. Dwiliński, R.; Baranowski, J.M.; Kamińska, M.; Doradziński, R.; Garczyński, J.; Sierzputowski, L. On GaN Crystallization by Ammonothermal Method. *Acta Phys. Pol. A* **1996**, *90*, 763–766. [CrossRef]
97. Grabianska, K.; Kucharski, R.; Puchalski, A.; Sochacki, T.; Bockowski, M. Recent progress in basic ammonothermal GaN crystal growth. *J. Cryst. Growth* **2020**, *547*, 125804. [CrossRef]
98. Boćkowski, M.; Grzegory, I. Recent Progress in Crystal Growth of Bulk GaN. *Acta Phys. Pol. A* **2022**, *141*, 167–174. [CrossRef]
99. Grabianska, K.; Jaroszynski, P.; Sidor, A.; Bockowski, M.; Iwinska, M. GaN single crystalline substrates by ammonothermal and HVPE methods for electronic devices. *Electronics* **2020**, *9*, 1342. [CrossRef]
100. Pimputkar, S. Gallium nitride 11. In *Single Crystals of Electronic Materials: Growth and Properties*; Elsevier: London, UK, 2018; p. 351. [CrossRef]
101. Dwiliński, R.; Doradziński, R.; Garczyński, J.; Sierzputowski, L.P.; Puchalski, A.; Kanbara, Y.; Yagi, K.; Minakuchi, H.; Hayashi, H. Excellent crystallinity of truly bulk ammonothermal GaN. *J. Cryst. Growth* **2008**, *310*, 3911–3916. [CrossRef]
102. Bao, Q.; Saito, M.; Hazu, K.; Kagamitani, Y.; Kurimoto, K.; Tomida, D.; Qiao, K.; Ishiguro, T.; Yokoyama, C.; Chichibu, S.F. Ammonothermal growth of GaN on a self-nucleated GaN seed crystal. *J. Cryst. Growth* **2014**, *404*, 168–171. [CrossRef]
103. Bao, Q.; Saito, M.; Hazu, K.; Furusawa, K.; Kagamitani, Y.; Kayano, R.; Tomida, D.; Qiao, K.; Ishiguro, T.; Yokoyama, C.; et al. Ammonothermal Crystal Growth of GaN Using an NH4F Mineralizer. *Cryst. Growth Des.* **2013**, *13*, 4158–4161. [CrossRef]
104. D'Evelyn, M.P.; Hong, H.C.; Park, D.-S.; Lu, H.; Kaminsky, E.; Melkote, R.R.; Perlin, P.; Lesczynski, M.; Porowski, S.; Molnar, R.J. Bulk GaN crystal growth by the high-pressure ammonothermal method. *J. Cryst. Growth* **2007**, *300*, 11–16. [CrossRef]
105. Zajac, M.; Kucharski, R.; Grabianska, K.; Gwardys-Bak, A.; Puchalski, A.; Wasik, D.; Litwin-Staszewska, E.; Piotrzkowski, R.; Domagala, J.Z.; Bockowski, M. Basic ammonothermal growth of Gallium Nitride–State of the art, challenges, perspectives. *Prog. Cryst. Growth Charact. Mater.* **2018**, *64*, 63–74. [CrossRef]
106. Bockowski, M.; Iwinska, M.; Amilusik, M.; Fijalkowski, M.; Lucznik, B.; Sochacki, T. Challenges and future perspectives in HVPE-GaN growth on ammonothermal GaN seeds. *Semicond. Sci. Technol.* **2016**, *31*, 93002. [CrossRef]

107. Kirste, L.; Grabianska, K.; Kucharski, R.; Sochacki, T.; Lucznik, B.; Bockowski, M. Structural Analysis of Low Defect Ammonothermally Grown GaN Wafers by Borrmann Effect X-ray Topography. *Materials* **2021**, *14*, 5472. [CrossRef] [PubMed]
108. Fujikura, H.; Yoshida, T.; Shibata, M.; Otoki, Y. Recent progress of high-quality GaN substrates by HVPE method. In *Gallium Nitride Materials and Devices XII, Proceedings of the SPIE OPTO, San Francisco, CA, USA, 28 January 2017*; Chyi, J.-I., Fujioka, H., Morkoç, H., Nanishi, Y., Schwarz, U.T., Shim, J.-I., Eds.; SPIE: Bellingham, WA, USA, 2017; p. 1010403.
109. Wang, B.; Callahan, M.J.; Rakes, K.D.; Bouthillette, L.O.; Wang, S.-Q.; Bliss, D.F.; Kolis, J.W. Ammonothermal growth of GaN crystals in alkaline solutions. *J. Cryst. Growth* **2006**, *287*, 376–380. [CrossRef]
110. Kucharski, R.; Zajac, M.; Puchalski, A.; Sochacki, T.; Bockowski, M.; Weyher, J.L.; Iwinska, M.; Serafinczuk, J.; Kudrawiec, R.; Siemiątkowski, Z. Ammonothermal growth of GaN crystals on HVPE-GaN seeds prepared with the use of ammonothermal substrates. *J. Cryst. Growth* **2015**, *427*, 1–6. [CrossRef]
111. Dooley, J.; Stoddard, N.; Landskron, K.; Pimputkar, S. On the solubility of boron nitride in supercritical ammonia-sodium solutions. *J. Cryst. Growth* **2023**, *621*, 127381. [CrossRef]
112. Zajac, M.; Doradziński, R.; Gosk, J.; Szczytko, J.; Lefeld-Sosnowska, M.; Kamińska, M.; Twardowski, A.; Palczewska, M.; Grzanka, E.; Gębicki, W. Magnetic and optical properties of GaMnN magnetic semiconductor. *Appl. Phys. Lett.* **2001**, *78*, 1276–1278. [CrossRef]
113. Ehrentraut, D.; Hoshino, N.; Kagamitani, Y.; Yoshikawa, A.; Fukuda, T.; Itoh, H.; Kawabata, S. Temperature effect of ammonium halogenides as mineralizers on the phase stability of gallium nitride synthesized under acidic ammonothermal conditions. *J. Mater. Chem.* **2007**, *17*, 886–893. [CrossRef]
114. Makiuchi, K.; Kawamura, F.; Jia, J.; Song, Y.; Yata, S.; Tampo, H.; Murata, H.; Yamada, N. Pressure-Induced Transition from Wurtzite and Epitaxial Stabilization for Thin Films of Rocksalt $MgSnN_2$. *Chem. Mater.* **2023**, *35*, 2095–2106. [CrossRef]
115. Yassine, M.; Nair, A.; Fammels, J.; Wade, E.; Fu, Z.; Yassine, A.; Kirste, L.; Ambacher, O. Influence of structural properties on the ferroelectric behavior of hexagonal AlScN. *J. Appl. Phys.* **2022**, *132*, 114101. [CrossRef]
116. Gorczyca, I.; Suski, T.; Perlin, P.; Grzegory, I.; Staszczak, G.; Aktas, M. Special role of indium nitride in the properties of related compounds and quantum structures. *AIP Adv.* **2024**, *14*, 040704. [CrossRef]
117. Karpiński, J.; Jun, J.; Porowski, S. Equilibrium pressure of N_2 over GaN and high pressure solution growth of GaN. *J. Cryst. Growth* **1984**, *66*, 1–10. [CrossRef]
118. Leszczynski, M.; Grzegory, I.; Teisseyre, H.; Suski, T.; Bockowski, M.; Jun, J.; Baranowski, J.M.; Porowski, S.; Domagala, J. The microstructure of gallium nitride monocrystals grown at high pressure. *J. Cryst. Growth* **1996**, *169*, 235–242. [CrossRef]
119. Richter, T.; Niewa, R. Chemistry of Ammonothermal Synthesis. *Inorganics* **2014**, *2*, 29–78. [CrossRef]
120. Pritula, I.; Sangwal, K. Fundamentals of Crystal Growth from Solutions. In *Handbook of Crystal Growth*, 2nd ed.; Rudolph, P., Ed.; Elsevier: Amsterdam, The Netherlands, 2015; pp. 1185–1227, ISBN 978-0-444-63303-3.
121. Byrappa, K. *Handbook of Hydrothermal Technology*, 2nd ed.; William Andrew: Oxford, UK, 2013; ISBN 9781437778366.
122. Hervey, P.R.; Foise, J.W. Synthetic quartz crystal—A review. *Min. Metall. Explor.* **2001**, *18*, 1–4. [CrossRef]
123. Ehrentraut, D.; Fukuda, T. The Ammonothermal Crystal Growth of Gallium Nitride—A Technique on the Up Rise. *Proc. IEEE* **2010**, *98*, 1316–1323. [CrossRef]
124. Eckert, C.A.; Knutson, B.L.; Debenedetti, P.G. Supercritical fluids as solvents for chemical and materials processing. *Nature* **1996**, *383*, 313–318. [CrossRef]
125. Brennecke, J.F.; Eckert, C.A. Phase equilibria for supercritical fluid process design. *AIChE J.* **1989**, *35*, 1409–1427. [CrossRef]
126. Knox, D.E. Solubilities in supercritical fluids. *Pure Appl. Chem.* **2005**, *77*, 513–530. [CrossRef]
127. Schimmel, S.; Kobelt, I.; Heinlein, L.; Kimmel, A.-C.L.; Steigerwald, T.G.; Schlücker, E.; Wellmann, P. Flow Stability, Convective Heat Transfer and Chemical Reactions in Ammonothermal Autoclaves—Insights by In Situ Measurements of Fluid Temperatures. *Crystals* **2020**, *10*, 723. [CrossRef]
128. Schimmel, S.; Tomida, D.; Ishiguro, T.; Honda, Y.; Chichibu, S.; Amano, H. Numerical Simulation of Ammonothermal Crystal Growth of GaN—Current State, Challenges, and Prospects. *Crystals* **2021**, *11*, 356. [CrossRef]
129. Rendel, P.M.; Mountain, B.W. Solubility of quartz in supercritical water from 375 °C to 600 °C and 200–270 bar. *J. Supercrit. Fluids* **2023**, *196*, 105883. [CrossRef]
130. Reinsch, T.; Dobson, P.; Asanuma, H.; Huenges, E.; Poletto, F.; Sanjuan, B. Utilizing supercritical geothermal systems: A review of past ventures and ongoing research activities. *Geotherm. Energy* **2017**, *5*, 16. [CrossRef]
131. Lemmon, E.W.; McLinden, M.O.; Friend, D.G. Thermophysical Properties of Fluid Systems: NIST Chemistry WebBook. NIST Standard Reference Database Number 69. Available online: http://webbook.nist.gov/chemistry (accessed on 5 June 2017).
132. Niewa, R. Significance of Ammonothermal Synthesis for Nitride Materials. In *Ammonothermal Synthesis and Crystal Growth of Nitrides*; Meissner, E., Niewa, R., Eds.; Springer International Publishing: Cham, Switzerland, 2021; pp. 3–12, ISBN 978-3-030-56304-2.
133. Zhang, S.; Hintze, F.; Schnick, W.; Niewa, R. Intermediates in Ammonothermal GaN Crystal Growth under Ammonoacidic Conditions. *Eur. J. Inorg. Chem.* **2013**, *2013*, 5387–5399. [CrossRef]
134. Zhang, S.; Alt, N.S.; Schlücker, E.; Niewa, R. Novel alkali metal amidogallates as intermediates in ammonothermal GaN crystal growth. *J. Cryst. Growth* **2014**, *403*, 22–28. [CrossRef]
135. Glasson, D.R.; Jayaweera, S.A.A. Formation and reactivity of nitrides I. Review and introduction. *J. Appl. Chem.* **1968**, *18*, 65–77. [CrossRef]

136. Pimputkar, S.; Nakamura, S. Decomposition of supercritical ammonia and modeling of supercritical ammonia–nitrogen–hydrogen solutions with applicability toward ammonothermal conditions. *J. Supercrit. Fluids* **2016**, *107*, 17–30. [CrossRef]
137. Bao, Q.; Hashimoto, T.; Sato, F.; Hazu, K.; Saito, M.; Kagamitani, Y.; Ishinabe, T.; Kayano, R.; Tomida, D.; Qiao, K.; et al. Acidic ammonothermal growth of GaN crystals using GaN powder as a nutrient. *CrystEngComm* **2013**, *15*, 5382. [CrossRef]
138. Zak, M.; Kempisty, P.; Lucznik, B.; Grabianska, K.; Kucharski, R.; Iwinska, M.; Bockowski, M. Modeling of convective transport in crystallization of gallium nitride by basic ammonothermal method. *J. Cryst. Growth* **2024**, *627*, 127525. [CrossRef]
139. Grabianska, K.; Kucharski, R.; Sochacki, T.; Weyher, J.L.; Iwinska, M.; Grzegory, I.; Bockowski, M. On Stress-Induced Polarization Effect in Ammonothermally Grown GaN Crystals. *Crystals* **2022**, *12*, 554. [CrossRef]
140. Pimputkar, S.; Kawabata, S.; Speck, J.S.; Nakamura, S. Improved growth rates and purity of basic ammonothermal GaN. *J. Cryst. Growth* **2014**, *403*, 7–17. [CrossRef]
141. Hertrampf, J.; Becker, P.; Widenmeyer, M.; Weidenkaff, A.; Schlücker, E.; Niewa, R. Ammonothermal Crystal Growth of Indium Nitride. *Cryst. Growth Des.* **2018**, *18*, 2365–2369. [CrossRef]
142. Ehrentraut, D.; Bockowski, M. High-pressure, high-temperature solution growth and ammonothermal synthesis of gallium nitride crystals. In *Handbook of Crystal Growth*; Elsevier: Amsterdam, The Netherlands, 2015; pp. 577–619.
143. Häusler, J.; Neudert, L.; Mallmann, M.; Niklaus, R.; Kimmel, A.-C.L.; Alt, N.S.A.; Schlücker, E.; Oeckler, O.; Schnick, W. Ammonothermal Synthesis of Novel Nitrides: Case Study on $CaGaSiN_3$. *Chemistry* **2017**, *23*, 2583–2590. [CrossRef]
144. Becker, P.; Cekovski, T.B.; Niewa, R. Two Intermediates in Ammonothermal InN Crystal Growth: $[In(NH_3)_5Cl]Cl_2$ and $InF_2(NH_2)$. *Z. Anorg. Allg. Chem.* **2021**, *647*, 2006–2014. [CrossRef]
145. Sintonen, S.; Kivisaari, P.; Pimputkar, S.; Suihkonen, S.; Schulz, T.; Speck, J.S.; Nakamura, S. Incorporation and effects of impurities in different growth zones within basic ammonothermal GaN. *J. Cryst. Growth* **2016**, *456*, 43–50. [CrossRef]
146. Sintonen, S.; Wahl, S.; Richter, S.; Meyer, S.; Suihkonen, S.; Schulz, T.; Irmscher, K.; Danilewsky, A.N.; Tuomi, T.O.; Stankiewicz, R.; et al. Evolution of impurity incorporation during ammonothermal growth of GaN. *J. Cryst. Growth* **2016**, *456*, 51–57. [CrossRef]
147. Pimputkar, S.; Malkowski, T.F.; Griffiths, S.; Espenlaub, A.; Suihkonen, S.; Speck, J.S.; Nakamura, S. Stability of materials in supercritical ammonia solutions. *J. Supercrit. Fluids* **2016**, *110*, 193–229. [CrossRef]
148. Schimmel, S.; Künecke, U.; Meisel, M.; Hertweck, B.; Steigerwald, T.G.; Nebel, C.; Alt, N.S.; Schlücker, E.; Wellmann, P. Chemical stability of carbon-based inorganic materials for in situ X-ray investigations of ammonothermal crystal growth of nitrides. *J. Cryst. Growth* **2016**, *456*, 33–42. [CrossRef]
149. Schimmel, S.; Koch, M.; Macher, P.; Kimmel, A.-C.L.; Steigerwald, T.G.; Alt, N.S.; Schlücker, E.; Wellmann, P. Solubility and dissolution kinetics of GaN in supercritical ammonia in presence of ammonoacidic and ammonobasic mineralizers. *J. Cryst. Growth* **2017**, *479*, 59–66. [CrossRef]
150. Schimmel, S.; Lindner, M.; Steigerwald, T.G.; Hertweck, B.; Richter, T.M.; Künecke, U.; Alt, N.S.; Niewa, R.; Schlücker, E.; Wellmann, P.J. Determination of GaN solubility in supercritical ammonia with NH_4F and NH_4Cl mineralizer by in situ X-ray imaging of crystal dissolution. *J. Cryst. Growth* **2015**, *418*, 64–69. [CrossRef]
151. Griffiths, S.; Pimputkar, S.; Speck, J.S.; Nakamura, S. On the solubility of gallium nitride in supercritical ammonia–sodium solutions. *J. Cryst. Growth* **2016**, *456*, 5–14. [CrossRef]
152. Tomida, D.; Kuroda, K.; Nakamura, K.; Qiao, K.; Yokoyama, C. Temperature dependent control of the solubility of gallium nitride in supercritical ammonia using mixed mineralizer. *Chem. Cent. J.* **2018**, *12*, 127. [CrossRef]
153. Yoshikawa, A.; Ohshima, E.; Fukuda, T.; Tsuji, H.; Oshima, K. Crystal growth of GaN by ammonothermal method. *J. Cryst. Growth* **2004**, *260*, 67–72. [CrossRef]
154. Tomida, D.; Kuribayashi, T.; Suzuki, K.; Kagamitani, Y.; Ishiguro, T.; Fukuda, T.; Yokoyama, C. Effect of halogen species of acidic mineralizer on solubility of GaN in supercritical ammonia. *J. Cryst. Growth* **2011**, *325*, 52–54. [CrossRef]
155. Tomida, D.; Kagamitani, Y.; Bao, Q.; Hazu, K.; Sawayama, H.; Chichibu, S.F.; Yokoyama, C.; Fukuda, T.; Ishiguro, T. Enhanced growth rate for ammonothermal gallium nitride crystal growth using ammonium iodide mineralizer. *J. Cryst. Growth* **2012**, *353*, 59–62. [CrossRef]
156. Tomida, D.; Bao, Q.; Saito, M.; Osanai, R.; Shima, K.; Kojima, K.; Ishiguro, T.; Chichibu, S.F. Ammonothermal growth of 2 inch long GaN single crystals using an acidic NH_4F mineralizer in a Ag-lined autoclave. *Appl. Phys. Express* **2020**, *13*, 55505. [CrossRef]
157. Purdy, A.P.; Case, S.; George, C. Ammonothermal Crystal Growth of Germanium and Its Alloys: Synthesis of a Hollow Metallic Crystal. *Cryst. Growth Des.* **2003**, *3*, 121–124. [CrossRef]
158. Tomida, D.; Bao, Q.; Saito, M.; Kurimoto, K.; Sato, F.; Ishiguro, T.; Chichibu, S.F. Effects of extra metals added in an autoclave during acidic ammonothermal growth of m -plane GaN single crystals using an NH_4F mineralizer. *Appl. Phys. Express* **2018**, *11*, 91002. [CrossRef]
159. Lan, Y.; Chen, X.; Cao, Y.; Xu, Y.; Xun, L.; Xu, T.; Liang, J. Low-temperature synthesis and photoluminescence of AlN. *J. Cryst. Growth* **1999**, *207*, 247–250. [CrossRef]
160. Peters, D. Preparation of aluminium nitride by ammonolysis of hexammine aluminium iodide: Ceramic Powder Processing Science. In *Ceramic Powder Processing Science, Proceedings of the Second International Conference, FRG, Berchtesgaden, Germany, 12–14 October 1988*; Hausner, H., Messing, G.L., Hirano, S., Eds.; Deutsche Keramische Gesellschaft: Köln, Germany, 2015; pp. 181–188.
161. Ketchum, D.R.; Schimek, G.L.; Pennington, W.T.; Kolis, J.W. Synthesis of new Group III fluoride–ammonia adducts in supercritical ammonia: Structures of $AlF_3(NH_3)_2$ and $InF_2(NH_2)(NH_3)$. *Inorganica Chim. Acta* **1999**, *294*, 200–206. [CrossRef]

162. Peters, D.; Bock, J.; Jacobs, H. Hexaaminaluminiumiodidmonoammoniakat—$[Al(NH_3)_6]I_3NH_3$—Darstellung und kristallstruktur. *J. Less Common Met.* **1989**, *154*, 243–250. [CrossRef]
163. Kobayashi, T. Solvent effects of fluorides in cubic BN high pressure synthesis. *Mater. Res. Bull.* **1979**, *14*, 1541–1551. [CrossRef]
164. Malkowski, T.F.; Pimputkar, S.; Speck, J.S.; DenBaars, S.P.; Nakamura, S. Acidic ammonothermal growth of gallium nitride in a liner-free molybdenum alloy autoclave. *J. Cryst. Growth* **2016**, *456*, 21–26. [CrossRef]
165. Hertweck, B.; Steigerwald, T.G.; Alt, N.S.; Schluecker, E. Different corrosion behaviour of autoclaves made of nickel base alloy 718 in ammonobasic and ammonoacidic environments. *J. Supercrit. Fluids* **2014**, *95*, 158–166. [CrossRef]
166. Kimmel, A.-C.; Malkowski, T.F.; Griffiths, S.; Hertweck, B.; Steigerwald, T.G.; Freund, L.P.; Neumeier, S.; Göken, M.; Speck, J.S.; Schluecker, E. High-temperature corrosion of Inconel®Alloy 718, Haynes®282®Alloy and CoWAlloy1&2 in supercritical ammonia/ammonium chloride solution. *J. Cryst. Growth* **2018**, *498*, 289–300. [CrossRef]
167. Linde, G.; Juza, R. Amidometallate von Lanthan und Gadolinium und Umsetzung von Lanthan, Gadolinium und Scandium mit Ammoniak. *Z. Anorg. Allg. Chem.* **1974**, *409*, 191–198. [CrossRef]
168. Stuhr, A.; Jocobs, H.; Juza, R. Amide des Yttriums. *Z. Anorg. Allg. Chem.* **1973**, *395*, 291–300. [CrossRef]
169. Dwiliński, R.; Doradziński, R.; Garczyński, J.; Sierzputowski, L.; Baranowski, J.; Kamińska, M. AMMONO method of GaN and AlN production. *Diam. Relat. Mater.* **1998**, *7*, 1348–1350. [CrossRef]
170. Adekore, B.T.; Callahan, M.J.; Bouthillette, L.; Dalmau, R.; Sitar, Z. Synthesis of erbium-doped gallium nitride crystals by the ammonothermal route. *J. Cryst. Growth* **2007**, *308*, 71–79. [CrossRef]
171. Bäucker, C.; Bauch, S.; Niewa, R. Synthesis and Characterization of the Amidomanganates $Rb_2[Mn(NH_2)_4]$ and $Cs_2[Mn(NH_2)_4]$. *Crystals* **2021**, *11*, 676. [CrossRef]
172. Hertrampf, J.; Alt, N.; Schlücker, E.; Knetzger, M.; Meissner, E.; Niewa, R. Ammonothermal synthesis of GaN using Ba(NH2)2 as mineralizer. *J. Cryst. Growth* **2016**, *456*, 2–4. [CrossRef]
173. Peters, D. Ammonothermal synthesis of aluminum nitride. *J. Cryst. Growth* **1990**, *104*, 411–418. [CrossRef]
174. Adekore, B.T.; Rakes, K.; Wang, B.; Callahan, M.J.; Pendurti, S.; Sitar, Z. Ammonothermal synthesis of aluminum nitride crystals on group III-nitride templates. *J. Electron. Mater.* **2006**, *35*, 1104–1111. [CrossRef]
175. Jacobs, H.; Juza, R. Darstellung und Eigenschaften von Berylliumamid und -imid. *Z. Anorg. Allg. Chem.* **1969**, *370*, 248–253. [CrossRef]
176. Xu, L.; Li, T.; Ren, G.; Su, X.; Gao, X.; Zheng, S.; Wang, H.; Xu, K. Study of lateral growth regions in ammonothermal c-plane GaN. *J. Cryst. Growth* **2021**, *556*, 125987. [CrossRef]
177. Mallmann, M.; Maak, C.; Schnick, W. Ammonothermal Synthesis and Crystal Growth of the Chain-type Oxonitridosilicate $Ca_{1+x}Y_{1-x}SiN_{3-x}O_x$ (x > 0). *Z. Anorg. Allg. Chem.* **2020**, *646*, 1539–1544. [CrossRef]
178. Häusler, J.; Eisenburger, L.; Oeckler, O.; Schnick, W. Ammonothermal Synthesis and Crystal Structure of the Nitridoalumogermanate $Ca_{1-x}Li_xAl_{1-x}Ge_{1+x}N_3$ ($x \approx 0.2$). *Eur. J. Inorg. Chem.* **2018**, *2018*, 759–764. [CrossRef]
179. Bäucker, C.; Niewa, R. Exchange of ammine- and fluorido-ligands in complex salts: The series $[Cr(NH_3)_6][AlF_6]$, $[Cr(NH_3)_5F][SiF_6]$ and $K_2[Cr(NH_3)_4F_2][Si(NH_3)_{0.5}F_{5.5}]_2$. *Z. Anorg. Allg. Chem.* **2022**, *648*, e2200209. [CrossRef]
180. Jacobs, H.; Juza, R. Darstellung und Eigenschaften von Magnesiumamid und -imid. *Z. Anorg. Allg. Chem.* **1969**, *370*, 254–261. [CrossRef]
181. Juza, R.; Jacobs, H.; Gerke, H. Ammonothermalsynthese von Metallamiden und Metallnitriden. *Berichte Bunsenges. Phys. Chem.* **1966**, *70*, 1103–1105. [CrossRef]
182. Kreiner, G.; Jacobs, H. Magnetische struktur von η-Mn_3N_2. *J. Alloys Compd.* **1992**, *183*, 345–362. [CrossRef]
183. Heikkinen, T.; Pavlov, J.; Ceponis, T.; Gaubas, E.; Zając, M.; Tuomisto, F. Effect of Mn and Mg dopants on vacancy defect formation in ammonothermal GaN. *J. Cryst. Growth* **2020**, *547*, 125803. [CrossRef]
184. Leineweber, A.; Jacobs, H.; Hull, S. ChemInform Abstract: Ordering of Nitrogen in Nickel Nitride Ni_3N Determined by Neutron Diffraction. *ChemInform* **2002**, *33*, 201004. [CrossRef]
185. Wendl, S.; Mallmann, M.; Strobel, P.; Schmidt, P.J.; Schnick, W. Ammonothermal Synthesis of Ba_2PO_3 N—An Oxonitridophosphate with Non-Condensed PO_3N Tetrahedra. *Eur. J. Inorg. Chem.* **2020**, *2020*, 841–846. [CrossRef]
186. Cordes, N.; Niklaus, R.; Schnick, W. Ammonothermal Crystal Growth of $ATaN_2$ with A = Na, K, Rb, and Cs and Their Optical and Electronic Properties. *Cryst. Growth Des.* **2019**, *19*, 3484–3490. [CrossRef]
187. Jacobs, H.; Kockelkorn, J. Über kalium- und rubidiumamidometallate des Europiums, Yttriums und Ytterbiums, $K_3M(NH_2)_6$ und $Rb_3M(NH_2)_6$. *J. Less Common Met.* **1982**, *85*, 97–110. [CrossRef]
188. Richter, T.T.; Zhang, S.; Niewa, R. Ammonothermal synthesis of dimorphic $K_2[Zn(_{NH2})_4]$. *Z. Krist.-Cryst. Mater.* **2013**, *228*, 351–358. [CrossRef]
189. Zajac, M.; Konczewicz, L.; Litwin-Staszewska, E.; Iwinska, M.; Kucharski, R.; Juillaguet, S.; Contreras, S. p-type conductivity in GaN:Zn monocrystals grown by ammonothermal method. *J. Appl. Phys.* **2021**, *129*, 135702. [CrossRef]
190. Jacobs, H.; Schröder, F.O. Penta-Ammoniates of Aluminium Halides: The Crystal Structures of $AlX_3{\cdot}5NH_3$ with X = Cl, Br, IProfessor Joachim Strähle zum 65. Geburtstag gewidmet. *Z. Anorg. Allg. Chem.* **2002**, *628*, 951. [CrossRef]
191. Juza, R.; Jacobs, H. Ammonothermal Synthesis of Magnesium and Beryllium Amides. *Angew. Chem. Int. Ed. Engl.* **1966**, *5*, 247. [CrossRef]
192. Jacobs, H.; Schmidt, D. ChemInform Abstract: High-pressure ammonolysis in solid-state chemistry. *Chem. Informationsdienst* **1983**, *14*, 381–427. [CrossRef]

193. Zachwieja, U.; Jacobs, H. Ammonothermalsynthese von kupfernitrid, Cu_3N. *J. Less Common Met.* **1990**, *161*, 175–184. [CrossRef]
194. Jacobs, H.; Birkenbeul, J.; Kockelkorn, J. Darstellung und eigenschaften des caesiumbarium-amids, $CsBa(NH_2)_3$: Strukturverwandtschaft zum NH4CdCl3-Typ. *J. Less Common Met.* **1982**, *85*, 71–78. [CrossRef]
195. Jacobs, H.; Kockelkorn, J.; Birkenbeul, J. Struktur und eigenschaften der ternären metallamide $NaCa(NH_2)_3$, $KBa(NH_2)_3$, $RbBa(NH_2)_3$, $RbEu(NH_2)_3$ und $RbSr(NH_2)_3$. *J. Less Common Met.* **1982**, *87*, 215–224. [CrossRef]
196. Jacobs, H.; Fink, U. Über natrium- und kaliumamidometallate des calciums, strontiums und europiums. *J. Less Common Met.* **1979**, *63*, 273–286. [CrossRef]
197. Jacobs, H.; Fink, U. Darstellung und Kristallstruktur von $KCa(NH_2)_3$. *Z. Anorg. Allg. Chem.* **1977**, *435*, 137–145. [CrossRef]
198. Jacobs, H.; Kockelkorn, J. Darstellung und KristallStruktur des Rubidiumcalciumamids, $RbCa(NH_2)_3$. *Z. Anorg. Allg. Chem.* **1979**, *456*, 147–154. [CrossRef]
199. Jacobs, H.; Kockelkorn, J. Über cäsiumamidometallate ($CsM(NH_2)_3$) des calciums, strontiums und europiums; verbindungen mit der struktur "hexagonaler perowskite". *J. Less Common Met.* **1981**, *81*, 143–154. [CrossRef]
200. Becker, P.; Niewa, R. Progress in ammonothermal crystal growth of indium nitride. *J. Cryst. Growth* **2022**, *581*, 126480. [CrossRef]
201. Jacobs, H.; Fink, U. Untersuchung des Systems Kalium/Europium/Ammoniak. *Z. Anorg. Allg. Chem.* **1978**, *438*, 151–159. [CrossRef]
202. Jacobs, H.; Rechenbach, D.; Zachwieja, U. Structure determination of γ'-Fe_4N and ϵ-Fe_3N. *J. Alloys Compd.* **1995**, *227*, 10–17. [CrossRef]
203. Jacobs, H.; Peters, D.; Hassiepen, K. Caesiumamidometallate des gadoliniums, ytterbiums und yttriums mit perowskitverwandten Atomanordnungen $Cs_3M_2(NH_2)_9$. *J. Less Common Met.* **1986**, *118*, 31–41. [CrossRef]
204. Hadenfeldt, C.; Gieger, B.; Jacobs, H. Darstellung und Kristallstruktur von $K_3La(NH_2)_6$. *Z. Anorg. Allg. Chem.* **1974**, *403*, 319–326. [CrossRef]
205. Zając, M.; Gosk, J.; Grzanka, E.; Stelmakh, S.; Palczewska, M.; Wysmołek, A.; Korona, K.; Kamińska, M.; Twardowski, A. Ammonothermal synthesis of GaN doped with transition metal ions (Mn, Fe, Cr). *J. Alloys Compd.* **2008**, *456*, 324–338. [CrossRef]
206. Jacobs, H.; Harbrecht, B. Schichtenweise Substitution von Kationen im Lithiumamid: Kaliumtrilithiumamid, $KLi_3(NH_2)_4$ und Kaliumheptalithiumamid, $KLi_7(NH_2)_8$. *Z. Anorg. Allg. Chem.* **1984**, *518*, 87–100. [CrossRef]
207. Jacobs, H.; Birkenbeul, J.; Schmitz, D. Strukturverwandtschaft des dicaesiumamidomagnesats, $Cs_2[Mg(NH_2)_4]$, zum β-K_2SO_4-Typ. *J. Less Common Met.* **1982**, *85*, 79–86. [CrossRef]
208. Jacobs, H.; Birkenbeul, J.; Kockelkorn, J. Darstellung und eigenschaften der amidomagnesate des kaliums und rubidiums $K_2[Mg(NH_2)_4]$- und $Rb_2[Mg(NH_2)_4]$-Verbindungen mit isolierten $[Mg(NH_2)_4]_{2-}$-tetraedern. *J. Less Common Met.* **1984**, *97*, 205–214. [CrossRef]
209. Jacobs, H.; Kistrup, H. Über das System Kalium/Samarium/Ammoniak. *Z. Anorg. Allg. Chem.* **1977**, *435*, 127–136. [CrossRef]
210. Hadenfeldt, C.; Jacobs, H.; Juza, R. Über die Amide des Europiums und Ytterbiums. *Z. Anorg. Allg. Chem.* **1970**, *379*, 144–156. [CrossRef]
211. Froehling, B.; Kreiner, G.; Jacobs, H. ChemInform Abstract: Synthesis and Crystal Structure of Manganese(II) and Zinc Amides, $Mn(NH_2)_2$ and $Zn(NH_2)_2$. *ChemInform* **1999**, *30*, 211–216. [CrossRef]
212. Juza, R. Über die Amide der 1. und 2. Gruppe des periodischen Systems. Metallamide. I. Mitteilung. *Z. Anorg. Allg. Chem.* **1937**, *231*, 121–135. [CrossRef]
213. Sun, D.; Liu, L.; Wang, G.; Yu, J.; Li, Q.; Tian, G.; Wang, B.; Xu, X.; Zhang, L.; Wang, S. Research Progress in Liquid Phase Growth of GaN Crystals. *Chem.–A Eur. J.* **2023**, *30*, e202303710. [CrossRef]
214. Kucharski, R.; Sochacki, T.; Lucznik, B.; Amilusik, M.; Grabianska, K.; Iwinska, M.; Bockowski, M. Ammonothermal and HVPE Bulk Growth of GaN. *Wide Bandgap Semicond. Power Electron. Mater. Devices Appl.* **2021**, *2*, 529–554.
215. Scholz, F.; Bockowski, M.; Grzanka, E. GaN-Based Materials: Substrates, Metalorganic Vapor-Phase Epitaxy, and Quantum Well Properties. In *Nitride Semiconductor Technology: Power Electronics and Optoelectronic Devices*; Wiley-VCH: Weinheim, Germany, 2020; pp. 41–98.
216. Xu, K.; Wang, J.-F.; Ren, G.-Q. Progress in bulk GaN growth. *Chin. Phys. B* **2015**, *24*, 66105. [CrossRef]
217. Brodt-Giles, D.; Gomach, S.; American-Made Team. *Spring 2024 Alumni Book of the AMERICAN-MADE*; U.S. Department of Energy: Washington, DC, USA, 2024.
218. Kirste, L.; Tran Thi Caliste, T.N.; Weyher, J.L.; Smalc-Koziorowska, J.; Zajac, M.A.; Kucharski, R.; Sochacki, T.; Grabianska, K.; Iwinska, M.; Detlefs, C.; et al. Large-Scale Defect Clusters with Hexagonal Honeycomb-like Arrangement in Ammonothermal GaN Crystals. *Materials* **2022**, *15*, 6996. [CrossRef]
219. Meissner, E.; Niewa, R. *Ammonothermal Synthesis and Crystal Growth of Nitrides*; Springer International Publishing: Cham, Switzerland, 2021; Volume 10, ISBN 978-3-030-56305-9.
220. Dwili, R.; Doradzi, R.; Garczy, J.; Sierzputowski, L.; Palczewska, M.; Wysmolek, A.; Kami, M. AMMONO method of BN, AlN and GaN synthesis and crystal growth. *MRS Internet J. Nitride Semicond. Res.* **1998**, *3*, e25. [CrossRef]
221. Cao, Y.; Chen, X.; Lan, Y.; Li, J.; Xu, Y.; Xu, T.; Liu, Q.; Liang, J. Blue emission and Raman scattering spectrum from AlN nanocrystalline powders. *J. Cryst. Growth* **2000**, *213*, 198–202. [CrossRef]
222. Cao, Y.G.; Chen, X.L.; Lan, Y.C.; Li, J.Y.; Zhang, Y.; Yang, Z.; Liang, J.K. Synthesis and Raman characteristics of hexagonal Al*x*Ga1-*x*N alloy nanocrystalline solids through ammonothermal routes. *Appl. Phys. A* **2001**, *72*, 125–127. [CrossRef]

223. Motchanyy, A.I.; Reu, A.A.; Kovalenko, V.S.; Balakirev, V.G. Ammonothermal Crystallization of AlN Crystals. *MRS Proc.* **2005**, *878*, Y3.3. [CrossRef]
224. Dwilinski, R.; Doradzinski, R.; Garczynski, J.; Sierzputowski, L.; Kanbara, Y. Method and Equipment for Manufacturing Aluminum Nitride Bulk Single Crystal. U.S. Patent No. 7,374,615, 20 May 2008.
225. Wang, B.; Callahan, M.J. Ammonothermal Synthesis of III-Nitride Crystals. *Cryst. Growth Des.* **2006**, *6*, 1227–1246. [CrossRef]
226. Dwilinski, R.; Doradzinski, R.; Garczynski, J.; Sierzputowski, L.P.; Kanbara, Y. Process for Obtaining of Bulk Monocrystalline Gallium-Containing Nitride. U.S. Patent 7,364,619, 29 April 2008.
227. Kukushkin, S.A.; Sharofidinov, S.S. A New Method of Growing AlN, GaN, and AlGaN Bulk Crystals Using Hybrid SiC/Si Substrates. *Phys. Solid State* **2019**, *61*, 2342–2347. [CrossRef]
228. Jander, G.; Spandau, H.; Addison, C.C. *Chemistry in Nonaequeous Ionizing Solvents*; Friedr. Vieweg & Sohn: Braunschweig, Germany, 1966.
229. Izyumskaya, N.; Demchenko, D.O.; Das, S.; Özgür, Ü.; Avrutin, V.; Morkoç, H. Recent Development of Boron Nitride towards Electronic Applications. *Adv. Electron. Mater.* **2017**, *3*, 1600485. [CrossRef]
230. Taniguchi, T.; Watanabe, K. Synthesis of high-purity boron nitride single crystals under high pressure by using Ba–BN solvent. *J. Cryst. Growth* **2007**, *303*, 525–529. [CrossRef]
231. Chilleri, J.; Siddiqua, P.; Shur, M.S.; O'Leary, S.K. Cubic boron nitride as a material for future electron device applications: A comparative analysis. *Appl. Phys. Lett.* **2022**, *120*, 122105. [CrossRef]
232. Naclerio, A.E.; Kidambi, P.R. A Review of Scalable Hexagonal Boron Nitride (h-BN) Synthesis for Present and Future Applications. *Adv. Mater.* **2023**, *35*, e2207374. [CrossRef]
233. Yankowitz, M.; Ma, Q.; Jarillo-Herrero, P.; LeRoy, B.J. van der Waals heterostructures combining graphene and hexagonal boron nitride. *Nat. Rev. Phys.* **2019**, *1*, 112–125. [CrossRef]
234. Maruyama, Y.; Kurozumi, T.; Omori, K.; Otsubo, H.; Sato, T.; WATANABE, T. Ammonothermal synthesis of rhombohedral boron nitride. *Mater. Lett.* **2018**, *232*, 110–112. [CrossRef]
235. Jacobs, H.; Pinkowski, E. von. Synthese ternärer nitride von alkalimetallen: Verbindungen mit tantal, $MTaN_2$ MIT M $\equiv$ Na, K, RbundCs. *J. Less Common Met.* **1989**, *146*, 147–160. [CrossRef]
236. Brokamp, T.; Jacobs, H. Synthese und Kristallstruktur eines gemischtvalenten Lithium-Tantalnitrids $Li_2Ta_3N_5$. *J. Alloys Compd.* **1991**, *176*, 47–60. [CrossRef]
237. Jacobs, H. Preparation and crystal structure of a sodium silicon nitride, $NaSi_2N_3$. *Eur. J. Solid State Inorg. Chem.* **1993**, *30*, 45–53.
238. Jacobs, H.; Nymwegen, R. Darstellung und Kristallstruktur eines Kaliumnitridophosphats, $K_3P_6N_{11}$. *Z. Anorg. Allg. Chem.* **1997**, *623*, 429–433. [CrossRef]
239. Mallmann, M.; Wendl, S.; Strobel, P.; Schmidt, P.J.; Schnick, W. $Sr_3P_3N_7$: Complementary Approach by Ammonothermal and High-Pressure Syntheses. *Chemistry* **2020**, *26*, 6257–6263. [CrossRef] [PubMed]
240. Tuomisto, F. Vacancy Defects in Bulk and Quasi-Bulk GaN Crystals. *Crystals* **2022**, *12*, 1112. [CrossRef]
241. Suihkonen, S.; Pimputkar, S.; Speck, J.S.; Nakamura, S. Infrared absorption of hydrogen-related defects in ammonothermal GaN. *Appl. Phys. Lett.* **2016**, *108*, 202105. [CrossRef]
242. Shima, K.; Kurimoto, K.; Bao, Q.; Mikawa, Y.; Saito, M.; Tomida, D.; Uedono, A.; Ishibashi, S.; Ishiguro, T.; Chichibu, S.F. Improved midgap recombination lifetimes in GaN crystals grown by the low-pressure acidic ammonothermal method. *Appl. Phys. Lett.* **2024**, *124*, 181103. [CrossRef]
243. Zajac, M.; Kaminski, P.; Kozlowski, R.; Litwin-Staszewska, E.; Piotrzkowski, R.; Grabianska, K.; Kucharski, R.; Jakiela, R. Formation of Grown-In Nitrogen Vacancies and Interstitials in Highly Mg-Doped Ammonothermal GaN. *Materials* **2024**, *17*, 1160. [CrossRef] [PubMed]
244. Chichibu, S.F.; Shima, K.; Uedono, A.; Ishibashi, S.; Iguchi, H.; Narita, T.; Kataoka, K.; Tanaka, R.; Takashima, S.; Ueno, K.; et al. Impacts of vacancy complexes on the room-temperature photoluminescence lifetimes of state-of-the-art GaN substrates, epitaxial layers, and Mg-implanted layers. *J. Appl. Phys.* **2024**, *135*, 185701. [CrossRef]
245. Freysoldt, C.; Grabowski, B.; Hickel, T.; Neugebauer, J.; Kresse, G.; Janotti, A.; van de Walle, C.G. First-principles calculations for point defects in solids. *Rev. Mod. Phys.* **2014**, *86*, 253–305. [CrossRef]
246. Lyons, J.L.; Wickramaratne, D.; van de Walle, C.G. A first-principles understanding of point defects and impurities in GaN. *J. Appl. Phys.* **2021**, *129*, 111101. [CrossRef]
247. Baker, J.N.; Bowes, P.C.; Harris, J.S.; Collazo, R.; Sitar, Z.; Irving, D.L. Complexes and compensation in degenerately donor doped GaN. *Appl. Phys. Lett.* **2020**, *117*, 102109. [CrossRef]
248. van de Walle, C.G. Interactions of hydrogen with native defects in GaN. *Phys. Rev. B* **1997**, *56*, R10020. [CrossRef]
249. Wright, A.F. Interaction of hydrogen with gallium vacancies in wurtzite GaN. *J. Appl. Phys.* **2001**, *90*, 1164–1169. [CrossRef]
250. Dreyer, C.E.; Alkauskas, A.; Lyons, J.L.; Speck, J.S.; van de Walle, C.G. Gallium vacancy complexes as a cause of Shockley-Read-Hall recombination in III-nitride light emitters. *Appl. Phys. Lett.* **2016**, *108*, 141101. [CrossRef]
251. Reshchikov, M.A.; Demchenko, D.O.; Ye, D.; Andrieiev, O.; Vorobiov, M.; Grabianska, K.; Zajac, M.; Nita, P.; Iwinska, M.; Bockowski, M. The effect of annealing on photoluminescence from defects in ammonothermal GaN. *J. Appl. Phys.* **2022**, *131*, 035704. [CrossRef]
252. Jiang, W.; Nolan, M.; Ehrentraut, D.; D'Evelyn, M.P. Electrical and optical properties of gallium vacancy complexes in ammonothermal GaN. *Appl. Phys. Express* **2017**, *10*, 75506. [CrossRef]

253. Tuomisto, F.; Mäki, J.-M.; Zając, M. Vacancy defects in bulk ammonothermal GaN crystals. *J. Cryst. Growth* **2010**, *312*, 2620–2623. [CrossRef]
254. Tuomisto, F.; Kuittinen, T.; Zając, M.; Doradziński, R.; Wasik, D. Vacancy–hydrogen complexes in ammonothermal GaN. *J. Cryst. Growth* **2014**, *403*, 114–118. [CrossRef]
255. Uedono, A.; Tsukada, Y.; Mikawa, Y.; Mochizuki, T.; Fujisawa, H.; Ikeda, H.; Kurihara, K.; Fujito, K.; Terada, S.; Ishibashi, S. Vacancies and electron trapping centers in acidic ammonothermal GaN probed by a monoenergetic positron beam. *J. Cryst. Growth* **2016**, *448*, 117–121. [CrossRef]
256. Reshchikov, M.A.; Demchenko, D.O.; McNamara, J.D.; Fernández-Garrido, S.; Calarco, R. Green luminescence in Mg-doped GaN. *Phys. Rev. B* **2014**, *90*, 035207. [CrossRef]
257. Myers, S.M.; Wright, A.F.; Petersen, G.A.; Seager, C.H.; Wampler, W.R.; Crawford, M.H.; Han, J. Equilibrium state of hydrogen in gallium nitride: Theory and experiment. *J. Appl. Phys.* **2000**, *88*, 4676–4687. [CrossRef]
258. Myers, S.M.; Wright, A.F.; Petersen, G.A.; Wampler, W.R.; Seager, C.H.; Crawford, M.H.; Han, J. Diffusion, release, and uptake of hydrogen in magnesium-doped gallium nitride: Theory and experiment. *J. Appl. Phys.* **2001**, *89*, 3195–3202. [CrossRef]
259. Saarinen, K.; Suski, T.; Grzegory, I.; Look, D.C. Thermal stability of isolated and complexed Ga vacancies in GaN bulk crystals. *Phys. Rev. B* **2001**, *64*, 233201. [CrossRef]
260. Horibuchi, K.; Yamaguchi, S.; Kimoto, Y.; Nishikawa, K.; Kachi, T. Formation of helical dislocations in ammonothermal GaN substrate by heat treatment. *Semicond. Sci. Technol.* **2016**, *31*, 34002. [CrossRef]
261. Tuomisto, F.; Mäki, J.-M.; Rauch, C.; Makkonen, I. On the formation of vacancy defects in III-nitride semiconductors. *J. Cryst. Growth* **2012**, *350*, 93–97. [CrossRef]
262. Tomida, D.; Chichibu, S.F.; Kagamitani, Y.; Bao, Q.; Hazu, K.; Simura, R.; Sugiyama, K.; Yokoyama, C.; Ishiguro, T.; Fukuda, T. Improving the purity of GaN grown by the ammonothermal method with in-autoclave gas-phase acidic mineralizer synthesis. *J. Cryst. Growth* **2012**, *348*, 80–84. [CrossRef]
263. Shim, J.B.; Kim, G.H.; Lee, Y.K. Basic ammonothermal growth of bulk GaN single crystal using sodium mineralizers. *J. Cryst. Growth* **2017**, *478*, 85–88. [CrossRef]
264. Neugebauer, J.; van de Walle, C.G. Chemical trends for acceptor impurities in GaN. *J. Appl. Phys.* **1999**, *85*, 3003–3005. [CrossRef]
265. Götz, W.; Johnson, N.M.; Chen, C.; Liu, H.; Kuo, C.; Imler, W. Activation energies of Si donors in GaN. *Appl. Phys. Lett.* **1996**, *68*, 3144–3146. [CrossRef]
266. Matsubara, M.; Bellotti, E. A first-principles study of carbon-related energy levels in GaN. II. Complexes formed by carbon and hydrogen, silicon or oxygen. *J. Appl. Phys.* **2017**, *121*, 195702. [CrossRef]
267. Lee, J.-H.; Hahm, S.-H.; Lee, J.-H.; Bae, S.-B.; Lee, K.-S.; Cho, Y.-H.; Lee, J.-L. Effect of Al doping in GaN films grown by metalorganic chemical vapor deposition. *Appl. Phys. Lett.* **2003**, *83*, 917–919. [CrossRef]
268. Lyons, J.L.; Janotti, A.; van de Walle, C.G. Shallow versus deep nature of Mg acceptors in nitride semiconductors. *Phys. Rev. Lett.* **2012**, *108*, 156403. [CrossRef] [PubMed]
269. Neugebauer, J.; van de Walle, C.G. Hydrogen in GaN: Novel aspects of a common impurity. *Phys. Rev. Lett.* **1995**, *75*, 4452–4455. [CrossRef] [PubMed]
270. Lyons, J.L.; van de Walle, C.G. Computationally predicted energies and properties of defects in GaN. *npj Comput. Mater.* **2017**, *3*, 12. [CrossRef]
271. Lee, D.; Mitchell, B.; Fujiwara, Y.; Dierolf, V. Thermodynamics and Kinetics of Three Mg−H−VN Complexes in Mg:GaN from Combined First-Principles Calculation and Experiment. *Phys. Rev. Lett.* **2014**, *112*, 205501. [CrossRef]
272. Wickramaratne, D.; Shen, J.-X.; Dreyer, C.E.; Alkauskas, A.; van de Walle, C.G. Electrical and optical properties of iron in GaN, AlN, and InN. *Phys. Rev. B* **2019**, *99*, 205202. [CrossRef]
273. Reshchikov, M.A. Photoluminescence from defects in GaN. In *Gallium Nitride Materials and Devices XVIII, Proceedings of the Gallium Nitride Materials and Devices XVIII, San Francisco, CA, USA, 28 January–3 February 2023*; Morkoç, H., Fujioka, H., Schwarz, U.T., Eds.; SPIE: Bellingham, WA, USA, 2023; p. 17, ISBN 9781510659476.
274. Lyons, J.L.; Janotti, A.; van de Walle, C.G. Impact of Group-II Acceptors on the Electrical and Optical Properties of GaN. *Jpn. J. Appl. Phys.* **2013**, *52*, 08JJ04. [CrossRef]
275. Cui, X.Y.; Delley, B.; Freeman, A.J.; Stampfl, C. Neutral and charged embedded clusters of Mn in doped GaN from first principles. *Phys. Rev. B* **2007**, *76*, 045201. [CrossRef]
276. Korotkov, R.Y.; Wessels, B.W. Electrical Properties of Oxygen Doped GaN Grown by Metalorganic Vapor Phase Epitaxy. *MRS Internet J. Nitride Semicond. Res.* **2000**, *5*, 301–307. [CrossRef]
277. Wickramaratne, D.; Dreyer, C.E.; Monserrat, B.; Shen, J.-X.; Lyons, J.L.; Alkauskas, A.; van de Walle, C.G. Defect identification based on first-principles calculations for deep level transient spectroscopy. *Appl. Phys. Lett.* **2018**, *113*, 192106. [CrossRef]
278. Dwiliński, R.; Doradziński, R.; Garczyński, J.; Sierzputowski, L.; Baranowski, J.; Kamińska, M. Exciton photo-luminescence of GaN bulk crystals grown by the AMMONO method. *Mater. Sci. Eng. B* **1997**, *50*, 46–49. [CrossRef]
279. Friedrich, J.; von Ammon, W.; Müller, G. 2—Czochralski Growth of Silicon Crystals. In *Handbook of Crystal Growth*, 2nd ed.; Rudolph, P., Ed.; Elsevier: Amsterdam, The Netherlands, 2015; pp. 45–104, ISBN 978-0-444-63303-3.
280. Lu, W.; Li, T.; Ren, G.; Xia, Z.; Xie, K.; Li, S.; Shen, L.; Xu, K. Study on Raman scattering spectroscopy of Mn-doped GaN grown by the ammonothermal method. *CrystEngComm* **2024**, *26*, 2166–2171. [CrossRef]

281. Wang, B.; Bliss, D.; Suscavage, M.; Swider, S.; Lancto, R.; Lynch, C.; Weyburne, D.; Li, T.; Ponce, F.A. Ammonothermal growth of high-quality GaN crystals on HVPE template seeds. *J. Cryst. Growth* **2011**, *318*, 1030–1033. [CrossRef]
282. Pimputkar, S.; Suihkonen, S.; Imade, M.; Mori, Y.; Speck, J.S.; Nakamura, S. Free electron concentration dependent sub-bandgap optical absorption characterization of bulk GaN crystals. *J. Cryst. Growth* **2015**, *432*, 49–53. [CrossRef]
283. Chyi, J.-I.; Fujioka, H.; Morkoç, H. (Eds.) *Gallium Nitride Materials and Devices X*; SPIE OPTO; SPIE: San Francisco, CA, USA, 2015.
284. Mikawa, Y.; Ishinabe, T.; Kawabata, S.; Mochizuki, T.; Kojima, A.; Kagamitani, Y.; Fujisawa, H. Ammonothermal growth of polar and non-polar bulk GaN crystal. In *Gallium Nitride Materials and Devices X*; Chyi, J.-I., Fujioka, H., Morkoç, H., Eds.; SPIE OPTO; SPIE: San Francisco, CA, USA, 2015; p. 936302.
285. Yoshida, K.; Aoki, K.; Fukuda, T. High-temperature acidic ammonothermal method for GaN crystal growth. *J. Cryst. Growth* **2014**, *393*, 93–97. [CrossRef]
286. Letts, E.; Hashimoto, T.; Hoff, S.; Key, D.; Male, K.; Michaels, M. Development of GaN wafers via the ammonothermal method. *J. Cryst. Growth* **2014**, *403*, 3–6. [CrossRef]
287. Amilusik, M.; Zajac, M.; Fijalkowski, M.; Iwinska, M.; Sochacki, T.; Wlodarczyk, D.; Somakumar, A.K.; Jakiela, R.; Suchocki, A.; Bockowski, M. Role of carbon in n-type bulk GaN crystals. *J. Cryst. Growth* **2024**, *632*, 127641. [CrossRef]
288. Jiang, W.; Ehrentraut, D.; Downey, B.C.; Kamber, D.S.; Pakalapati, R.T.; Yoo, H.D.; D'Evelyn, M.P. Highly transparent ammonothermal bulk GaN substrates. *J. Cryst. Growth* **2014**, *403*, 18–21. [CrossRef]
289. Trivellin, N.; Meneghini, M.; Zanoni, E.; Orita, K.; Yuri, M.; Tanaka, T.; Ueda, D.; Meneghesso, G. A review on the reliability of GaN-based laser diodes. In Proceedings of the 2010 IEEE International Reliability Physics Symposium, Garden Grove (Anaheim), CA, USA, 2–6 May 2010; IEEE: Piscataway, NJ, USA, 2010.
290. Wen, P.Y.; Zhang, S.M.; Li, D.Y.; Liu, J.P.; Zhang, L.Q.; Zhou, K.; Feng, M.X.; Tian, A.Q.; Zhang, F.; Gao, X.D.; et al. Identification of degradation mechanisms of blue InGaN/GaN laser diodes. *J. Phys. D Appl. Phys.* **2015**, *48*, 415101. [CrossRef]
291. Leszczynski, M.; Prystawko, P.; Suski, T.; Lucznik, B.; Domagala, J.; Bak-Misiuk, J.; Stonert, A.; Turos, A.; Langer, R.; Barski, A. Lattice parameters of GaN single crystals, homoepitaxial layers and heteroepitaxial layers on sapphire. *J. Alloys Compd.* **1999**, *286*, 271–275. [CrossRef]
292. Krysko, M.; Sarzynski, M.; Domagała, J.; Grzegory, I.; Łucznik, B.; Kamler, G.; Porowski, S.; Leszczyński, M. The influence of lattice parameter variation on microstructure of GaN single crystals. *J. Alloys Compd.* **2005**, *401*, 261–264. [CrossRef]
293. Letts, E.; Key, D.; Hashimoto, T. Reduction of crack density in ammonothermal bulk GaN growth. *J. Cryst. Growth* **2016**, *456*, 27–32. [CrossRef]
294. Van de Walle, C.G. Effects of impurities on the lattice parameters of GaN. *Phys. Rev. B* **2003**, *68*, 165209. [CrossRef]
295. Tanaka, A.; Nagamatsu, K.; Usami, S.; Kushimoto, M.; Deki, M.; Nitta, S.; Honda, Y.; Bockowski, M.; Amano, H. V-shaped dislocations in a GaN epitaxial layer on GaN substrate. *AIP Adv.* **2019**, *9*, 095002. [CrossRef]
296. Erlekampf, J.; Seebeck, J.; Savva, P.; Meissner, E.; Friedrich, J.; Alt, N.; Schlücker, E.; Frey, L. Numerical time-dependent 3D simulation of flow pattern and heat distribution in an ammonothermal system with various baffle shapes. *J. Cryst. Growth* **2014**, *403*, 96–104. [CrossRef]
297. Alt, N.S.A. *Entwicklung Spezieller Hochdruckapparate und Methoden zur Durchführung von In-Situ Messungen Beim Ammonothermalen Prozess*; FAU Erlangen-Nürnberg: Erlangen, Germany, 2010.
298. Alt, N.; Meissner, E.; Schlücker, E.; Frey, L. In situ monitoring technologies for ammonthermal reactors. *Phys. Status Solidi C* **2012**, *9*, 436–439. [CrossRef]
299. Schimmel, S. *In Situ Visualisierung des Ammonothermalen Kristallisationsprozesses Mittels Röntgenmesstechnik*; Friedrich-Alexander-Universität Erlangen-Nürnberg (FAU): Erlangen, Germany, 2018.
300. Schimmel, S.; Wellmann, P. In Situ Visualization of the Ammonothermal Crystallization Process by X-ray Technology. In *Ammonothermal Synthesis and Crystal Growth of Nitrides*; Meissner, E., Niewa, R., Eds.; Springer International Publishing: Cham, Switzerland, 2021; pp. 171–190, ISBN 978-3-030-56304-2.
301. Schlücker, E.; Kimmel, A.-C.L. Technological Challenges of Autoclave Design for Ammonothermal Syntheses. In *Ammonothermal Synthesis and Crystal Growth of Nitrides*; Meissner, E., Niewa, R., Eds.; Springer International Publishing: Cham, Switzerland, 2021; pp. 27–44, ISBN 978-3-030-56304-2.
302. Schlücker, E. High-Pressure Technology. In *Ullmann's Encyclopedia of Industrial Chemistry*; Wiley-VCH Verlag GmbH & Co. KGaA: Weinheim, Germany, 2000; pp. 1–49, ISBN 9783527306732.
303. Hertweck, B.; Steigerwald, T.G.; Alt, N.S.A.; Schluecker, E. Corrosive Degeneration of Autoclaves for the Ammonothermal Synthesis: Experimental Approach and First Results. *Chem. Eng. Technol.* **2014**, *37*, 1903–1906. [CrossRef]
304. Schimmel, S.; Salamon, M.; Tomida, D.; Neumeier, S.; Ishiguro, T.; Honda, Y.; Chichibu, S.F.; Amano, H. High-Energy Computed Tomography as a Prospective Tool for In Situ Monitoring of Mass Transfer Processes inside High-Pressure Reactors-A Case Study on Ammonothermal Bulk Crystal Growth of Nitrides including GaN. *Materials* **2022**, *15*, 6165. [CrossRef] [PubMed]
305. Mallmann, M.; Wendl, S.; Schnick, W. Crystalline Nitridophosphates by Ammonothermal Synthesis. *Chemistry* **2020**, *26*, 2067–2072. [CrossRef]
306. Cordes, N.; Schnick, W. Ammonothermal Synthesis of Crystalline Oxonitride Perovskites $LnTaON_2$ (Ln = La, Ce, Pr, Nd, Sm, Gd). *Chemistry* **2017**, *23*, 11410–11415. [CrossRef]
307. NeoNickel. A 286 Technical Datasheet. Available online: https://www.neonickel.com/alloys/all-alloys/a286 (accessed on 25 April 2024).

308. Larson, F.R.; Miller, J. A Time-Temperature Relationship for Rupture and Creep Stresses. *J. Fluids Eng.* **1952**, *74*, 765–771. [CrossRef]
309. Guštin, A.Z.; Žužek, B.; Podgornik, B. Creep Life Prediction of 10CrMo9-10 Steel by Larson-Miller Model. *Materials* **2022**, *15*, 4431. [CrossRef] [PubMed]
310. Tamura, M.; Esaka, H.; Shinozuka, K. Stress and Temperature Dependence of Time to Rupture of Heat Resisting Steels. *ISIJ Int.* **1999**, *39*, 380–387. [CrossRef]
311. Sherby, O.D.; Burke, P.M. Mechanical behavior of crystalline solids at elevated temperature. *Prog. Mater. Sci.* **1968**, *13*, 323–390. [CrossRef]
312. Metzger, F.; Rienzi, V.; Mascetti, C.; Nguyen, T.; Pimputkar, S. Properties of Titanium Zirconium Molybdenum Alloy after Exposure to Indium at Elevated Temperatures. *Materials* **2022**, *15*, 5270. [CrossRef]
313. Jiang, W.; Ehrentraut, D.; Cook, J.; Kamber, D.S.; Pakalapati, R.T.; D'Evelyn, M.P. Transparent, conductive bulk GaN by high temperature ammonothermal growth. *Phys. Status Solidi (b)* **2015**, *252*, 1069–1074. [CrossRef]
314. Kimmel, A.-C.L.; Schlücker, E. Corrosive Degeneration of Process Equipment and Technical Solutions for Corrosion Protection Under Ammonothermal Conditions. In *Ammonothermal Synthesis and Crystal Growth of Nitrides*; Meissner, E., Niewa, R., Eds.; Springer International Publishing: Cham, Switzerland, 2021; pp. 191–201, ISBN 978-3-030-56304-2.
315. Hertweck, B.; Schimmel, S.; Steigerwald, T.G.; Alt, N.S.; Wellmann, P.J.; Schluecker, E. Ceramic liner technology for ammonoacidic synthesis. *J. Supercrit. Fluids* **2015**, *99*, 76–87. [CrossRef]
316. Becker, P.; Wonglakhon, T.; Zahn, D.; Gudat, D.; Niewa, R. Approaching Dissolved Species in Ammonoacidic GaN Crystal Growth: A Combined Solution NMR and Computational Study. *Chemistry* **2020**, *26*, 7008–7017. [CrossRef] [PubMed]
317. Hashimoto, T.; Wu, F.; Speck, J.S.; Nakamura, S. Growth of Bulk GaN Crystals by the Basic Ammonothermal Method. *Jpn. J. Appl. Phys.* **2007**, *46*, L889. [CrossRef]
318. Kagamitani, Y.; Ehrentraut, D.; Yoshikawa, A.; Hoshino, N.; Fukuda, T.; Kawabata, S.; Inaba, K. Ammonothermal Epitaxy of Thick GaN Film Using NH 4 Cl Mineralizer. *Jpn. J. Appl. Phys.* **2006**, *45*, 4018. [CrossRef]
319. Tomida, D.; Kuroda, K.; Hoshino, N.; Suzuki, K.; Kagamitani, Y.; Ishiguro, T.; Fukuda, T.; Yokoyama, C. Solubility of GaN in supercritical ammonia with ammonium chloride as a mineralizer. *J. Cryst. Growth* **2010**, *312*, 3161–3164. [CrossRef]
320. Purdy, A.P.; Jouet, R.J.; George, C.F. Ammonothermal Recrystallization of Gallium Nitride with Acidic Mineralizers. *Cryst. Growth Des.* **2002**, *2*, 141–145. [CrossRef]
321. Hertweck, B.; Zhang, S.; Steigerwald, T.G.; Alt, N.S.A.; Niewa, R.; Schluecker, E. Applicability of Metals as Liner Materials for Ammonoacidic Crystal Growth. *Chem. Eng. Technol.* **2014**, *37*, 1835–1844. [CrossRef]
322. Malkowski, T.F.; Speck, J.S.; DenBaars, S.P.; Nakamura, S. An exploratory study of acidic ammonothermal growth in a TZM autoclave at high temperatures. *J. Cryst. Growth* **2018**, *499*, 85–89. [CrossRef]
323. Pimputkar, S.; Speck, J.S.; Nakamura, S. Basic ammonothermal GaN growth in molybdenum capsules. *J. Cryst. Growth* **2016**, *456*, 15–20. [CrossRef]
324. Griffiths, S.; Pimputkar, S.; Kearns, J.; Malkowski, T.F.; Doherty, M.F.; Speck, J.S.; Nakamura, S. Growth kinetics of basic ammonothermal gallium nitride crystals. *J. Cryst. Growth* **2018**, *501*, 74–80. [CrossRef]
325. Steigerwald, T.G.; Schlücker, E. Direct Determination of Viscosity of Supercritical Solutions. In *Ammonothermal Synthesis and Crystal Growth of Nitrides*; Meissner, E., Niewa, R., Eds.; Springer International Publishing: Cham, Switzerland, 2021; pp. 117–153, ISBN 978-3-030-56304-2.
326. Schlücker, E.; Hertweck, B.; Schimmel, S.; Wellmann, P. Special Equipment for Ammonothermal Processes. In *Ammonothermal Synthesis and Crystal Growth of Nitrides*; Meissner, E., Niewa, R., Eds.; Springer Series in Materials Science; Springer: Cham, Switzerland, 2021; Volume 304. [CrossRef]
327. MacKay, R.A.; Gabb, T.P.; Smialek, J.L.; Nathal, M.V. A new approach of designing superalloys for low density. *JOM* **2010**, *62*, 48–54. [CrossRef]
328. Schlücker, E. Technical Solutions for In Situ Monitoring of Ammonothermal Processes. In *Ammonothermal Synthesis and Crystal Growth of Nitrides*; Meissner, E., Niewa, R., Eds.; Springer International Publishing: Cham, Switzerland, 2021; pp. 47–64, ISBN 978-3-030-56304-2.
329. Schimmel, S.; Tomida, D.; Ishiguro, T.; Honda, Y.; Chichibu, S.F.; Amano, H. Temperature Field, Flow Field, and Temporal Fluctuations Thereof in Ammonothermal Growth of Bulk GaN-Transition from Dissolution Stage to Growth Stage Conditions. *Materials* **2023**, *16*, 2016. [CrossRef] [PubMed]
330. Schimmel, S.; Duchstein, P.; Steigerwald, T.G.; Kimmel, A.-C.; Schlücker, E.; Zahn, D.; Niewa, R.; Wellmann, P. In situ X-ray monitoring of transport and chemistry of Ga-containing intermediates under ammonothermal growth conditions of GaN. *J. Cryst. Growth* **2018**, *498*, 214–223. [CrossRef]
331. William, S. The viscosity of gases and molecular force. *Lond. Edinb. Dublin Philos. Mag. J. Sci.* **1893**, *36*, 507–531. [CrossRef]
332. Lencka, M.M.; Anderko, A.; Sanders, S.J.; Young, R.D. Modeling Viscosity of Multicomponent Electrolyte Solutions. *Int. J. Thermophys.* **1998**, *19*, 367–378. [CrossRef]
333. Jones, G.; Dole, M. The viscosity of aqueous solutions of strong electrolytes with special reference to barium chloride. *J. Am. Chem. Soc.* **1929**, *51*, 2950–2964. [CrossRef]
334. Saeed, R.; Uddin, F.; Masood, S.; Asif, N. Viscosities of ammonium salts in water and ethanol+water systems at different temperatures. *J. Mol. Liq.* **2009**, *146*, 112–115. [CrossRef]

335. Jenkins, H.D.B.; Marcus, Y. Viscosity B-Coefficients of Ions in Solution. *Chem. Rev.* **1995**, *95*, 2695–2724. [CrossRef]
336. Steigerwald, T.G.; Balouschek, J.; Hertweck, B.; Kimmel, A.-C.L.; Alt, N.S.; Schluecker, E. In situ investigation of decomposing ammonia and ammonobasic solutions under supercritical conditions via UV/vis and Raman Spectroscopy. *J. Supercrit. Fluids* **2018**, *134*, 96–105. [CrossRef]
337. Baser, H.; Schwieger, W.; Freitag, D.; Steigerwald, T.G.; Schluecker, E. Solubility Studies of Sodium Azide in Liquid Ammonia by In Situ Ultrasonic Velocity Measurement. *Chem. Eng. Technol.* **2017**, *40*, 1101–1106. [CrossRef]
338. Schimmel, S.; Tomida, D.; Saito, M.; Bao, Q.; Ishiguro, T.; Honda, Y.; Chichibu, S.; Amano, H. Boundary Conditions for Simulations of Fluid Flow and Temperature Field during Ammonothermal Crystal Growth—A Machine-Learning Assisted Study of Autoclave Wall Temperature Distribution. *Crystals* **2021**, *11*, 254. [CrossRef]
339. Chervin, J.C.; Syfosse, G.; Besson, J.M. Mechanical strength of sapphire windows under pressure. *Rev. Sci. Instrum.* **1994**, *65*, 2719–2725. [CrossRef]
340. Ockenfels, T.; Vewinger, F.; Weitz, M. Sapphire optical viewport for high pressure and temperature applications. *Rev. Sci. Instrum.* **2021**, *92*, 65109. [CrossRef] [PubMed]
341. Grunwaldt, J.-D.; Wandeler, R.; Baiker, A. Supercritical Fluids in Catalysis: Opportunities of In Situ Spectroscopic Studies and Monitoring Phase Behavior. *Catal. Rev.* **2003**, *45*, 1–96. [CrossRef]
342. Alt, N.S.; Meissner, E.; Schluecker, E. Development of a novel in situ monitoring technology for ammonothermal reactors. *J. Cryst. Growth* **2012**, *350*, 2–4. [CrossRef]
343. Spieß, L.; Teichert, G.; Schwarzer, R.; Behnken, H.; Genzel, C. (Eds.) Erzeugung und Eigenschaften von Röntgenstrahlung. In *Moderne Röntgenbeugung*; Vieweg+Teubner: Wiesbaden, Germany, 2009; pp. 5–40, ISBN 978-3-8351-0166-1.
344. Schimmel, S.; Künecke, U.; Baser, H.; Steigerwald, T.G.; Hertweck, B.; Alt, N.S.A.; Schlücker, E.; Schwieger, W.; Wellmann, P. Towards X-ray in-situ visualization of ammonothermal crystal growth of nitrides. *Phys. Status Solidi C* **2014**, *11*, 1439–1442. [CrossRef]
345. *Advances in Imaging and Electron Physics Volume 171*; Elsevier: Amsterdam, The Netherlands, 2012; ISBN 9780123942975.
346. Berger, M.J.; Hubbell, J.H.; Seltzer, S.M.; Chang, J.; Coursey, J.S.; Sukumar, R.; Zucker, D.S.; Olsen, K. XCOM: Photon Cross Sections Database, NIST Standard Reference Database 8 (XGAM). Available online: https://physics.nist.gov/PhysRefData/Xcom/html/xcom1.html (accessed on 13 May 2022).
347. Richter, T.M.M.; Alt, N.S.A.; Schlücker, E.; Niewa, R. Ammonothermal Synthesis and Characterization of $Li_4[Zn(NH_2)_4](NH_2)_2$. *Z. Anorg. Allg. Chem.* **2015**, *641*, 1016–1023. [CrossRef]
348. Richter, T.M.M.; Alt, N.S.A.; Schlücker, E.; Niewa, R. Ammonothermal Synthesis and Characterization of $Cs_2[Zn(NH_2)_4]$. *Z. Anorg. Allg. Chem.* **2016**, *642*, 1207–1211. [CrossRef]
349. Braeuer, A. *In Situ Spectroscopic Techniques at High Pressure*; Elsevier: Amsterdam, The Netherlands, 2015; ISBN 9780444634207.
350. Jones, R.R.; Hooper, D.C.; Zhang, L.; Wolverson, D.; Valev, V.K. Raman Techniques: Fundamentals and Frontiers. *Nanoscale Res. Lett.* **2019**, *14*, 231. [CrossRef] [PubMed]
351. Das, R.S.; Agrawal, Y.K. Raman spectroscopy: Recent advancements, techniques and applications. *Vib. Spectrosc.* **2011**, *57*, 163–176. [CrossRef]
352. Klein, T.; Wu, W.; Rausch, M.H.; Giraudet, C.; Koller, T.M.; Fröba, A.P. Influence of Liquid Structure on Fickian Diffusion in Binary Mixtures of n-Hexane and Carbon Dioxide Probed by Dynamic Light Scattering, Raman Spectroscopy, and Molecular Dynamics Simulations. *J. Phys. Chem. B* **2018**, *122*, 7122–7133. [CrossRef] [PubMed]
353. Stratmann, A.; Schweiger, G. Fluid Phase Equilibria of Binary Mixtures with Supercritical Solvents with in-situ Concentration Measurements by Raman Spectroscopy. In *Supercritical Fluids as Solvents and Reaction Media*; Elsevier: Amsterdam, The Netherlands, 2004; pp. 85–120, ISBN 9780444515742.
354. Braeuer, A. Chapter 3—Raman Spectroscopy from an Engineering Point of View. In *Supercritical Fluid Science and Technology: Spectroscopic Techniques at High Pressure*; Braeuer, A., Ed.; Elsevier: Amsterdam, The Netherlands, 2015; pp. 193–281, ISBN 2212-0505.
355. Stegemann, D.; Raj, B.; Bhaduri, A. NDT for Analysis of Microstructures and Mechanical Properties of Metallic Materials. In *Reference Module in Materials Science and Materials Engineering*; Elsevier: Amsterdam, The Netherlands, 2016; ISBN 9780128035818. [CrossRef]
356. Schwieger, W.; Baser, H. Ultrasound Measurement as a Tool for In Situ Determination of Filling Degree Under Extreme Conditions. In *Ammonothermal Synthesis and Crystal Growth of Nitrides*; Meissner, E., Niewa, R., Eds.; Springer International Publishing: Cham, Switzerland, 2021; pp. 105–116, ISBN 978-3-030-56304-2.
357. Viswanath, D.S.; Ghosh, T.K.; Prasad, D.H.L.; Dutt, N.V.; Rani, K.Y. Viscometers. In *Viscosity of Liquids*; Springer: Dordrecht, The Netherlands, 2007; pp. 9–107, ISBN 978-1-4020-5481-5.
358. Gupta, S.V. Oscillating Viscometers. In *Viscometry for Liquids*; Gupta, S.V., Ed.; Springer International Publishing: Cham, Switzerland, 2014; pp. 107–136, ISBN 978-3-319-04857-4.
359. Gupta, S.V. New Trends in Viscometers. In *Viscometry for Liquids*; Gupta, S.V., Ed.; Springer International Publishing: Cham, Switzerland, 2014; pp. 137–170, ISBN 978-3-319-04857-4.
360. Cohen-Tenoudji, F.; Pardee, W.J.; Tittmann, B.R.; Ahlberg, L.A.; Elsley, R.K. A shear wave rheology sensor. *IEEE Trans. Ultrason. Ferroelectr. Freq. Control* **1987**, *34*, 263–269. [CrossRef]

361. Kleis, S.J.; Sanchez, L.A. Dependence of speed of sound on salinity and temperature in concentrated NaCl solutions. *Sol. Energy* **1990**, *45*, 201–206. [CrossRef]
362. Natarajan, S.; Randolph, T.W. Ultrasonic velocity measurements in supercritical jet fuel. *J. Supercrit. Fluids* **1997**, *10*, 65–70. [CrossRef]
363. Sheen, S.H.; Chien, H.-T.; Raptis, A.C. An in-Line Ultrasonic Viscometer. In *Review of Progress in Quantitative Nondestructive Evaluation*; Thompson, D.O., Chimenti, D.E., Eds.; Springer: New York, NY, USA, 1995; pp. 1151–1158, ISBN 978-1-4613-5819-0.
364. McSkimin, H.J.; Andreatch, P. Measurement of Dynamic Shear Impedance of Low Viscosity Liquids at Ultrasonic Frequencies. *J. Acoust. Soc. Am.* **1967**, *42*, 248–252. [CrossRef]
365. Balasubramaniam, K.; Shah, V.V.; Costley, R.D.; Boudreaux, G.; Singh, J.P. High temperature ultrasonic sensor for the simultaneous measurement of viscosity and temperature of melts. *Rev. Sci. Instrum.* **1999**, *70*, 4618–4623. [CrossRef]
366. Moore, R.S.; McSkimin, H.J. Dynamic Shear Properties of Solvents and Polystyrene Solutions from 20 to 300 MHz. In *Physical Acoustics*; Mason, W.P., Thurston, R.N., Eds.; Elsevier: Amsterdam, The Netherlands, 1970; pp. 167–242, ISBN 9780123956668. [CrossRef]
367. Carmichael, L.T.; Sage, B.H. Viscosity of Liquid Ammonia at High Pressures. *Ind. Eng. Chem.* **1952**, *44*, 2728–2732. [CrossRef]
368. Dobson, D.P.; Jones, A.P.; Rabe, R.; Sekine, T.; Kurita, K.; Taniguchi, T.; Kondo, T.; Kato, T.; Shimomura, O.; Urakawa, S. In-situ measurement of viscosity and density of carbonate melts at high pressure. *Earth Planet. Sci. Lett.* **1996**, *143*, 207–215. [CrossRef]
369. Suzuki, A.; Ohtani, E.; Funakoshi, K.; Terasaki, H.; Kubo, T. Viscosity of albite melt at high pressure and high temperature. *Phys. Chem. Miner.* **2002**, *29*, 159–165. [CrossRef]
370. Funakoshi, K.; Nozawa, A. Development of a method for measuring the density of liquid sulfur at high pressures using the falling-sphere technique. *Rev. Sci. Instrum.* **2012**, *83*, 103908. [CrossRef]
371. Steigerwald, T.G.; Alt, N.S.; Hertweck, B.; Schluecker, E. Feasibility of density and viscosity measurements under ammonothermal conditions. *J. Cryst. Growth* **2014**, *403*, 59–65. [CrossRef]
372. Schlücker, E.; Hertweck, B.; Schimmel, S.; Wellmann, P. Special Equipment for Ammonothermal Processes. In *Ammonothermal Synthesis and Crystal Growth of Nitrides*; Meissner, E., Niewa, R., Eds.; Springer International Publishing: Cham, Switzerland, 2021; pp. 317–328, ISBN 978-3-030-56304-2.
373. Richter, T.; Strobel, S.; Alt, N.; Schlücker, E.; Niewa, R. Ammonothermal Synthesis and Crystal Structures of Diamminetriamidodizinc Chloride $[Zn_2(NH_3)_2(NH_2)_3]Cl$ and Diamminemonoamidozinc Bromide $[Zn(NH_3)_2(NH_2)]Br$. *Inorganics* **2016**, *4*, 41. [CrossRef]
374. Hashimoto, T.; Saito, M.; Fujito, K.; Wu, F.; Speck, J.S.; Nakamura, S. Seeded growth of GaN by the basic ammonothermal method. *J. Cryst. Growth* **2007**, *305*, 311–316. [CrossRef]
375. Ehrentraut, D.; Kagamitani, Y.; Yokoyama, C.; Fukuda, T. Physico-chemical features of the acid ammonothermal growth of GaN. *J. Cryst. Growth* **2008**, *310*, 891–895. [CrossRef]
376. Jouet, R.J.; Purdy, A.P.; Wells, R.L.; Janik, J.F. Preparation of Phase Pure Cubic Gallium Nitride, c-GaN, by Ammonothermal Conversion of Gallium Imide, {Ga(NH)3/2} n. *J. Clust. Sci.* **2002**, *13*, 469–486. [CrossRef]
377. Bao, Q.; Sawayama, H.; Hashimoto, T.; Sato, F.; Hazu, K.; Kagamitani, Y.; Ishinabe, T.; Saito, M.; Kayano, R.; Tomida, D.; et al. Powder synthesis and ammonothermal crystal growth of GaN from metallic Ga in the presence of NH4I. *CrystEngComm* **2012**, *14*, 3351. [CrossRef]
378. Harbrecht, B.; Jacobs, H. Hochdrucksynthese von Caesiumamidazid, $Cs_2(NH_2)N_3$ aus Caesiummetall und Ammoniak. *Z. Anorg. Allg. Chem.* **1983**, *500*, 181–187. [CrossRef]
379. Hertrampf, J.; Alt, N.S.A.; Schlücker, E.; Niewa, R. Three Solid Modifications of $Ba[Ga(NH_2)_4]_2$: A Soluble Intermediate in Ammonothermal GaN Crystal Growth. *Eur. J. Inorg. Chem.* **2017**, *2017*, 902–909. [CrossRef]
380. Hashimoto, T.; Wu, F.; Saito, M.; Fujito, K.; Speck, J.S.; Nakamura, S. Status and perspectives of the ammonothermal growth of GaN substrates. *J. Cryst. Growth* **2008**, *310*, 876–880. [CrossRef]
381. Ketchum, D.R.; Kolis, J.W. Crystal growth of gallium nitride in supercritical ammonia. *J. Cryst. Growth* **2001**, *222*, 431–434. [CrossRef]
382. Schwieger, W.; Baser, H. Determination of Solubility of GaN in Ammonobasic Systems. In *Ammonothermal Synthesis and Crystal Growth of Nitrides*; Meissner, E., Niewa, R., Eds.; Springer International Publishing: Cham, Switzerland, 2021; pp. 155–170, ISBN 978-3-030-56304-2.
383. Bäucker, C.; Niewa, R. A New Modification of Rb[Al(NH2)4] and Condensation in Solid State. *Crystals* **2020**, *10*, 1018. [CrossRef]
384. Jacobs, H.; Stüve, C. Hochdrucksynthese der η-phase im system Mn-N:Mn_3N_2. *J. Less Common Met.* **1984**, *96*, 323–329. [CrossRef]
385. Peters, D.; Jacobs, H. Ammonothermalsynthese von kristallinem siliciumnitridimid, Si_2N_2NH. *J. Less Common Met.* **1989**, *146*, 241–249. [CrossRef]
386. Pust, P.; Schmiechen, S.; Hintze, F.; Schnick, W. Ammonothermal Synthesis and Crystal Structure of $BaAl_2(NH_2)_8{\cdot}2NH_3$. *Z. Anorg. Allg. Chem.* **2013**, *639*, 1185–1187. [CrossRef]
387. Hadenfeldt, C.; Jacobs, H. Darstellung, Eigenschaften und Kristallstruktur von $Na_3[Yb(NH_2)_6]$. *Z. Anorg. Allg. Chem.* **1972**, *393*, 111–125. [CrossRef]
388. Jacobs, H.; Hadenfeldt, C. Die Kristallstruktur von Bariumamid, $Ba(NH_2)_2$. *Z. Anorg. Allg. Chem.* **1975**, *418*, 132–140. [CrossRef]

389. Senker, J.; Jacobs, H.; Müller, M.; Press, W.; Müller, P.; Mayer, H.M.; Ibberson, R.M. Reorientational Dynamics of Amide Ions in Isotypic Phases of Strontium and Calcium Amide. 1. Neutron Diffraction Experiments. *J. Phys. Chem. B* **1998**, *102*, 931–940. [CrossRef]
390. Jacobs, H.; Kablitz, D. Untersuchung des Systems Kalium/Cer/Ammoniak. *Z. Anorg. Allg. Chem.* **1979**, *454*, 35–42. [CrossRef]
391. Jacobs, H.; Schmidt, D. Struktur und eigenschaften von perowskitartigen cäsiumamidometallaten des cers, neodyms und samariums $Cs_3Ln_2(NH_2)_9$. *J. Less Common Met.* **1980**, *76*, 227–244. [CrossRef]
392. Jacobs, H.; SCHMIDT, D.; Schmitz, D.; Fleischhauer, J.; Schleker, W. Struktur und eigenschaften der cäsiumamidolanthanat-monoammoniakate $Cs_3La(NH_2)_6{\cdot}NH_3$ und $Cs_4La(NH_2)_7{\cdot}NH_3$. *J. Less Common Met.* **1981**, *81*, 121–133. [CrossRef]
393. Jacobs, H.; Zachwieja, U. Kupferpalladiumnitride, Cu3PdxN mit x = 0,020 und 0,989, Perowskite mit "bindender 3d10-4d10-Wechselwirkung". *J. Less Common Met.* **1991**, *170*, 185–190. [CrossRef]
394. Jacobs, H.; Bock, J.; Stüve, C. Röntgenographische strukturbestimmung und IR-spektroskopische Untersuchungen an Hexaammindiiodiden, [M(NH3)6]I2, von Eisen und Mangan. *J. Less Common Met.* **1987**, *134*, 207–214. [CrossRef]
395. Jacobs, H.; Bock, J. Einkristallzüchtung von γ'-Fe4N in überkritischem ammoniak. *J. Less Common Met.* **1987**, *134*, 215–220. [CrossRef]
396. Jacobs, H.; Nöcker, B. Neubestimmung von Struktur und Eigenschaften isotyper Natriumtetraamidometallate des Aluminiums und Galliums. *Z. Anorg. Allg. Chem.* **1993**, *619*, 381–386. [CrossRef]
397. Becker, P.; Cekovski, T.B.; Niewa, R. Indium Ammoniates from Ammonothermal Synthesis: $InAlF_6(NH_3)_2$, $[In(NH_3)_6][AlF_6]$, and $[In_2F(NH_3)_{10}]2[SiF_6]_5{\cdot}2NH_3$. *Crystals* **2021**, *11*, 679. [CrossRef]
398. Hadenfeldt, C.; Gieger, B.; Jacobs, H. Darstellung und Kristallstruktur von $KLa_2(NH_2)_7$. *Z. Anorg. Allg. Chem.* **1974**, *408*, 27–36. [CrossRef]
399. Jacobs, V.H.; Scholze, H. Untersuchung des Systems $Na/La/NH_3$. *Z. Anorg. Allg. Chem.* **1976**, *427*, 8–16. [CrossRef]
400. Jacobs, H.; Stüve, C. Rubidiumhexaamidolanthanat und -neodymat, $Rb_3[La(NH_2)_6]$ und $Rb_3[Nd(NH_2)_6]$; Strukturverwandtschaft zu $K_3[Cr(OH_6)]$ und K_4CdCl_6. *Z. Anorg. Allg. Chem.* **1987**, *546*, 42–47. [CrossRef]
401. Jacobs, H.; Schmidt, D. Über ein cäsiumheptaamidodilanthanat $CsLa_2(NH_2)_7$. *J. Less Common Met.* **1981**, *78*, 51–59. [CrossRef]
402. Kraus, F.; Korber, N. $K_2Li(NH_2)_3$ and $K_2Na(NH_2)_3$—Synthesis and crystal structure of two crystal-chemically isotypic mixed-cationic amides. *J. Solid State Chem.* **2005**, *178*, 1241–1246. [CrossRef]
403. Jakiela, R.; Sierakowski, K.; Sochacki, T.; Iwinska, M.; Fijalkowski, M.; Barcz, A.; Bockowski, M. Investigation of diffusion mechanism of beryllium in GaN. *Phys. B Condens. Matter* **2020**, *594*, 412316. [CrossRef]
404. Schipper, F.; Kubo, S.; Fellinger, T.-P. Nitrogen-doped porous carbon via ammonothermal carbonization for supercapacitors. *J. Sol-Gel Sci. Technol.* **2019**, *89*, 101–110. [CrossRef]
405. Blunck, H.; Juza, R. über Verbindungen des Thoriums mit Stickstoff und Wasserstoff. *Z. Anorg. Allg. Chem.* **1974**, *410*, 9–20. [CrossRef]
406. Jacobs, H.; Peters, D. Übergang von dichter anionenpackung zu perowskitartiger struktur bei kalium- und rubidiumamidoyttriat, $KY(NH_2)_4$ und $RbY(NH_2)_4$. *J. Less Common Met.* **1986**, *119*, 99–113. [CrossRef]
407. Blunck, H.; Juza, R. Über Ein Nitridamid des Zirkoniums. *Z. Anorg. Allg. Chem.* **1974**, *406*, 145–152. [CrossRef]

materials

Article

Temperature Field, Flow Field, and Temporal Fluctuations Thereof in Ammonothermal Growth of Bulk GaN—Transition from Dissolution Stage to Growth Stage Conditions

Saskia Schimmel [1,2,3,*], Daisuke Tomida [2,*], Tohru Ishiguro [4], Yoshio Honda [2], Shigefusa F. Chichibu [4] and Hiroshi Amano [2]

[1] Crystal Growth Lab, Materials for Electronics and Energy Technology (i-MEET), Friedrich-Alexander-Universität Erlangen-Nürnberg, 91058 Erlangen, Germany
[2] Institute of Materials and Systems for Sustainability, Nagoya University, Nagoya 464-8601, Japan
[3] Electron Devices (LEB), Friedrich-Alexander-Universität Erlangen-Nürnberg, 91058 Erlangen, Germany
[4] Institute of Multidisciplinary Research for Advanced Materials, Tohoku University, Sendai 980-8577, Japan
* Correspondence: saskia.schimmel@fau.de (S.S.); tomida@imass.nagoya-u.ac.jp (D.T.)

Abstract: With the ammonothermal method, one of the most promising technologies for scalable, cost-effective production of bulk single crystals of the wide bandgap semiconductor GaN is investigated. Specifically, etch-back and growth conditions, as well as the transition from the former to the latter, are studied using a 2D axis symmetrical numerical model. In addition, experimental crystal growth results are analyzed in terms of etch-back and crystal growth rates as a function of vertical seed position. The numerical results of internal process conditions are discussed. Variations along the vertical axis of the autoclave are analyzed using both numerical and experimental data. During the transition from quasi-stable conditions of the dissolution stage (etch-back process) to quasi-stable conditions of the growth stage, significant temperature differences of 20 K to 70 K (depending on vertical position) occur temporarily between the crystals and the surrounding fluid. These lead to maximum rates of seed temperature change of 2.5 K/min to 1.2 K/min depending on vertical position. Based on temperature differences between seeds, fluid, and autoclave wall upon the end of the set temperature inversion process, deposition of GaN is expected to be favored on the bottom seed. The temporarily observed differences between the mean temperature of each crystal and its fluid surrounding diminish about 2 h after reaching constant set temperatures imposed at the outer autoclave wall, whereas approximately quasi-stable conditions are reached about 3 h after reaching constant set temperatures. Short-term fluctuations in temperature are mostly due to fluctuations in velocity magnitude, usually with only minor variations in the flow direction.

Keywords: ammonothermal; gallium nitride; crystal growth; numerical simulation; computational fluid dynamics; natural convection; buoyancy; conjugated heat transfer; solvothermal; hydrothermal

Citation: Schimmel, S.; Tomida, D.; Ishiguro, T.; Honda, Y.; Chichibu, S.F.; Amano, H. Temperature Field, Flow Field, and Temporal Fluctuations Thereof in Ammonothermal Growth of Bulk GaN—Transition from Dissolution Stage to Growth Stage Conditions. *Materials* **2023**, *16*, 2016. https://doi.org/10.3390/ma16052016

Academic Editor: Fabrizio Roccaforte

Received: 20 January 2023
Revised: 23 February 2023
Accepted: 27 February 2023
Published: 28 February 2023

1. Introduction

The low-pressure acidic ammonothermal (LPAAT) method is considered one of the most promising technologies for the scalable production of bulk GaN substrates with high structural quality and at low cost [1–3]. Therefore, it represents a promising route for overcoming the lack of native substrates, which currently limits the development of GaN-based electronic devices [4]. Although blue and white light-emitting diodes realized on foreign substrates have already revolutionized lighting technology [5,6], further applications among others in power electronics [7] demand better structural quality and benefit more from the use of native substrates. The lower defect densities achievable with native substrates enable the realization of lower off-state leakage currents, which in turn improves the energy efficiency in energy conversion applications [8]. However, for widespread application to tap the full potential of GaN-on-GaN technology, wafer prices would need to decrease significantly [8] in relation to the cost of the saved energy, and the availability needs to improve. For both, efficient scale-up of bulk GaN growth technologies such as the ammonothermal method is essential. To provide an avenue for efficient scale-up and high yield in production, the development of numerical models of the growth process is expected to be very important, especially considering the great technical difficulty of experimental access to ammonothermal autoclaves. Purely data-based empirical models or hybrid models may eventually become effective for process improvement and control once the method is used industrially at a large scale and enough experimental data are generated. At the present stage, however, physics-based numerical simulations [9] and in situ monitoring technologies [10–12] are more feasible and are also particularly suitable for advancing fundamental understanding. Besides the application to GaN crystal growth, an improved fundamental understanding of ammonothermal crystal growth conditions would also facilitate the targeted application of the ammonothermal method as a tool for explorative synthesis [13] of various nitride materials and the development of single crystal growth processes for these materials. In other crystal growth processes, data generated via physics-based numerical simulations have also proven highly valuable as input data for machine learning models. In SiC solution growth, for instance, the tremendously accelerated prediction of growth conditions has been demonstrated [14]. In the directional solidification of silicon, the accelerated optimization of growth conditions has been presented [15]. Therefore, once the physics have been clarified and sufficiently accurate models have been developed on that basis, machine learning from numerical simulation data may provide a pathway to realize computationally efficient compact models for ammonothermal crystal growth.

In this study, we investigate the quasi-stationary conditions in ammonothermal growth of GaN in the geometrical configuration used for growth under retrograde solubility conditions, which applies to the use of NH_4F mineralizer [1] as well as ammonobasic growth [16]. Contrary to most prior studies, we do not use heater-long fixed temperatures as a thermal boundary condition. Instead, a temperature distribution based on a simulation, including the growth furnace, is used, which is closer to the experimental conditions according to recent studies [17,18].

A detailed analysis of flow stability and temperature fluctuations at different probing locations is presented. A lack of spatial and temporal uniformity of material transport to the growth front is suspected of causing the formation of defects, and therefore the spatial and temporal uniformity of fluid flow is also thought to affect defect formation [19].

To shine a first light on the conditions during the initial stages of the crystal growth process, the effect of inverted set temperatures is investigated approximately. Such an inverted temperature gradient is applied during temperature ramp-up by experimentalists and is intended for back-etching of the seed crystals [20], hereby likely removing impurities as well as sub-surface damage from cutting and grinding (if present) and modifying the surface morphology due to defect-selective etching, which occurs in both ammonobasic [21] and ammonoacidic [22] solutions. Besides quasi-stable temperature and fluid flow conditions during etch-back and growth, we also analyze the conditions occurring temporarily while changing set temperatures from etch-back to growth conditions. In addition, we analyze how quasi-stable conditions are re-established after the set temperatures of the growth stage have been reached and are kept constant.

2. Materials and Methods

2.1. Numerical Simulation

In our previous study [17], we examined the temperature distribution along the outer wall of an ammonothermal autoclave inside a resistively heated furnace with two separately controlled heaters (the thermocouples for heater power control being located at the outer autoclave wall, one for each heater). Regarding the geometry, materials properties, and the use of the LVEL (Length-VELocity) turbulence model, the models in this study were set up in full analogy to the type C cases in the mentioned previous study. Note that although the occurrence of turbulent flow was not prevented by the selection of the flow model, laminar flow can likewise occur, depending on the location and the rate of thermal boundary condition change. In preliminary numerical experiments with this model, we observed that the flow cell structure between the baffle and uppermost seed does change if the flow is forced to be laminar (also see our previous publications [9] and [17] for further discussion of the question of laminar versus turbulent flow in ammonothermal crystal growth). In addition, we intentionally kept the model very similar to the one in the previous study because otherwise, it would not be reasonable to use the results of the previous study for boundary condition definition. Minor modifications with regard to the type C cases in our previous study [17] and model variants are detailed below. Note that two versions with different seed thicknesses were studied. While we applied a seed thickness of 3 mm (thereof 1.5 mm inside the simulation domain) in our previous simulation [17], a seed thickness of 10 mm (thereof 5 mm inside the simulation domain) was used in the present study unless otherwise stated. The reason for this is that we preferred to first obtain results for the larger seed thickness because these can easily be used as initial conditions for further studies with thinner seeds. This does not work as easily if performed in the opposite order (likely due to issues with non-zero velocity values inside solids if enlarging the seed size rather than shrinking it).

In Table 1, the geometry is described by the positions of the centers of the components and their dimensions.

Table 1. Geometry of the model: Components, radial, and vertical positions referring to the center of each component, and radial width as well as vertical height of the components. For the seeds, the thickness refers to the model version with the thicker seeds.

Component	Position r/mm	Δr/mm	Position z/mm	Δz/mm
Autoclave	0.0	35.0	185.0	370.0
Nutrient	0.0	10.0	250.0	150.0
Baffle	8.3	3.4	170.0	1.0
Top seed	0.0	5.0	110.0	20.0
Middle seed	0.0	5.0	80.0	20.0
Bottom seed	0.0	5.0	50.0	20.0
Inner volume of autoclave	0.0	12.0	185.0	300.0

The geometry, alongside information on probing locations for the evaluation of temporal fluctuations, is depicted in Figure 1.

To complement the information in Figure 1, the domain probe locations are listed and described in Table 2.

A list of the components labeled with numbers in Figure 1, together with a description of the materials they consist of, can be found in Table 3. Note that, in the case of GaN, bulk GaN properties were assumed for the seeds, whereas the GaN nutrient was represented as a porous medium with a porosity of 0.7 and a permeability of 0.6352 mm^2.

Table 2. Detailed description of domain probes, including the labels used, radial and vertical position, and description of locations.

Label	Position		Description of Location: Radial Position, Vertical Position (Further Description)
	r/mm	z/mm	
DP0	0.00	369.00	Center, 34 mm from top inner wall (near upper outer autoclave wall)
DP1	0.00	330.00	Center, above nutrient halfway between inner upper autoclave wall and top of nutrient

Table 2. *Cont.*

Label	Position		Description of Location: Radial Position, Vertical Position (Further Description)
	r/mm	z/mm	
DP2	0.00	250.00	Center of the nutrient
DP3	0.00	170.00	Center, at half baffle height
DP4	0.00	145.00	Center, halfway between bottom edge of the baffle and upper edge of the top seed
DP5	0.00	110.00	Center of the top seed
DP6	0.00	95.00	Center, in the middle between top and middle seed
DP7	0.00	80.00	Center of the middle seed
DP8	0.00	65.00	Center, in the middle between middle and bottom seed
DP9	0.00	50.00	Center of the bottom seed
DP10	0.00	35.00	Center, between bottom seed and bottom inner wall
DP11	0.00	1.00	Center, 34 mm from bottom inner wall (near outer autoclave wall)
DP12	11.00	250.00	Middle of gap between nutrient and inner wall, half nutrient height
DP13	11.00	170.00	Middle of gap between baffle and inner autoclave wall, half baffle height
DP14	6.75	110.00	Halfway between edge of top seed and inner autoclave wall, center of top seed
DP15	6.75	80.00	Halfway between edge of middle seed and inner autoclave wall, center of middle seed
DP16	6.75	50.00	Halfway between edge of bottom seed and inner autoclave wall, center of bottom seed
DP17	13.00	252.50	1 mm from inner autoclave wall, center of upper zone (near inner autoclave wall)
DP18	13.00	102.50	1 mm from inner autoclave wall, center of lower zone (near inner autoclave wall)
DP19	34.00	280.00	1 mm from outer autoclave wall, control thermocouple of upper zone
DP20	34.00	252.50	1 mm from outer autoclave wall, center of upper zone
DP21	34.00	102.50	1 mm from outer autoclave wall, center of lower zone
DP22	34.00	70.00	1 mm from outer autoclave wall, control thermocouple of lower zone
DP23	0.00	22.00	Center, 1 mm from inner autoclave wall (near bottom inner wall)
DP24	0.00	336.00	Center, 1 mm from inner autoclave wall (near top inner wall)

Table 3. Components with the numerical label used in Figure 1 and description of the respective materials.

Component (Number)	Description
NH_3 (1)	At 426.6 °C, 100 MPa, treated as incompressible
Autoclave wall (2)	Ni-Cr superalloy
Baffle (3)	Corrosion-resistant metal
Bulk GaN (4)	Wurtzite GaN

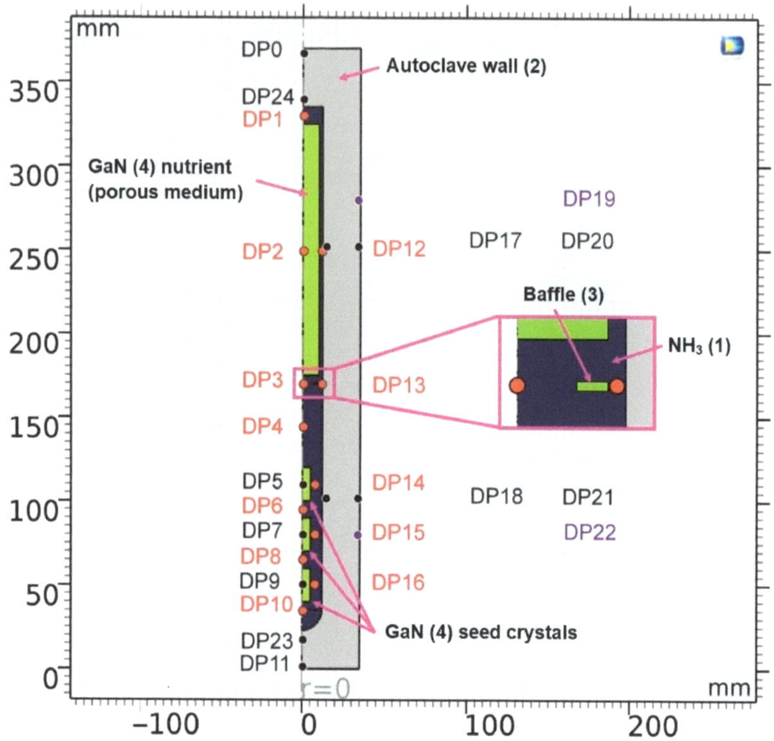

Figure 1. Geometry of the model with location of domain point probes (see Table 2 for coordinates thereof and for a description of the chosen locations). The color coding of the labels is as follows: probes in the fluid (red), probes in solids (black and purple), and probes at the location of control thermocouples in an experimental setup (purple). The components are also labeled, including a number that refers to the material (see Tables 3–5 for further information on materials).

Note that ammonia is in its supercritical state during all studied stages of a growth run. The amount of ammonia inside the autoclave does not undergo any externally controlled changes during a growth run. While chemical reactions, as well as leakages, can cause changes in ammonia density during a growth run in experimental work, neither of these possible effects was implemented in the numerical model. For these reasons, the same mean density of the fluid was used regardless of the temperatures applied (that is, neglecting possible effects of solutes on the thermophysical properties of the fluid, thus also neglecting solutal convection). The temperature-induced variations of ammonia density were, therefore, the only driving force for convection and implemented using Boussinesq approximation (Equation (1)), with a reference temperature T_{ref} = 293.15 K and a reference density ρ_{ref} = 233.950 kg/m^3. Therein, α_v is the volumetric thermal expansion coefficient.

$$\rho - \rho_{ref} = -\rho_{ref}\alpha_v\left(T - T_{ref}\right) \tag{1}$$

The density of 233.950 kg/m^3 corresponds to the density of supercritical ammonia at 100 MPa at a temperature of 554.0 °C (mean value of fluid temperatures inside the autoclave at growth stage conditions). These conditions mimic those typically used in the experimental LPAAT method. Temperature-dependent data of supercritical ammonia taken from the NIST database [23] were used. Beyond 426.9 °C, data for thermal conductivity, heat capacity at constant pressure, density, and the ratio of specific heats were extrapolated linearly, whereas a constant extrapolation was used for the dynamic viscosity. The materials properties of supercritical ammonia at 426.9 °C are listed in Table 4.

Table 4. Fluid properties of supercritical ammonia at 426.9 °C as obtained from NIST database [23].

Material	ρ/kg/m^3	ν/m^2/s	c$_p$/J/(kg·K)	k/W/(m·K)	α/1/K
NH_3 (1)	233.950	1.570×10^{-7}	4216.5	0.16224	2.631×10^{-3}

Table 5. Materials properties of the solid components. For the autoclave walls, the in-build materials property definitions for the nickel base superalloy Haynes R-41 of the Comsol Multiphysics materials library were used.

Material	ρ/kg/m^3	c$_p$/J/(kg·K)	k/W/(m·K)
Autoclave wall (2)	8301.112+0.1012931*T^1-0.001639246*T^2+8.064631E-7*T^3+1.085603E-8*T^4-1.915217E-11*T^5 (from 33 to 366 K) 8335.344-0.3077835*T^1+6.941174E-5*T^2-9.845641E-8*T^3 (from 366.0 to 1366.0 K)	474.575639-0.12868256*T^1+1.2284019E-4*T^2+2.01913857E-7*T^3-4.98757486E-11*T^4 (from 293.0 to 1483.0 K)	3.843196+0.01841176*T^1 (from 293.0 to 1173.0 K)
Baffle (3)	10200.00	250.00	138.00
Bulk GaN (4)	6150.00	518.41	100.13

For all solid components, the properties of the materials are listed in Table 5.

The remaining governing equations are given in the following, with the subscript w indicating quantities at interfaces to solids (walls).

The conservation of mass is expressed by the continuity equation (Equation(2)):

$$\rho \nabla \cdot \vec{u} = 0 \tag{2}$$

Equation (3) describes the conservation of momentum for the flow of an incompressible fluid. The term $\left(\rho - \rho_{ref}\right)\vec{g}$ therein represents the buoyancy force per unit volume, and the pressure gradient is equal to $\nabla \cdot \left(-p\vec{I}\right)$. The vector $\vec{u}$ contains the velocity components u and w, t is the time, p the pressure, $\vec{I}$ the identity matrix, μ the dynamic viscosity of the fluid, μ_T the turbulent viscosity.

$$\begin{gathered}\rho\frac{\partial \vec{u}}{\partial t} + \rho\left(\vec{u}\cdot\nabla\right)\vec{u} = \nabla\cdot\left[-p\vec{I} + \vec{K}\right] + \left(\rho - \rho_{ref}\right)\vec{g} \\ \text{with } \vec{K} = (\mu + \mu_T)\left(\nabla\vec{u} + \left(\nabla\vec{u}\right)^T\right) \\ \text{with } \mu_T = \mu\left(\frac{dl_w^+}{du^+} - 1\right)\end{gathered} \tag{3}$$

The dimensionless distance from the wall in the LVEL model l_w^+ is defined by Spalding's law of the wall [24] (Equation (4)):

$$l_w^+ = u^+ + \frac{1}{E}\left(e^{\kappa u^+} - 1 - \kappa u^+ - \frac{\left(\kappa u^+\right)^2}{2} - \frac{\left(\kappa u^+\right)^3}{6} - \frac{\left(\kappa u^+\right)^4}{24}\right) \tag{4}$$

Equation (5) describes the local Reynolds number at the interfaces to solids, with the dimensionless quantity u^+ being the local flow speed divided by the friction velocity and l_w^+ being the dimensionless distance from the wall.

$$Re_w = \frac{\rho\left|\vec{u}\right| l_w}{\mu} = \frac{\left|\vec{u}\right|}{u_\tau} \cdot \frac{\rho u_\tau l_w}{\mu} = u^+ l_w^+ \tag{5}$$

The wall distance l_w is given by Equation (6):

$$l_w = \frac{1}{G} - \frac{l_{ref}}{2} \tag{6}$$

A modified version of the Eikonal equation (in which the dependent variable is $G = 1/l_w$ instead of the exact distance to the closest wall l_w) is used to describe the reciprocal wall distance G (Equation (7)), with the smoothing parameter σ_w being equal to 0.2:

$$\nabla G \cdot \nabla G + \sigma_w G(\nabla \cdot \nabla G) = (1 + 2\sigma_w)G^4 \tag{7}$$

The dimensionless effective viscosity ν^+ can be obtained from Equation (8):

$$\nu^+ = 1 + \frac{\kappa}{E}\left(e^{\kappa u^+} - 1 - \kappa u^+ - \frac{(\kappa u^+)^2}{2} - \frac{(\kappa u^+)^3}{6}\right) \tag{8}$$

No slip wall conditions are used (Equation (9)):

$$\vec{u}\Big|_{l_w=0} = \vec{0} \tag{9}$$

The flow in the porous medium is described as follows by the Brinkman equations (Equations (10) and (11)), with Q_m being the mass source term:

$$\rho \nabla \cdot \vec{u} = Q_m = \vec{0} \tag{10}$$

With the porosity ϵ_p and the local permeability κ:

$$\frac{1}{\epsilon_p}\rho\frac{\partial \vec{u}}{\partial t} + \frac{1}{\epsilon_p}\rho\left(\vec{u}\cdot\nabla\right)\vec{u}\frac{1}{\epsilon_p} = \nabla\cdot\left[-p\vec{I} + \vec{K}\right] - \left(\mu\kappa^{-1} + \beta\rho\left|\vec{u}\right| + \frac{Q_m}{\epsilon_p^2}\right)\vec{u} + \left(\rho - \rho_{ref}\right)\vec{g} \tag{11}$$

With $K = \mu\frac{1}{\epsilon_p}\left(\nabla\vec{u} + \left(\nabla\vec{u}\right)^T\right) - \frac{2}{3}\mu\frac{1}{\epsilon_p}\left(\nabla\cdot\vec{u}\right)\vec{I}$

The Forchheimer coefficient β is calculated from the Forchheimer parameter c_F and the local permeability κ (Equation (12)):

$$\beta = \frac{c_F}{\sqrt{\kappa}} \tag{12}$$

The energy equation in the temperature form (Equation (13)) is as follows. Therein, Q represents the heat supplied to the system and the vector $\vec{q}$ describes the conductive heat flux, C_p is the specific heat capacity and k is the thermal conductivity.

$$\rho C_p \frac{\partial T}{\partial t} + \rho C_p \vec{u} \cdot \nabla T + \nabla \cdot \vec{q} = Q \tag{13}$$

The conductive heat flux vector therein is defined by Equation (14), Fourier's law of conduction:

$$\vec{q} = -k\nabla T = 0 \tag{14}$$

Two model versions with thermal boundary conditions constant over time were used to investigate the quasi-stable conditions during the etch-back and growth stages of a crystal growth experiment. Model 550_650 uses the temperature distributions at the outer autoclave wall as obtained in model A3 in the previous study [17] (model A3 included the furnace as well as internal solids). It represents an ammonothermal setup with an uninsulated autoclave head, with the top heater set to 550 °C and the bottom heater set to 650 °C.

To obtain a first insight into the conditions in the case of inverted thermal gradient in a computationally efficient way, the following assumptions were used to obtain approximate but reasonably realistic thermal boundary conditions for this case, which we term Model 550_450. We assume that since the set temperature of the top heater, as well as the uninsulated condition of the autoclave head, is identical to model A3 in our previous study [17], autoclave wall temperatures of the upper part of the autoclave down to the position of the control thermocouple of the top heater can be approximated as the outer wall temperature distribution extracted from the results of model A3. This is likely not perfectly accurate, as a measurable influence of convective heat transfer on the wall temperatures of the autoclave head has been observed in an experimental study of the first author [10]. However, this influence is likely less pronounced in the configuration studied here, as the presence of the porous medium is expected to lead to lower flow velocities. For the vertical temperature distribution from the location of the control thermocouple of the top heater to the bottom of the setup, we assume that a near-linear temperature profile will develop in analogy to the one obtained in case A3 in our previous study [17], albeit with different absolute temperatures to account for the different set temperature of the bottom heater (450 °C). The temperature distributions imposed along the vertical autoclave wall are plotted in Figure 2 for Model 550_650 as well as Model 550_450. The temperature distribution along the bottom autoclave wall is assumed to show the same characteristics as in A3 but was shifted so that the vertical and horizontal autoclave wall temperature distributions share the same temperature at the bottom outer edge of the autoclave wall.

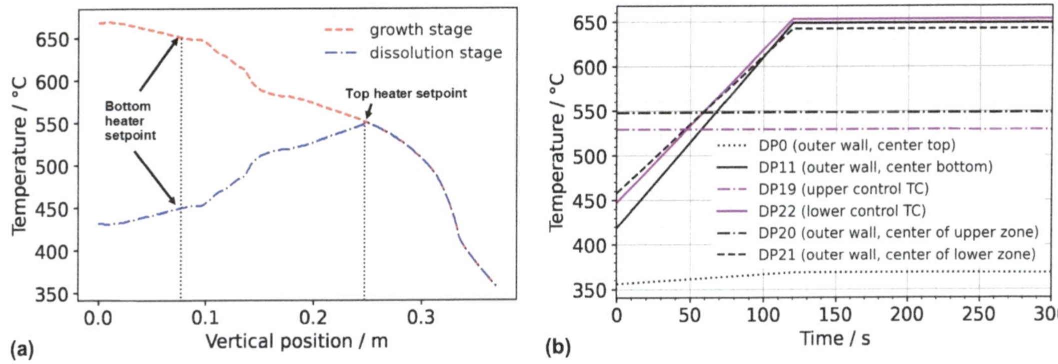

Figure 2. (**a**) Thermal boundary conditions imposed along the vertical outer wall of the autoclave for Model 550_650 (growth stage) and Model 550_450 (dissolution stage of the growth experiment). (**b**) Temperatures at domain probes located at the outer wall over time. The abbreviation TC stands for thermocouple, referring to the position thereof in the experimental setup.

The boundary conditions were based on the temperature profiles extracted from the previous study and defined as a function of coordinates, the time t, and the duration of the set temperature inversion process Δt as follows. For the upper outer autoclave wall (z = 370 mm), the thermal boundary conditions were defined by the function in Equation (15):

$$T_W(r, t, z = 370\ \text{mm}) = T_w\_GS(r, z = 370\ \text{mm}) - \frac{T_w\_GS(r,\ z = 370\ \text{mm}) - T_w\_DS(r,\ z = 370\ \text{mm})}{\Delta \text{t}} |t - \Delta \text{t}| \quad (15)$$

In the same way, the function describing the thermal boundary condition at the bottom outer autoclave wall was defined by Equation (16):

$$T_W(r,t,\ z=0\ \text{mm}) = T_w\_GS(r,\ z=0\ \text{mm}) - \frac{T_w\_GS(r,\ z=0\ \text{mm}) - T_w\_DS(r,\ z=0\ \text{mm})}{\Delta t}|t-\Delta t| \tag{16}$$

The function describing the time-dependent thermal boundary conditions for the vertical outer wall is defined in the same way (Equation (17)):

$$T_W(z,t,\ r=35\ \text{mm}) = T_w\_GS(z,\ r=35\ \text{mm}) - \frac{T_w\_GS(z,\ r=35\ \text{mm}) - T_w\_DS(z,\ r=35\ \text{mm})}{\Delta t}|t-\Delta t| \tag{17}$$

The resulting thermal boundary conditions as a function of the coordinates and time are visualized by the graphs in Figure 3, where subfigures (a), (b), and (c) represent the thermal boundary conditions at the upper outside wall, the vertical outside wall, and the lower outside wall, respectively.

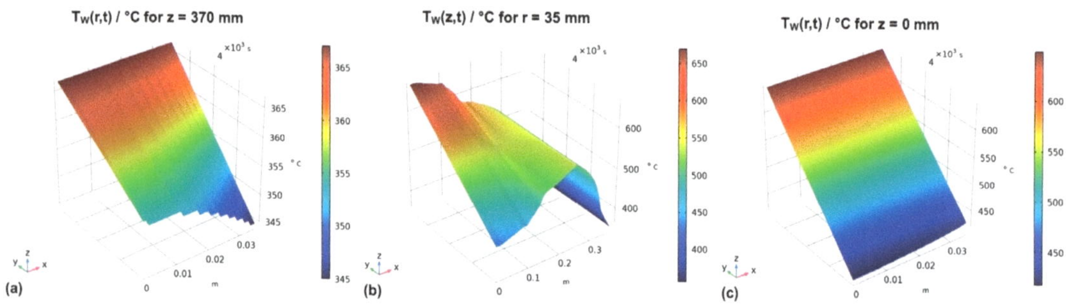

Figure 3. Thermal boundary conditions over time for (**a**) the horizontal upper outer autoclave wall, (**b**) the vertical outer autoclave wall, and (**c**) the horizontal bottom outer autoclave wall.

Calculations were done using the commercial software Comsol Multiphysics. A combination of structured and unstructured mesh regions was used, as depicted in Figure 4.

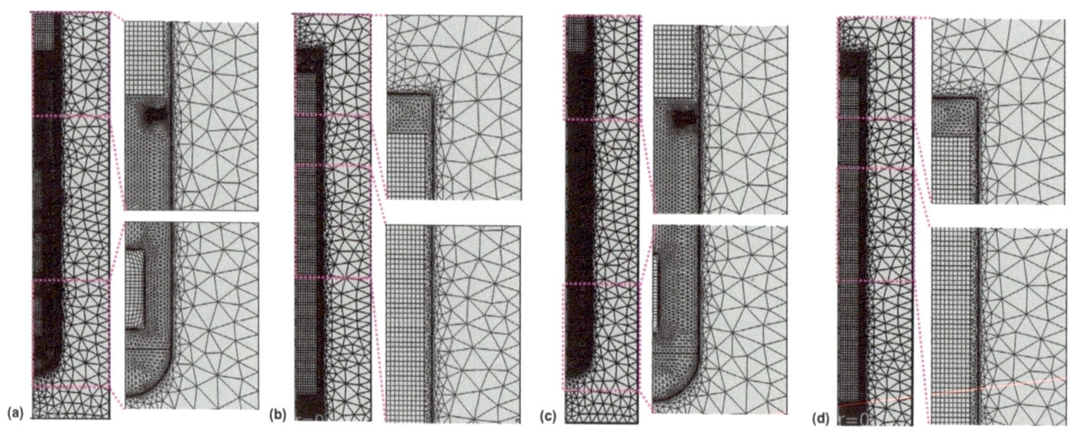

Figure 4. Mesh in (**a**,**c**) the lower and (**b**,**d**) the upper part of the autoclave for the model versions with thick seeds (**a**,**b**) and thin seeds (**c**,**d**).

The parameters used to create the mesh are given in Table 6. Four boundary layers with a stretching factor of 1.2 were employed at the interface of the fluid with the autoclave walls, baffle, and seeds, respectively. The thickness of the first boundary layer was set to 0.02 mm. The relative tolerance was set to 0.002 for Model 550_650, whereas a relative tolerance of 0.050 was applied for the inverted temperature Model 550_450 to facilitate convergence. The convergence difficulties observed when initially using the same relative tolerance are thought to be due to the less accurate estimation of thermal boundary conditions in the case of the inverted temperature version of the Model 550_450.

Table 6. Parameters applied for mesh generation.

	Maximum Element Size/mm	Minimum Element Size/mm	Maximum Element Growth Rate	Curvature Factor	Resolution of Narrow Regions
Autoclave walls	122.00	18.5000	2.00	1.0	0.9
Free flow	1.00	0.1000	1.15	0.6	0.8
Baffle	0.25	0.0074	1.10	0.2	1.0
Nutrient	1.23	0.0350	1.13	0.3	1.0
Seeds	0.80	0.0074	1.13	0.2	1.0

Time steps taken by the solver were constrained to a maximum of 0.005 s to ensure sufficient temporal resolution to resolve fluid flow fluctuations and short-lived small eddies on small time scales. As for the time-stepping method, the implicit backward differentiation formula (BDF) was applied. The model was solved using a fully coupled approach, employing a direct solver (PARallel DIrect Solver, PARDISO for short). Newton's method was used as the nonlinear method (specifically, Constant (Newton)), with the Jacobian update on every iteration. To avoid memory-related issues as well as the creation of excessively large recovery files, the computation was split into sections of 1000 s of real time. To ensure that any disturbances upon restart would not be part of the evaluated data, the last 900 s of each section were used for evaluation.

To identify whether a quasi-stable state had been reached, the temperatures, as well as the velocity components (where applicable), were plotted over time for all domain point probe locations specified in Figure 1. It should be noted that while it is computationally feasible to analyze most of the temporal changes in temperature and velocity fields using our current model and resources, the following limitations exist and are expected to lead to minor inaccuracies. Firstly, as a physical effect, the rates of temperature and velocity changes (time-averaged, so far as domain probes are located inside the fluid and therefore experience short-term fluctuations) decrease while the system approaches a quasi-stable state. This is expected because the thermal gradients between adjacent system components decrease, especially those between crystals and surrounding fluid. Even after the average temperature of the surrounding fluid has been established inside each seed crystal, slow temperature changes are still observed. This includes sluggish, small changes of fluid temperatures at various locations within the fluid, which also lead to a slow change of temperatures inside the seed crystals. Since the rates of temperature change decrease as quasi-stable conditions are approached, the computational cost in relation to the improvement in accuracy increases. To balance computational cost and accuracy, quasi-stable conditions were determined approximately. The term quasi-stable conditions in this study refer to a state at which the rates of temperature changes have decreased below 0.15 K/min at all domain probe locations. Note that the rates of temperature change at conditions termed quasi-stable were virtually constant at most domain probe locations and below 0.03 K/min at center and gap locations at the vertical position of baffle and center of the nutrient (i.e., DP2, DP12, DP3, DP13). The remaining temperature differences between the centers of the seeds and the domain probe inside the fluid at the same vertical height were 2.2 K, 0.9 K, and 3.0 K for the top, middle, and bottom seeds, respectively. The loss of accuracy due to using an approximately quasi-stable solution is estimated to be in the order of a few Kelvin.

For visualization using animations, arrows with logarithmic scaling were used for both models (arrow placement: Gauss points with the maximum number of 5000, arrow base: tail). For Model 550_450, a scale factor of 300 was used for the arrows visualizing fluid flow. The scale factor was adapted to 100 for Model 550_650 to maintain distinguishable arrows despite the higher flow velocities. In both cases, one frame was exported every 0.02 s, and a frame rate of 50 frames per second was used. Thus, the videos are a real-time representation of the results. The Videos S1–S5 represent a 30 s timespan at quasi-stable dissolution stage conditions for the uppermost region of the autoclave, the middle region of the autoclave (between the baffle and top seed), the region around the top seed, the region around the middle seed, and the region around the bottom seed, respectively. The Videos S6–S10 represent a 30 s timespan at quasi-stable growth stage conditions for the uppermost region of the autoclave, the middle region of the autoclave (between the baffle and top seed), the region around the top seed, the region around the middle seed, and the region around the bottom seed, respectively.

For studying the transition from dissolution to the growth stage, slight modifications were made to improve computational efficiency while maintaining stable convergence (constant time step size taken by the solver). Specifically, the relative tolerance was increased to 0.3, and the maximum time step was increased to 0.01 s. The growth zone set temperature change imposed was 200 K over a time period of 2 h, i.e., 100 K/h.

2.2. Experimental Crystal Growth under Similar Conditions

To provide at least some comparison to experimental data, results of an ammonoacidic crystal growth experiment using an NH_4F mineralizer are reported. The mineralizer concentration was 2.00% (amount of substance: 0.02714 mol) and the crystal growth time (at growth stage set temperature) was 24 h. The initial pressure upon reaching growth stage temperatures was 98.5 MPa and the final pressure before cool-down was 100.3 MPa. The initial amount of polycrystalline nutrients was 35.59 g. GaN seed crystals grown by hydride vapor phase epitaxy (HVPE) were purchased from Mitsubishi Chemical Corporation. Each seed was 10 mm wide and 20 mm long (for thicknesses, see the results section). The vertical positions of seed centers were 15 mm, 35 mm, and 58 mm above the inner bottom wall. The baffle position was 100 mm above the bottom inner wall. The outer diameter of the ring-shaped baffle was 20 mm, and the diameter of the center hole was 4.0 mm. The dimensions of the autoclave were the same as in the numerical model. During etch-back, set temperatures were 500 °C (top heater) and 450 °C (bottom heater), respectively. Etch-back set temperatures were kept constant for 5 h. Within this time, a saturation of the solution can be expected so far as not prevented by transport and redeposition following solubility gradients: In situ monitoring experiments conducted with near-isothermal temperature field showed a saturation of the solution within about 2 h while passing through a temperature range from about 220 °C to 500 °C [21]. Rates of set temperature increase were 100 K/h as in the numerical model. During the growth stage, set temperatures were 575 °C (top heater) and 625 °C (bottom heater), respectively. The thickness of the seed crystal before the experiment was measured with a micrometer. The seed crystal and growth layer thicknesses after the experiments were determined from cross-sectional UV fluorescent micrographs. The post-etch-back mass was estimated based on the remaining seed thickness using the density of GaN (6.15 g/cm^3) and the initial seed area (200 mm^2).

3. Results and Discussion

To confirm that quasi-stable conditions were reached, plots of temperature and velocity probes at various locations are displayed in Figure A1. For both quasi-stable conditions, the most pronounced fluctuations in flow velocities and temperatures are observed for domain probe DP1 located in-between the top inner wall and the upper end of the nutrient. The general characteristics regarding both temperature and fluid flow fields were confirmed to be similar for both model versions differing in the thickness of seeds (compare the left subfigures of Figures 5 and 6 to Figure A2). This suggests that the internal temperature and fluid flow characteristics are relatively insensitive to moderate changes in crystal dimensions, such as those during etch-back and early crystal growth.

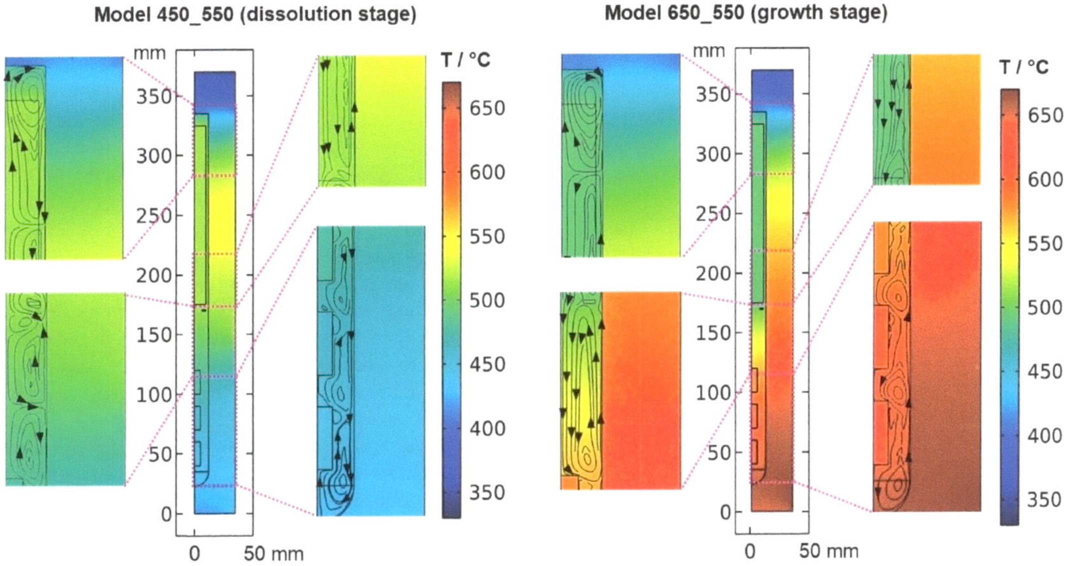

Figure 5. Thermal field in the quasi-stable state during dissolution and growth stage, together with the respective velocity fields visualized by streamlines.

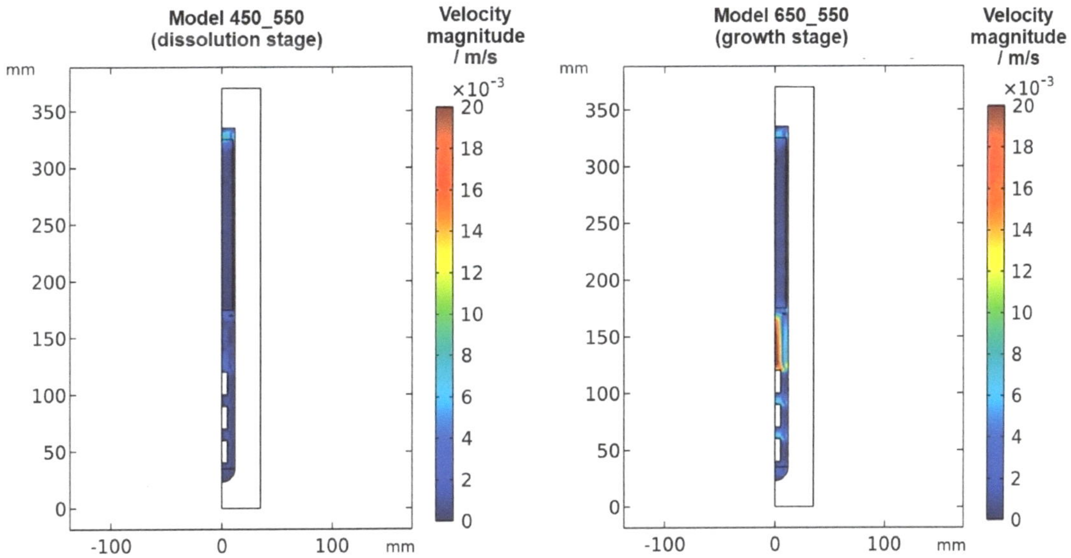

Figure 6. Velocity magnitude at dissolution stage (**left**) and growth stage (**right**) in a quasi-stable state.

The resulting thermal fields and flow fields for both quasi-stable cases are shown in Figures 5 and 6, respectively. As expected, the inversion of the direction of the thermal gradient along the vertical autoclave wall carries over into the interior, though with noticeable temperature differences between the fluid in the growth zone and the seeds compared to the autoclave wall. Likewise, at the growth stage, a noticeable temperature difference occurs between the nutrient and the autoclave wall for most locations within the nutrient. These effects are thought to be caused by the pronounced natural convection that is present at the growth stage due to the warmer temperatures at the bottom. The conditions observed for both quasi-stable conditions is discussed in Sections 3.1. and 3.3, respectively. Regardless of the different stages of the process, the temperature is relatively uniform within each seed. This is likely due to the high thermal conductivity of GaN (100.13 $W{\cdot}m^{-1}{\cdot}K^{-1}$) compared to that of supercritical ammonia of the considered density (about 0.16 $W{\cdot}m^{-1}{\cdot}K^{-1}$), which is also a likely additional reason for the rather uniform temperature within the nutrient.

Expectably, the temperature of each seed is an average of the temperatures within the surrounding fluid (see the times 0 min and 300 min in Figure 7a as well). Consequently, the dissolving and growing GaN surfaces are immersed in a fluid of higher or lower temperature depending not only on the process stage but also on the position within a seed. The question arises of how this inhomogeneity of the thermal gradient between seed and fluid influences crystal growth, especially if seeds with a large dimension in the vertical direction are used. In experimental ammonothermal growth of GaN, the remaining seed after etch-back is regularly found to be thinner at the lower side of the seed, which is in qualitative agreement with this numerical result. Note that there is also a possibility that this effect might be altered if fluid flow in 3D and its time-averaged effects are considered: The degrees of freedom of the fluid flow will have an influence on convective heat transfer within the fluid, and different velocity magnitudes at relevant surfaces such as the autoclave walls will influence the heat transfer coefficient and hence the heat transfer between solids and fluid.

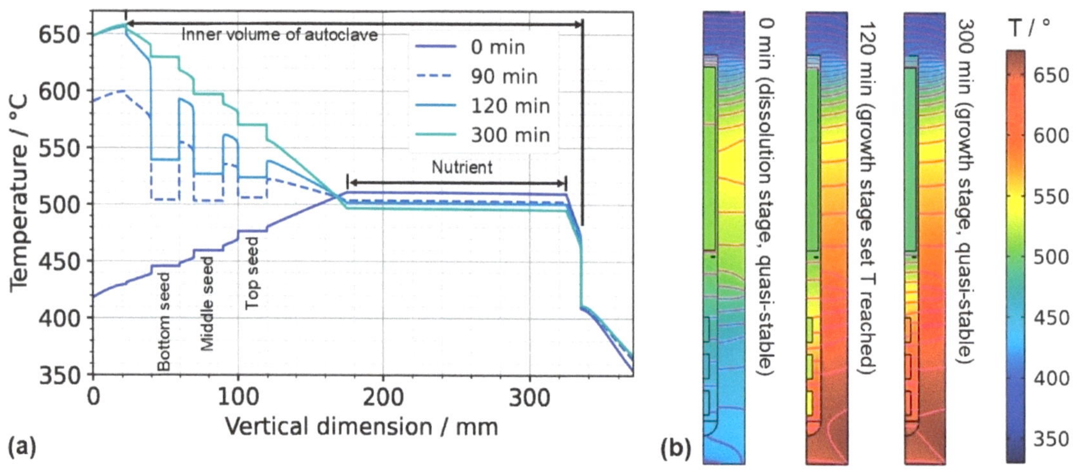

Figure 7. (**a**) Temperature distribution along the centerline for selected points in time. The data labeled with 0 min correspond to quasi-stable conditions at dissolution stage, whereas the data labeled with 120 min correspond to the time at which growth stage set temperatures were reached. The data labeled with 300 min correspond to the time at which quasi-stable conditions at growth stage were reached. The time t = 90 min is also shown, as it represents the time at which all seeds and nutrients have nearly the same temperature. (**b**) Temperature distribution with isotherms for quasi-stable dissolution stage conditions (0 min), at the end of set temperature change (120 min), and for quasi-stable conditions at growth stage (300 min). The temperature difference between isotherms is 10 K, and the range of temperatures is 330 °C to 670 °C.

The quasi-stable conditions of etch-back and growth, as well as the conditions occurring temporarily in between, are considered to be important characteristics of the growth process. In Figure 7a, the temperature distribution along the centerline is shown for selected times corresponding to quasi-stable conditions of etch-back, reaching of set temperatures of growth stage, and quasi-stable conditions of growth stage, whereas Figure 7b) shows the respective temperature distributions and isotherms in 2D. Although details are discussed in the respective subsections, this overview already reveals a couple of relevant characteristics.

As for the temperature within the nutrient, it is noteworthy that it *decreases* as the set temperature of the bottom heater is changed from its etch-back value to its (higher) growth stage value, in spite of the constant thermal boundary conditions in the upper part of the model. This can be explained considering the changes in the flow field: By establishing a boundary condition with the bottom of the autoclave being warmer than the nutrient and top region, a strong driving force for natural convection is established, which also leads to a pronounced downward flow through the baffle and towards the top seed (Figure 6). This enhanced convection facilitates heat transfer from the cold top to the nutrient, which apparently dominates over the increased heat transfer from the warm bottom to the nutrient zone.

Within the growth zone, rather high thermal gradients occur at the seed–fluid and the fluid–wall interface under certain conditions. These are most pronounced under transient conditions during fast changes in thermal boundary conditions (t = 120 min in Figure 7b). Such high thermal gradients occur only at interfaces and only while a significant driving force for natural convection is present (t = 120 min and 300 min in Figure 7b). They are therefore thought to be closely related to natural convection. In the case of the fluid–wall interface, the inner wall of the autoclave is heated by heat conduction in the autoclave wall and thereby keeps a temperature similar to the temperature imposed at the outer wall. The natural convection, however, transports rather cool fluid from the nutrient region and top of the autoclave into the growth zone, which temporarily overcompensates the heating of the fluid by the warm inner wall. The high thermal gradient at the fluid-wall interface is, therefore, explainable.

Even higher thermal gradients are temporarily observed at the seed–fluid interfaces. Major temperature differences occur around the time at which set temperatures of growth stage have been reached, but the respective quasi-stable conditions have not been established yet. In addition to the effect of natural convection described for the fluid–wall interface in the paragraph above, this is likely related to the heat capacity of the seeds (see Section 3.2. for discussion). Although the fluid temperature increases relatively quickly, the temperature increase in the seeds lags behind, leading to a temporary condition of relatively cool seeds immersed in a warmer fluid (t = 120 min in Figure 7). Under quasi-stable conditions, the temperature inside the seeds is similar to that of the surrounding fluid, as expected.

For considering the effects of temperature distribution on supersaturation, it is also important to account for the temperature differences between each seed and the nutrient. A driving force for transport away from the seeds can be assumed under the conditions of the etch-back stage even after an initial, approximate saturation of the solution (assuming that solubility is retrograde for the entire temperature range studied). Under the assumption of retrograde solubility, the direction of transport should transition to the opposite direction after about 90 min, as temperatures of seeds and the nutrient become rather similar around this time, and thereafter, the temperatures of seeds increase further, whereas the temperature of the nutrient decreases further.

3.1. Quasi-Stable Conditions at Dissolution Stage

At the dissolution stage (left subfigures of Figures 5 and 6) as well as at other times investigated, the temperature distribution is relatively uniform within the nutrient, whereas a noticeable thermal gradient is observable along the vertical direction throughout the growth zone and more generally, in the major free flow regions. This suggests that those thermal gradients are related to convective heat transfer and, consequently, to the flow field. At the top of the autoclave, a smaller flow cell develops, which extends into the upper third of the nutrient and exhibits an upward flow at the center and a downward flow near the vertical autoclave wall. Within the remaining part of the nutrient height, a larger flow cell develops, which exhibits downward flow inside the nutrient and upward flow in the gap between the nutrient and autoclave wall. A circular flow cell is found around the baffle, i.e., the downward flow through the center changes its direction already in the vicinity of the baffle and joins an upward flow passing through the gap between the baffle and autoclave wall. This is likely related to the upward flow near the center that is observed in the growth zone and probably originates from the cold wall in the growth zone at the dissolution stage.

The temperatures of all the seeds, as well as the fluid surrounding them, are in a range of about 440 to 480 °C, which is well above the minimum parameters required for the dissolution of GaN with NH_4F mineralizer (about 220 °C, 21 MPa) [21]. Therefore, back etching of the seeds can be expected to occur at this stage as anticipated. Due to the higher temperature and thus faster reaction kinetics and earlier onset of etch-back, the amount of etch-back can be expected to be highest for the top seed and lowest for the bottom seed. In the nutrient region, the temperature is significantly higher (slightly below 520 °C). Therefore, the dissolution rates in the nutrient region are likely higher than in the growth zone until a saturated solution is established within the upper zone of the autoclave, which is also fostered by the larger surface area of the nutrient. The inverted flow direction in the lower zone, as well as the overall slow convection speeds during etch-back, are thought to prevent the transport of the saturating and eventually saturated solution to the seeds, hereby ensuring sufficient back etching of the seeds prior to growth. Note, however, that the here-studied convection driven by thermal gradients will be superimposed by solutal convection in an actual growth experiment. Due to the presumably faster dissolution in the upper zone, the density of the solute-containing fluid there will likely increase faster compared to the density of the fluid in the growth zone. This inequal density increase is expected to increase the driving force for downward flow into the growth zone. However, the increasing concentration of Ga-containing solutes formed with NH_4F mineralizer is also suspected to increase the viscosity [22], which might counteract the mentioned effect of density increase.

The velocity magnitudes at the dissolution stage are below 0.007 m/s, with much lower velocity magnitudes in most regions (see Figures 6 and A1). Only in the uppermost region within and especially above the nutrient, velocity magnitudes comparable to those at the growth stage are found. This is expectable because the nutrient zone has a higher temperature than the top of the autoclave, and therefore, a driving force for natural convection exists in the uppermost part of the autoclave. The temperature of the growth zone, however, is lower than that of the nutrient region above, and therefore there is no local driving force to cause pronounced natural convection within the growth zone at the dissolution stage.

3.2. *Transition from Quasi-Stable Conditions at Dissolution Stage to Growth Stage*

Following the gradual change of set temperatures during the temperature inversion process, the downward flow from baffle to first seed gradually develops, mostly during the second half of the temperature inversion process (velocity magnitude subfigures in Figure 8). The strengthened downward flow in the center brings cooler fluid into the growth zone, which counteracts the fluid heating effect of the increasingly warm autoclave walls there. As a result of these effects, increasing temperature differences between the seeds and the surrounding fluid develop, as well as increasing temperature differences between the autoclave wall and the surrounding fluid. Both can be seen by comparing the thermal fields at different times of the temperature inversion process shown in the temperature subfigures in Figure 8. These effects are likely exacerbated by differences in volumetric heat capacity (GaN: 3.19 $MJ \cdot K^{-1} \cdot kg^{-3}$, Ni-Cr superalloy: 4.4 $MJ \cdot K^{-1} \cdot kg^{-3}$, supercritical NH_3: 0.99 $MJ \cdot K^{-1} \cdot kg^{-3}$, all based on values for specific heat capacity and density as stated in [17]). That is, the lower volumetric heat capacity of ammonia in comparison to GaN leads to relatively quick changes in fluid temperature followed by a sluggish response of the GaN components immersed in the fluid. Note that the fluid in the growth zone exhibits a temperature that is in-between the temperature of seeds and the autoclave wall at the same vertical position. That is, autoclave walls already show higher temperatures than the fluid, but the temperature inside the seeds is temporarily lower than that of the surrounding fluid. Therefore, parasitic nucleation on the autoclave walls is likely to occur, especially during the first 90 min of the set temperature inversion process. Starting from 90 min, the temperatures of the seeds rise above that of the nutrient, which is expected to lead to deposition on the seeds. However, parasitic nucleation or growth on the growth zone inner walls is likely to continue simultaneously as the temperature of the autoclave walls is even higher.

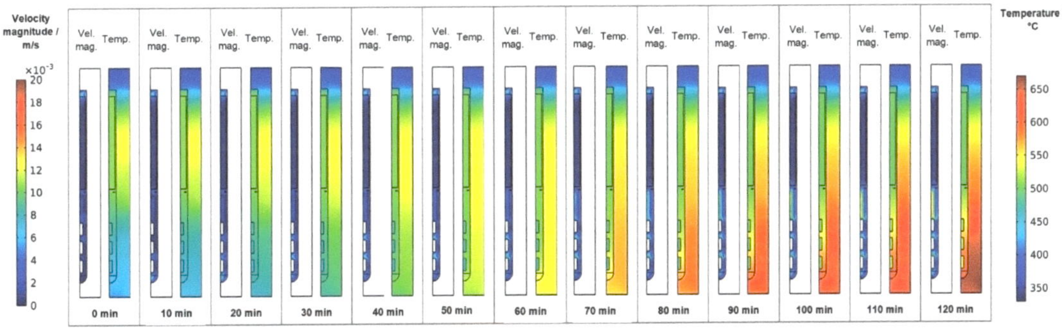

Figure 8. Velocity magnitude and temperature fields occurring during set temperature change for temperature inversion (0 min: quasi-stable etch-back condition, 120 min: reaching of growth stage set temperatures).

For all domain probe locations, the resulting temperatures plotted over time are presented in Figure 9. Significant short-term fluctuations are visible for DP1 (at the center halfway between the top of the nutrient and the top inner wall) and DP13 (in the gap between the baffle and autoclave wall). Note that the temperature data for DP3 (in the center of the center hole of the baffle) are hidden by the data of DP13 but exhibit significantly fewer short-term fluctuations, similar to DP4 (between the baffle and top seed in the center). At present, it is not entirely clear whether the oscillations in temperatures in the gap regions represent a physical effect or whether numerical artifacts related to the mesh in this narrow region play a noticeable role. The absence of significant temperature fluctuations for the other domain probes deviates from previously published experimental results of one of the authors [10], which is, however, explainable due to significant differences in the experimental setup and procedures. Firstly, the experimental setup used in that previous study [10] contained a much smaller volume fraction of solids in the nutrient zone, which represented much less of an obstacle to natural convection. Secondly, the control thermocouples in that setup were located inside the furnace insulation and consequently without thermal contact with the outer autoclave wall (contrary to the here-considered setup). In the previous experimental study [10], fluctuations in autoclave wall temperature (that is, thermal boundary conditions) may therefore have occurred due to possible oscillations in heater powers.

In addition, Figure 9 reveals that the fluid in the growth zone follows the changes in thermal boundary conditions rather swiftly. The bulk material of the seeds in contact with the fluid, however, shows a more sluggish, damped response to the imposed boundary condition change (see DP5 to DP10). As elaborated before, this is likely related to the relatively high heat capacity of the crystals slowing down their temperature increase.

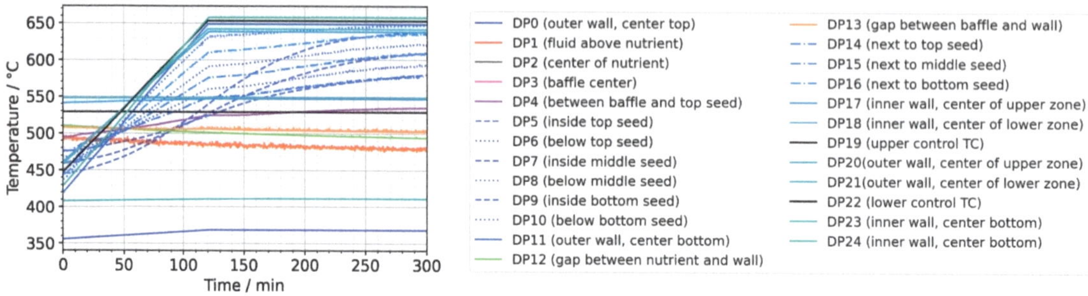

Figure 9. Temperatures over time for the entire set of domain probes.

To further ease quantitative interpretation at various times of the set temperature inversion process, a vertical temperature profile along the centerline of the autoclave is presented in Figure 10 for the respective times. Horizontal temperature profiles through the center of each seed are presented in Figure 11. At quasi-stable etch-back condition (0 min), the temperature of each seed is an average of the temperatures of the surrounding fluid. Upon reaching growth stage set temperatures (120 min), that is, before re-establishing quasi-stable conditions, the temperatures of the seeds are significantly *lower* than the temperatures of the surrounding fluid: about 70 K for the bottom seed, 45 K for the middle seed, and 20 K for the top seed, respectively. On the contrary, the temperatures of the inner autoclave wall are significantly *higher* than the fluid temperatures: 48 K for the bottom seed, 74 K for the middle seed, and 80 K for the top seed, respectively. That is, the driving force for parasitic growth should be highest at the bottom of the autoclave, as well as the driving force for a possibly ongoing dissolution of the seed (if parasitic deposition temporarily leads to an undersaturated fluid there). The increasing amount of parasitic deposition towards the bottom of the autoclave is in line with typical experimental observations.

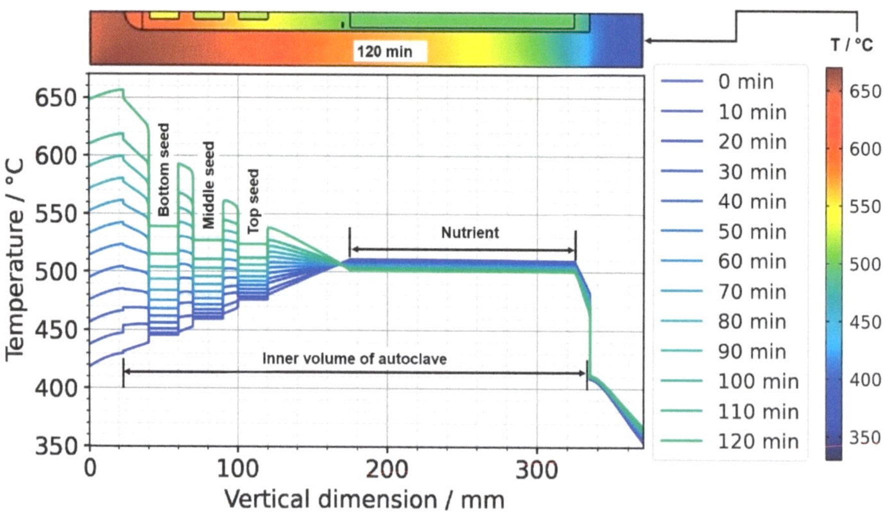

Figure 10. Temperature distribution in vertical direction in the center (0 min: quasi-stable etch-back condition, 120 min: reaching of growth stage set temperatures).

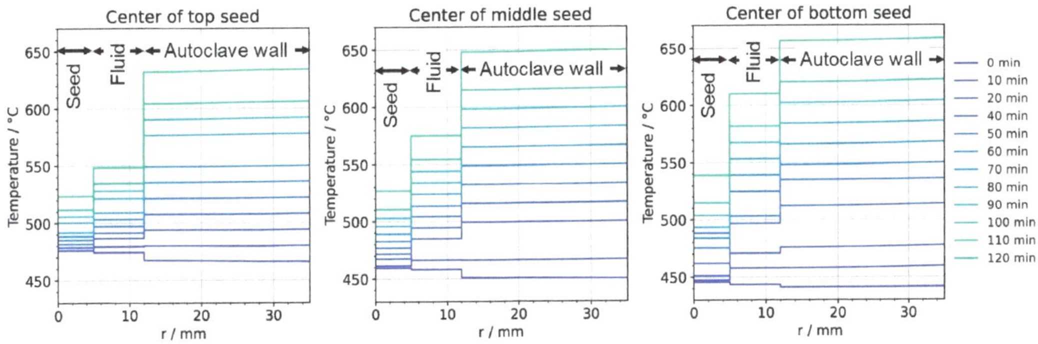

Figure 11. Temperature distribution in horizontal direction through the center of each seed (0 min: quasi-stable etch-back condition, 120 min: reaching of growth stage set temperatures).

Note also that the thermal gradient from one seed to another (slope of the temperature in fluid sections within the growth zone in Figure 10) is also inverted during a set temperature inversion. This occurs after all seeds temporarily reach nearly the same temperature of about 500 °C after about 90 min. The maximum rate of temperature change of seeds is observed for the bottom seed directly before growth stage set temperatures are reached: From the time 110 min to 120 min, rates of temperature change are about 2.5 K/min for the bottom seed, 1.8 K/min for the middle seed, and 1.2 K/min for the top seed, respectively. The fastest rates of seed temperature change correlate with the highest temperature differences between each seed and its fluid surrounding, which is in accordance with expectations because the temperature difference drives these temperature changes.

3.3. Re-Establishment of Quasi-Stable Internal Conditions after Set Temperature Inversion

After the process of set temperature inversion is completed, it takes about 3 h for the system to reach quasi-stable conditions again. The respective temporarily occurring temperature distributions, as well as the one under quasi-stable conditions of the growth stage, are shown in Figure 12, alongside the quasi-stable growth stage temperature distribution. Expectably, the rates of temperature change decrease gradually as quasi-stable conditions are approached. After about 2 h, the seed temperatures become an average of the temperatures of the surrounding fluid (Figures 12 and 13).

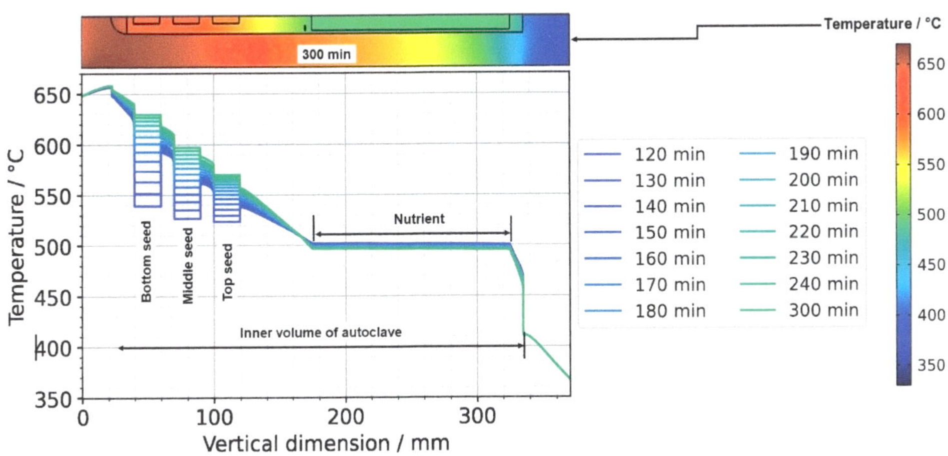

Figure 12. Temperature distribution in vertical direction in the center (120 min: reaching of growth stage set temperatures, 300 min quasi-stable conditions of growth stage).

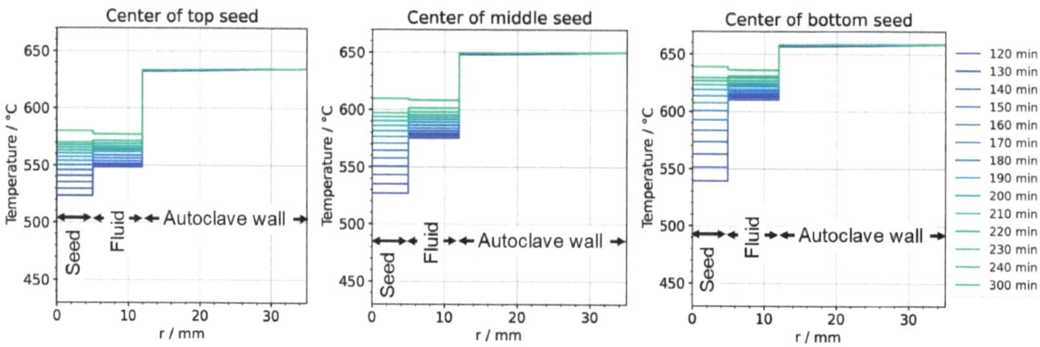

Figure 13. Temperature distribution in horizontal direction through the center of each seed (120 min: reaching of growth stage set temperatures, 300 min quasi-stable conditions of growth stage).

Figure 13 also reveals the respective temperature differences between the autoclave wall and the interior of the autoclave at the same vertical position. The temperature differences between the autoclave wall and fluid do not decrease nearly as much as the temperature difference between seeds and fluid. However, the significant downward flow of saturated fluid in the center region of the autoclave is thought to favor the deposition on the seeds already at this stage of the process.

A comprehensive representation containing the temperature distribution data over the entire investigated timespan can be found in Figure A3 (along the centerline) and Figure A4 (horizontally through the seeds).

3.4. *Quasi-Stable Conditions at Growth Stage*

At the growth stage (right subfigures of Figures 5 and 6), a relatively uniform temperature distribution is maintained within the nutrient, and the clockwise flow direction of the upper as well as the counterclockwise flow direction of the main flow cell are maintained. The downward flow extending through the center hole of the baffle is much stronger at the growth stage. In the growth zone, the uppermost flow cell extends from the baffle to the first seed, joining the downward flow from the center hole of the baffle. Along and between the seeds, the downward flow near the center persists. Circular counterclockwise flow cells are observed in between the seeds and the autoclave wall, with the center of the flow cells residing at vertical positions between the seeds, similar to the dissolution stage but with the opposite flow direction.

The thermal gradient along the autoclave wall translates into a thermal gradient inside the autoclave that is modified and shifted by the effects of convection. Consequently, each seed at a different vertical position experiences a different temperature, with temperature differences between seeds of about 15 K. A different temperature of seeds placed at different vertical positions may at least partially explain the experimental observation of growth results depending on the vertical position of seeds. Variations of growth rate depending on the vertical position of seeds have been observed not only in our own experiments using the ammonoacidic mineralizer NH_4F but were also reported in a detailed analysis of ammonobasic experiments by Grabianska et al. [20]. Under quasi-stable conditions of the growth stage, temperature differences between fluid and seed reach up to about 10 K at the upper and lower ends of the bottom seed (Figure 12), with the seeds being in contact with warmer fluid at their lower end and with cooler fluid at their upper end. While temperature differences between seed and fluid are negligible at the center of the seed, there are significant temperature differences between fluid and autoclave wall. These differ significantly at the vertical positions of each seed (about 60 K next to the top seed, 50 K next to the middle seed, and 30 K next to the bottom seed, with fluid temperatures being lower than the autoclave wall temperatures in all cases, see Figure 13). However, the direction of flow cells supports the deposition on the seeds, as fluid coming from the nutrient zone tends to reach the seeds before coming in contact with the autoclave walls.

Velocity magnitudes at the growth stage reach up to about 0.02 m/s. The highest velocity magnitudes are found around the centerline in-between the baffle and the uppermost seed. At this location, the velocity magnitude at the growth stage is about ten times the velocity magnitude at the dissolution stage. The lowest velocity magnitudes are observed in the nutrient region and partially in the growth zone (see Figure 6).

3.5. Lateral Temperature Distribution within the Autoclave Wall

In the following, the impact of convective heat transfer on the temperatures of the inner autoclave walls shall be evaluated for quasi-stable conditions as well as for transient conditions with typical rates of set temperature change. For this purpose, temperatures near the inner and outer autoclave wall along the vertical dimension of the simulation domain are displayed in Figure 14. In addition, the respective temperature profile in the fluid near the inner wall is shown for comparison. The radial position of r = 11 mm corresponds to the middle of the gap between the nutrient and the inner wall. Inside the autoclave wall, the differences between temperatures near the outer and inner walls are small, suggesting that meaningful results could likely also be achieved if imposing the outer wall thermal boundary conditions at or much closer to the inner autoclave wall. From our previous results obtained with a simulation of the entire furnace [17], this had not been obvious, as more significant temperature differences between outer and inner walls had been observed there. This was likely an inaccuracy in the previous simulation that originated from a less precise simulation of the convective flow in the sense of not resolving flow phenomena at small time scales (a time resolution comparable to the present study was not feasible in the previous study due to the large domain size and need to perform the calculation for a high number of heater power combinations). In agreement with this interpretation of the deviation, the flow cells observed in the previous study are also somewhat different from those found in the present study. The time resolution was improved in the present simulation using much smaller timesteps. We conclude that while it is necessary to use continuous temperature distributions reflecting realistic thermal losses, it is likely not necessary to include the full thickness of the autoclave wall in the simulation domain if set temperature changes occur gradually.

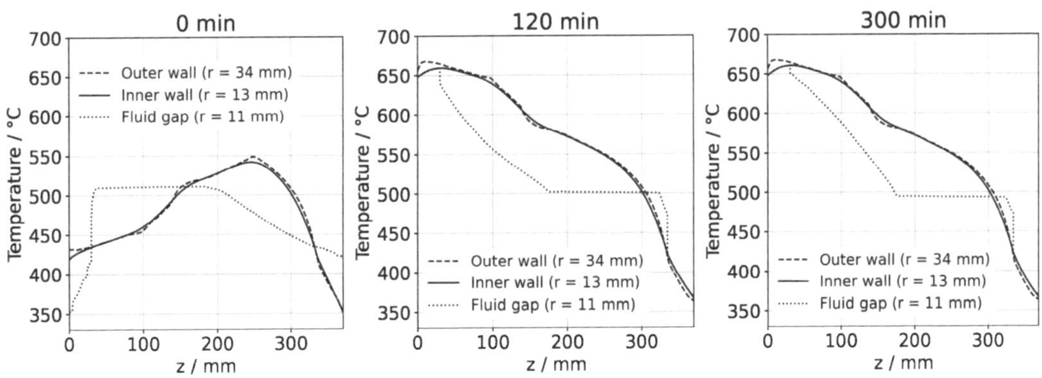

Figure 14. Temperatures inside the autoclave wall near the outer (dashed lines) and inner (solid lines) walls and inside the fluid near the inner autoclave wall (dotted lines). The data labeled with 0 min correspond to quasi-stable conditions at dissolution stage, whereas the data labeled with 120 min correspond to the time at which growth stage set temperatures were reached. The data labeled with 300 min correspond to quasi-stable conditions at growth stage.

3.6. Experimental Results Obtained Using Conditions Similar to the Numerical Simulation

The UV fluorescent micrographs of the cleaved crystals are presented in Figure A5. The interfaces of seeds and grown layers are clearly visible as intended. The experimental results in terms of GaN mass transport are listed in Tables 7 and 8 in terms of mass changes, dimensional changes, and related quantities, respectively. As expected, the overall mass changes indicate transport from the nutrient to the growth zone. The top seed shows the smallest increase in mass; however, this seed has also experienced the largest loss in thickness by etch-back and hence the largest (estimated) mass loss by etch-back. The effects of seed positioning will therefore be discussed based on the remaining seed thicknesses and the effective growth rates (that is, considering growth from the seed interface rather than the thickness change during the overall experiment).

Table 7. Masses of GaN components before and after the ammonothermal growth experiment. Changes with negative values represent a decrease in mass. Note that the bottom edges of the bottom seed (position 15 mm) dissolved completely so that the surface area available for growth on the bottom seed was reduced.

GaN Component	Pre-Run Mass/g	Etch-Back Mass Loss (Estimate)/g	Post-Run Mass/g	Mass Change/g	Mass Change/%
Nutrient	35.59	Not determined	14.14	−21.45	−60.3
Top seed (58 mm)	0.4394	−0.1098	0.9517	0.5123	116.6
Middle seed (35 mm)	0.4397	−0.0892	1.1323	0.6926	157.5
Bottom seed (15 mm)	0.4399	−0.0795	1.0679	0.6280	142.8

Table 8. Results of the growth experiment. Note that face-specific growth rates refer to the polarity of the seed, which could potentially differ from the polarity of the grown layer in case of polarity inversion. Note as well that the bottom edges of the bottom seed (position 15 mm) dissolved. Note that growth rates were determined based on the duration of constant growth stage set temperatures, which is only an approximation of the actual timespan during which growth occurs.

Seed Position/mm	Seed Thickness Loss/µm	Effective Ga-Face Growth Rate/µm/d	Effective N-Face Growth Rate/µm/d	N:Ga Face Growth Rate Ratio	Effective Cumulative Growth Rate/µm/d
58	99	14.3	86.5	6.1	130.3
35	83	30.5	107.8	3.5	168.8
15	74	36.5	113.8	3.1	176.5

The amount of seed thickness loss increases from the bottom to the top seed. This may be due to the different seed temperatures during etch-back, as the seed temperatures increase from the bottom to the top seed. The different temperatures are thought to influence dissolution kinetics: dissolution is generally expected to be faster at higher temperatures. This should lead to a larger amount of thickness loss at the upper seeds within a given timespan. Assuming a retrograde solubility also in the temperature range used for etch-back, mass transport should occur from bottom to top seed, which could slow down the etch-back of the upper seeds and facilitate the etch-back of the lower seeds. This may contribute to the experimentally observed dissolution of the bottom part of the bottom seed. Since seed mass loss is highest for the uppermost seed, etch-back appears to be dominated by dissolution kinetics rather than by the temperature dependence of solubility, which is expectable because, initially, there is no saturated solution.

The effective cumulative growth rates increase towards the bottom seed position. This trend has also been seen in other experiments with similar configurations. The increase in effective cumulative growth rates coincides with the increase in seed temperatures from the top to the bottom position at the growth stage. This is likely due to the increasing temperature difference between the nutrient and the respective seed. Effective face-specific growth rates fit this trend on both the Ga- and N-face. On all seeds, growth on the N-face was faster than on the Ga-face. However, the ratio of face-specific growth rates decreased towards the bottom of the autoclave.

On the N-face of each crystal, two sublayers can be distinguished within the grown layer (see UV fluorescent micrographs in Figure A5). The ratio of sublayer thicknesses changes with the vertical position of the seed in the autoclave as follows. For the uppermost seed, a ratio of complete N-face layer thickness to dark N-face sublayer thickness of about 6.0 is observed. This ratio decreases to about 4.8 for the middle and about 3.4 for the bottom seed. The absolute thickness of the dark sublayer also increases towards the bottom of the autoclave (top seed: 61 μm, middle seed: 93 μm, bottom seed: 140 μm). Since the distinguishable sublayers must be expected to represent distinguishable growth conditions occurring within distinguishable timeframes during the experiment, we suspect that the dark sublayer may represent the material grown while seed temperatures were still significantly lower than at quasi-stable conditions of the growth stage. Since the duration of etch-back was likely sufficient to establish a saturated solution also within the growth zone (as discussed in Section 2.2), we expect growth to be initiated by the rise of fluid temperature, which begins early during the set temperature inversion process (according to the numerical simulation, see Figure 10). The timespan with significant differences between the temperature of each seed and its fluid surrounding can be estimated to be about 4 h according to the simulation (2 h for the actual set temperature inversion process and another 2 h with diminishing thermal gradients between crystals and surrounding fluid), whereas the quasi-stable conditions of growth stage are established after about 5 h in total.

All in all, the experimental and numerical results can be interpreted in a coherent way, supporting the correctness of the main characteristics determined via numerical simulations.

4. Conclusions

As expected, both numerical and experimental results indicate a systematic variation of growth conditions along the vertical axis of the autoclave. According to the analysis of cleavage surfaces of the grown crystals under UV illumination, the effective cumulative growth rates increase towards the bottom of the autoclave. According to the numerical model, this correlates with an increasing temperature at the growth stage. Moreover, according to the analysis of the cleavage surfaces, the amount of etch-back (seed thickness loss) decreases towards the bottom of the autoclave. According to the numerical model, this correlates with a decreasing temperature at the dissolution stage towards the bottom of the autoclave. Therefore, the transport and recrystallization process of crystal growth appears to be governed mostly by a retrograde solubility, whereas the process of etch-back appears to be strongly influenced by reaction kinetics, provided that the solubility of GaN is retrograde under both the dissolution stage and growth stage conditions investigated.

According to the numerical results, short-term temperature fluctuations originate primarily from fluctuations in velocity magnitude, while the flow direction usually exhibits only minor changes within short time scales. However, it should be noted that flow directions might be stabilized by the symmetrical nature of the axis of this 2D model.

Methodically, it can be concluded that if all set temperature changes (more generally: boundary condition changes) are applied gradually and if the initial values are close enough to the solution for the initial boundary condition, it should be feasible to exclude most of the autoclave wall thickness from the simulation domain to reduce computation times.

The numerical model is found to be particularly insightful in terms of analyzing temporary effects, including temperature differences between crystals and the surrounding fluid, which would be extremely difficult to observe experimentally.

The response of the fluid to changes in the autoclave wall temperatures is relatively swift, whereas the response of the crystals is relatively sluggish. This is likely related to the low volumetric heat capacity of the fluid compared to both GaN and the nickel base alloy. The rates of seed temperature change range from 2.5 K/min to 1.2 K/min and are thus closely related to the rate of set temperature change (100 K/h, i.e., about 1.7 K/min). Due to the relatively swift response of the fluid temperature compared to the seeds, the following occurs during the set temperature inversion process: Depending on the vertical position of the seed, significant temperature differences of 20 K to 70 K occur temporarily between a seed and its fluid surrounding during the set temperature inversion process, so that the fluid temperature in the growth zone is between that of the seeds and that of the neighboring inner autoclave walls. Therefore, in a retrograde solubility scenario, parasitic nucleation and growth on the autoclave walls may already occur while the seeds are not growing yet or may still experience back-etching. This may explain why the bottom part of the lowest seed tends to dissolve preferentially: If the seed is temporarily colder than the fluid and the inner autoclave wall is warmer, then mass transport from the bottom of the lowest seed to the bottom inner wall is expectable. In addition, the relatively high surface area in the vicinity of the lowest seed may exacerbate this effect, as parasitic deposition at the bottom autoclave wall may lead to an undersaturated solution in its vicinity. These insights also pose the question of whether ammonothermal growth with normal and retrograde solubility differ fundamentally in the conditions during the early stages of growth. If a normal solubility configuration is used, the positions of seeds and nutrients are inverted. In this case, it would be the nutrient rather than the seeds that would approach growth condition temperatures in a sluggish way, as the seeds would be residing in the zone that does not experience major changes in thermal boundary conditions. This suggests that if a similar temperature ramp-up process is used for growth under normal solubility conditions, then nucleation on the seeds may occur earlier in the set temperature inversion process. In addition, crystal growth could occur already while the crystals approach their quasi-stable growth condition temperatures, and significant fluid-crystal temperature differences could result in fast growth during the early stages of the process. This might affect structural quality as well as impurity incorporation. Further research is needed to clarify if this is the case and whether there is a general advantage of retrograde solubility process variants due to such effects.

About 90 min into the process of set temperature inversion, the temperatures of the seeds begin to exceed those of the nutrient. Therefore, crystal growth is likely initiated around that time due to the inversion of the solubility gradient. When the process of set temperature inversion is completed after 120 min, the most pronounced thermal gradients between seeds, fluid, and autoclave wall are observed. Within another 120 min (120 min after reaching the growth stage set temperatures), the mean temperature of each seed approaches the average temperature of its fluid surrounding. It takes an additional 60 min until approximately quasi-stable conditions of the growth stage are established.

Supplementary Materials: The following supporting information can be downloaded at: https://zenodo.org/record/7555506#.Y8rq5HbMKUk, Video S1: dissolution stage (upper part of nutrient and above), Video S2: dissolution stage (baffle region), Video S3: dissolution stage (top seed), Video S4: dissolution stage (middle seed), Video S5: dissolution stage (bottom seed), Video S6: growth stage (upper part of nutrient and above), Video S7: growth stage (baffle region), Video S8: growth stage (top seed), Video S9: growth stage (middle seed), Video S10: growth stage (bottom seed).

Author Contributions: Conceptualization, S.S. and D.T.; methodology, S.S. and D.T.; validation, D.T.; formal analysis, S.S. and D.T.; investigation, S.S. and D.T.; resources, S.S.; data curation, S.S. and D.T.; writing—original draft preparation, S.S.; writing—review and editing, S.S., D.T., T.I., Y.H., S.F.C. and H.A.; visualization, S.S.; supervision, D.T., Y.H., H.A. and S.F.C., project administration, S.S., D.T., Y.H. and H.A.; funding acquisition, S.S., D.T., Y.H. and H.A. All authors have read and agreed to the published version of the manuscript.

Funding: The first author would like to thank the Japan Society for the Promotion of Science (JSPS), grant number P19752 (Postdoctoral Fellowships for Research in Japan (Standard)), and the Alexander von Humboldt Foundation (Feodor Lynen Return Fellowship) for funding. This research was furthermore supported by a Casiozaidan research grant.

Institutional Review Board Statement: Not applicable.

Informed Consent Statement: Not applicable.

Data Availability Statement: The data presented in this study are available upon reasonable request from the corresponding author.

Acknowledgments: The first author would like to thank Peter Wellmann as the return fellowship host and Jörg Schulze as the current host.

Conflicts of Interest: The authors declare no conflict of interest. The funders had no role in the design of the study; in the collection, analyses, or interpretation of data; in the writing of the manuscript; or in the decision to publish the results.

Appendix A

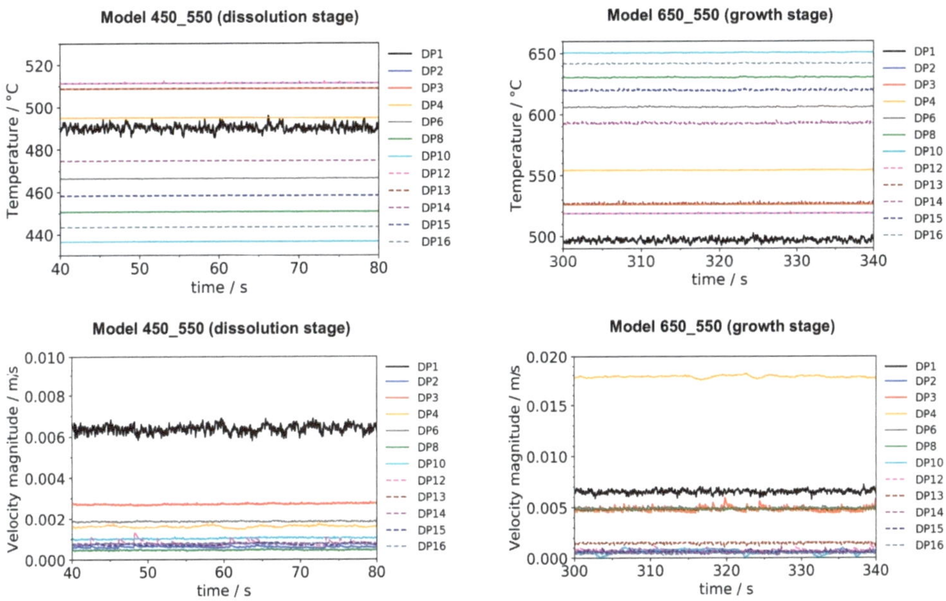

Figure A1. Temperature (**top**) and velocity magnitude (**bottom**) over time for all probe locations in the fluid at dissolution stage (**left**) and growth stage (**right**), respectively. Note the different y-axis scales at growth and dissolution stage. Data were exported with a temporal resolution of 0.05 s. Note that the times shown here should be interpreted only in terms of time range, as they do not relate to the time values elsewhere in this work.

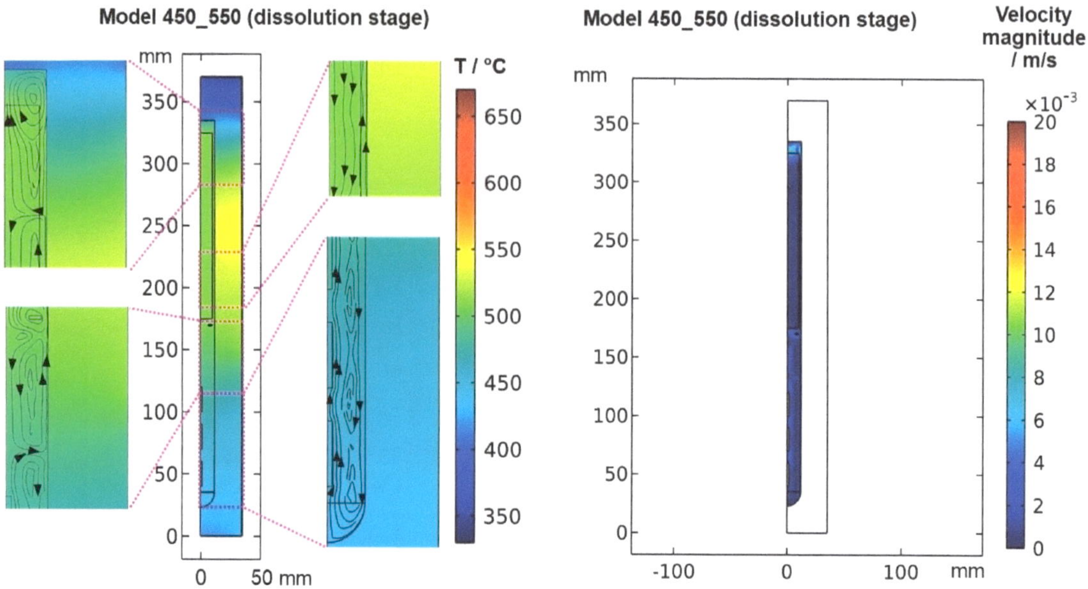

Figure A2. Quasi-stable dissolution stage conditions for the model version with slim seeds: Temperature field with streamlines visualizing fluid flow directions (**left**) and velocity magnitude field (**right**).

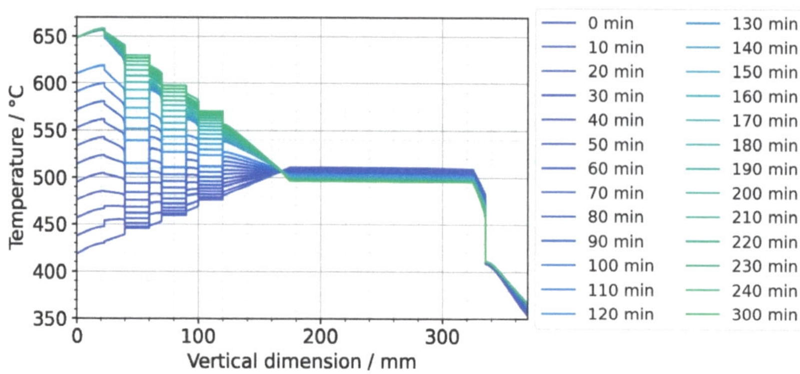

Figure A3. Temperature distribution along the centerline for the entire set temperature inversion process. The quasi-stable conditions of etch-back stage are present at the time 0 min. The time 120 min represents the moment at which set temperatures have reached their values for growth stage. The time 300 min corresponds to quasi-stable conditions of growth stage.

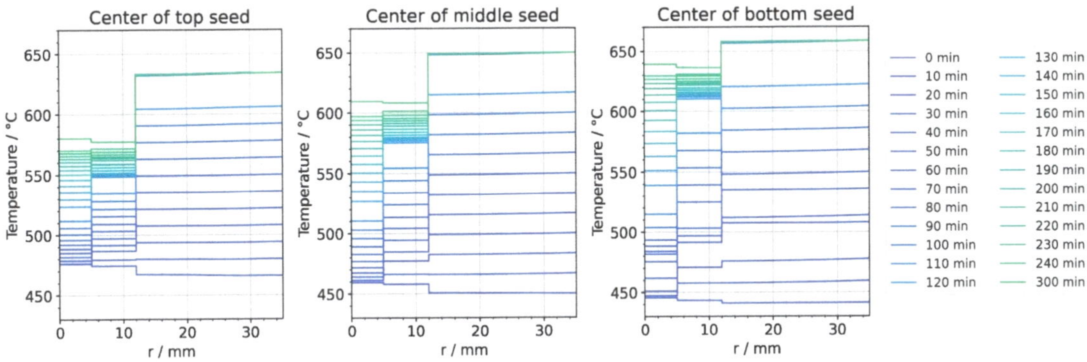

Figure A4. Temperature distribution in horizontal direction through the center of each seed for the entire set temperature inversion process. The quasi-stable conditions of etch-back stage are present at the time 0 min. The time 120 min represents the moment at which set temperatures have reached their values for growth stage. The time 300 min corresponds to quasi-stable conditions of growth stage.

Appendix B

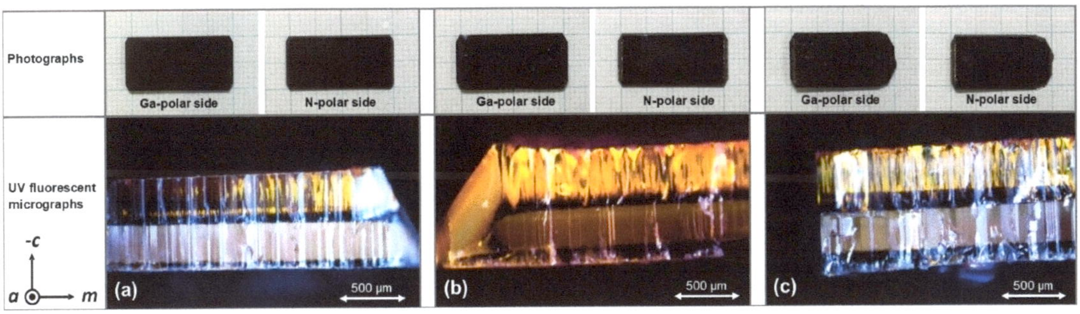

Figure A5. Photographs (top) and UV fluorescent micrographs of cleavage planes (bottom) of crystals grown at (**a**) top, (**b**) middle, and (**c**) bottom position. The lack of optical transparency of the crystals visible in the photographs is related to the cost-effective choice of seeds (HVPE) and nutrient due to the different goals of this study. The UV micrographs were used to quantify seed thickness loss by etch-back as well as effective growth rates stated in Table 8.

References

1. Bao, Q.; Saito, M.; Hazu, K.; Furusawa, K.; Kagamitani, Y.; Kayano, R.; Tomida, D.; Qiao, K.; Ishiguro, T.; Yokoyama, C.; et al. Ammonothermal Crystal Growth of GaN Using an NH4F Mineralizer. *Cryst. Growth Des.* **2013**, *13*, 4158–4161. [CrossRef]
2. Tomida, D.; Bao, Q.; Saito, M.; Osanai, R.; Shima, K.; Kojima, K.; Ishiguro, T.; Chichibu, S.F. Ammonothermal growth of 2 inch long GaN single crystals using an acidic NH 4 F mineralizer in a Ag-lined autoclave. *Appl. Phys. Express* **2020**, *13*, 55505. [CrossRef]
3. Kurimoto, K.; Bao, Q.; Mikawa, Y.; Shima, K.; Ishiguro, T.; Chichibu, S.F. Low-pressure acidic ammonothermal growth of 2-inch-diameter nearly bowing-free bulk GaN crystals. *Appl. Phys. Express* **2022**, *15*, 55504. [CrossRef]
4. Kucharski, R.; Sochacki, T.; Lucznik, B.; Amilusik, M.; Grabianska, K.; Iwinska, M.; Bockowski, M. Ammonothermal and HVPE Bulk Growth of GaN. In *Wide Bandgap Semiconductors for Power Electronics*; Wiley-VCH: Weinheim, Germany, 2021; pp. 529–554.
5. Wasisto, H.S.; Prades, J.D.; Gülink, J.; Waag, A. Beyond solid-state lighting: Miniaturization, hybrid integration, and applications of GaN nano- and micro-LEDs. *Appl. Phys. Rev.* **2019**, *6*, 41315. [CrossRef]
6. Amano, H. Nobel Lecture: Growth of GaN on sapphire via low-temperature deposited buffer layer and realization of p-type GaN by Mg doping followed by low-energy electron beam irradiation. *Rev. Mod. Phys.* **2015**, *87*, 1133–1138. [CrossRef]
7. Amano, H.; Baines, Y.; Beam, E.; Borga, M.; Bouchet, T.; Chalker, P.R.; Charles, M.; Chen, K.J.; Chowdhury, N.; Chu, R.; et al. The 2018 GaN power electronics roadmap. *J. Phys. D* **2018**, *51*, 163001. [CrossRef]
8. Zhang, Y.; Dadgar, A.; Palacios, T. Gallium nitride vertical power devices on foreign substrates: A review and outlook. *J. Phys. D* **2018**, *51*, 273001. [CrossRef]

9. Schimmel, S.; Tomida, D.; Ishiguro, T.; Honda, Y.; Chichibu, S.; Amano, H. Numerical Simulation of Ammonothermal Crystal Growth of GaN—Current State, Challenges, and Prospects. *Crystals* **2021**, *11*, 356. [CrossRef]
10. Schimmel, S.; Kobelt, I.; Heinlein, L.; Kimmel, A.-C.L.; Steigerwald, T.G.; Schlücker, E.; Wellmann, P. Flow Stability, Convective Heat Transfer and Chemical Reactions in Ammonothermal Autoclaves—Insights by In Situ Measurements of Fluid Temperatures. *Crystals* **2020**, *10*, 723. [CrossRef]
11. Schimmel, S.; Wellmann, P. In Situ Visualization of the Ammonothermal Crystallization Process by X-ray Technology. In *Ammonothermal Synthesis and Crystal Growth of Nitrides: Chemistry and Technology*; Meissner, E., Niewa, R., Eds.; Springer International Publishing: Cham, Switzerland, 2021; pp. 171–190. ISBN 978-3-030-56305-9.
12. Griffiths, S.; Pimputkar, S.; Kearns, J.; Malkowski, T.F.; Doherty, M.F.; Speck, J.S.; Nakamura, S. Growth kinetics of basic ammonothermal gallium nitride crystals. *J. Cryst. Growth* **2018**, *501*, 74–80. [CrossRef]
13. Häusler, J.; Schnick, W. Ammonothermal Synthesis of Nitrides: Recent Developments and Future Perspectives. *Chem.—A Eur. J.* **2018**, *24*, 11864–11879. [CrossRef]
14. Tsunooka, Y.; Kokubo, N.; Hatasa, G.; Harada, S.; Tagawa, M.; Ujihara, T. High-speed prediction of computational fluid dynamics simulation in crystal growth. *CrystEngComm* **2018**, *20*, 6546–6550. [CrossRef]
15. Liu, X.; Dang, Y.; Tanaka, H.; Fukuda, Y.; Kutsukake, K.; Kojima, T.; Ujihara, T.; Usami, N. Data-Driven Optimization and Experimental Validation for the Lab-Scale Mono-Like Silicon Ingot Growth by Directional Solidification. *ACS Omega* **2022**, *7*, 6665–6673. [CrossRef] [PubMed]
16. Zajac, M.; Kucharski, R.; Grabianska, K.; Gwardys-Bak, A.; Puchalski, A.; Wasik, D.; Litwin-Staszewska, E.; Piotrzkowski, R.; Domagala, J.Z.; Bockowski, M. Basic ammonothermal growth of Gallium Nitride—State of the art, challenges, perspectives. *Prog. Cryst. Growth Charact.* **2018**, *64*, 63–74. [CrossRef]
17. Schimmel, S.; Tomida, D.; Saito, M.; Bao, Q.; Ishiguro, T.; Honda, Y.; Chichibu, S.; Amano, H. Boundary Conditions for Simulations of Fluid Flow and Temperature Field during Ammonothermal Crystal Growth—A Machine-Learning Assisted Study of Autoclave Wall Temperature Distribution. *Crystals* **2021**, *11*, 254. [CrossRef]
18. Han, P.; Gao, B.; Song, B.; Yu, Y.; Tang, X.; Liu, B. Large-Sized GaN Crystal Growth Analysis in an Ammonothermal System Based on a Well-Developed Numerical Model. *Materials* **2022**, *15*, 4137. [CrossRef]
19. Kirste, L.; Grabianska, K.; Kucharski, R.; Sochacki, T.; Lucznik, B.; Bockowski, M. Structural Analysis of Low Defect Ammonothermally Grown GaN Wafers by Borrmann Effect X-ray Topography. *Materials* **2021**, *14*, 5472. [CrossRef]
20. Grabianska, K.; Kucharski, R.; Sochacki, T.; Weyher, J.L.; Iwinska, M.; Grzegory, I.; Bockowski, M. On Stress-Induced Polarization Effect in Ammonothermally Grown GaN Crystals. *Crystals* **2022**, *12*, 554. [CrossRef]
21. Schimmel, S.; Koch, M.; Macher, P.; Kimmel, A.-C.L.; Steigerwald, T.G.; Alt, N.S.A.; Schlücker, E.; Wellmann, P. Solubility and dissolution kinetics of GaN in supercritical ammonia in presence of ammonoacidic and ammonobasic mineralizers. *J. Cryst. Growth* **2017**, *479*, 59–66. [CrossRef]
22. Schimmel, S.; Duchstein, P.; Steigerwald, T.G.; Kimmel, A.C.L.; Schlücker, E.; Zahn, D.; Niewa, R.; Wellmann, P. In situ X-ray monitoring of transport and chemistry of Ga-containing intermediates under ammonothermal growth conditions of GaN. *J. Cryst. Growth* **2018**, *498*, 214–223. [CrossRef]
23. Lemmon, E.W.; McLinden, M.O.; Friend, D.G. Thermophysical Properties of Fluid Systems: NIST Chemistry WebBook. NIST Standard Reference Database Number 69. Available online: http://webbook.nist.gov/chemistry (accessed on 5 June 2017).
24. Spalding, D.B. A single formula for the "law of the wall". *J. Appl. Mech. Trans. ASME* **1960**, *28*, 455–458. [CrossRef]

materials

MDPI

Article

Electronic Properties of Group-III Nitride Semiconductors and Device Structures Probed by THz Optical Hall Effect

Nerijus Armakavicius [1,2,*], Philipp Kühne [1,2], Alexis Papamichail [1,2], Hengfang Zhang [1,2], Sean Knight [1,2], Axel Persson [1,2], Vallery Stanishev [1,2], Jr-Tai Chen [1,3], Plamen Paskov [1,2], Mathias Schubert [1,4] and Vanya Darakchieva [1,2,5,*]

[1] Center for III-Nitride Technology (C3NiT-Janzén), Linköping University, 581 83 Linköping, Sweden; philipp.kuhne@liu.se (P.K.); alexis.papamichail@liu.se (A.P.); vallery.stanishev@liu.se (V.S.); plamen.paskov@liu.se (P.P.); mschubert4@unl.edu (M.S.)

[2] Department of Physics, Chemistry and Biology (IFM), Linköping University, 581 83 Linköping, Sweden

[3] SweGaN AB, 582 78 Linköping, Sweden

[4] Department of Electrical and Computer Engineering, University of Nebraska-Lincoln, Lincoln, NB 68588, USA

[5] NanoLund, Center for III-Nitride Technology (C3NiT-Janzén), Terahertz Materials Analysis Center, THeMAC and Solid State Physics Division, Lund University, 221 00 Lund, Sweden

* Correspondence: nerijus.armakavicius@liu.se (N.A.); vanya.darakchieva@liu.se or vanya.darakchieva@ftf.lth.se (V.D.)

Abstract: Group-III nitrides have transformed solid-state lighting and are strategically positioned to revolutionize high-power and high-frequency electronics. To drive this development forward, a deep understanding of fundamental material properties, such as charge carrier behavior, is essential and can also unveil new and unforeseen applications. This underscores the necessity for novel characterization tools to study group-III nitride materials and devices. The optical Hall effect (OHE) emerges as a contactless method for exploring the transport and electronic properties of semiconductor materials, simultaneously offering insights into their dielectric function. This non-destructive technique employs spectroscopic ellipsometry at long wavelengths in the presence of a magnetic field and provides quantitative information on the charge carrier density, sign, mobility, and effective mass of individual layers in multilayer structures and bulk materials. In this paper, we explore the use of terahertz (THz) OHE to study the charge carrier properties in group-III nitride heterostructures and bulk material. Examples include graded AlGaN channel high-electron-mobility transistor (HEMT) structures for high-linearity devices, highlighting the different grading profiles and their impact on the two-dimensional electron gas (2DEG) properties. Next, we demonstrate the sensitivity of the THz OHE to distinguish the 2DEG anisotropic mobility parameters in N-polar GaN/AlGaN HEMTs and show that this anisotropy is induced by the step-like surface morphology. Finally, we present the temperature-dependent results on the charge carrier properties of 2DEG and bulk electrons in GaN with a focus on the effective mass parameter and review the effective mass parameters reported in the literature. These studies showcase the capabilities of the THz OHE for advancing the understanding and development of group-III materials and devices.

Keywords: electrical properties; free charge carriers; optical Hall effect; terahertz; group-III nitrides; gallium nitride; aluminum nitride; aluminum gallium nitride; high-electron-mobility transistor; HEMT

check for updates

Citation: Armakavicius, N.; Kühne, P.; Papamichail, A.; Zhang, H.; Knight, S.; Persson, A.; Stanishev, V.; Chen, J.-T.; Paskov, P.; Schubert, M.; et al. Electronic Properties of Group-III Nitride Semiconductors and Device Structures Probed by THz Optical Hall Effect. *Materials* **2024**, *17*, 3343. https://doi.org/10.3390/ma17133343

Academic Editor: Spiros Gardelis

Received: 30 April 2024
Revised: 11 June 2024
Accepted: 27 June 2024
Published: 5 July 2024

1. Introduction

The advent of group-III nitride semiconductor materials has revolutionized solid-state lighting [1,2] and radio frequency (RF) wireless communication [3,4] and it is already advancing power electronics in a critical and impactful way [5,6]. Indium nitride (InN), gallium nitride (GaN), and aluminum nitride (AlN) compounds and their alloys are direct bandgap semiconductors with bandgap energies ranging from 0.7 to 6.0 eV, facilitating their utilization in optoelectronics and photonics [7]. The GaN-based blue light-emitting diodes (LEDs) have enabled efficient and energy-saving lighting, for which the Nobel Prize in Physics 2014 was awarded. GaN and AlN have high critical electric fields, high saturation carrier velocities, and high thermal conductivity parameters, which make them promising candidates for replacing silicon (Si) in next-generation power electronic devices. The polarization-induced two-dimensional electron gas (2DEG), formed at the interface of AlGaN and GaN has enabled GaN-based high-electron-mobility transistors (HEMTs) [3,8]. These devices are suitable for high-power switching, power amplification, and high-frequency applications in the millimeter-wave range such as next-generation 5G and 6G communication systems, radars, and satellites [4].

Understanding the fundamental material parameters unambiguously drives the development of device technology forward. Especially for electronic and optoelectronic devices, a deep understanding of optical and electrical parameters and their behavior under various conditions is crucial. Moreover, revealing the underlying physics can lead to new and unforeseen applications of the materials. Both electrical and optical methods have been exploited to assess the charge carrier properties of group-III nitrides and their heterostructures. Commonly, 2DEG mobility and sheet carrier density are obtained from electrical direct-current (DC) Hall effect measurements, which require the fabrication of electrical contacts. Nondestructive and non-invasive measurement of the free charge carriers is key to advance modern materials and device design, but it constitutes a significant challenge. Contactless methods permit the testing of structures without the need for full device processing [9–12]. The optical Hall effect (OHE) is a contactless method for studying the properties of free and confined charge carriers in semiconductor materials. In addition, it simultaneously provides the dielectric function spectra, which are of high importance for electronic and optoelectronic devices. The significance of the OHE is that it allows obtaining quantitative information about the charge carrier parameters in both bulk materials and heterostructures in a fully contactless manner. In addition, the charge carrier effective mass parameter can be obtained in a wide temperature range, including room temperature and above, along with density and mobility parameters, providing a complete understanding of 2DEG and free carrier transport within the heterostructures. OHE studies of charge carriers in the THz range are often realized employing frequency domain systems employing the continuous-wave THz sources that were widely used to study free charge carrier properties of various electronic materials and device structures [13–18]. Studies involving THz time-domain systems [19,20] have also been used, primarily for the measurement of Faraday rotation and Kerr effect [21–23].

In this paper, we explore the application of THz OHE to study the charge carrier properties in group-III nitride heterostructures and epitaxial layers. Firstly, we demonstrate the THz OHE results for graded AlGaN channel HEMT structures with different grading profiles and compare their 2DEG properties with the case of conventional (non-graded) AlGaN/GaN HEMT. Secondly, we investigate the anisotropic mobility parameters in a N-polar GaN/AlGaN HEMT structure and discuss the origin of the anisotropy. In the third part, we present a study of the temperature-dependent effective mass parameters of 2DEGs in GaN-based HEMT structures and bulk electrons in GaN epitaxial layers, and review the effective mass parameters determined by OHE reported in the literature.

2. Materials and Methods

The optical Hall effect is produced by the motion of the free charge carriers under the influence of the external magnetic field as exemplified in Figure 1. An incident electromagnetic wave with linear polarization parallel to the plane of incidence interacting with the conductive sample subjected to an external magnetic field causes displacement of the free charge carriers along the direction of the electric field oscillations. As a result of the Lorentz force, the motion of the free charge carrier deviates from a straight line and adopts a small circular component (Figure 1). The circular component depends on the cyclotron frequency (and as a result on the effective mass parameter), and the Fermi velocity of the free charge carriers. By measuring this change in the polarization state by generalized ellipsometry, access to the free charge carrier sign, the density, mobility, and effective mass, including their anisotropy, is enabled in bulk [11,24–27] and 2D systems [17,18,28–31]. In contrast to classical cyclotron resonance and Shubnikov–de Haas oscillations, which require measurements in low-defect density materials to reduce impurity potential scattering and at very low temperatures to reduce phonon scattering, THz OHE provides access to the free charge carrier properties even at room temperature and in materials with a high concentration of defects.

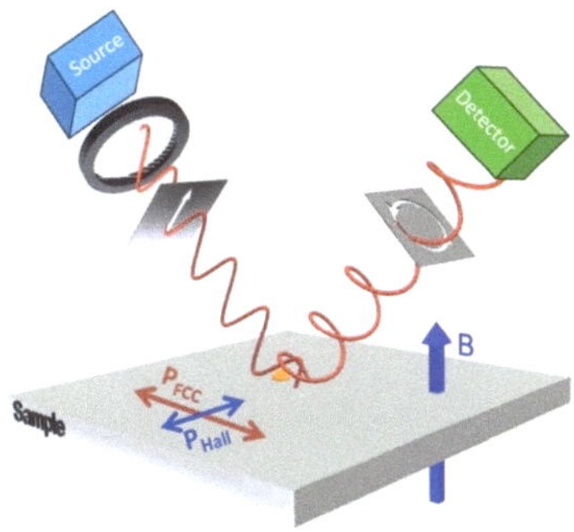

Figure 1. Schematic of the optical Hall effect in a reflection configuration. Free charge carriers produce a material polarization P_{FCC} following the electric field of an incident electromagnetic waves (analogous to the longitudinal voltage in electrical Hall effect case). The induced polarization P_{FCC} produces P_{Hall} due to the Lorentz force, oriented perpendicular to the external magnetic field and the incident electric field vector (analogous to the Hall voltage). P_{FCC} + P_{Hall} are the source of the reflected light, which acquires a small circular polarization component due to P_{Hall}, which mainly carries information about the type of the charge carriers, and the cyclotron frequency parameter, while the reflected wave induced by P_{FCC} is mainly determined by the plasma frequency and broadening parameters. These parameters can be recalculated into carrier density, mobility, and effective mass. If multiple layers are thin enough against the penetration depth of the wave, light interacts with multiple layers and reveals, for example, free carrier properties within buried layers otherwise inaccessible to direct electrical measurements. The figure is adapted from Ref. [11] under CC-BY 4.0 license.

Spectroscopic ellipsometry provides access to the dielectric function tensors of materials by measuring the light polarization change upon reflection (or transmission) from (through) a sample composed of, e.g., a single bulk substrate or multilayered structure [32]. The intrinsic material parameters can be evaluated from the obtained dielectric functions via analysis based on physical models. For example, in the infrared range, phonon excitations contribute to the dielectric function, while in FIR and THz ranges, contributions from free charge carriers become significant for conductive materials. Standard spectroscopic ellipsometry is commonly used to study samples when no conversion between *s*- and *p*-polarization modes (perpendicular and parallel to the plane of incidence, respectively) occurs upon reflection or transmission [32]. In more general cases, when the studied sample is anisotropic and causes conversion between the *s*- and *p*-polarization modes, standard ellipsometry is not sufficient, and generalized ellipsometry must be used, which measures the sample's optical response in terms of at least 6 frequency-dependent parameters.

Mueller matrix ellipsometry is a powerful technique categorized as generalized ellipsometry. It employs Stokes vector formalism to describe the polarization state of the light beam in terms of 4 real-valued parameters, S_0, S_1, S_2, and S_3, defined in terms of intensities measured for different light polarization state components [33]:

$$S = \begin{pmatrix} S_0 \\ S_1 \\ S_2 \\ S_3 \end{pmatrix} = \begin{pmatrix} I_x + I_y \\ I_x - I_y \\ I_{+45^\circ} - I_{-45^\circ} \\ I_R - I_L \end{pmatrix} \quad (1)$$

where I_x and I_y are intensities of linearly polarized light components along the x and y directions, which are commonly chosen to match with s- and p-polarization orientations, respectively (I_x=I_s and I_y=I_p); I_{+45°, I_{-45° are the intensities of light polarized at polarization planes rotated by 45° clockwise and anti-clockwise, respectively; and I_R and I_L are the intensities of right-hand and left-hand circular polarization components. Then, the light interaction with the sample is described as the transformation of the incoming beam's Stokes vector S_{in} into the outgoing beam's Stokes vector S_{out} by a 4×4 Mueller matrix:

$$S_{out} = \begin{pmatrix} M_{11} & M_{12} & M_{13} & M_{14} \\ M_{21} & M_{22} & M_{23} & M_{24} \\ M_{31} & M_{32} & M_{33} & M_{34} \\ M_{41} & M_{42} & M_{43} & M_{44} \end{pmatrix} S_{in} \quad (2)$$

Typically, the Mueller matrix is normalized by the M_{11} element. The primary advantages of Mueller matrix ellipsometry include obtaining ellipsometry data in real-valued parameters derived from directly measured intensity parameters, and providing information about depolarization effects that may arise during the measurement. Moreover, the formalism grants access to the optical anisotropy of measured samples, which can be readily inferred from the symmetry properties of the Mueller matrix elements.

Measured Mueller matrix data of the THz OHE require model-based data analysis to extract quantitative information about the studied sample. For that reason, an optical model is employed that allows calculation of the Mueller matrix spectra and fitting them to the experimental data by varying model parameters. The optical model is commonly described as a set of layers separated by parallel interfaces representing the sample structure. Each layer is described by its thickness parameter and complex dielectric function tensor $\tilde{\varepsilon}$. The ellipsometric data are calculated using 4×4 transfer matrix formalism [32,34,35]. Fitting is performed iteratively by adjusting the model parameters using a multivariate regression algorithm (the Levenberg–Marquardt method) until the mean-square error function is minimized. This results in the best-match model, which is used to extract the parameters of interest. Error bars are derived from the covariance matrix of the fit parameters simultaneously during the analysis and represent the standard 90% confidence limits. The data analysis presented in this work is performed using the WVase software (version 3) developed by J.A. Woollam Co Inc.

The most technologically relevant crystal structure of group-III nitride materials is the thermodynamically stable wurtzite crystal structure possessing uniaxial optical anisotropy. In such a case, the optical properties can be described by a frequency-dependent dielectric function tensor:

$$\tilde{\varepsilon} = \begin{bmatrix} \varepsilon_\perp & 0 & 0 \\ 0 & \varepsilon_\perp & 0 \\ 0 & 0 & \varepsilon_\parallel \end{bmatrix}, \quad (3)$$

where $\epsilon_\perp$ and $\epsilon_\parallel$ denotes the dielectric function for polarization perpendicular and parallel to the c-axis of the wurtzite crystal structure. The dielectric function of group-III nitrides in the THz spectral range is described by the contribution of free charge carriers $\tilde{\varepsilon}^{FCC}$, the tail of the phonon contributions in the mid-infrared (MIR) range $\tilde{\varepsilon}^L$, and an offset related to high-energy electronic excitations $\tilde{\varepsilon}_\infty$ [11]. It can be written as:

$$\tilde{\varepsilon} = \tilde{\varepsilon}_\infty(\tilde{\varepsilon}^L + \tilde{\varepsilon}^{FCC}) . \quad (4)$$

The phonon contribution is described by a factorized four-parameter semi-quantum model containing transverse and longitudinal optical phonon frequencies, (ω_{TO}, ω_{LO}) and broadenings (γ_{TO}, γ_{LO}) as parameters:

$$\tilde{\varepsilon}^{L} = \begin{pmatrix} \varepsilon_{\perp}^{L} & 0 & 0 \\ 0 & \varepsilon_{\perp}^{L} & 0 \\ 0 & 0 & \varepsilon_{\parallel}^{L} \end{pmatrix}, \tag{5a}$$

$$\varepsilon_{j}^{L} = \prod_{i} \frac{\omega_{\mathrm{LO},i,j}^{2} - \omega^{2} - i\omega\gamma_{\mathrm{LO},i,j}}{\omega_{\mathrm{TO},i,j}^{2} - \omega^{2} - i\omega\gamma_{\mathrm{TO},i,j}} \qquad (j = \perp, \parallel), \tag{5b}$$

with ω being the angular frequency, j denoting polarization direction, and i indicating different active optical phonon mode pairs. The effect of free charge carriers on the dielectric function is described by the classical Drude model, which is extended to include the contribution of an external static magnetic field and is often referred to as the magneto-optic Drude model. This model contains plasma frequency and broadening parameters, as well as the cyclotron frequency, which depends on carrier density, mobility, and effective mass parameters:

$$\tilde{\varepsilon}^{FCC} = -\tilde{\omega}_{p}^{2} \left(\omega^{2} + i\omega\tilde{\gamma}_{p} - i\omega\tilde{\omega}_{c} \begin{pmatrix} 0 & -b_{z} & b_{y} \\ b_{z} & 0 & -b_{x} \\ -b_{y} & b_{x} & 0 \end{pmatrix} \right)^{-1}, \tag{6}$$

with screened plasma frequency tensor $\tilde{\omega}_{p}$, effective mass tensor $\tilde{m}^{*}$, plasma broadening tensor $\tilde{\gamma}_{p}$, and cyclotron frequency tensor $\tilde{\omega}_{c}$ defined as:

$$\tilde{\omega}_{p}^{2} = \frac{Nq^{2}}{\varepsilon_{0}} \tilde{\varepsilon}_{\infty}^{-1} \tilde{m}^{*-1}, \quad \tilde{m}^{*} = \begin{bmatrix} m_{x}^{*} & 0 & 0 \\ 0 & m_{y}^{*} & 0 \\ 0 & 0 & m_{z}^{*} \end{bmatrix}, \quad \tilde{\gamma}_{p} = \begin{bmatrix} \tau_{x}^{-1} & 0 & 0 \\ 0 & \tau_{y}^{-1} & 0 \\ 0 & 0 & \tau_{z}^{-1} \end{bmatrix}, \quad \tilde{\omega}_{c} = q|\vec{B}|\tilde{m}^{*-1}. \tag{7}$$

Here, N, q, ε_{0} are the carrier concentration, charge carrier unit charge, and vacuum permittivity parameter, while τ is the scattering time. The mobility parameter can be obtained from the scattering time and effective mass parameters using the relation $\tilde{\mu} = q\tilde{\tau}\tilde{m}^{*-1}$. The magnetic field vector is defined as $\vec{B} = |\vec{B}|(b_{x}, b_{y}, b_{z})$. Normally, the effective mass and scattering time parameters are isotropic for polarization orthogonal to the c-axis; therefore, they can be written as $m_{\perp}^{*} = m_{x}^{*} = m_{y}^{*}$ and $\tau_{\perp} = \tau_{x} = \tau_{y}$, while $m_{\parallel}^{*} = m_{z}^{*}$ and $\tau_{\parallel} = \tau_{z}$. In case of lack of detectable anisotropy, all the effective mass and scattering time parameters can be replaced with the two scalar parameters m^{*} and τ. Regardless of whether the intrinsic anisotropy of the charge carrier properties is present or not, the free charge carrier contribution to the dielectric function tensor, $\tilde{\varepsilon}^{FCC}$, gives rise to non-zero off-diagonal block Mueller matrix elements (M_{ij}, $ij = 13, 14, 23, 24, 31, 32, 41, 42$) whenever the external magnetic field is present, resulting in optical birefringence that we refer to as the OHE.

The OHE measurements were performed using a THz frequency domain ellipsometer that was designed and built using a stealth technology approach to reduce the formation of standing waves. The system is described in detail in Reference [36] and covers a spectral range from 100 GHz to 1 THz. The THz ellipsometer operates in a polarizer-sample-rotating analyzer arrangement. In the Mueller matrix ellipsometry mode, it is capable of measuring the 3×3 upper left block ($\mathrm{M}_{ij}/\mathrm{M}_{11}$, where $i, j = 1, 2, 3$) of the 4×4 Mueller matrix ($\mathrm{M}_{ij}/\mathrm{M}_{11}$, where $i, j = 1, 2, 3, 4$). The THz system has a continuous wave solid-state source with a base frequency from 100 GHz to 170 GHz, which is augmented by Schottky-diode frequency multipliers, $\times 2$, $\times 3$, and $\times 6$ (Virgin Diodes, Charlottesville, VA, USA), providing the accessible range of 100–1000 GHz with a few narrow gaps that are not covered. The ellipsometer can perform cavity-enhanced OHE measurements at room temperature in reflection geometry using a permanent neodymium magnet (0.6 T) [37]. In addition, the system is augmented by a superconducting magnet with a cryostat (Cryogenics Ltd., London, UK), which allows performing OHE measurements in reflection or transmission modes at magnetic fields up to 8 T and sample temperatures from 2 K to 400 K. Conductive layers with thickness parameters much smaller than the wavelength of THz radiation (in our case ranging from 3 mm to 300 μm in the frequency range of 100 GHz–1 THz) behave as 2D conductive channels. Therefore, only the sheet carrier density ($N_s = Nd$, where N is carrier concentration and d is layer thickness) can be obtained from the THz OHE data analysis. To obtain the bulk carrier concentration, the thickness parameters of conductive layers must be measured independently and fixed in the model analysis. For the studies presented in this paper, the thickness parameters of thin layers were measured independently using Vis-UV ellipsometry (RC2, J.A. Woollam Co. Inc., Lincoln, NE, USA).

The studied group-III nitride epitaxial layers and device heterostructures were grown on semi-insulating 4H-SiC substrates using a hot-wall metal-organic chemical vapor deposition (MOCVD) reactor [38–40]. The standard characterization of the grown layers included X-ray diffraction (XRD) for compositional and crystal quality studies and electrical sheet resistance, capacitance–voltage (C-V) and Lehighton mobility measurements for carrier density and mobility parameters of 2DEG in HEMTs, and electrons in doped layers. For selected samples, the Al-content profiles and the structural quality were determined by scanning transmission electron microscopy (STEM) combined with energy-dispersive X-ray spectroscopy (EDS) [40,41]. The measurements were performed using the double corrected Linköping FEI Titan^3 60–300 microscope (Linköping, Sweden), operated at 300 kV. The built-in Super-X/Quantax EDS system (Bruker, Billerica, MA, USA) was employed, and the absolute quantification was made using the Esprit software (https://www.espritcam.com/, accessed on 29 December 2023) and its built-in calibrations for TEM-EDS.

3. Results

3.1. Graded AlGaN Channel HEMTs

Incorporating a thin AlGaN layer with Al composition grading as a channel at the interface between the GaN buffer and AlGaN barrier layers in AlGaN/GaN HEMTs modifies the distribution of the 2DEG carriers formed at the interface. The graded layer induces delocalization of the electrons' wavefunction which is now allowed to extend across the graded layer [40,42]. Here, we refer to the 2DEG channel formed at such graded channel HEMTs as a three-dimensional electron gas (3DEG) channel. Such spatially broader distribution in the conductive channel can improve the HEMT device linearity at high powers [40,42–45]. Here, we present THz OHE results for graded channel AlGaN/GaN HEMT structures with different grading profiles [40]. The HEMT structures consist of the following stack of layers on semi-insulating 4H-SiC substrate: AlN (60 nm)/GaN buffer (1100 nm)/graded $Al_xGa_{1-x}N$ channel (x = 0–0.1, 10 nm)/$Al_xGa_{1-x}N$ barrier ($x = 0.3$). Three different grading profiles were implemented: exponential, linear, and hybrid (hyperbolic tangent-linear). The growth conditions can be found in Ref. [40]. The studied structures with nominal Al grading profiles of the channel layer are shown in Figure 2a–c. The linearly graded channel HEMT also contains a GaN cap (2–3 nm) layer on top of the structure. The intended Al grading in the channel was confirmed by STEM and EDS measurements. Figure 2d,e show a representative STEM image and the respective EDS map of the Al content. Due to the very low thickness, it is difficult to resolve the GaN cap layer in the STEM image and the respective EDS map in Figure 2e. The high Al content layer that appears on top of the structure is an artifact related to the TEM sample preparation and is associated with the Al oxide that can form on top of the structure. The surface roughness (RMS) of the HEMT structures determined from 5 × 5 μm atomic force microscopy (AFM) images are 0.32 nm, 0.45 nm, and 0.28 nm for the exponential, hybrid, and linear grading channel HEMTs, respectively.

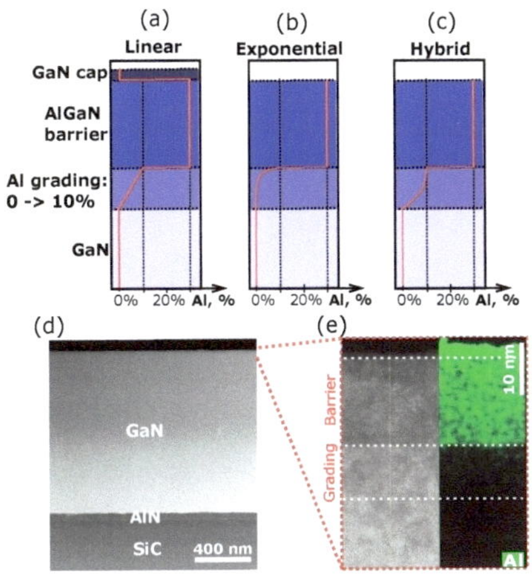

Figure 2. Graded channel AlGaN/GaN HEMT structures: (**a**–**c**) schematic drawings of the heterostructures with the nominal Al grading profiles in the channel, (**d**) STEM image of the linearly graded HEMT structure and (**e**) the EDS map of Al distribution in the vicinity of the AlGaN/GaN interface.

The THz OHE measurements were performed at room temperature in reflection geometry using the permanent neodymium magnet (0.6 T) and employing a backside cavity for OHE enhancement [36,37]. The measurements were carried out on both sides of the magnet, resulting in Mueller matrix datasets obtained at two opposing magnetic field directions ($\mathbf{M}[\pm 0.6\ T]$). The experimental difference Mueller matrix spectra ($\Delta\mathbf{M}[\pm 0.6\ T] = \mathbf{M}[+0.6\ T] - \mathbf{M}[-0.6\ T]$) and the best-match model spectra for the exponentially graded channel HEMT are shown in Figure 3a, and the spectra for the selected off-diagonal-block Mueller matrix elements ($\Delta M_{13,23}$) are shown for all HEMT structures in Figure 3b. Specifically, the off-diagonal block elements are proportional to the cyclotron frequency ω_c and provide sensitivity for this parameter. The experimental difference Mueller matrix spectra reveal OHE signatures following the Fabry–Perot oscillations pattern within the sample–air gap–mirror optical system [13,37], with the strongest OHE signatures in the spectral ranges where the reflectivity is the smallest. The lineshapes and amplitudes of these features are governed by the free electron properties. More specifically, the features observed in the Mueller matrix difference spectra are caused by differences between left- and right-hand circularly polarized light propagation due to the magneto-optic anisotropy of the free charge carriers. Therefore, the differences between the Mueller matrix spectra among the samples indicate different free-charge carrier properties of the 3DEG channel. In addition, the spectral position of the observed features is also affected by the substrate and cavity thickness parameters. Note that the Mueller matrix difference spectra are equal to zero in cases where there are no free-charge carriers in the structure. The 3DEG parameters extracted from the best-match model analysis are shown in Table 1.

The sensitivity of the THz OHE to the out-of-plane anisotropy, i.e., to the parameters $\tau_{\|}$ and $m^*_{\|}$, is very low for such thin conductivity channels, therefore these parameters are always assumed to be equal to the in-plane parameters. Here, the model used for data analysis further assumes the effective mass and mobility parameters being in-plane isotropic. The very good agreement between the experimental data and the model shows that no detectable in-plane anisotropy of the 3DEG properties for the studied structures can be inferred.

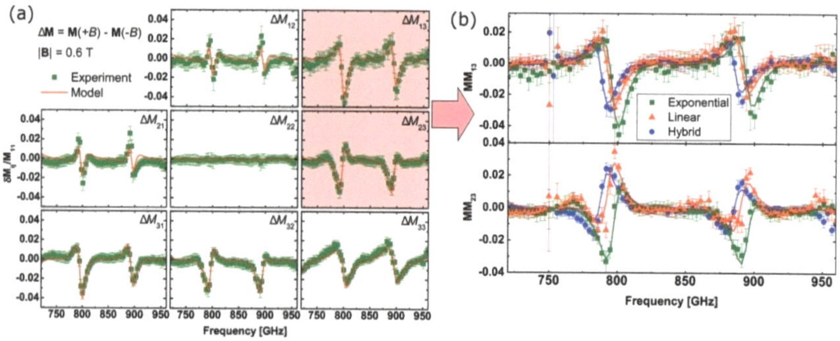

Figure 3. (**a**) The experimental (squares) and the best-match model (solid line) Mueller matrix difference spectra ($\Delta\mathbf{M}(\pm 0.6\ T) = \mathbf{M}(+0.6\ T) - \mathbf{M}(-0.6\ T)$) for the exponentially graded channel AlGaN/GaN-HEMT. (**b**) Comparison of the selected experimental (squares, triangles, circles) and best-match model (solid lines) off-diagonal-block Mueller matrix element spectra ($\Delta M_{13,23}$) for the graded channel AlGaN/GaN HEMTs with different grading profile in the channel.

Table 1. The 3DEG parameters of the graded-channel AlGaN/GaN HEMTs with different grading profiles obtained from the THz OHE best-match model.

Channel Grading	Sheet Density, 10^{13} cm^{-2}	Mobility, cm^{2}V^{-1}s^{-1}	Effective Mass, m_0
Exponential	1.05 ± 0.18	960 ± 60	0.40 ± 0.03
Hybrid	1.11 ± 0.09	730 ± 50	0.44 ± 0.03
Linear	1.27 ± 0.15	720 ± 40	0.43 ± 0.02

For all graded HEMT structures, the obtained sheet density parameters are in the range of $N_s = 1.1 \times 10^{13} - 1.3 \times 10^{13}$ cm^{-2} and can be considered to be the same within the 90% confidence limits. Mobility values for all the samples are significantly lower than those in conventional (non-graded channel) AlGaN/GaN HEMT structures with sharp interfaces and typical mobilities around 2000 cm^{2}V^{-1}s^{-1} and above [18,40]. This can be anticipated since the alloy scattering in the graded channel is expected to be significantly stronger as compared to the case of non-graded conventional HEMTs. We find that the exponentially graded channel-HEMT has slightly higher mobility ($\sim$30%), $\mu^{exp} = 960(\pm 60)$ cm^{2}V^{-1}s^{-1}, in comparison with the hybrid and linear grading structures with mobility values of $\mu^{hybrid} = 730(\pm 50)$ cm^{2}V^{-1}s^{-1} and $\mu^{linear} = 720(\pm 40)$ cm^{2}V^{-1}s^{-1}, respectively. This could be explained by the effectively sharper potential barrier for the 3DEG and as a result of a lower Al concentration in the channel region where most of the electrons are localized, resulting in a lower alloy scattering. The exponential grading channel HEMT structure has the lowest 3DEG effective mass parameter. That further supports the statement that electrons in the exponentially graded channel are more localized in the region with lower Al concentration. Note that this leads to a lower effective mass since the latter increases with Al content [25]. As the mobility parameter inversely depends on the effective mass ($\mu = e\tau m^{*-1}$), the lower effective mass partly explains the higher mobility for this sample. However, the difference in effective masses is only around 10%. Therefore, this cannot fully account for the higher 3DEG mobility parameter in the exponential grading channel, showing that a longer scattering time is associated with the HEMT. We note that the results obtained from the THz OHE on mobility parameters and sheet density are in good agreement with the results from C-V, Lehighton mobility, and Eddy current measurements [46]. More information about the graded-channel AlGaN/GaN HEMT structures, their structural and electrical properties, and the respective device performance can be found in Ref. [40].

3.2. *Anisotropic Mobility in N-Polar GaN/AlGaN HEMT*

Nitrogen-polar III-nitride heterostructures offer advantages over metal-polar structures in high-frequency and high-power applications [47]. Due to the inverted polarization field in N-polar structures as compared to their Ga-polar counterparts, the 2DEG channel forms at the interface between the top GaN layer and the AlGaN back barrier. This enables enhanced 2DEG confinement and reduction in the on-resistance, resulting in improved power efficiency. The N-polar GaN/AlGaN HEMT structure studied here was grown on 4H-SiC (000$\bar{1}$) substrate with miscut angle of 4° toward the m-plane (10$\bar{1}$0). The high-temperature AlN nucleation layer was grown on that substrate, which was followed by the growth of the GaN buffer layer using a multi-step temperature process [48,49], the $Al_{0.31}Ga_{0.69}N$ back barrier and the GaN channel layer on top. The surface roughness (RMS) determined from 5 × 5 µm AFM images was 2 nm.

The THz OHE measurements were performed in reflection geometry (angle of incidence 45°) using the permanent neodymium magnet (0.6 T) and employing a backside cavity [37]. The measurements were taken at both sides of the neodymium magnet, resulting in two opposing magnetic field directions ($\mathbf{M}[\pm 0.6\ T]$). The Mueller matrix spectra for positive magnetic field ($\mathbf{M}[+0.6\ T]$) are shown in Figure 4. Most of the Mueller matrix spectra contain distinct spectral features modulated by the Fabry–Perot oscillations within the mirror(magnet)–cavity–sample optical system. The red solid lines indicate the best-match model data, which can be seen to be in perfect agreement with the experimental data points.

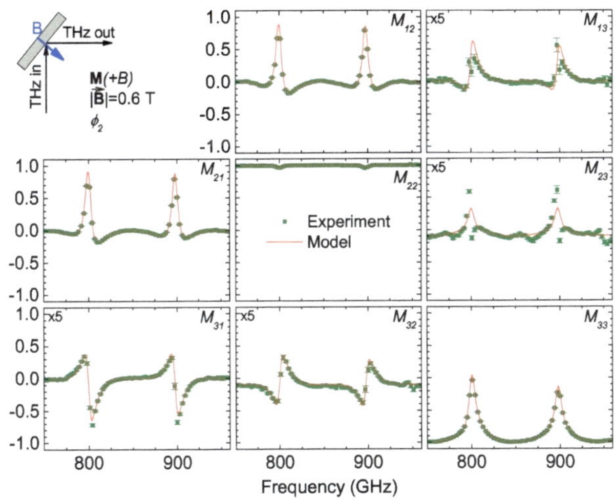

Figure 4. THz-OHE Mueller matrix spectra ($\mathbf{M}[+0.6\ T]$) for N-polar GaN/AlGaN HEMT structure measured in reflection geometry (at angle of incidence of 45°) with the sample mounted on the permanent neodymium magnet and implemented backside cavity.

The THz OHE measurements revealed a significant dependence of the Mueller matrix spectra on the azimuth orientation (in-plane rotation of the sample with respect to the surface normal) of the N-polar HEMT sample. The measured Mueller matrix spectra of the block-off diagonal elements ($M_{13,23,31,32}$) for three different azimuth orientations (ϕ_1, ϕ_2, ϕ_3) are shown in Figure 5. Each column represents the different azimuth orientation with a nominal 45° rotation in between. Such azimuth dependence of the Mueller matrix spectra is a direct fingerprint of the in-plane optical anisotropy of the sample. Without the presence of in-plane anisotropy, the OHE induces symmetric off-diagonal block Mueller matrix elements, i.e., $M_{ij} = M_{ij}$, with $ij = 13, 23, 31, 32$. Any in-plane anisotropy causes antisymmetric Mueller matrix spectra $M_{ij} = -M_{ij}$, with $ij = 13, 23, 31, 32$ at B = 0 T, whenever the anisotropy axis is not parallel to the plane of incidence. In Figure 5a,b,d,e,g,h, one can observe that different azimuth angles result in different symmetry properties of the off-diagonal block Mueller matrix elements. At the azimuth angles ϕ_1 and ϕ_3, the spectra are nearly symmetric ($M_{ij} \approx M_{ji}$); however, the spectra are not the same for the two different orientations. Meanwhile, the spectra at ϕ_2 show more complicated asymmetric behavior as a result of the convolution of the magneto-optical and in-plane anisotropy effects, which correspond to symmetric and antisymmetric parts.

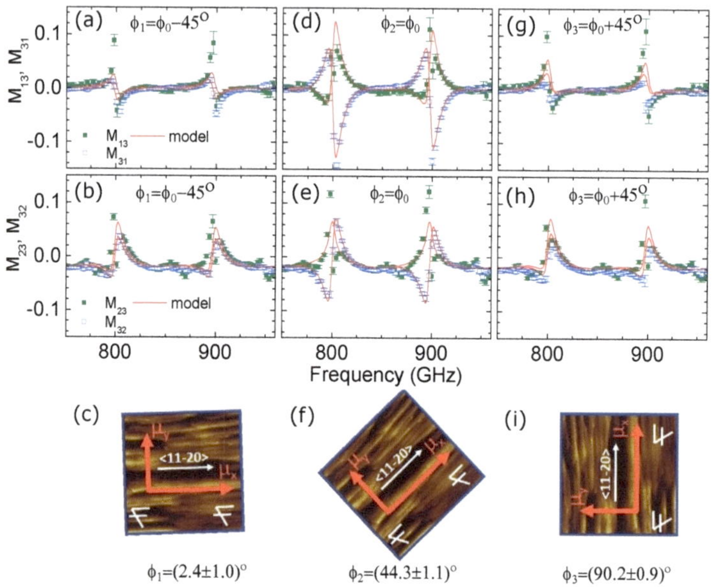

Figure 5. Experimental (symbols) and the best-match model (solid lines) THz-OHE off-diagonal block Mueller matrix elements spectra (M_{13}, M_{23}, M_{31}, M_{32}) for N-polar GaN/AlGaN HEMT structure measured at 3 different azimuth orientations (**a**,**b**—ϕ_1, **d**,**e**—ϕ_2, and **g**,**h**—ϕ_3). The azimuth angles obtained from the best-match optical model representing the angle between the fast mobility axis μ_x and the plane of incidence, and the AFM images showing the absolute sample orientation for each measurement are depicted in (**c**,**f**,**i**).

Our model-based data analysis revealed that the observed anisotropy is induced by the directionally dependent mobility parameter of the 2DEG channel. Two different in-plane mobility parameters (μ_x and μ_y) for the 2DEG in the magneto-optic Drude model had to be implemented in the model in order to fit the experimental data with the model. The out-of-plane mobility parameter μ_z was set to zero due to the limited sensitivity to the out-of-plane optical properties for such a thin layer (2DEG is localized within a few nanometers). Therefore, in the best-match model, the dielectric function for the 2DEG channel contains the Drude contribution, with the mobility parameter expressed as a second-rank tensor with the on-diagonal elements representing the mobilities in the three different orthogonal directions being different. The azimuth angle ϕ indicates the angle between the plane of incidence and the μ_x mobility axis. It is a fitting parameter in the model analysis and therefore was extracted from the best-match model. When mobility axes are nearly aligned with the plane of incidence (at $\phi_1 = 2.4°$ and $\phi_3 = 90.2°$), the Mueller matrix spectra are nearly symmetric and therefore, the spectral features in Figure 5 are mainly induced by the OHE. At the azimuth angle $\phi_2 = 44.3°$, the μ_x mobility axis makes an angle with the plane of incidence, and therefore the Mueller matrix spectra (in Figure 5d,e) reflect the combined magneto-optic and in-plane anisotropy-induced effects.

The in-plane mobility parameters extracted from the best-match model are $\mu_x = 1170 \pm 20$ and $\mu_y = 820 \pm 10$. The higher mobility parameter along the fast axis μ_x is around 40% larger than the mobility along the slow axis. This results in significant anisotropy as observed in the Mueller matrix spectra variation with the azimuth orientation of the sample (Figure 5). The determined sheet carrier density was $N_s = 7.8(\pm 0.3) \times 10^{12}$ cm^{-2}, while the extracted effective mass parameter was $0.258(\pm 0.002)\ m_0$.

Additional AFM measurements of the N-polar HEMT structure were performed in order to correlate the THz OHE results with the surface morphology (see Figure 5c,f,i). The absolute azimuth orientation of the sample in the AFM measurements was recorded and correlated with the azimuth angles of the fast and slow mobility axis extracted from the THz OHE measurements. The AFM shows a step-like surface morphology, typical for the growth on the off-axis SiC substrate [48]. The AFM results further reveal that the mobility axes follow the orientation of the steps with the fast mobility axis μ_x being oriented along the steps parallel to the [11$\bar{2}$0], while the slow mobility axis μ_y is oriented across the steps i.e., along the [1$\bar{1}$00] (see Figure 5c,f,i). This suggests that the step-like morphology also translates into the GaN/AlGaN interface where 2DEG is located and introduces the anisotropic mobility. The results indicate that steps introduce additional scattering of the 2DEG for electrons moving in the direction perpendicular to the steps. A similar effect is observed for another 2D electronic system on SiC substrate, namely, epitaxial graphene , where the step-like surface morphology of the substrate introduces additional carrier scattering, leading to significant in-plane optical anisotropy observed by OHE measurements [17].

3.3. *Temperature Dependence of Electronic Properties*

The OHE measurements allow accessing the charge carrier properties in a wide temperature range, which is mainly limited by the capabilities of the current experimental setup (2 K to 400 K). A major advantage of OHE is the access to the effective mass parameter at elevated temperatures along with the free charge carrier density and mobility (or scattering time). The OHE data are sensitive to the cyclotron frequency parameter which depends on the effective mass even when the former is outside the measured frequency range. This allows determination of the effective mass parameter from model-based analysis of the OHE data. Commonly used methods for effective mass determination, such as Shubnikov–de Haas oscillations or cyclotron resonance measurements, require high mobilities, low temperatures, and high magnetic fields. In contrast, OHE has been demonstrated to provide effective mass parameters even at room temperature or above for various electronic materials at magnetic fields as low as 0.5 T [13,18,24–28,36,37,50].

Bulk free charge carriers in group-III nitrides are generated by a donor, i.e., Si, and an acceptor, i.e., Mg doping. On the other hand, AlGaN/GaN HEMT structures contain 2DEG channel at the interface between two nominally undoped layers with different bandgaps, and different spontaneous and piezoelectric polarizations [51,52]. Bulk electrons and 2DEG have very different electrical properties, which stipulate their different temperature dependencies. The 2DEG channel demonstrates much higher mobility values in comparison to the bulk carriers. Two HEMT structures containing 2DEG and an n-type doped GaN substrate containing bulk (3D) free electrons were selected for temperature dependence study. The HEMT structures were grown epitaxially on semi-insulating 4H-SiC (0001) substrates using an AlN nucleation layer with a nominal thickness of 60 nm followed by a nominally 1.9 um-thick relaxed GaN buffer layer, and by a 10–20 nm thick pseudomorphic AlGaN (nominal Al content of 29%) or AlN barrier layers [40]. The n-type GaN substrate (MSE Supplies) contains a nominal Si concentration of $N_{\rm Si} = 1 \times 10^{17}$ cm^{-3}[31]. The THz OHE measurements of the samples were performed in the spectral range of 700–1000 GHz. The charge carrier properties of bulk 3D electrons in the GaN substrate, and 2DEGs in the AlGaN/GaN and AlN/GaN HEMT structures obtained from THz OHE in a wide temperature range from 4.5 K to 400 K are shown in Figure 6.

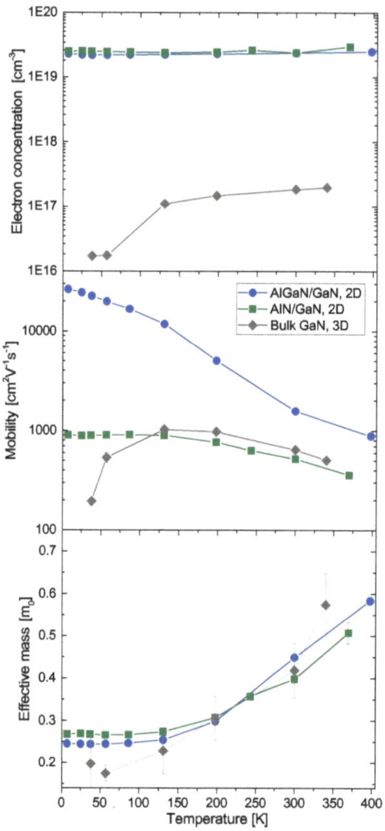

Figure 6. Temperature dependence of electron concentration, mobility and effective mass parameters of 2DEG in AlGaN/GaN (nominal Al content of 29%) and AlN/GaN HEMT structures, and bulk carriers (denoted as 3D) in GaN substrate doped with Si (nominal Si concentration $N_{Si} = 1 \times 10^{17}$ cm^{-3}) [31]. The electron concentration for the AlGaN/GaN and AlN/GaN HEMTs structures is calculated from the sheet density parameters N_s, obtained from the THz OHE analysis, and assuming a 4 nm layer thickness and uniform distribution across the channel ($N = N_s/d$).

For comparative purposes, in Figure 6, the electron concentrations for AlGaN/GaN and AlN/GaN HEMTs are calculated from the obtained sheet density parameters N_s, assuming a 4 nm layer thickness and uniform distribution across the channel ($N = N_s/d$, where $d = 4$ nm). As expected, the 2DEG density remains virtually unchanged with temperature, while the carrier density of the n-type doped GaN substrate gradually increases with temperature due to the thermal activation of the donor states. The mobility behavior is also markedly different. For the 2DEG, the mobility decreases with increasing temperature. In contrast, for n-type GaN, the electron mobility parameter first increases, reaching a maximum value of around 150–200 K, after which it decreases with increasing temperature. For both bulk carriers and 2DEG, the mobility at high temperatures is limited by optical phonon scattering, while at low temperatures, the scattering mechanisms limiting the mobility are different [53–55]. For bulk (3D) carriers, the low-temperature mobility is limited by impurity scattering, which is negligible for the 2DEG [53,54,56]. For the 2DEG, the low-temperature mobility is limited by either alloy disorder or interface roughness scattering. The AlN/GaN HEMT has significantly lower mobility compared to the AlGaN/GaN HEMT, which could be related to higher alloy disorder scattering. We observed in similar structures grown by MOCVD that achieving a pure AlN barrier layer is very difficult. Instead, it commonly results in a compositionally graded Al-rich AlGaN layer, causing strong delocalization of the electron wavefunction and high penetration into the barrier layer (results to be published elsewhere).

Despite the differences in the temperature dependencies of the bulk and 2DEG carrier density and mobility parameters, the results for the respective effective mass parameters are very similar. For all of the studied samples, at low temperatures, the effective mass parameter value is close to the well-established literature value of $0.23m_0$ [57,58], while above 150 K, it shows an increase with temperature. Initially, such a temperature dependence of the effective mass parameter was observed for 2DEG in AlGaN/GaN HEMT structures without an AlN inter-layer between the buffer and the barrier layers and assigned to the 2DEG wavefunction hybridization due to a penetration into the AlGaN barrier layer [18,28]. However, such a strong increase with temperature cannot be explained by hybridization alone, nor can it be explained by conduction band non-parabolicity. Polaronic effects due to the coupling of electrons to the optical phonons could also be excluded as a sole reason for the effective mass enhancement, as they only account for about 10 % increase. A detailed study of the electron effective mass parameter increase at high temperatures in bulk GaN was recently reported in Ref. [31]. It was suggested that possible deviation of the free charge carrier behavior from the classical Drude model commonly used to describe conductivity in group-III nitrides (leading to the renormalization of the effective mass parameter) may play an important role for the observed increase in the electron effective mass parameter at high temperatures and THz frequencies [31].

Table 2 summarizes the published effective mass parameters obtained via Drude model-based analysis of the THz OHE measurements of various group-III nitride heterostructures containing 2DEG conductivity channels. The RT electron effective mass parameter of 2DEG is close to 0.30 m_0 for HEMTs with a GaN channel. This value is higher than the well-accepted value of 0.23 m_0 for bulk GaN. The latter, however, is primarily determined by infrared spectroscopy, cyclotron resonance or Shubnikov–de Haas oscillations techniques at low temperatures. The 2DEG effective mass parameter of high Al content AlGaN channel HEMT is even higher than 0.60 m_0. Figure 7 summarizes the bulk effective mass parameters obtained from MIR OHE for group-III nitride materials [24–26,50,59–64]. The bulk effective mass parameter for GaN obtained from THz-OHE is included for comparison and can be seen to be significantly higher [31].

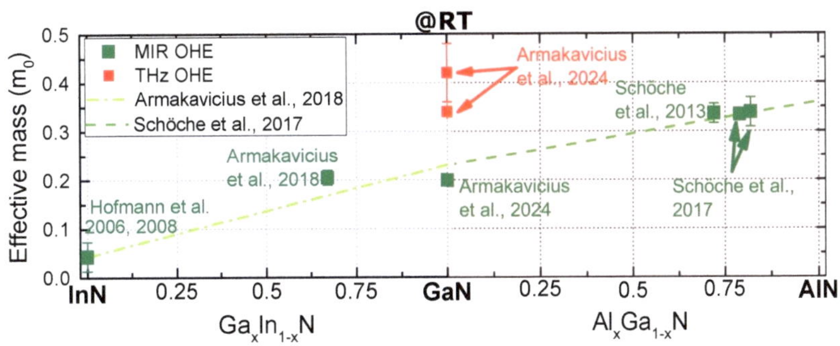

Figure 7. Summary of the published bulk (3D) electron effective mass parameters (isotropically averaged) at room temperature (RT) in group-III nitrides determined from MIR OHE [24–26,50,59]. The bulk effective mass of GaN obtained from THz-OHE at RT for n-type bulk GaN substrate and epitaxial layer on SiC are also included for comparison [31]. The dashed lines indicate linear dependencies of the electron effective mass parameters for $Ga_xIn_{1-x}N$ (conduction band minima) and $Al_xGa_{1-x}N$ published in references [25,26] reporting MIR OHE results.

Table 2. Summary of the effective mass parameters for the 2DEG in HEMT structures obtained from THz-OHE.

Structure	Substrate	Temp., K	Mag. Field, T	Exp. Eff. Mass, m_0	Ref.
1) $Al_{0.82}In_{0.18}N/AlN/GaN$	Al_2O_3	RT	0.55	0.27(±0.01)	[13]
2) $Al_{0.75}Ga_{0.25}N/GaN$	4H-SiC	RT	0.55	0.32(±0.01)	[36]
3) $Al_{0.19}Ga_{0.81}N/GaN$	4H-SiC	RT	0.55	0.30(±0.01)–0.32(±0.01)	[18]
4) $Al_{0.82}In_{0.18}N/AlN/GaN$	Al_2O_3	RT	0.55	0.24(±0.02)	[37]
5) $Al_{0.25}Ga_{0.75}N/GaN$	4H-SiC	1.5–300	3	0.22(±0.01)–0.36(±0.03)	[28]
6) * $AlN/Al_{0.78}Ga_{0.22}N$	4H-SiC	5	8	0.63(±0.04)	[30]
7) $Al_{0.29}Ga_{0.71}N/GaN$	4H-SiC	7–400	8	0.25(±0.01)–0.58(±0.01)	This work
8) AlN/GaN	4H-SiC	7–370	8	0.27(±0.01)–0.51(±0.03)	This work

* Note that this structure contains 2DEG in $Al_{0.78}Ga_{0.22}N$, while all other HEMTs contain 2DEG in GaN.

3.4. Conclusions

In summary, we present the application of THz OHE to study charge carrier properties of group-III nitride heterostructures and bulk material using several examples of contemporary research significance. We show that in graded channel AlGaN/GaN HEMT structures for linear power amplifiers and receivers, exponential grading of the Al content over 10 nm of the channel thickness leads to higher mobility and a lower effective mass parameter as compared to the cases of linear and hybrid Al grading profiles. The results suggest that exponential grading causes free electrons to be localized in the channel region with lower Al content, resulting in reduced alloy scattering. THz OHE study of an N-polar GaN/AlGaN HEMT structure grown on a SiC substrate with off-cut angle toward the m-plane reveals strong optical anisotropy inferred from the symmetry properties of the measured Mueller matrix spectra. Data analysis reveals in-plane anisotropic electron mobility due to directionally dependent 2DEG scattering induced by the step-like surface morphology, with higher mobility along the steps parallel to the $[11\bar{2}0]$ plane. We also compare free-charge carrier properties over a wide temperature range from 4.5 K to 400 K between 2DEG and bulk electrons in GaN. An increase in the electron effective mass parameter with the temperature is observed in both cases independently of the differences between 2D and bulk carriers. This observation is consistent with the prevalence of the reported GaN electron effective mass parameters in the literature. Such behavior could be potentially explained by a deviation of the charge carrier behavior from the Drude model, commonly used to describe conductivity in group-III nitride materials, in addition to the non-porabolicity, polaronic, and hybridization effects. These studies showcase THz-OHE as a powerful tool for studying the free and confined charge carrier properties of group-III nitride materials and heterostructures at different temperatures, with high sensitivity to their anisotropy.

Exciting new opportunities are presented by the future development of OHE measurements of group-III-nitride device structures under different conditions relevant to the device operation. Further development of the THz OHE technique can enable in situ studies of the charge carrier properties under electric gating, and high temperatures in fully contactless manner, which can further advance our understanding of the factors affecting device performance and can provide routes to improve it beyond the current state of the art.

Author Contributions: Conceptualization, N.A., P.K., A.P. (Alexis Papamichail), H.Z. and V.D.; methodology, N.A., P.K., S.K., A.P. (Alexis Papamichail), H.Z., V.S., J.-T.C., P.P., M.S. and V.D.; validation, N.A., P.K., A.P. (Alexis Papamichail), H.Z., A.P. (Axel Persson) and V.D.; formal analysis, N.A., P.K., S.K. and A.P. (Axel Persson); investigation, N.A., P.K., S.K. and V.S.; resources, J.-T.C., V.D. and M.S.; data curation, N.A. and P.K.; writing—original draft, N.A.; writing—review and editing, N.A. and V.D.; visualization, N.A., P.K. and A.P. (Axel Persson); supervision, M.S. and V.D.; project administration, P.P., M.S. and V.D.; funding acquisition, M.S. and V.D. All authors have read and agreed to the published version of the manuscript.

Funding: This work is performed within the framework of the competence center for III-Nitride technology, C3Nit-Janzén supported by the Swedish Governmental Agency for Innovation Systems (VINNOVA) under the Competence Center Program Grant No. 2022-03139, Lund University, Linköping University, Chalmers University of Technology, Ericsson, Epiluvac, FMV, Gotmic, Hexagem, Hitachi Energy, On Semiconductor, Region Skåne SAAB, SweGaN, Volvo Cars and UMS. We further acknowledge support from the Swedish Research Council VR under Awards No. 2016-00889 and 2022-04812, Swedish Foundation for Strategic Research under Grants No. RIF14-055 and No. EM16-0024, and the Swedish Government Strategic Research Area NanoLund, Lund University and in Materials Science on Functional Materials at Linköping University, Faculty Grant SFO Mat LiU No. 2009-00971. M.S. acknowledges support by the National Science Foundation under awards ECCS 2329940 and EPSCoR RII Track-1: Emergent Quantum Materials and Technologies (EQUATE), OIA-2044049, by Air Force Office of Scientific Research under awards FA9550-19-S-0003, FA9550-21-1-0259, and FA9550-23-1-0574 DEF, and by the University of Nebraska Foundation. M.S. also acknowledges support from the J. A. Woollam Foundation.

Institutional Review Board Statement: Not applicable.

Informed Consent Statement: Not applicable.

Data Availability Statement: The data presented in this study are available on request from the corresponding author. The data are not publicly available due to privacy or ethical restrictions.

Acknowledgments: We acknowledge Steffen Richter for his support.

Conflicts of Interest: Author Jr-Tai Chen was employed by the company SweGaN AB. The remaining authors declare that the research was conducted in the absence of any commercial or financial relationships that could be construed as a potential conflict of interest.

Abbreviations

The following abbreviations are used in this manuscript:

THz	Terahertz
GHz	Gigahertz
OHE	Optical Hall Effect
HEMT	High-electron-mobility transistor
2D	Two-dimensional
3D	Three-dimensional
2DEG	Two-dimensional electron gas
3DEG	Three-dimensional electron gas
RT	Room temperature
GaN	Gallium nitride
AlN	Aluminum nitride
AlGaN	Aluminum gallium nitride
AlINN	Aluminum indium nitride
SiC	Silicon carbide
Vis	Visible
UV	Ultraviolet
MOCVD	Metal-organic chemical vapor deposition
XRD	X-ray diffraction

References

1. Pimputkar, S.; Speck, J.; DenBaars, S.; Nakamura, S. Prospects for LED lighting. *Nat. Photon.* **2009**, *3*, 180. [CrossRef]
2. DenBaars, S.P.; Feezell, D.; Kelchner, K.; Pimputkar, S.; Pan, C.C.; Yen, C.C.; Tanaka, S.; Zhao, Y.; Pfaff, N.; Farrell, R.; et al. Development of gallium-nitride-based light-emitting diodes (LEDs) and laser diodes for energy-efficient lighting and displays. *Acta Mater.* **2013**, *61*, 945–951. [CrossRef]
3. Mishra, U.K.; Shen, L.; Kazior, T.E.; Wu, Y.F. GaN-Based RF Power Devices and Amplifiers. *Proc. IEEE* **2008**, *96*, 287–305. [CrossRef]
4. Lu, H.; Zhang, M.; Yang, L.; Hou, B.; Martinez, R.P.; Mi, M.; Du, J.; Deng, L.; Wu, M.; Chowdhury, S.; et al. A review of GaN RF devices and power amplifiers for 5G communication applications. *Fundam. Res.* **2023**. [CrossRef]
5. Amano, H.; Baines, Y.; Beam, E.; Borga, M.; Bouchet, T.; Chalker, P.R.; Charles, M.; Chen, K.J.; Chowdhury, N.; Chu, R.; et al. The 2018 GaN power electronics roadmap. *J. Phys. D Appl. Phys.* **2018**, *51*, 163001. [CrossRef]

6. Tsao, J.Y.; Chowdhury, S.; Hollis, M.A.; Jena, D.; Johnson, N.M.; Jones, K.A.; Kaplar, R.J.; Rajan, S.; Van de Walle, C.G.; Bellotti, E.; et al. Ultrawide-bandgap semiconductors: Research opportunities and challenges. *Adv. Electron. Mater.* **2018**, *4*, 1600501. [CrossRef]
7. Amano, H.; Collazo, R.; Santi, C.D.; Einfeldt, S.; Funato, M.; Glaab, J.; Hagedorn, S.; Hirano, A.; Hirayama, H.; Ishii, R.; et al. The 2020 UV emitter roadmap. *J. Phys. D Appl. Phys.* **2020**, *53*, 503001. [CrossRef]
8. Mishra, U.; Parikh, P.; Wu, Y.F. AlGaN/GaN HEMTs-an overview of device operation and applications. *Proc. IEEE* **2002**, *90*, 1022–1031. [CrossRef]
9. Huang, Y.S.; Pollak, F.H. Non-destructive, room temperature characterization of wafer-sized III–V semiconductor device structures using contactless electromodulation and wavelength-modulated surface photovoltage spectroscopy. *Phys. Status Solidi A* **2005**, *202*, 1193. [CrossRef]
10. Kudrawiec, R.; Syperek, M.; Motyka, M.; Misiewicz, J.; Paszkiewicz, R.; Paszkiewicz, B.; Tlaczala, M. Contactless electromodulation spectroscopy of AlGaN/GaN heterostructures with a two-dimensional electron gas: A comparison of photoreflectance and contactless electroreflectance. *J. Appl. Phys.* **2006**, *100*, 013501. [CrossRef]
11. Schubert, M.; Kühne, P.; Darakchieva, V.; Hofmann, T. Optical Hall effect—Model description: Tutorial. *J. Opt. Soc. Am. A* **2016**, *33*, 1553. [CrossRef]
12. Turkulets, Y.; Shalish, I. Contactless Method to Measure 2DEG Charge Density and Band Structure in HEMT Structures. *IEEE J. Electron Devices Soc.* **2018**, *6*, 703. [CrossRef]
13. Knight, S.; Schöche, S.; Kühne, P.; Hofmann, T.; Darakchieva, V.; Schubert, M. Tunable cavity-enhanced terahertz frequency-domain optical Hall effect. *Rev. Sci. Instruments* **2020**, *91*, 083903 [CrossRef] [PubMed]
14. Hofmann, T.; Herzinger, C.; Tedesco, J.; Gaskill, D.; Woollam, J.; Schubert, M. Terahertz ellipsometry and terahertz optical-Hall effect. *Thin Solid Film.* **2011**, *519*, 2593–2600.
15. Kühne, P.; Hofmann, T.; Herzinger, C.; Schubert, M. Terahertz optical-Hall effect for multiple valley band materials: N-type silicon. *Thin Solid Film.* **2011**, *519*, 2613–2616. [CrossRef]
16. Armakavicius, N.; Bouhafs, C.; Stanishev, V.; Kühne, P.; Yakimova, R.; Knight, S.; Hofmann, T.; Schubert, M.; Darakchieva, V. Cavity-enhanced optical Hall effect in epitaxial graphene detected at terahertz frequencies. *Appl. Surf. Sci.* **2017**, *421*, 357–360. [CrossRef]
17. Armakavicius, N.; Kühne, P.; Eriksson, J.; Bouhafs, C.; Stanishev, V.; Ivanov, I.G.; Yakimova, R.; Zakharov, A.A.; Al-Temimy, A.; Coletti, C.; et al. Resolving mobility anisotropy in quasi-free-standing epitaxial graphene by terahertz optical Hall effect. *Carbon* **2021**, *172*, 248–259. [CrossRef]
18. Armakavicius, N.; Chen, J.T.; Hofmann, T.; Knight, S.; Kühne, P.; Nilsson, D.; Forsberg, U.; Janzén, E.; Darakchieva, V. Properties of two-dimensional electron gas in AlGaN/GaN HEMT structures determined by cavity-enhanced THz optical Hall effect. *Phys. Status Solidi C* **2016**, *13*, 369–373. [CrossRef]
19. Mazaheri, Z.; Koral, C.; Andreone, A. Accurate THz ellipsometry using calibration in time domain. *Sci. Rep.* **2022**, *12*, 7342. [CrossRef] [PubMed]
20. Mazaheri, Z.; Koral, C.; Andreone, A.; Marino, A. Terahertz time-domain ellipsometry: Tutorial. *J. Opt. Soc. Am. A* **2022**, *39*, 1420–1433. [CrossRef]
21. Miyagawa, K.; Nagai, M.; Ashida, M.; Kim, C.; Akiyama, H. Enhanced magneto-optical Kerr effect of GaAs-based pn junctions in the terahertz range. *J. Infrared Millimeter Terahertz Waves* **2021**, *42*, 325–337. [CrossRef]
22. Hu, J.; Liu, J.; Wang, K. Cross-polarization coupling terahertz time-domain spectroscopy in a semiconductor based on the Hall effect. *Sci. Rep.* **2017**, *7*, 11464. [CrossRef]
23. Ino, Y.; Shimano, R.; Svirko, Y.; Kuwata-Gonokami, M. Terahertz time domain magneto-optical ellipsometry in reflection geometry. *Phys. Rev. B* **2004**, *70*, 155101. [CrossRef]
24. Hofmann, T.; Darakchieva, V.; Monemar, B.; Lu, H.; Schaff, W.J.; Schubert, M. Optical Hall Effect in Hexagonal InN. *J. Electron. Mater.* **2008**, *37*, 611–615. [CrossRef]
25. Schöche, S.; Hofmann, T.; Nilsson, D.; Kakanakova-Georgieva, A.; Janzén, E.; Kühne, P.; Lorenz, K.; Schubert, M.; Darakchieva, V. Infrared dielectric functions, phonon modes, and free-charge carrier properties of high-Al-content $Al_xGa_{1-x}N$ alloys determined by mid infrared spectroscopic ellipsometry and optical Hall effect. *J. Appl. Phys.* **2017**, *121*, 205701. [CrossRef]
26. Armakavicius, N.; Stanishev, V.; Knight, S.; Kühne, P.; Schubert, M.; Darakchieva, V. Electron effective mass in $In_{0.33}Ga_{0.67}N$ determined by mid-infrared optical Hall effect. *Appl. Phys. Lett.* **2018**, *112*, 082103. [CrossRef]
27. Knight, S.; Mock, A.; Korlacki, R.; Darakchieva, V.; Monemar, B.; Kumagai, Y.; Goto, K.; Higashiwaki, M.; Schubert, M. Electron effective mass in Sn-doped monoclinic single crystal */beta*-gallium oxide determined by mid-infrared optical Hall effect. *Appl. Phys. Lett.* **2018**, *112*, 012103. [CrossRef]
28. Hofmann, T.; Kühne, P.; Schöche, S.; Chen, J.T.; Forsberg, U.; Janzén, E.; Ben Sedrine, N.; Herzinger, C.M.; Woollam, J.A.; Schubert, M.; et al. Temperature dependent effective mass in AlGaN/GaN high electron mobility transistor structures. *Appl. Phys. Lett.* **2012**, *101*, 192102. [CrossRef]
29. Knight, S.; Hofmann, T.; Bouhafs, C.; Armakavicius, N.; Kühne, P.; Stanishev, V.; Ivanov, I.G.; Yakimova, R.; Wimer, S.; Schubert, M.; et al. In-situ terahertz optical Hall effect measurements of ambient effects on free charge carrier properties of epitaxial graphene. *Sci. Rep* **2017**, *7*, 5151. [CrossRef] [PubMed]

30. Kühne, P.; Armakavicius, N.; Papamichail, A.; Tran, D.Q.; Stanishev, V.; Schubert, M.; Paskov, P.P.; Darakchieva, V. Enhancement of 2DEG effective mass in AlN/Al0.78Ga0.22N high electron mobility transistor structure determined by THz optical Hall effect. *Appl. Phys. Lett.* **2022**, *120*, 253102. [CrossRef]
31. Armakavicius, N.; Knight, S.; Kühne, P.; Stanishev, V.; Tran, D.Q.; Richter, S.; Papamichail, A.; Stokey, M.; Sorensen, P.; Kilic, U.; et al. Electron effective mass in GaN revisited: New insights from terahertz and mid-infrared optical Hall effect. *APL Mater.* **2024**, *12*, 021114. [CrossRef]
32. Fujiwara, H., Principles of Spectroscopic Ellipsometry. In *Spectroscopic Ellipsometry*; John Wiley & Sons, Ltd: Hoboken, NJ, USA, 2015; pp. 81–146. [CrossRef]
33. Hilfiker, J.N.; Hong, N.; Schoeche, S. Mueller matrix spectroscopic ellipsometry. *Adv. Opt. Technol.* **2022**, *11*, 59–91. [CrossRef]
34. Schubert, M. Polarization-dependent optical parameters of arbitrarily anisotropic homogeneous layered systems. *Phys. Rev. B* **1996**, *53*, 4265–4274. [CrossRef]
35. Schubert, M. *Infrared Ellipsometry on Semiconductor Layer Structures: Phonons, Plasmons, and Polaritons*; Springer Science & Business Media: Berlin/Heidelberg, Germany, 2004; Volume 209.
36. Kühne, P.; Armakavicius, N.; Stanishev, V.; Herzinger, C.M.; Schubert, M.; Darakchieva, V. Advanced Terahertz Frequency-Domain Ellipsometry Instrumentation for In Situ and Ex Situ Applications. *IEEE Trans. Terahertz Sci. Technol.* **2018**, *8*, 257–270. [CrossRef]
37. Knight, S.; Schöche, S.; Darakchieva, V.; Kühne, P.; Carlin, J.F.; Grandjean, N.; Herzinger, C.M.; Schubert, M.; Hofmann, T. Cavity-enhanced optical Hall effect in two-dimensional free charge carrier gases detected at terahertz frequencies. *Opt. Lett.* **2015**, *40*, 2688–2691. [CrossRef] [PubMed]
38. Chen, J.T.; Forsberg, U.; Janzén, E. Impact of residual carbon on two-dimensional electron gas properties in $Al_xGa_{1-x}N$/GaN heterostructure. *Appl. Phys. Lett.* **2013**, *102*, 193506. [CrossRef]
39. Papamichail, A.; Kakanakova-Georgieva, A.; Sveinbjörnsson, E.O.; Persson, A.R.; Hult, B.; Rorsman, N.; Stanishev, V.; Le, S.P.; Persson, P.O.Å.; Nawaz, M.; et al. Mg-doping and free-hole properties of hot-wall MOCVD GaN. *J. Appl. Phys.* **2022**, *131*, 185704. [CrossRef]
40. Papamichail, A.; Persson, A.; Richter, S.; Kühne, P.; Stanishev, V.; Persson, P.Å.; Ferrand-Drake Del Castillo, R.; Thorsell, M.; Hjelmgren, H.; Paskov, P.; et al. Tuning composition in graded AlGaN channel HEMTs toward improved linearity for low-noise radio-frequency amplifiers. *Appl. Phys. Lett.* **2023**, *122*, 153501. [CrossRef]
41. Persson, A.R.; Papamichail, A.; Darakchieva, V.; Persson, P.O.Å. Mg segregation at inclined facets of pyramidal inversion domains in GaN:Mg. *Sci. Rep.* **2022**, *12*, 17987. [CrossRef] [PubMed]
42. Jena, D.; Heikman, S.; Green, D.; Buttari, D.; Coffie, R.; Xing, H.; Keller, S.; DenBaars, S.; Speck, J.S.; Mishra, U.K.; et al. Realization of wide electron slabs by polarization bulk doping in graded III–V nitride semiconductor alloys. *Appl. Phys. Lett.* **2002**, *81*, 4395–4397. [CrossRef]
43. Bajaj, S.; Yang, Z.; Akyol, F.; Park, P.S.; Zhang, Y.; Price, A.L.; Krishnamoorthy, S.; Meyer, D.J.; Rajan, S. Graded AlGaN Channel Transistors for Improved Current and Power Gain Linearity. *IEEE Trans. Electron Devices* **2017**, *64*, 3114–3119. [CrossRef]
44. Ancona, M.G.; Calame, J.P.; Meyer, D.J.; Rajan, S.; Downey, B.P. Compositionally Graded III-N HEMTs for Improved Linearity: A Simulation Study. *IEEE Trans. Electron Devices* **2019**, *66*, 2151–2157. [CrossRef]
45. Sohel, S.H.; Xie, A.; Beam, E.; Xue, H.; Razzak, T.; Bajaj, S.; Campbell, S.; White, D.; Wills, K.; Cao, Y.; et al. Improved DC-RF dispersion with epitaxial passivation for high linearity graded AlGaN channel field effect transistors. *Appl. Phys. Express* **2020**, *13*, 036502. [CrossRef]
46. Knight, S.; Richter, S.; Papamichail, A.; Kühne, P.; Armakavicius, N.; Guo, S.; Persson, A.R.; Stanishev, V.; Rindert, V.; Persson, P.O.Å.; et al. Room temperature two-dimensional electron gas scattering time, effective mass, and mobility parameters in AlxGa1-xN/GaN heterostructures (0.07 < x < 0.42). *J. Appl. Phys.* **2023**, *134*, 185701. [CrossRef]
47. Wong, M.H.; Keller, S.; Nidhi, S.D.; Denninghoff, D.J.; Kolluri, S.; Brown, D.F.; Lu, J.; Fichtenbaum, N.A.; Ahmadi, E.; Singisetti, U.; et al. N-polar GaN epitaxy and high electron mobility transistors. *Semicond. Sci. Technol.* **2013**, *28*, 074009. [CrossRef]
48. Zhang, H.; Chen, T.; Papamichail, A.; Persson, I.; Paskov, P.P.; Darakchieva, V. High-quality N-polar GaN optimization by multi-step temperature growth process. *J. Cryst. Growth* **2023**, *603*, 127002. [CrossRef]
49. Zhang, H.; Persson, I.; Chen, J.T.; Papamichail, A.; Tran, D.Q.; Persson, P.O.A.; Paskov, P.P.; Darakchieva, V. Polarity Control by Inversion Domain Suppression in N-Polar III-Nitride Heterostructures. *Cryst. Growth Des.* **2023**, *23*, 1049. [CrossRef]
50. Schöche, S.; Kühne, P.; Hofmann, T.; Schubert, M.; Nilsson, D.; Kakanakova-Georgieva, A.; Janzén, E.; Darakchieva, V. Electron effective mass in Al0.72Ga0.28N alloys determined by mid-infrared optical Hall effect. *Appl. Phys. Lett.* **2013**, *103*, 212107. [CrossRef]
51. Ambacher, O.; Smart, J.; Shealy, J.R.; Weimann, N.G.; Chu, K.; Murphy, M.; Schaff, W.J.; Eastman, L.F.; Dimitrov, R.; Wittmer, L.; et al. Two-dimensional electron gases induced by spontaneous and piezoelectric polarization charges in N- and Ga-face AlGaN/GaN heterostructures. *J. Appl. Phys.* **1999**, *85*, 3222. [CrossRef]
52. Yoo, S.H.; Todorova, M.; Neugebauer, J.; Van de Walle, C.G. Microscopic Origin of Polarization Charges at GaN/(Al, Ga)N Interfaces. *Phys. Rev. Appl.* **2023**, *19*, 064037. [CrossRef]
53. Arakawa, Y.; Ueno, K.; Imabeppu, H.; Kobayashi, A.; Ohta, J.; Fujioka, H. Electrical properties of Si-doped GaN prepared using pulsed sputtering. *Appl. Phys. Lett.* **2017**, *110*, 042103. [CrossRef]

54. Ueno, K.; Fudetani, T.; Arakawa, Y.; Kobayashi, A.; Ohta, J.; Fujioka, H. Electron transport properties of degenerate n-type GaN prepared by pulsed sputtering. *APL Mater.* **2017**, *5*, 126102. [CrossRef]
55. Lisesivdin, S.B.; Yildiz, A.; Balkan, N.; Kasap, M.; Ozcelik, S.; Ozbay, E. Scattering analysis of two-dimensional electrons in AlGaN/GaN with bulk related parameters extracted by simple parallel conduction extraction method. *J. Appl. Phys.* **2010**, *108*, 013712. [CrossRef]
56. Teke, A.; Gokden, S.; Tulek, R.; Leach, J.H.; Fan, Q.; Xie, J.; Ozgur, U.; Morkoc, H.; Lisesivdin, S.B.; Ozbay, E. The effect of AlN interlayer thicknesses on scattering processes in lattice-matched AlInN/GaN two-dimensional electron gas heterostructures. *New J. Phys.* **2009**, *11*, 063031. [CrossRef]
57. Perlin, P.; Litwin-Staszewska, E.; Suchanek, B.; Knap, W.; Camassel, J.; Suski, T.; Piotrzkowski, R.; Grzegory, I.; Porowski, S.; Kaminska, E.; et al. Determination of the effective mass of GaN from infrared reflectivity and Hall effect. *Appl. Phys. Lett.* **1996**, *68*, 1114. [CrossRef]
58. Kasic, A.; Schubert, M.; Einfeldt, S.; Hommel, D.; Tiwald, T.E. Free-carrier and phonon properties of n- and p-type hexagonal GaN films measured by infrared ellipsometry. *Phys. Rev. B* **2000**, *62*, 7365–7377. [CrossRef]
59. Hofmann, T.; Chavdarov, T.; Darakchieva, V.; Lu, H.; Schaff, W.J.; Schubert, M. Anisotropy of the Gamma-point effective mass and mobility in hexagonal InN. *Phys. Status Solidi C* **2006**, *3*, 1854–1857. [CrossRef]
60. Schubert, M.; Woollam, J.; Kasic, A.; Rheinländer, B.; Off, J.; Kuhn, B.; Scholz, F. Free-Carrier Response and Lattice Modes of Group III-Nitride Heterostructures Measured by Infrared Ellipsometry. *Phys. Status Solidi (b)* **1999**, *216*, 655–658. [CrossRef]
61. Kasic, A.; Schubert, M.; Rheinländer, B.; Riede, V.; Einfeldt, S.; Hommel, D.; Kuhn, B.; Off, J.; Scholz, F. Effective carrier mass and mobility versus carrier concentration in p-and n-type α-GaN determined by infrared ellipsometry and Hall resistivity measurements. *Mater. Sci. Eng. B* **2001**, *82*, 74–76. [CrossRef]
62. Kasic, A.; Schubert, M.; Saito, Y.; Nanishi, Y.; Wagner, G. Effective electron mass and phonon modes in n-type hexagonal InN. *Phys. Rev. B* **2002**, *65*, 115206. [CrossRef]
63. Schubert, M.; Kasic, A.; Tiwald, T.; Off, J.; Kuhn, B.; Scholz, F. Optical phonons and free-carrier effects in MOVPE grown AlxGa1-xN measured by Infrared Ellipsometry. *Mater. Res. Soc. Internet J. Nitride Semicond. Res.* **1999**, *4*, e11. [CrossRef]
64. Kasic, A.; Schubert, M.; Off, J.; Kuhn, B.; Scholz, F.; Einfeldt, S.; Böttcher, T.; Hommel, D.; As, D.J.; Köhler, U.; et al. Phonons and free-carrier properties of binary, ternary, and quaternary group-III nitride layers measured by Infrared Spectroscopic Ellipsometry. *Phys. Status Solidi (c)* **2003**, *6*, 1750–1769. [CrossRef]

materials

MDPI

Article

Features of the Carrier Concentration Determination during Irradiation of Wide-Gap Semiconductors: The Case Study of Silicon Carbide

Alexander A. Lebedev [1,*], Vitali V. Kozlovski [2], Klavdia S. Davydovskaya [1], Roman A. Kuzmin [1], Mikhail E. Levinshtein [1] and Anatolii M. Strel'chuk [1]

[1] Ioffe Institute, Politekhnicheskaya Street 26, St. Petersburg 194021, Russia
[2] Department of Experimental Physics, St. Petersburg State Polytechnic University, Polytekhnicheskaya 29, St. Petersburg 195251, Russia
* Correspondence: shura.lebe@mail.ioffe.ru; Tel.: +7-812-292-71-25

Abstract: In this paper, the features of radiation compensation of wide-gap semiconductors are discussed, considering the case study of silicon carbide. Two classical methods of concentration determination are compared and analyzed: capacitance-voltage (*C–V*) and current-voltage (*I–V*) characteristics. The dependence of the base resistance in high-voltage 4H-SiC Schottky diodes on the dose of irradiation by electrons and protons is experimentally traced in the range of eight orders of magnitude. It is demonstrated that the dependence of the carrier concentration on the irradiation dose can be determined unambiguously and reliably in a very wide range of compensation levels, based on the results of measuring the *I–V* characteristics. It is shown that the determination of the carrier removal rate using the *I–V* characteristics is more correct than using the *C–V* characteristics, especially in the case of high radiation doses.

Keywords: radiation compensation; wide-gap semiconductors; deep levels; protons; electrons; current-voltage characteristics; capacitance-voltage characteristics

Citation: Lebedev, A.A.; Kozlovski, V.V.; Davydovskaya, K.S.; Kuzmin, R.A.; Levinshtein, M.E.; Strel'chuk, A.M. Features of the Carrier Concentration Determination during Irradiation of Wide-Gap Semiconductors: The Case Study of Silicon Carbide. *Materials* **2022**, *15*, 8637. https://doi.org/10.3390/ma15238637

Academic Editor: Fabrizio Roccaforte

Received: 1 November 2022
Accepted: 30 November 2022
Published: 3 December 2022

1. Introduction

The study of the wide-gap materials properties is one of the most dynamically developing areas of semiconductor physics. As a rule, materials with a band gap (Eg) > (2.5–3.0) eV are referred to wide-band semiconductors. Compared to the classic semiconductor materials, Si and GaAs, wide-gap semiconductors make it possible to create radiation-resistant devices, operating at significantly higher temperatures [1–3]. Such devices can be used to improve the reliability of nuclear power plants, thermonuclear power plants already under design, and space technology devices that require the use of radiation-resistant semiconductor electronics. Such electronics are the subject of increased requirements for preserving the initial properties (and/or changing properties within acceptable limits) under the effect of various radiation types: protons, electrons, neutrons, alpha and gamma particles, as well as heavy high-energy particles.

Despite numerous studies in this area, many unresolved issues remain in the choice of options for the radiation resistance of wide-band semiconductors and in establishing possible ways to improve it. This paper considers one of the most important parameters characterizing radiation resistance: the change in the carrier concentration depending on the level of sample compensation due to irradiation with electrons and protons.

2. Statement of the Problem

To estimate the radiation resistance of semiconductors at a relatively low level of compensation, the parameter η_e is often used, that is, the removal rate of carriers under the irradiation influence (see, for example, [4,5]):

$$\eta_e = (n_0 - n)/\Phi, \tag{1}$$

where n_0 is the electron concentration in the semiconductor before irradiation, n is the electron concentration after irradiation, and Φ is the fluence (for certainty, we will consider the n-type conductivity material).

In wide-gap semiconductors (i.e., SiC, GaN, AlGaN, Ga_2O_3), at the current level of technology, there are always deep centers (levels) that are practically not ionized at room temperature. When heated, the degree of their ionization increases, which leads to a temperature dependence of the measured η_e value [6]. It should also be noted that, in determining the η_e value, the concentration of uncompensated donors, $N_d - N_a$, is often used instead of the n value. In current n-type device-quality SiC, prior to irradiation, the concentration of carriers at room temperature can be considered equal with acceptable accuracy to the concentration of shallow donor N_d (nitrogen). Irradiation with protons and/or electrons creates acceptor-type levels in the band gap [7,8]. As long as the total concentration of acceptor levels created by irradiation, N_a^Σ, is less than the initial carrier concentration n_0, the residual electron concentration in the conduction band can be considered equal to $n = n_0 - N_a^\Sigma$. However, at $N_a^\Sigma > n_0$, the shallow donor level responsible for the initial electron concentration in the non-irradiated semiconductor turns out to be completely empty, and the "residual" electron concentration n is determined by the generation of electrons from deep levels.

The η value can critically depend on the method of measurement. In the case of measuring current-voltage (I–V) characteristics or the Hall effect, the electron concentration after irradiation n is directly determined from the measurements. The value of n is determined directly, in this case, at any levels of compensation, including large fluences Φ, when expression (1) becomes inapplicable. When measuring capacitance-voltage characteristics (C–V measurements), carried out on reverse-biased Schottky diodes or p-n junctions, the $N_d - N_a$ value is measured. In this case, the removal rate η_N (also at a relatively low level of compensation) is defined as:

$$\eta_N = [(N_d - N_a)_0 - (N_d - N_a)_1]/\Phi \tag{2}$$

where $(N_d - N_a)_0$ is the initial value of the concentration of uncompensated donors, and $(N_d - N_a)_1$ is the $(N_d - N_a)$ value after irradiation.

Properly speaking, the degree of deep centers filling in the I–V (or the Hall effect) measurements and C–V measurements is always different. The quasi-neutral situation is realized when measuring the I–V characteristics. Then, the filling of the level is determined by two processes: thermal emission of electrons from the corresponding center, and recapture from the conduction band (C-band) to the level. If the level at the measurement temperature lies above the Fermi level, then it is empty; below it, it is filled.

In C–V measurements, the levels at which electrons are captured as a result of irradiation are located in the space charge layer, where there are practically no electrons in the C-band, and the carrier concentration at the level is determined by thermal emission from the level. At long measurement times, all levels lying in the upper half of the band gap should be emptied. However, for sufficiently deep levels, the time for complete depletion can be extremely long and significantly exceed the practically acceptable measurement time.

If the acceptor level is formed during irradiation in the lower half of the band gap, it will be filled with electrons, charged negatively, and will contribute to both a decrease in the $N_d - N_a$ value and a decrease in the concentration n.

If an acceptor level is formed in the upper half of the band gap, then it will reduce the n value. However, the decrease in the value of $N_d - N_a$ will depend on its depth (ionization energy). If the level ionization energy is large enough (deep level), then in a reasonable time for C–V measurements, it can remain almost completely filled with electrons. In this case, this level will contribute to the measured $N_d - N_a$ value. However, a relatively shallow level in C–V measurements can be completely emptied in a time much shorter than the measurement time already at room temperature. In this case, such a level will not contribute to the measured $N_d - N_a$ value.

Measurements of current-voltage and C–V characteristics of SiC Schottky diodes irradiated with protons and electrons were made in the Ref. [9]. Both types of irradiation lead to an increase in the resistance of the diode base. With large radiation doses, the resistance of the sample base increases extremely sharply, and measuring the I–V characteristics allows to trace the change in electron concentration by many orders of magnitude.

The value of $N_d - N_a$, determined from the capacitance–voltage characteristics at the same irradiation level, decreases at a much weaker rate [10,11]. In addition, since levels with different ionization empty at different rates, the measured C–V characteristics significantly depend on the measurement frequency and are characterized by a noticeable frequency dispersion at a given irradiation level [11].

The above considerations can be clearly illustrated using a simple qualitative model.

3. Qualitative Model

It is known that a number of deep levels are formed in silicon carbide under the actions of electron and proton irradiation [9–15]. Within the framework of the problem under consideration, these levels can be divided into three groups. Next, we will consider a simple model where each of these groups will be presented by one most characteristic level (Figure 1):

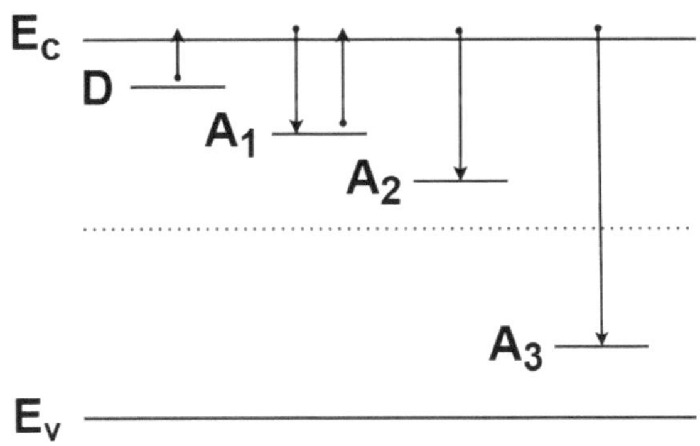

Figure 1. Schematic presentation of the main types of levels in the band gap of SiC. The arrows show the dominant electron transitions corresponding to the level. The dotted line corresponds to the middle of the band gap. Here, D is the shallow donor level; a relatively shallow acceptor level A1 associates with the $Z_{1/2}$ level; an A2 level associates with the $E_{6/7}$ level; and the A3 level represents the group of levels lying in the lower half of the band gap.

1. Shallow donor level D with N_d concentration that existed in the sample before irradiation. The concentration of this level does not depend on the radiation dose. An impurity nitrogen level with an ionization energy E_i = 0.1 eV is considered such a level. With a sufficiently large fluence Φ, this level will be almost completely depleted in the quasi-neutral region. In the space charge layer, the level is completely empty at any Φ.

2. Relatively shallow acceptor level A1 with a concentration of N_{A1}. Its characteristic relaxation time to the equilibrium value, $\tau = \frac{\tau_e \tau_c}{\tau_e + \tau_c}$, is determined by the time of electron emission from the level to the C-band τ_e and the capture time τ_c. Such a level will be filled in the quasi-neutral region and emptied in the space charge layer. It is expedient to choose the $Z_{1/2}$ level with the ionization energy E_i = 0.68 eV, as such a model level is in 4H-SiC [5,8].

The appearance of this level as a result of irradiation will reduce the electron concentration in the conduction band n, determined from the current–voltage characteristics. However, as will be shown below, during C–V measurements, this level will be completely empty and not affect the decrease in the $N_d - N_a$ value.

3. The A2 level is located in the upper half of the band gap, but its ionization energy ($E_c - E_{A2}$) is quite high (>1 eV), so that during the characteristic time of measuring the C–V characteristics, the ionization of electrons from this level can be neglected. It is expedient to choose the $E_{6/7}$ level with the ionization energy E_i ~ 1.5 eV, as such a model level in 4H-SiC [5,8].

4. Level A3 is located in the lower half of the band gap, and its ionization energy is even higher. The total concentration of very deep levels will be denoted as $N_A = N_{A2} + N_{A3}$. These levels will decrease both the n value and the $N_d - N_a$ value.

Due to the large difference in the ionization energy of the levels under consideration, the concentration of electron n_j, supplied to the conduction band from each level, can be considered with good accuracy independent of the filling of all other levels. Then (see, for example, [16]):

$$n_j = \frac{2(N_d - N_a)}{1 + \frac{gN_a}{N_c}\exp\varepsilon_d + \left[\left(1 + \frac{gN_a}{N_c}\exp\varepsilon_d\right)^2 + \frac{4g(N_d - N_a)\exp\varepsilon_d}{N_c}\right]^{1/2}} \quad (3)$$

where N_c is the density of states in the conduction band, $\varepsilon_d = \frac{E_c - E_t}{kT}$ (here, $E_c - E_t$ is the ionization energy of the corresponding level), and g is the degeneracy factor of the corresponding level. In calculations, the g was taken to be equal to unity (g = 1).

At low radiation doses ($N_d \geq N_{A1} + N_A$), we took the total electron concentration n in the conduction band as the sum $n_1 + n_2$, where n_1 and n_2 are the concentrations of electrons delivered to the conduction band from levels D and A_1, respectively, calculated by Formula (3).

When calculating the n_1 according to Formula (3), the $N_d - N_a$ will obviously be equal to $N_d - \Phi(a_{A1} + a_{NA})$, where a_{A1} is the rate of center A_1 generation; a_{NA} is the total rate of all other acceptor levels' generation. Note that the A1 level plays the role of an electron source in the C-zone, especially at high temperatures [17]. When calculating the value $n_j = n_2$ according to (3), the $N_d - N_a$ will be equal to $a_{A1}\Phi$.

At high radiation doses ($N_d < N_{A1} + N_A$), the D level is completely empty. The residual conductivity of the base is due to the carrier's ionization from level A1. Note that, in accordance with the adopted model, the probability of ionization from levels with higher ionization energies (A2 and A3 levels) is assumed to be zero.

In calculations, the rate of the center $Z_{1/2}$ generation, determined from the analysis of DLTS spectra, was used.

Irradiation with 15 MeV energy protons was carried out in a portable MGTs-20 cyclotron. Irradiation with 0.9 MeV electrons was carried out on a resonant transformer accelerator. The electron beam's current density was 12.5 μA cm^{-2}. The pulse duration was 330 μs, and the pulse repetition rate was 490 Hz.

The carrier removal rates and generation rates of radiation defects used in the calculations are presented in Table 1.

Table 1. Generation rates of deep levels.

	Electron Irradiation	Proton Irradiation
$a_{NA} = \eta_N$, cm^{-1}	0.14	70
$a_{Z1/2}$, cm^{-1}	0.034	13.2

4. Results and Discussion

Let us consider the situation using an example of irradiation with protons and electrons of 4H-SiC Schottky diodes with a blocking voltage of 600 V. The initial concentration of donors in the base, that is, the concentration corresponding to the D level, N_d, is taken to be equal to 7×10^{15} cm^{-3} [6]. The density of states, N_c at room temperature in the 4H-SiC is $N_c \approx 9 \times 10^{18}$ cm^{-3}. Electron mobility μ_e is assumed to be independent of fluence Φ and equal to 800 cm^2/Vs [18].

Figure 2 presents the experimental dependences of the diode base resistance R_b = L/(enμS) on the doses with electron (Figure 2a) and proton (Figure 2b) irradiation [11,17]. Here, e is the electron charge, n is the electron concentration in the conduction band calculated in accordance with (3), L = 10 μm, and $S = 6.75 \times 10^{-4}$ cm^{-2} is the base length and diode area, respectively. These dependences are compared with the results of calculations in accordance with Equation (3).

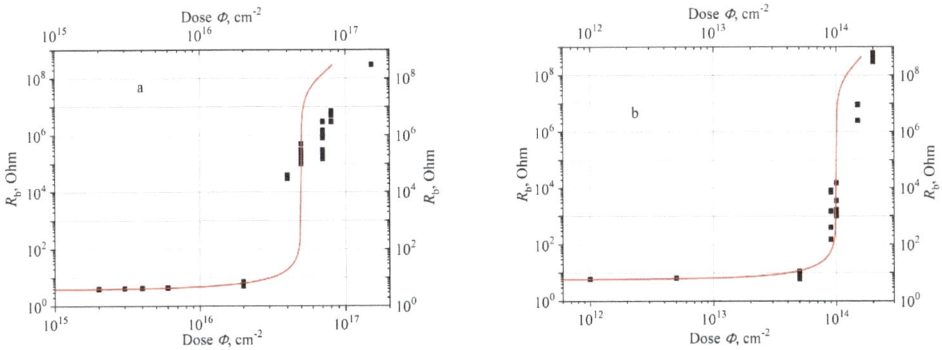

Figure 2. Experimental [9] and calculated (Equation (3)) dependences of the base resistance of 600 V 4H-SiC Schottky diodes, R_b, on the dose Φ when irradiated with electrons with an energy of 0.9 MeV (**a**) and protons with an energy of 15 MeV (**b**). Points present experimental data, lines correspond to calculations.

Taking into account the fact that the change in the resistance R_b was traced in Figure 2 by eight orders of magnitude, the agreement between the experimental data and the calculations should be recognized as satisfactory. The obtained experimental results qualitatively agree with the data obtained in [7] during irradiation of 4H-SiC structures with an initial electron concentration of 7.2×10^{15} cm^{-3} by electrons with an energy of 400 keV.

Two characteristic parts can be distinguished in Figure 2.

1. At radiation doses of $\Phi \leq 3 \times 10^{16}$ cm^{-2} (electrons) and $\Phi \leq 6 \times 10^{13}$ cm^{-2} (protons), the resistance R_b increases relatively weakly (and practically linearly) with the radiation dose growth. This situation corresponds to the condition $N_d \geq N_{A1} + N_A$. Conductivity is due to electrons ionized from the shallow donor level D. In this case, the carrier removal rates calculated by Formulas (1) and (2) will be close.
2. At radiation doses $\Phi \geq 3 \times 10^{16}$ cm^{-2} (electrons) and $\Phi \geq 6 \times 10^{13}$ cm^{-2} (protons), there is a sharp increase in the base resistance with increasing Φ. The shallow level D is almost completely compensated, and the residual conductivity is due to the generation of electrons to the conduction band from the A1 level. (In more accurate calculations, it should be taken into account that, in addition to the $Z_{1/2}$ level, which is presented in the model by the A1 level, electron irradiation also creates shallow levels with ionization energies of 0.25, 038, and 0.43 eV [5]).

It should be emphasized that the type of dependences $R_b(\Phi)$, presented in Figure 2a,b, is of a general nature and characteristic of all studied SiC Schottky diodes.

As mentioned above, when measuring the C–V characteristics of reverse-biased Schottky diodes and p-n junctions, it is not the carrier concentration n that is measured, but the $N_d - N_a$ value. In this case, when calculating the value of n from the data of such measurements, it is assumed that all acceptor centers created by irradiation with a concentration of N_a remain negatively charged in the process of measuring the C–V characteristics. However, when a reverse voltage is applied, the acceptor centers in the space charge region become empty (neutral), and do not contribute to the measured $N_d - N_a$ value. In this case, the characteristic time of electron emission τ_e from the corresponding level after the application of the reverse voltage is of fundamental importance. This time can be estimated as [19]:

$$\tau_e = \exp[(E_c - E_t)/kT]/\sigma v_t N_c \quad (4)$$

where σ is the capture cross-section, v_t is the electron thermal velocity, and N_c is the density of states in the conduction band.

Taking for the $Z_{1/2}$ level (level A1 in Figure 1) $(E_c - E_t) = 0.68$ eV, $\sigma = 6 \times 10^{-14}$ cm^2, $v_T \approx 2 \times 10^7$ cm/s, and $N_c = 1.7 \times 10^{19}$ cm^{-3} [5, 18), we get $\tau_e \approx 4 \times 10^{-2}$ s. For shallower levels with ionization energy 0.25, 038, and 0.43 eV [5], the time of depletion will be even shorter. Such times are significantly shorter than the practical time of measurement of the C–V characteristics. Thus, the value of $N_d - N_a$ obtained from C–V measurements is not, strictly speaking, equal to the decrease in the electron concentration n as a result of irradiation.

Note that for a deeper level A2 with an ionization energy $E_i \sim 1.5$ eV (Figure 1), identified with the $E_{6/7}$ level [5,8], the emission time will be ~5×10^{12} s, which is many orders of magnitude greater than any practically acceptable time for measuring C–V characteristics. Thus, the $E_{6/7}$ level and, of course, the levels lying in the lower half of the band gap will contribute to the measured $N_d - N_a$ value.

When estimating the n value from C–V measurements, a factor that makes it difficult to unambiguously interpret the experimental results is also a noticeable frequency dispersion of the capacitance of reverse-biased junctions in SiC structures.

Figure 3 shows the frequency dependence of the capacitance of a reverse-biased high-voltage 4H-SiC Schottky diode (CPW4-1200S002B diode with 1200 V blocking voltage), measured in the Ref. [11] over a wide frequency range.

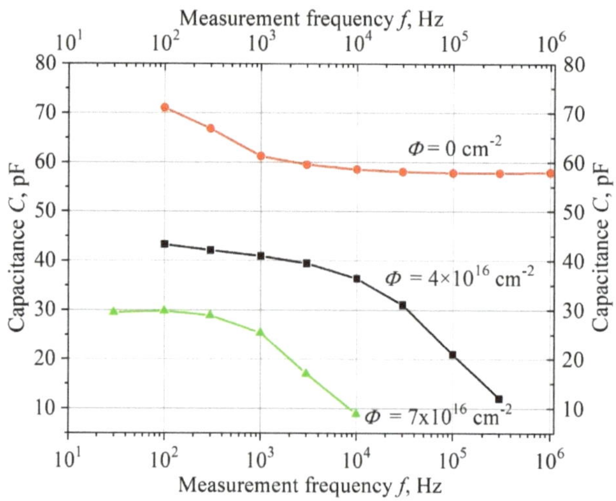

Figure 3. Frequency dependences of 4H-SiC Schottky diode capacitance at reverse bias $V = -10$ V for different values of fluencies Φ (irradiation with 0.9 MeV electrons).

As can be seen from Figure 3, the frequency dispersion in a non-radiated diode is rather insignificant, and it can be neglected at frequencies $f \geq 10$ Hz. As the dose (fluence) Φ increases, the dispersion increases greatly, reflecting the appearance of relatively shallow centers, rapidly emptying when reverse bias is applied. The frequency dependence of the capacitance is due to the fact that at high frequencies, the deep centers do not have time to ionize, and at low and high frequencies, they find themselves in different charge states, giving a different contribution to the value of the measured capacitance.

It is important to emphasize that when measuring the carrier concentration from *I–V* characteristics, the directly measured value (base resistance R_b) at an irradiation dose $\Phi = 7 \times 10^{16}$ cm^{-2} changes with respect to R_b in a non-irradiated structure by ~5 orders of magnitude (Figure 2a). With *C–V* measurements, the capacitance under the same conditions only changes several times (Figure 3). It is obvious that the determination of the concentration from the *I–V* characteristics has an incomparably higher resolution.

5. Conclusions

In this study, the features of wide-band semiconductors radiation compensation were considered using the example of silicon carbide. The base resistance dependence of the high-voltage (blocking voltage 600 V) 4H-SiC Schottky diodes on the dose of irradiation with electrons (energy 0.9 MeV) and protons (energy 15 MeV) in the range of eight orders of magnitude was experimentally traced. It has been shown that the observed experimental dependences can be qualitatively interpreted on the basis of a simple three-level model that takes into account the formation of acceptor levels during SiC irradiation in the upper half of the band gap. It has been demonstrated that the measurement of current–voltage (*I–V*) characteristics makes it possible to uniquely and reliably determine the dependence of the carrier concentration (base resistance) on the radiation dose over a very wide range. It has been shown that determination of the carrier concentration from measurement of *I–V* characteristics has an incomparably higher resolution than *C–V* measurements.

Author Contributions: Methodology, M.E.L.; Investigation, K.S.D.; Data curation, V.V.K., R.A.K. and A.M.S.; Writing—review & editing, A.A.L. All authors have read and agreed to the published version of the manuscript.

Funding: This work was supported by the Russian Science Foundation through project № 22-12-00003.

Conflicts of Interest: The authors declare that they have no known competing financial interests or personal relationships that could have appeared to influence the work reported in this paper.

References

1. Levinshtein, M.E.; Ivanov, P.A.; Boltovets, M.S.; Krivutsa, V.A.; Palmour, J.W.; Das, M.K.; Hull, B.A. High-temperature (up to 773 K) operation of 6-kV 4H-SiC junction diodes. *Solid State Electron.* **2005**, *49*, 1228–1232. [CrossRef]
2. Elahipanah, H.; Kargarrazi, S.; Salemi, A.; Ostling, M.; Zetterling, C.-M. 500 °C High Current 4H-SiC Lateral BJTs for High-Temperature Integrated Circuits. *IEEE Electron. Device Lett.* **2017**, *38*, 1429–1432. [CrossRef]
3. Kaneko, M.; Nakajima, M.; Jin, Q.; Kimoto, T. SiC Complementary Junction Field-Effect Transistor Logic Gate Operation at 623 K. *IEEE Electron. Device Lett.* **2022**, *43*, 997–1000. [CrossRef]
4. Omotoso, E.; Meyer, W.E.; Auret, F.D.; Paradzah, A.T.; Diale, M.; Coelho, S.M.M.; Janse van Rensburg, P.J. The influence of high energy electron irradiation on the Schottky barrier height and the Richardson constant of Ni/4H-SiC Schottky diodes. *Mater. Sci. Semicond. Process.* **2015**, *39*, 112–118. [CrossRef]
5. Hazdra, P.; Vobecký, J. Radiation Defects Created in n-Type 4H-SiC by electron Irradiation in the Energy Range of 1–10 MeV. *Phys. Stat. Solidi A* **2019**, *216*, 1900312. [CrossRef]
6. Lebedev, A.A.; Kozlovski, V.V. Comparison of the radiation hardness of silicon and silicon carbide. *Semiconductors* **2014**, *48*, 1293–1295. [CrossRef]
7. Kaneko, H.; Kimoto, T. Formation of a semi-insulating layer in n-type 4H-SiC by electron irradiation. *Appl. Phys. Lett.* **2011**, *98*, 262106. [CrossRef]
8. Castaldini, A.; Cavallini, A.; Rigutti, L.; Nava, F. Low temperature annealing of electron irradiation induced defects in 4H-SiC. *Appl. Phys. Lett.* **2004**, *85*, 3780–3782. [CrossRef]
9. Strel'chuk, A.M.; Kozlovski, V.V.; Lebedev, A.A. Radiation-Induced Damage of Silicon-Carbide Diodes by High-Energy Particles. *Semiconductors* **2014**, *52*, 1752–1756. [CrossRef]
10. Kozlovski, V.V.; Lebedev, A.A.; Bogdanova, E.V. Model for conductivity compensation of moderately doped n- and p- 4H-SiC by high-energy electron bombardment. *J. Appl. Phys.* **2015**, *117*, 155702. [CrossRef]
11. Kozlovski, V.V.; Lebedev, A.; Levinshtein, M.E.; Rumyantsev, S.L.; Palmour, J.W. Impact of high energy electron irradiation on high voltage Ni/4H-SiC Schottky diodes. *Appl. Phys. Lett.* **2017**, *110*, 083503. [CrossRef]
12. Danno, K.; Kimoto, T. Investigation of deep levels in n-type 4H-SiC epilayers irradiated with low-energy electrons. *J. Appl. Phys.* **2006**, *100*, 113728. [CrossRef]
13. Kozlovski, V.; Lebedev, A.; Lomasov, V.; Bogdanova, E.; Seredova, N. Conductivity compensation in n-4H-SiC (CVD) under irradiation with 0.9-MeV electrons. *Semiconductors* **2014**, *48*, 1006–1009. [CrossRef]

14. Hazdra, P.; Popelka, S. Lifetime control in SiC PiN diodes Using Radiation Defects. *Mater. Sci. Forum.* **2017**, *897*, 463–466. [CrossRef]
15. Lebedev, A.A.; Kozlovski, V.V.; Levinshtein, M.E.; Ivanov, A.E.; Strel'chuk, A.M.; Zubov, A.V.; Fursin, L. Impact of 0.9 MeV electron irradiation on main properties of high voltage vertical power 4H-SiC MOSFETs. *Radiat. Phys. Chem.* **2020**, *177*, 109200. [CrossRef]
16. Shalimova, K.V. *Semiconductor Physics;* Energoatom: Moscow, Russia, 1985; p. 122.
17. Kozlovski, V.V.; Lebedev, A.A.; Levinshtein, M.E.; Rumyantsev, S.L.; Palmour, J.W. Electrical and noise properties of proton irradiated 4H-SiC Schottky diodes. *J. Appl. Phys.* **2018**, *123*, 024502. [CrossRef]
18. Levinshtein, M.E.; Rumyantsev, S.L.; Shur, M.S. *Properties of Advanced Semiconductor Materials GaN, AlN, InN, BN, SiC, SiGe;* John Wiley & Sons Inc.: New York, NY, USA, 2001; p. 110.
19. Sze, S.M. *Physics of Semiconductor Devices;* Section 1.4; John Wiley & Sons Inc.: New York, NY, USA, 1981.

materials

MDPI

Article

Effects of High-Temperature Treatments in Inert Atmosphere on 4H-SiC Substrates and Epitaxial Layers

Francesca Migliore [1], Marco Cannas [1], Franco Mario Gelardi [1], Filippo Pasquali [2], Andrea Brischetto [2], Daniele Vecchio [2], Massimo Davide Pirnaci [2] and Simonpietro Agnello [1,3,*]

[1] Department of Physics and Chemistry Emilio Segrè, University of Palermo, Via Archirafi 36, 90123 Palermo, Italy; francesca.migliore02@unipa.it (F.M.); marco.cannas@unipa.it (M.C.); franco.gelardi@unipa.it (F.M.G.)

[2] STMicroelectronics, Stradale Primosole, 95121 Catania, Italy

[3] ATeN Center, University of Palermo, Viale delle Scienze Ed. 18, 90128 Palermo, Italy

* Correspondence: simonpietro.agnello@unipa.it

Abstract: Silicon carbide is a wide-bandgap semiconductor useful in a new class of power devices in the emerging area of high-temperature and high-voltage electronics. The diffusion of SiC devices is strictly related to the growth of high-quality substrates and epitaxial layers involving high-temperature treatment processing. In this work, we studied the thermal stability of substrates of 4H-SiC in an inert atmosphere in the range 1600–2000 °C. Micro-Raman spectroscopy characterization revealed that the thermal treatments induced inhomogeneity in the wafer surface related to a graphitization process starting from 1650 °C. It was also found that the graphitization influences the epitaxial layer successively grown on the wafer substrate, and in particular, by time-resolved photoluminescence spectroscopy it was found that graphitization-induced defectiveness is responsible for the reduction of the carrier recombination lifetime.

Keywords: 4H-SiC; thermal treatments; micro-Raman spectroscopy; graphitization; exciton recombination; time-resolved photoluminescence spectroscopy

check for updates

Citation: Migliore, F.; Cannas, M.; Gelardi, F.M.; Pasquali, F.; Brischetto, A.; Vecchio, D.; Pirnaci, M.D.; Agnello, S. Effects of High-Temperature Treatments in Inert Atmosphere on 4H-SiC Substrates and Epitaxial Layers. *Materials* **2024**, *17*, 5761. https://doi.org/10.3390/ma17235761

Academic Editors: Christof Schneider and Sergey Kukushkin

Received: 29 October 2024
Revised: 15 November 2024
Accepted: 21 November 2024
Published: 25 November 2024

1. Introduction

The interest in developing devices for applications in emerging fields such as high-voltage and high-temperature electronics is pushing the use of wide-bandgap semiconductors. Among them, silicon carbide stands out for its excellent physical properties; in particular, the polytype 4H-SiC, which has a bandgap value of 3.3 eV, has a three-times-higher thermal conductivity and a ten-times-higher breakdown electric field strength than silicon. This brings two significant advantages for electronic applications. First, devices made of SiC with the same dimensions as Si devices withstand ten-times-higher voltages before breakdown. Additionally, it is possible to produce SiC devices that operate at the same voltage as Si devices but that are ten times smaller. Therefore, the strong potentiality of SiC for the production of high-power devices is evident. SiC also exhibits high chemical inertness, mechanical strength, high conductivity along a direction, and high-saturation drift velocity. Additionally, it is easy to control either n-type or p-type doping during the growth of both substrates and epitaxial layers. Uniquely, SiC is the only compound semiconductor with SiO_2 as a native oxide, identical to that of silicon [1–4]. From an applicative perspective, it is important to highlight that SiC exhibits high radiation hardness due to its large threshold energy for defect formation. For this reason, SiC stands out as one of the most promising semiconductors to make devices that are able to operate in harsh environments and extreme conditions such as high levels of radiation, elevated temperatures, and environments with high chemical reactivity [5–19].

To fabricate a semiconductor device in 4H-SiC, it is generally necessary to grow an epitaxial layer with a specific thickness and doping level on a degenerate doped 4H-SiC substrate. This growth process, typically carried out via Chemical Vapor Deposition (CVD), involves high-temperature treatments that can reach up to 2000 °C [2,20,21]. Notably, it has been observed that thermal treatments in the 1200–1900 °C range, conducted in various atmospheres or ultra-high vacuum, can result in the formation of graphene layers on the surface of hexagonal or cubic SiC polytypes [22–26]. This highlights the potential vulnerability of the material to thermal treatments under different conditions, even during the epitaxial layer growth process. Consequently, the thermal stability of 4H-SiC at high temperatures remains a critical factor in the production of SiC devices.

In this study, we examined the stability of 4H-SiC substrates in an inert atmosphere within the temperature range of 1600–2000 °C, analyzing the wafer surface using micro-Raman spectroscopy. Traces of graphitization were observed on the thermally treated substrates. To evaluate the impact of this graphitization on subsequent epitaxial layer growth, a 12 μm thick epitaxial layer with n-type doping of 10^{16} cm^{-3} was grown on a treated substrate. The influence of graphitization in the epitaxial layer was investigated using time-resolved photoluminescence spectroscopy, focusing on the lifetime of the exciton recombination band.

2. Materials and Methods

STMicroelectronics (Catania, Italy) provided two substrates, named A and B hereafter, of 4H-SiC of 6-inch and 350 μm thickness. The substrates have the same nominal features of doping (n-type 10^{19} cm^{-3} realized using nitrogen) and thickness. They differ, however, in the extent of the facet region, a darker area on the wafer associated with substrate growth, which has higher doping than the rest of the wafer. Specifically, sample A has a circular facet, 9 cm in diameter, at the center of the wafer, while in sample B, the facet extends across the entire wafer. Both substrates, A and B, underwent thermal treatment within the temperature range of 1600–2000 °C. This process involved a rapid thermal ramp-up to the target temperature (below 2000 °C), a plateau of a few seconds at the peak temperature in Ar atmosphere, followed by natural cooling (approximately 3 h). Additionally, STMicroelectronics grew a 12 μm thick n-type 4H-SiC epitaxial layer with doping of 10^{16} cm^{-3} on two new thermally treated substrates.

The entire surfaces of the samples were characterized before and after the thermal process using micro-Raman spectroscopy to obtain information on crystal structure and electronic properties. Spectra were collected in back-scattering geometry with a Horiba HR-Evolution Micro-Raman system, equipped with a confocal microscope (100× magnification) and a laser excitation wavelength of 633 nm, filtered with a neutral density (ND) filter at 50% to prevent sample damage. The pinhole size was set to 200 or 50 μm for measurements on substrates and epitaxial layers, respectively. All spectra were acquired using a grating with 600 lines/mm and calibrated to the Si peak at 520.7 cm^{-1}. Notably, the system's vertical resolution is 2–3 μm, ensuring that measurements on epitaxial layers were unaffected by the underlying substrate.

The exciton recombination band lifetime at 3.2 eV (associated with nitrogen doping) was investigated through time-resolved photoluminescence spectroscopy, using excitation at 4.66 eV (266 nm). The penetration depth of the selected excitation wavelength (around 0.6 μm) enabled the generation of exciton pairs primarily within the epitaxial layer. Measurements were conducted with a tunable laser system consisting of an optical parametric oscillator (OPO) pumped by the third harmonic of a pulsed, Q-switched Nd:YAG laser, producing 5 ns pulses at a 10 Hz repetition rate. Emission spectra were collected with an intensified CCD camera at progressively increasing delays from the laser pulse and within a defined time window.

3. Results and Discussion

The micro-Raman spectra of substrates A and B in the range of the LO-PC (longitudinal optical phonon–plasmon coupled) band (925–1100 cm^{-1}) are reported, respectively, in Figure 1a,b together with a graphical representation of the samples themselves. The LOPC asymmetric shape is well-known in the literature and is related to doping concentration. In fact, as in other polar semiconductors, the free carriers interact with the LO phonons, giving rise to the longitudinal optical phonon–plasmon coupled mode. As a consequence, this Raman mode is strongly affected by the carrier density. In particular, when the carrier density increases, the LO peak shifts toward high Raman shift values and features peak broadening and asymmetry (LOPC) [21,27,28]. In sample A, it is evident that the doping in the facet and in the rest of the wafer is different, in particular, in the facet, the doping is higher. Instead, in sample B, the doping is quite homogeneous (reasonably, because the facet is extended across the whole wafer). It is worth noting that based on the spectral characteristics of the Raman LOPC band, the doping of sample B is the same as sample A in the facet area.

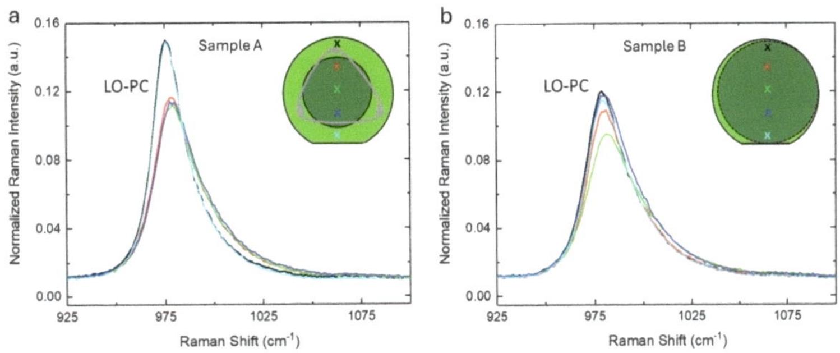

Figure 1. (**a**) Normalized micro−Raman spectra of substrate A along the diameter in the range 925–1100 cm^{-1}. (**b**) Normalized micro−Raman spectra of substrate B along the diameter in the range 925–1100 cm^{-1}. The spectra are normalized to the TO band at 780 cm^{-1} amplitude. The graphical representations of the substrates are inserted in both panels. The color of the spectra refer to the positions where they were acquired, which are marked by an x of the same color on the graphical representation.

In Figure 2a, micro-Raman spectra acquired in different radial positions, from the center to the edges, of sample A after the thermal process conducted at 1650 °C are shown. Each color refers to a particular area of the wafer: green for a circular area 3 cm in diameter centered at the center of the wafer; pink at 1–2 cm from the edges; red on the edges; and black on the remaining parts of the surface. It is evident that the surface of sample A is not homogeneous after the thermal process. In particular, the spectra are characterized by the typical bands of 4H-SiC together with specific shape changes (along the black dashed lines, as a guide for the eye) going from the center to the edges of the wafer [29,30].

To demonstrate the effects of the thermal treatment, the spectra of Figure 2a after the subtraction of the reference Raman spectrum of the same kind of substrate of 4H-SiC before any treatment are reported in Figure 2b. Furthermore, with blue dashed lines, the typical Raman bands D ($\sim$1350 cm^{-1}), G ($\sim$1580 cm^{-1}), and 2D ($\sim$2700 cm^{-1}) of graphene are indicated [31,32]. The green spectrum, collected in the center of sample A, has no difference in signal that would indicate that the corresponding part of the wafer is made of high quality, unmodified 4H-SiC. In the black spectrum, which refers to the most widespread region of the wafer, two bands around 1300 cm^{-1} and 1580 cm^{-1} are visible. In the pink spectrum, at 1–2 cm from the edges, it is possible to observe a shift in the bands toward high Raman shift values. Moreover, even if it is hardly visible for the magnification motif, a band around 2700 cm^{-1} is present. Lastly, in correspondence with the edges, for the red spectrum, the band around 2700 cm^{-1} is evident and the bands at 1350 cm^{-1} and 1600 cm^{-1} become narrower. The red spectrum matches with the Raman spectra of low-quality and damaged graphene, whereas the other spectra resemble those of amorphous carbon [29,31,33].

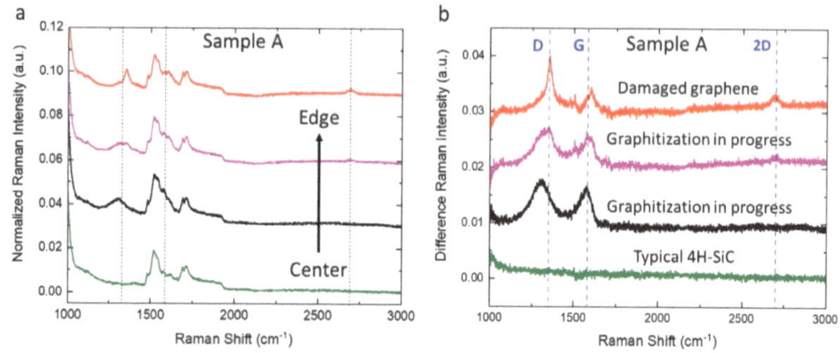

Figure 2. (**a**) Normalized micro−Raman spectra, with respect to TO amplitude band, of substrate A in different radial positions subsequent to thermal treatment at 1650 °C in the range of the second-order Raman peaks. Specifically, the green spectrum corresponds to a circular area with a 3 cm diameter centered on the wafer, the pink spectrum to positions 1–2 cm from the edges, the red spectrum to the edges, and the black spectrum to the remaining surface area. Dashed lines are reported as guides for the eye to mark relevant spectral positions. (**b**) Normalized micro-Raman spectra of Figure 2a after the subtraction of the reference spectrum of 4H-SiC. The dashed lines labeled D, G, and 2D indicate the Raman peaks associated with graphene.

The micro-Raman characterization conducted on sample B after the thermal process (data not reported) indicates that the effect of the treatment is homogeneous all over the wafer. To demonstrated these effects, in Figure 3, the recorded micro-Raman spectrum after the subtraction of the reference spectrum is reported in red. Furthermore, in this case, two bands at 1300 cm^{-1} and 1580 cm^{-1} related to amorphous carbon are evident. In addition, in Figure 3, the most common micro-Raman difference in spectrum of sample A is reported in black. It is evident that this one is similar to the spectra of sample B after the thermal treatment and the subtraction of the reference 4H-SiC spectrum. It is worth noting that the areas of sample A and sample B compared in Figure 3 have the same doping. For this reason, the observed effects seem to be related to the doping levels of the material.

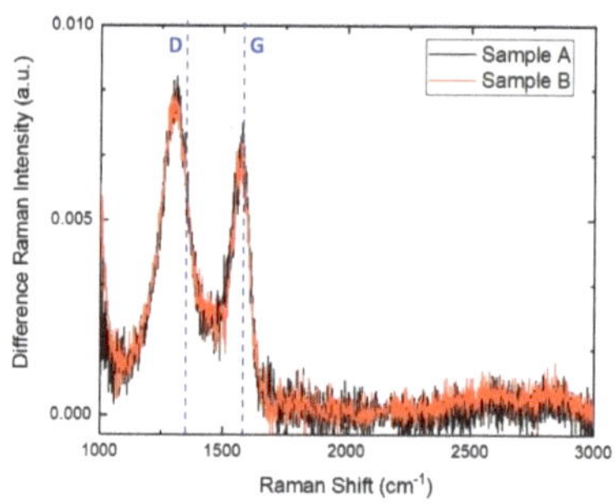

Figure 3. Normalized micro-Raman spectra, with respect to TO amplitude band, in the range of the second-order Raman peaks of sample A (black) and sample B (red) after the subtraction of the 4H-SiC reference spectrum. The spectra were acquired from positions with the same doping level. The dashed lines labeled D and G indicate the Raman peaks associated with graphene.

It is possible to explain the observed results as a graphitization process of the surface triggered by the desorption of Si from the SiC lattice and the subsequent reorganization of the C atoms. This is supported by the fact that this mechanism is exploited to produce high-quality graphene on 4H-SiC or on other SiC polytypes with hexagonal symmetry in an inert atmosphere or in vacuum starting from around 1000 °C up to 2000 °C (depending on other parameters such as gas pressure and growth time, which can reach 30 min) [22–26,31,34]. In this study, the substrates were treated for just a few seconds at 1650 °C, and this was sufficient to start the graphitization of the wafers. Therefore, this short treatment time range can explain the almost total presence of amorphous carbon on both sample A and sample B.

To further investigate the graphitized layer, a scan of the region underneath the surface of the substrate (z-profile) was carried out by micro-Raman spectroscopy. In particular, in Figure 4, the z-profile in a point of the edge of sample A (red spectrum in Figure 2a) is reported. Moving beneath the surface of the sample, the spectrum changes gradually until a depth of 4 μm below the surface, where the traces of graphitization completely disappear. Since the estimated vertical resolution of the micro-Raman setup is 2–3 μm, it is possible to conclude that the effect induced by the thermal treatment is limited to the surface of the sample within 3 μm of the surface.

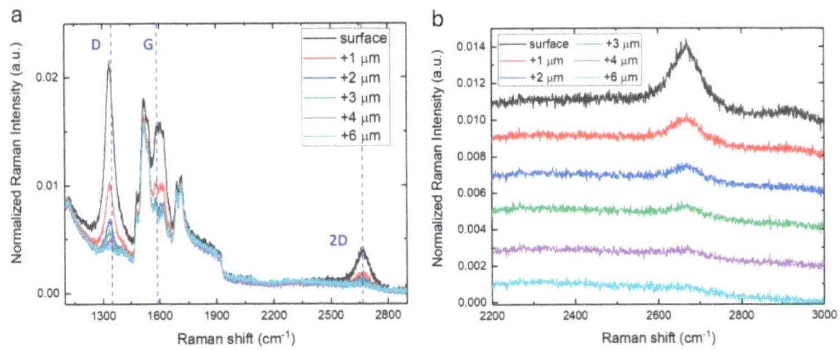

Figure 4. (**a**) Micro−Raman measurements in the range of the second−order Raman peaks, normalized with respect to the TO amplitude band, acquired on a point of the edge of sample A, varying the z-positions underneath the sample surface. (**b**) Zoom in the 2200–3000 cm^{-1} range of the spectra in panel (**a**). The spectra are vertically shifted for convenience.

To study how the graphitization revealed on the thermally treated substrates can affect the subsequent step of the device growth, an epitaxial layer of 12 μm thickness and n-type doping of 10^{16} cm^{-3} was grown on two new substrates, which presented the graphitization effects described above. In particular, the structure and the doping (qualitatively) of the wafers after the growth of the epitaxial layer were monitored by micro-Raman spectroscopy from the surface of the epitaxial layer down to the interface with the substrate, performing the z-profile reported in Figure 5a,b. It is possible to notice that no traces of graphitization are visible even at a depth of 12 μm below the surface, that is, at the interface between the substrate and the epitaxial layer. This indicates that the graphitized layer on the substrate does not affect the crystal structure of the epitaxial layers grown on it.

To further explore the effects of the graphitization of the substrate once an epitaxial layer is grown on it, time-resolved photoluminescence spectroscopy measurements were conducted on the epitaxial layer grown on a damaged substrate to study the lifetime of the excitonic band (at around 3.2 eV). In fact, the carrier lifetime is one useful parameter to evaluate the quality of semiconductor materials, and time-resolved photoluminescence spectroscopy can be used as a probe of the epitaxial layer quality due to its sensitivity to the presence of point defects and the possibility to obtain (qualitatively, by comparison) local information about their presence [35]. The normalized photoluminescence time decay of the excitonic band recorded in different positions on the epitaxial layer grown on the two damaged substrates and on an epitaxial layer grown (under the same conditions) on a high-quality substrate are, respectively, reported in Figure 5c,d. In this study, the lifetime is defined as the time required for the intensity of the excitonic band to decay to 1/e of its value measured 6 ns after the laser pulse peak, and the associated errors are related to the experimental uncertainty arising from the repeatability of the measurements. The lifetime values are different, in particular, they correspond to 12 ± 2 ns in the epitaxial layer grown on the damaged substrates and 25 ± 2 ns in the epitaxial layer grown on the high-quality substrate. Considering that the utilized substrates have the same nominal features (thickness, doping, amount of extended defects); that the epitaxial layers were grown in exactly the same way; that the crystal quality of the epitaxial layer seems to be not affected by the graphitization of the substrate underneath; and that a shorter lifetime is associated with an increase in the number of defects that influence the exciton recombination, our results demonstrate that a remembrance of the graphitization process is present even if the carbon-related species have been removed or masked during the growth of the epitaxial layer.

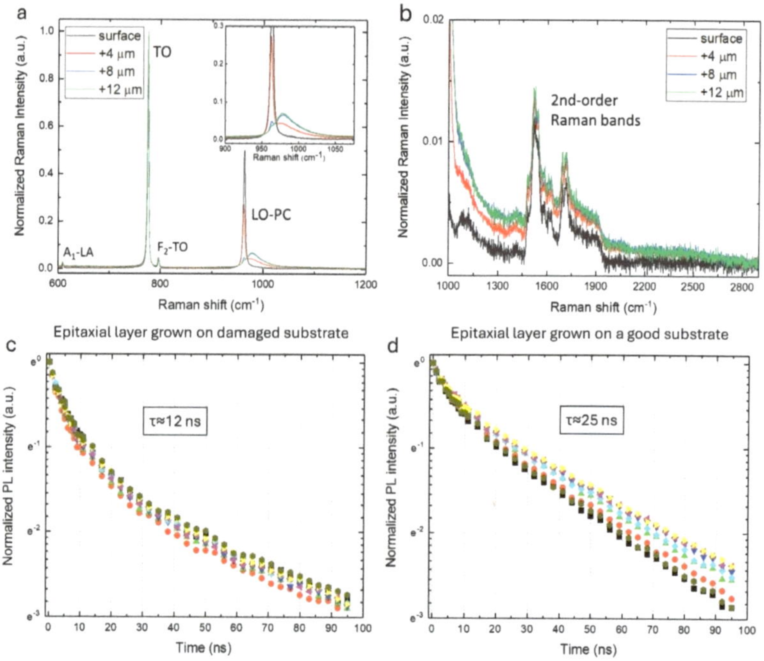

Figure 5. (**a**) Micro-Raman measurements, normalized with respect to the TO amplitude band, acquired at different positions underneath the epitaxial layer surface grown on one of the thermally treated damaged substrates. The spectrum collected at 12 μm is in correspondence with the interface between the substrate and epitaxial layer, and the inset reports a zoom of the LOPC band. (**b**) Zoom of the 1000–2900 cm^{-1} range of the spectra in panel (**a**). (**c**) Photoluminescence time decay (excitation wavelength 266 nm) of the excitonic band recorded at different positions (reported with different colors) on the epitaxial layer grown on the damaged substrates. (**d**) Photoluminescence time decay (excitation wavelength 266 nm) of the excitonic band recorded at different positions (reported with different colors) on the epitaxial layer grown on a high-quality substrate.

4. Conclusions

The thermal stability in an Ar atmosphere at a temperature in the range 1600–2000 °C of 4H-SiC substrates was studied. Amorphous carbon and traces of damaged graphene were observed by micro-Raman spectroscopy after the thermal treatment. Furthermore a dependence on doping was observed. The effect of graphitization on the surface of the substrates is attributed to the desorption of Si from the 4H-SiC lattice. The epitaxial layer grown on thermally treated damaged substrates was studied by micro-Raman and time-resolved photoluminescence spectroscopy. The crystal structure of the epitaxial layer was not affected by the graphitization damages of the substrate underneath, but despite that, it has an important influence on the recombination lifetime of the excitonic couple, shortening it.

Author Contributions: Conceptualization, F.M. and S.A.; Methodology, F.M., M.C., F.M.G., D.V., M.D.P., F.P., A.B. and S.A.; Validation, F.M. and S.A.; Formal analysis, F.M.; Investigation, F.M.; Resources, F.P., A.B., D.V. and M.D.P.; Data curation, F.M.; Writing—original draft, F.M.; Writing—review & editing, M.C., F.M.G., M.D.P. and S.A.; Supervision, S.A.; Project administration, S.A.; Funding acquisition, S.A. All authors have read and agreed to the published version of the manuscript.

Funding: This research was funded by STMicroelectronics.

Data Availability Statement: The original contributions presented in this study are included in the article. Further inquiries can be directed to the corresponding author.

Acknowledgments: The authors would like to thank STMicroelectronics for funding and ATeN Center for the micro-Raman measurements.

Conflicts of Interest: Authors Filippo Pasquali, Andrea Brischetto, Daniele Vecchio, Massimo Davide Pirnaci were employed by the company STMicroelectronics. The remaining authors declare that the research was conducted in the absence of any commercial or financial relationships that could be construed as a potential conflict of interest.

References

1. Physical Properties of Silicon Carbide. In *Fundamentals of Silicon Carbide Technology*; John Wiley and Sons, Ltd.: Hoboken, NJ, USA, 2014; Chapter 2, pp. 11–38. [CrossRef]
2. Bulk Growth of Silicon Carbide. In *Fundamentals of Silicon Carbide Technology*; John Wiley and Sons, Ltd.: Hoboken, NJ, USA, 2014; Chapter 3, pp. 39–74. [CrossRef]
3. Rost, H.J.; Schulz, D.; Siche, D. High Nitrogen Doping During Bulk Growth of SiC. In *Silicon Carbide: Recent Major Advances*; Choyke, W.J., Matsunami, H., Pensl, G., Eds.; Springer: Berlin/Heidelberg, Germany, 2004; pp. 163–178. [CrossRef]
4. Starke, U. Atomic Structure of SiC Surfaces. In *Silicon Carbide: Recent Major Advances*; Choyke, W.J., Matsunami, H., Pensl, G., Eds.; Springer: Berlin/Heidelberg, Germany, 2004; pp. 281–316. [CrossRef]
5. La Via, F.; Alquier, D.; Giannazzo, F.; Kimoto, T.; Neudeck, P.; Ou, H.; Roncaglia, A.; Saddow, S.E.; Tudisco, S. Emerging SiC Applications beyond Power Electronic Devices. *Micromachines* **2023**, *14*, 1200. [CrossRef] [PubMed]
6. Shakir, M.; Hou, S.; Hedayati, R.; Malm, B.G.; Östling, M.; Zetterling, C.M. Towards Silicon Carbide VLSI Circuits for Extreme Environment Applications. *Electronics* **2019**, *8*, 496. [CrossRef]
7. Neudeck, P.G.; Spry, D.J.; Chen, L.; Prokop, N.F.; Krasowski, M.J. Demonstration of 4H-SiC digital integrated circuits above 800 °C. *IEEE Electron Device Lett.* **2017**, *38*, 1082–1085.
8. Zetterling, C.M.; Hallén, A.; Hedayati, R.; Kargarrazi, S.; Lanni, L.; Malm, B.G.; Mardani, S.; Norström, H.; Rusu, A.; Suvanam, S.S.; et al. Bipolar integrated circuits in SiC for extreme environment operation. *Semicond. Sci. Technol.* **2017**, *32*, 034002. [CrossRef]
9. Altana, C.; Calcagno, L.; Ciampi, C.; La Via, F.; Lanzalone, G.; Muoio, A.; Pasquali, G.; Pellegrino, D.; Puglia, S.; Rapisarda, G.; et al. Radiation Damage by Heavy Ions in Silicon and Silicon Carbide Detectors. *Sensors* **2023**, *23*, 6522. [CrossRef]
10. Ciampi, C.; Pasquali, G.; Altana, C.; Bini, M.; Boscardin, M.; Calcagno, L.; Casini, G.; Cirrone, G.; Fazzi, A.; Giove, D.; et al. Nuclear fragment identification with ΔE-E telescopes exploiting silicon carbide detectors. *Nucl. Instrum. Methods Phys. Res. Sect. A Accel. Spectrometers Detect. Assoc. Equip.* **2019**, *925*, 60–69. [CrossRef]
11. Huang, X.; Zhang, X. Investigating the advanced characteristics of SiC based piezoresistive pressure sensors. *Mater. Today Commun.* **2020**, *25*, 101493. [CrossRef]
12. Kalinina, E.V.; Lebedev, A.; Kozlovski, V.V.; Zabrodski, V.; Strel'chuk, A.M.; Nikitina, I.P. Electrophysical and Optical Properties of 4H-SiC UV Detectors Irradiated with Electrons. *Mater. Sci. Forum* **2019**, *963*, 722–725. [CrossRef]
13. Tudisco, S.; La Via, F.; Agodi, C.; Altana, C.; Borghi, G.; Boscardin, M.; Bussolino, G.; Calcagno, L.; Camarda, M.; Cappuzzello, F.; et al. SiCILIA—Silicon Carbide Detectors for Intense Luminosity Investigations and Applications. *Sensors* **2018**, *18*, 2289. [CrossRef]
14. Chaudhuri, S.K.; Zavalla, K.J.; Mandal, K.C. High resolution alpha particle detection using 4H–SiC epitaxial layers: Fabrication, characterization, and noise analysis. *Nucl. Instrum. Methods Phys. Res. Sect. A Accel. Spectrometers Detect. Assoc. Equip.* **2013**, *728*, 97–101. [CrossRef]
15. Mandal, K.C.; Muzykov, P.G.; Russell Terry, J. Highly sensitive x-ray detectors in the low-energy range on n-type 4H-SiC epitaxial layers. *Appl. Phys. Lett.* **2012**, *101*, 051111. [CrossRef]
16. Nava, F.; Bertuccio, G.; Cavallini, A.; Vittone, E. Silicon carbide and its use as a radiation detector material. *Meas. Sci. Technol.* **2008**, *19*, 102001. [CrossRef]
17. Kalinina, E. The effect of irradiation on the properties of SiC and devices based on this compound. *Semiconductors* **2007**, *41*, 745–783. [CrossRef]
18. Moscatelli, F. Silicon carbide for UV, alpha, beta and X-ray detectors: Results and perspectives. *Nucl. Instrum. Methods Phys. Res. Sect. A Accel. Spectrometers Detect. Assoc. Equip.* **2007**, *583*, 157–161. [CrossRef]
19. Le Donne, A.; Binetti, S.; Acciarri, M.; Pizzini, S. Electrical characterization of electron irradiated X-rays detectors based on 4H-SiC epitaxial layers. *Diam. Relat. Mater.* **2004**, *13*, 414–418. [CrossRef]

20. Ohtani, N.; Katsuno, M.; Fujimoto, T.; Yashiro, H. Defect Formation and Reduction During Bulk SiC Growth. In *Silicon Carbide: Recent Major Advances*; Choyke, W.J., Matsunami, H., Pensl, G., Eds.; Springer: Berlin/Heidelberg, Germany, 2004; pp. 137–162. [CrossRef]
21. Burton, J.C.; Sun, L.; Pophristic, M.; Lukacs, S.J.; Long, F.H.; Feng, Z.C.; Ferguson, I.T. Spatial characterization of doped SiC wafers by Raman spectroscopy. *J. Appl. Phys.* **1998**, *84*, 6268–6273. [CrossRef]
22. Zebardastan, N.; Bradford, J.; Lipton-Duffin, J.; MacLeod, J.; Ostrikov, K.K.; Tomellini, M.; Motta, N. High quality epitaxial graphene on 4H-SiC by face-to-face growth in ultra-high vacuum. *Nanotechnology* **2022**, *34*, 105601. [CrossRef]
23. Mishra, N.; Boeckl, J.; Motta, N.; Iacopi, F. Graphene growth on silicon carbide: A review. *Phys. Status Solidi (A)* **2016**, *213*, 2277–2289. [CrossRef]
24. Emtsev, K.; Bostwick, A.; Horn, K.; Jobst, J.; Kellogg, G.L.; Ley, L.; McChesney, J.L.; Ohta, T.; Reshanov, S.A.; Röhrl, J.; et al. Towards wafer-size graphene layers by atmospheric pressure graphitization of silicon carbide. *Nat. Mater.* **2009**, *8*, 203–207. [CrossRef]
25. Hass, J.; de Heer, W.D.; Conrad, E.H. The growth and morphology of epitaxial multilayer graphene. *J. Phys. Condens. Matter* **2008**, *20*, 32. [CrossRef]
26. Tedesco, J.; Jernigan, G.; Culbertson, J.; Hite, J.; Yang, Y.; Daniels, K.; Myers-Ward, R.; Eddy, C.; Robinson, J.; Trumbull, K.; et al. Morphology characterization of argon-mediated epitaxial graphene on C-face SiC. *Appl. Phys. Lett.* **2010**, *96*, 222103. [CrossRef]
27. Nakashima, S.; Kitamura, T.; Mitani, T.; Okumura, H.; Katsuno, M.; Ohtani, N. Raman scattering study of carrier-transport and phonon properties of 4H-SiC crystals with graded doping. *Phys. Rev. B* **2007**, *76*, 245208. [CrossRef]
28. Nakashima, S.; Kitamura, T.; Kato, T.; Kojima, K.; Kosugi, R.; Okumura, H.; Tsuchida, H.; Ito, M. Determination of free carrier density in the low doping regime of 4H-SiC by Raman scattering. *Appl. Phys. Lett.* **2008**, *93*, 121913. [CrossRef]
29. Faugeras, C.; Nerrière, A.; Potemski, M.; Mahmood, A.; Dujardin, E.; Berger, C.; Heer, W. Few-layer graphene on SiC, pyrolitic graphene: A Raman scattering study. *Appl. Phys. Lett.* **2007**, *92*, 011914. [CrossRef]
30. Nakashima, S.; Harima, H. Raman Investigation of SiC Polytypes. *Phys. Status Solidi (A)* **1997**, *162*, 39–64. [CrossRef]
31. Zolyomi, V.; Koltai, J.; Kurti, J. Resonance Raman spectroscopy of graphite and graphene. *Phys. Status Solidi B* **2011**, *248*, 2435–2444. [CrossRef]
32. Klar, P.; Lidorikis, E.; Eckmann, A.; Verzhbitskiy, I.A.; Ferrari, A.C.; Casiraghi, C. Raman scattering efficiency of graphene. *Phys. Rev. B* **2013**, *87*, 205435. [CrossRef]
33. Rouchon, D.; Becerra, L.; Renault, O.; Kaja, K.; Mariolle, D.; Chevalier, N. Raman Spectra and Imaging of Graphene Layers Grown by SiC Sublimation. *AIP Conf. Proc.* **2010**, *1267*, 445–446. [CrossRef]
34. Zhao, J.; Ji, P.; Li, Y.; Li, R.; Zhang, K.; Tian, H.; Yu, K.; Bian, B.; Hao, L.; Xiao, X.; et al. Ultrahigh-mobility semiconducting epitaxial graphene on silicon carbide. *Nature* **2024**, *625*, 60–65. [CrossRef]
35. Migliore, F.; Cannas, M.; Gelardi, F.M.; Vecchio, D.; Brischetto, A.; Agnello, S. Defects in epitaxial 4H-SiC revealed by exciton recombination. *J. Phys. Condens. Matter* **2024**, *36*, 185601. [CrossRef]

materials

MDPI

Article

Charge Trap States of SiC Power TrenchMOS Transistor under Repetitive Unclamped Inductive Switching Stress

Juraj Marek [1,*], Jozef Kozarik [1,2], Michal Minarik [1], Aleš Chvála [1], Matej Matus [1], Martin Donoval [1], Lubica Stuchlikova [1] and Martin Weis [1]

[1] Institute of Electronics and Photonics, Slovak University of Technology in Bratislava, 812 19 Bratislava, Slovakia
[2] NanoDesign, s.r.o, Drotárska 19a, 811 04 Bratislava, Slovakia
* Correspondence: juraj.marek@stuba.sk

Abstract: Silicon carbide (SiC) has been envisioned as an almost ideal material for power electronic devices; however, device reliability is still a great challenge. Here we investigate the reliability of commercial 1.2-kV 4H-SiC MOSFETs under repetitive unclamped inductive switching (UIS). The stress invoked degradation of the device characteristics, including the output and transfer characteristics, drain leakage current, and capacitance characteristics. Besides the shift of steady-state electrical characteristics, a significant change in switching times points out the charge trapping phenomenon. Transient capacitance spectroscopy was applied to investigate charge traps in the virgin device as well as after UIS stress. The intrinsic traps due to metal impurities or $Z_{1,2}$ transitions were recognized in the virgin device. The UIS stress caused suppression of the second stage of the $Z_{1,2}$ transition, and only the first stage, Z_1^0, was observed. Hence, the UIS stress is causing the reduction of multiple charging of carbon vacancies in SiC-based devices.

Keywords: charge trap states; reliability; degradation; SiC MOSFET; TrenchMOSFET; repetitive UIS

Citation: Marek, J.; Kozarik, J.; Minarik, M.; Chvála, A.; Matus, M.; Donoval, M.; Stuchlikova, L.; Weis, M. Charge Trap States of SiC Power TrenchMOS Transistor under Repetitive Unclamped Inductive Switching Stress. *Materials* **2022**, *15*, 8230. https://doi.org/10.3390/ma15228230

Academic Editor: Fabrizio Roccaforte

Received: 17 October 2022
Accepted: 16 November 2022
Published: 19 November 2022

1. Introduction

Due to their superior properties, silicon carbide-based (SiC) devices have become the number one choice for power electronic conversion semiconductor (PECS) applications, where they operate under high power, high temperature, and high switching frequency conditions [1,2]. These applications typically employ a high quantity of large-area dies per system while demanding high system-level reliability under aggressive electrical and environmental operating conditions. Significant progress in recent years in the development of SiC transistors and, above all, in their reliability have contributed to the fact that SiC-based devices have become a viable alternative to silicon-based power devices [3]. Nevertheless, before SiC can completely replace silicon (Si) counterparts, robustness and reliability remain the main issue [4,5] for the devices under several extreme operational conditions, such as overcurrent, overtemperature, short circuit, and unclamped inductive switching (UIS) [6]. For example, sudden high currents and voltage pulses may occur during start-up events or load changes of electric motors. Even with a good inverter or power supply design, high-voltage peaks may occur due to inevitable parasitic inductances. As a result, voltage spikes can easily reach the value of the breakdown voltage of the switch, leading to an avalanche breakdown operation for a short time. This imposed stress might trigger the drift of electrical parameters that can limit the devices' operating range and lifetime. Even though the degradation of device electrical parameters is well studied, the relation to material defects is still unclear and represents a gap in knowledge of SiC-based devices.

Therefore, in our analysis, we focused on the impact of repetitive unclamped inductive switching (UIS) on the stability of electrical parameters of commercial 1.2-kV 4H-SiC MOSFETs with an asymmetric channel configuration proposed by R. Siemieniec [7]. The stress-induced shift of device characteristics and parameters, including the transfer, output, and diode characteristics, as well as parameters such as drain leakage current (I_{dss}), breakdown voltage (BV_{dss}), and on-resistance (R_{DSon}), was observed. Switching performance was verified using a double-pulse test, where a significant change in switching times was observed. These changes during the lifetime operation of the device pose considerable risks against the requirement for the long-term reliability of modern power devices. Here we aim for the application of well-accepted UIS stress on SiC-based TrenchMOSFET with a detailed study of trap states localized within the energy gap. The mutual comparison of device electrical parameters and trap states in the band diagram illustrates the impact of UIS stress.

2. Experimental

Tested devices were commercially available 1.2-kV SiC TrenchMOSFET devices produced by Infineon (Dresden, Germany) with rated current 13 A and a typical value of on-resistance, R_{DSon} = 220 mΩ. The structure of the device is vertical, with an asymmetric channel oriented along a plane, which gives the best channel mobility and lowest interface trap density [7]. Two deep p-wells serve as an anode of the body diode for freewheeling operation, while one of the p-wells also shields the gate structure from high potential and decreases the risk of gate dielectric breakdown. Devices are encapsulated in the PG-TO247-4 package.

Commercial UIS tester ITC55100 (Integrated Technology Corporation, Tempe, USA)was used for UIS testing. The entire testing procedure, as well as the layout of the test circuit, were defined by JEDEC standard—JESD24-5. Figure 1a,b show the circuit diagram and the typical voltage and current waveforms during the SiC MOSFET test. The device under test (DUT) is connected to the power supply through the inductor and the high-side switch (HSW). When HSW is shorted, and DUT is turned on, the current limited by the inductor starts to rise linearly. DUT and HSW are turned off when the current (I_D) reaches the required (pre-set) value in the experiments. The magnetic field in the inductor (L) induces a counter-electromotive force that can build up surprisingly high potentials across the switch (DUT). If no protective circuits are added to the switch, all the energy accumulated in the inductor is dissipated directly in the device switch. To correctly set parameters for repetitive UIS tests, the first single pulse UIS capability of tested devices was verified. Obtained values of destructive currents for single pulse operation, measured for different inductances, are shown in Figure 2. Observed results are in good agreement with basic UIS theory, where if the intrinsic temperature of the blocking PN junction is assumed as a limit value for the passive mode of destruction and tests start at a constant temperature, then we can write the relation between the inductance, L, and the critical value of the avalanche current, I_{Avcrit}, as [8]:

$$\Delta T_M = \frac{\sqrt{2}}{3} C_{th0} . I_{AV} . V_{BReff} . \sqrt[2.2]{\frac{L I_{AV}}{V_{BReff}}} \rightarrow I_{AVcrit} \sim \sqrt[3.2]{\frac{1}{L}} \quad (1)$$

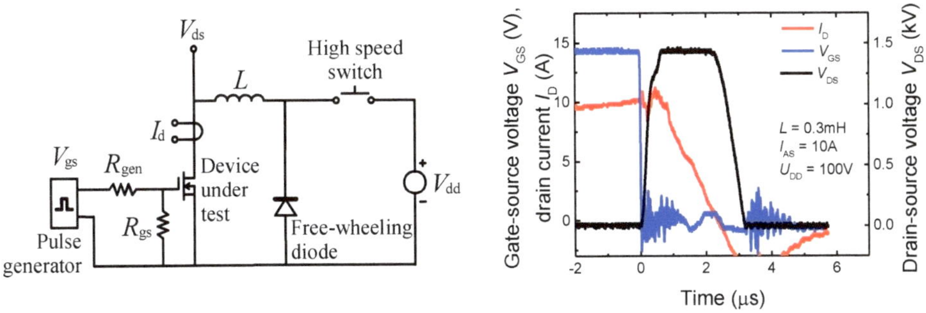

(a) (b)

Figure 1. UIS test: (**a**) Basic UIS test circuit. (**b**) Typical current and voltage waveforms of SiC MOSFET under UIS test conditions.

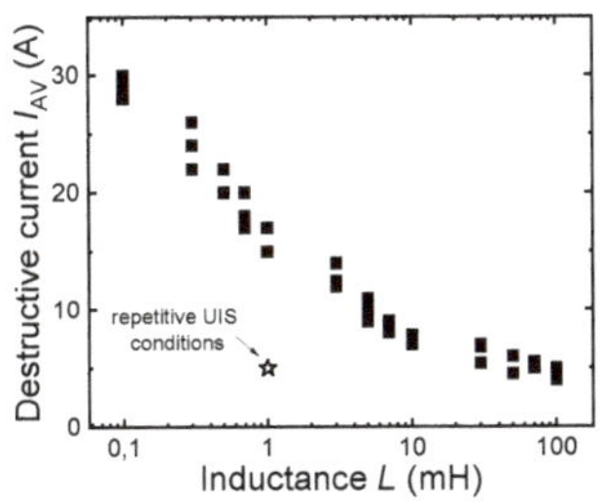

Figure 2. Measured destructive currents of investigated SiC TrenchMOS samples for different inductive loads. The star symbol represents the stress current induced during the repetitive UIS test.

3. Repetitive UIS Stress

3.1. Impact on I-V and C-V Characteristics

Based on the results from single pulse UIS measurements, conditions for the repetitive UIS stress were set with a significantly lower avalanche current than the critical value for the given inductance. Parameters were set as follows: load inductance (L) = 1 mH, supply voltage (V_{DD}) = 100 V, switched current (I_{AS}) = 5 A. These parameters have been determined in accordance with the recommendation given by the application note AN-1005 [9]. The switching period between two pulses was set to 5 ms, and an additional time of 800 ms was set after every 100 sets of pulses. The current and switching period was set according to the maximum non-destructive single pulse current for a given inductive load and to minimize heat accumulation in samples. However, during switching, significant heat is still generated in transistors; therefore, all I-V measurements were performed after the device cooled down for ten minutes to prevent the impact of temperature on device parameters shifting.

First, we analyzed the impact of repetitive UIS on static I-V characteristics. Device characteristics were measured before and after each sequence of pulses to observe the gradual effects of repetitive UIS stress. Measured I-V characteristics for virgin and stressed samples, up to 7 million pulses, were shown in Figure 3. Observations were as follows: Repetitive UIS pulses induced shifts in all analyzed I-V characteristics. The output and transfer characteristics' shifts were mostly uniform (Figure 3a,b). We observed a slight negative shift of the threshold voltage. This negative shift of the threshold voltage is obviously also responsible for the decline of the on-resistance, evident in the output characteristics. The on-resistance has decreased from the initial value of 207 mΩ, before the stress, to the value of 195 mΩ after 7 million pulses. The decrease in on-resistance clearly points to the fact that the characteristics are not affected by thermal cycling, but other mechanisms have to be present. It is a fact that during the avalanching phase of the UIS pulse, large concentrations of high energy carriers are generated close to blocking the PN junction. Therefore, a negative shift of threshold voltage indicates the trapping of a positive charge on the channel-gate interface or in its vicinity. The origin of this positive charge can be attributed to the high energy holes generated during the UIS pulse. Our assumption is confirmed by a significant change in the gate leakage current (Figure 3d), which must be caused by the formation or activation of electrically active defects in the gate dielectric. A significant change in the concentration of electrically active defects in the vicinity of the blocking PN junction is also indicated by a significant change in the drain-to-source leakage current (Figure 3c). However, no change in static breakdown voltage value was observed. The change in the leakage current characteristics (Figure 3c), which is caused by an increase in the concentration of electrically active traps, is manifested by two things, an increase in the leakage current itself and a change in the shape of the I-V curve, which is most likely related to a change in the distribution of the electric field due to a change in the charge distribution.

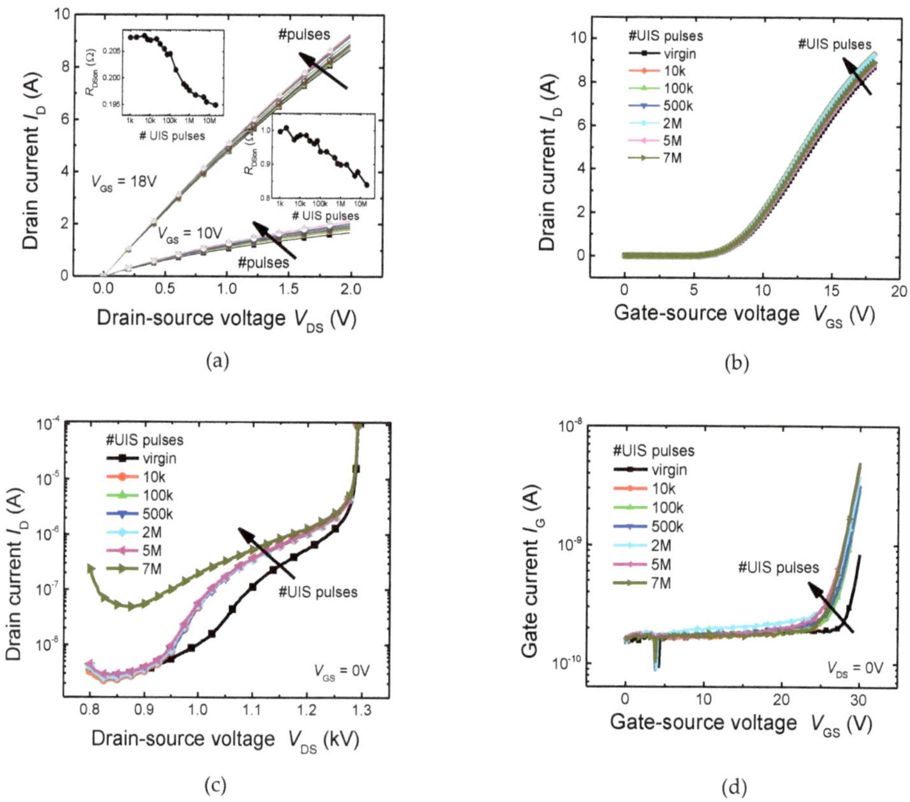

Figure 3. *Cont.*

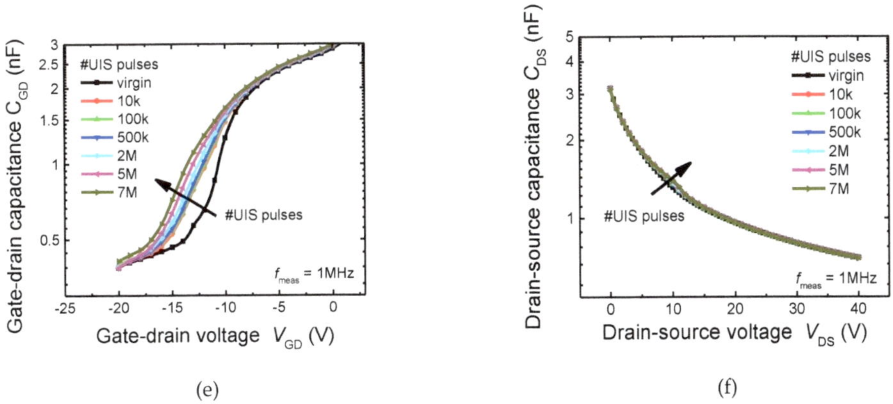

Figure 3. Measured the impact of repetitive UIS on the stability of electrical characteristics. (**a**) Output characteristics and on-resistance (inset graph), various colours of output characteristics stand for different numbers of UIS pulses, (**b**) transfer characteristics, (**c**) drain leakage current and breakdown, (**d**) gate leakage current, (**e**) gate–drain capacitance, (**f**) drain–source capacitance.

Another important observation was the shift of static C-V characteristics (Figure 3e,f). While only minimal change was observed in the drain-source capacitance characteristic, C_{DS}, which is primarily determined by the capacity of the blocking PN junction, a significant change in the gate-to-drain capacitance, C_{GD}, was observed. It is caused by the fact that the bottom part of the trench gate electrode, which determines the C_{GD} capacity, is strongly exposed to hot carriers generated during UIS stress.

3.2. Impact on Switching Performance

Next, switching processes (turn-on and -off) were analyzed using a double-pulse tester. Waveforms of the gate and drain voltages and drain current are shown in Figure 4. First, measurements were performed on virgin samples. Next, samples were exposed to 10 million UIS pulses. After repetitive UIS stress, a significant increase in switching time for the turn-on process was observed. However, no impact on switching time was observed for the turn-off process. The explanation may be that defects are generated close to or on the gate-drain interfaces. These defects are charged when the transistor is in an off state and high voltage is present between the gate and drain. Then, when the transistor is turned on, discharging of these defects slows down the drain voltage decrease. A similar change in the dynamic behaviour was previously observed and described by T. Laska et al. in [10] for IGBTs, where authors assumed that carriers accelerated by high dynamic electrical fields were injected into the gate oxide. As a result, charges at or near the interface Si/oxide over the whole trench shape below the p-body region are formed, leading to an increase in the turn-off delay time. The charge trapping takes an extremely short time (in the ns region), while the discharging process is drastically slower (in the µs and ms region). As a result, both the turn-on and turn-off processes are affected by the charge trap states; however, the observation of the switching process in the ns region can reveal the charge trapping only, while the charge detrapping is out of the observed time range. The charge detrapping (charge release) was analyzed in detail by the deep-level transient spectroscopy that is suitable for long-living states only.

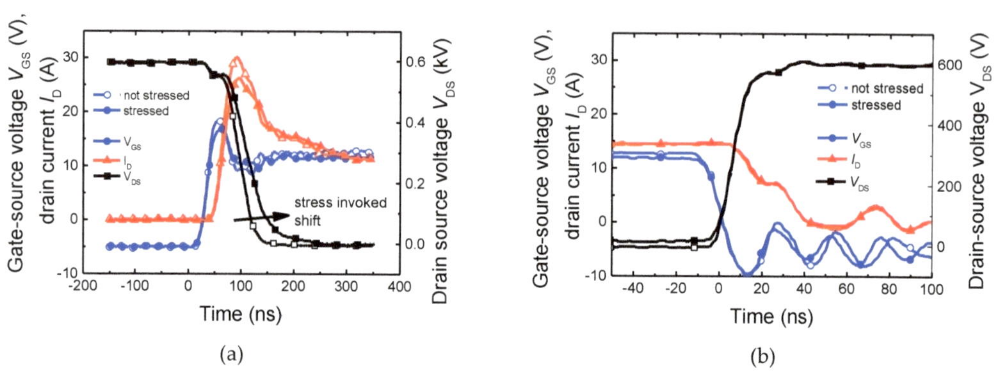

Figure 4. Impact of repetitive UIS on switching performance of the power SiC TrenchMOS. (**a**) Turn-on and (**b**) turn-off the process. A significant increase in turn-on time was observed after 10 million UIS pulses. No impact on the turn-off process was observed. Open symbols mean the values correspond to virgin samples, while the filled symbols represent the measurement after 10 million UIS pulses.

4. Defect Analysis

The structural defects originated in the device fabrication process or due to device stress, serve as charge carrier traps. The charge trap identification was performed using the deep-level transient Fourier spectroscopy (DLTFS) system DL8000 (BIO-RAD Micromeasurement, UK). The gate-drain capacitance transients, C_{GD}, were recorded since the channel region stands for the most affected capacitance contribution by repetitive UIS stress and different sets of parameters, and that is all. Note that the DLTFS method relies on the observation of a transient response on an excitation pulse applied at various temperatures. As a result, the temperature dependence of transients' relaxation time reveals the activation energies of hidden charge traps. The measurement conditions, such as excitation pulse amplitude or time, were conserved for all measurements to achieve fully comparable parameters for the virgin device and device after the stress. The Fourier transform analysis was employed to analyze the multiexponential transients. Fourier coefficients, a_1 and b_1, representing the cosine and sine coefficients of the first order, were estimated for the Arrhenius plots evaluation [11]. First, the virgin device was analyzed to map the traps' energy distribution before the stress application, as shown in Figure 5. Interestingly, four dominant electron trap states denoted as ET1 to ET4 were observed. The Arrhenius plot was applied to extract the traps' parameters, such as the activation energy and the capture cross-section, as shown in Table 1. A numerical simulation was used to verify the evaluated parameters of each trap state. All the measurements were summarized in the Supplementary Material, Figure S1.

Note that even though the activation energy derived from transient spectroscopy is not an actual position of the energy level within the energy band gap due to the temperature-dependent cross-section, many reports found relations between the activation energies and structural defects causing the trap states. For example, metal impurities are a common source of electron traps originating from the imperfection of the MOSFET fabrication technology. The electron traps, ET1 and ET2, are associated with the presence of Cr, Ti, or W, since these impurities create localized states, 0.13 to 0.18 eV from the bottom edge of the conduction band, E_c [12–15]. Electron trap, ET3, is a known intrinsic defect denoted as $Z_{1,2}$ and stands for the negative U-center, the defect that can exist in more than a single charge state. Even though it was assigned to the carbon vacancy as the most common structural defect in 4H-SiC, Hemmingsson et al. [15] had shown that it consists of two contributions corresponding to the emission of two electrons from the $Z_1^{0/+}$ and $Z_2^{0/+}$ levels [12,13,16–20]. These defects were discussed as $(=/0)$ transitions of carbon vacancy, V_C, located at the hexagonal $(-h)$ and pseudo-cubic $(-k)$ sites [12,13]. However, under common conditions, we observe only the $Z_{1,2}^{-/+}$ transition, while the $Z_1^{-/0}$ and $Z_2^{-/0}$ levels with activation energies of about 0.52 and 0.45 eV, respectively, are not observable [15]. As a result, the $Z_{1,2}^{-/+}$ transition is an origin of a strongly asymmetric peak in the spectrum that can be deconvoluted into two energy levels close to each other. Interestingly, the Laplace DLTS was used to separate the $Z_{1,2}^{-/+}$ transition into two emissions with different emission times since it represents a two-stage process ($V_C^{=} \rightarrow V_C^{-} + e \rightarrow V_C^{0} + 2e$), and conventional DLTS merges them into a single process. Here, the first stage, Z_1, reached an energy level of 0.58 eV, while the second stage, Z_2, was estimated at 0.65 eV. In other words, these two negative U-centres have one donor and one acceptor level. The donor levels, Z_1^0 and Z_2^0, have activation energies of 0.58 and 0.45 eV, respectively, whereas the acceptor levels, Z_1^- and Z_2^-, exhibit activation energies of 0.67 and 0.71 eV, respectively [16]. Therefore, we assume that the electron traps, ET3 and ET4, are related to the $Z_{1,2}$ transition and stand for Z_1^- and Z_2^- transitions.

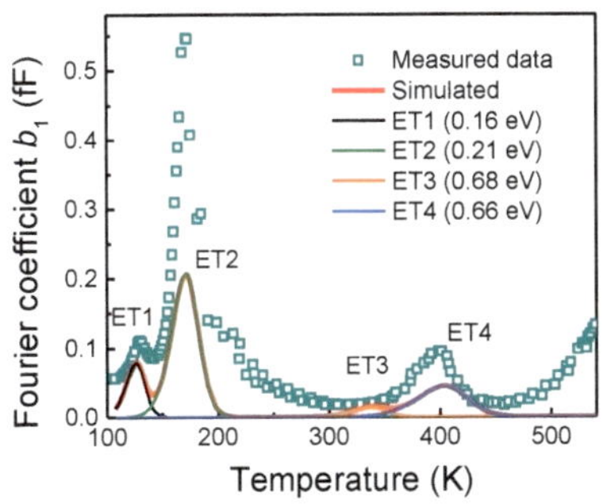

Figure 5. Fourier coefficient, b_1, spectrum before the stress. The position of evaluated deep levels is verified by the simulation (solid lines).

Table 1. Parameters of evaluated trap states. The asterisk (*) denotes the trap after stress conditions. The capture cross-section has an evaluated range of one standard deviation.

Trap	E_c-E_t (eV)	σ (cm^2)	Possible Origin
ET1	0.16 ± 0.01	1.27×10^{-17} ($0.39{\sim}3.91 \times 10^{-17}$)	metal impurity
ET1*	0.16 ± 0.00	1.11×10^{-17} ($0.81{\sim}1.52 \times 10^{-17}$)	
ET2	0.21 ± 0.01	4.06×10^{-18} ($2.44{\sim}6.75 \times 10^{-18}$)	metal impurity
ET2*	0.29 ± 0.00	1.12×10^{-15} ($0.88{\sim}1.44 \times 10^{-15}$)	
ET3	0.68 ± 0.01	1.37×10^{-14} ($0.86{\sim}2.18 \times 10^{-14}$)	$Z_{1,2}$ transition (Z_1^-)
ET4	0.66 ± 0.01	1.11×10^{-16} ($0.73{\sim}1.69 \times 10^{-16}$)	$Z_{1,2}$ transition (Z_2^-)
ET5*	0.57 ± 0.02	3.39×10^{-17} ($1.92{\sim}5.99 \times 10^{-17}$)	$Z_{1,2}$ transition (Z_1^0)

Next, the device was exposed to two different stress conditions: short repetitive UIS stress with a duration of 1 h and long repetitive UIS stress with a duration of 24 h. Finally, to distinguish stress impact, another device for trap state analysis was also prepared and exposed to static stress for 24 h. During this static stress, the device was kept in the drain-to-source breakdown with a constant drain current of 100 μA.

The Fourier coefficient b_1's spectra and Arrhenius plot, shown in Figure 6a, illustrated the static stress effect. As a result, we can summarize these observations: As the spectra for the unstressed virgin device and the device after static stress are almost identical, we can conclude that static stress had minimal impact on the device degradation. On the other hand, both repetitive UIS stress considerably affected the observed spectra. The concentration of the trap ET1 was slightly increasing with the length of UIS stress, while the increase in the trap state ET2 was observed only for the long-term stress (24 h). The highest impact of repetitive UIS was observed on trap states ET3 and ET4. First, after the short stress (1 h), the signal from trap state ET3 was increased. However, the long UIS stress had the opposite effect, and the signal from the trap ET3 was significantly suppressed. Similarly, the signal from trap ET4 is suppressed after the long-term UIS stress. Using different measurement conditions, we identified one newly generated electron trap state, ET5, with an activation energy of 0.57 eV. Since the emission from trap state ET5 is present without signals from ET3 and ET4, we can conclude that it is the first stage of the two-electron $Z_{1,2}$ process (Z_1^0 transition). Furthermore, we assume that the trap state was activated by the stress conditions because it was observable only after long-term UIS stress.

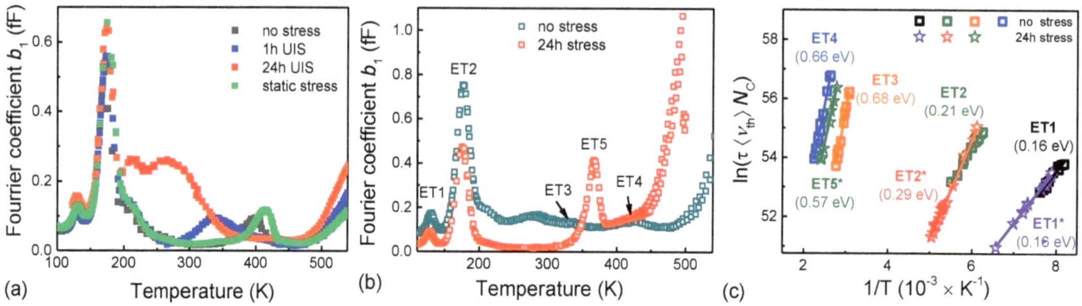

Figure 6. (**a**) Comparison of the Fourier coefficient b_1 spectra induced by different stress conditions and (**b**) for the virgin device and after repetitive UIS stress with a duration of 24 h. (**c**) Arrhenius plot of evaluated trap states.

5. Conclusions

The gradual degradation of commercial trench gate 1.2 kV 4H-SiC MOSFETs produced by Infineon, subjected to the repetitive avalanche pulses, was measured and analyzed. The ITC55100 UIS tester was utilized to generate the stress. Devices were exposed to up to 7 million stress pulses and stressed for up to 24 h. The strong impact of repetitive avalanching stress on the electrical performance of tested devices was observed. After repetitive UIS stress tests, devices showed a consistent degradation tendency characterized by a decreased threshold voltage (V_{th}), decreased on-resistance (R_{DSon}), and a significant increase in drain-source and gate-source leakage currents. Moreover, a significant shift of gate-to-drain capacitance (C_{DG}) was observed. Long-term repetitive UIS stress also invoked a significant increase in turn-on time. The trapping was also analyzed using transient capacitance spectroscopy. Four electron traps were evaluated from the measured spectra of the virgin device. Numerical simulations were applied to verify their parameters, and the trap states were associated with metal impurities or charge transitions of carbon vacancies. A significant shift of Fourier coefficients b_1 spectra and new electron trap state ET5 was observed after long-term repetitive UIS stress. The newly generated state was assumed to be a first-stage charging of the two-stage $Z_{1,2}$ transition related to the multiple charge states of carbon vacancy. The analysis confirmed the significant impact of repetitive avalanche stress on the overall stability of electrical parameters of the power SiC MOSFET and pointed to a possible negative impact on its overall long-term reliability due to the impact on traps capable of multiple charge states. The results point out that the device resistance to UIS stress can be improved by fabrication technology improvements that will suppress the presence of interface states and structural defects.

Supplementary Materials: The following supporting information can be downloaded at: https://www.mdpi.com/article/10.3390/ma15228230/s1, Figure S1: Deep-level transient spectroscopy data.

Author Contributions: Investigation, J.M., J.K., M.M. (Matej Matus) and L.S.; data curation, M.M. (Michal Minarik) and M.M. (Matej Matus); simulation, A.C.; funding acquisition, J.M. and M.D.; writing—original draft preparation, J.M., M.W., L.S., J.K. and M.M. (Michal Minarik) All authors have read and agreed to the published version of the manuscript.

Funding: This work was supported by the project HiEFFICIENT. The project has received funding from the Ecsel JU under grant agreement No. 101007281. The JU receives support from the European Union's Horizon 2020 research and innovation program and Austria, Germany, Slovenia, Netherlands, Belgium, Slovakia, France, Italy, and Turkey. This work was also supported by the Slovak Research and Development Agency under contract No. APVV-20-0266 and by grant VEGA 1/0727/19 and VEGA 1/0758/19 supported by the Ministry of Education, Science, Research and Sport of Slovakia.

Institutional Review Board Statement: Not applicable.

Informed Consent Statement: Not applicable.

Data Availability Statement: The data presented in this study are available in the Supplementary Materials.

Conflicts of Interest: The authors declare no conflict of interest.

References

1. Batunlu, C.; Alrweq, M.; Albarbar, A. Effects of power tracking algorithms on lifetime of power electronic devices used in solar systems. *Energies* **2016**, *9*, 884. [CrossRef]
2. Liu, S.; Gu, C.; Wei, J.; Qian, Q.; Sun, W.; Huang, A.Q. Repetitive unclamped-inductive-switching-induced electrical parameters degradations and simulation optimizations for 4H-SiC MOSFETs. *IEEE Trans. Electron. Devices* **2016**, *63*, 4331–4338. [CrossRef]
3. Martinella, C.; Alía, R.G.; Stark, R.; Coronetti, A.; Cazzaniga, C.; Kastriotou, M.; Kadi, Y.; Gaillard, R.; Grossner, U.; Javanainen, A. Impact of terrestrial neutrons on the reliability of SiC VD-MOSFET technologies. *IEEE Trans. Nucl. Sci.* **2021**, *68*, 634–641. [CrossRef]
4. Hazra, S.; Madhusoodhanan, S.; Moghaddam, G.K.; Hatua, K.; Bhattacharya, S. Design considerations and performance evaluation of 1200-V 100-A SiC MOSFET-based two-level voltage source converter. *IEEE Trans. Ind. Appl.* **2016**, *52*, 4257–4268. [CrossRef]
5. Lutz, J.; Baburske, R. Some aspects on ruggedness of SiC power devices. *Microelectron. Reliab.* **2014**, *54*, 49–56. [CrossRef]
6. Castellazzi, A.; Fayyaz, A.; Romano, G.; Yang, L.; Riccio, M.; Irace, A. SiC power MOSFETs performance, robustness and technology maturity. *Microelectron. Reliab.* **2016**, *58*, 164–176. [CrossRef]
7. Siemieniec, R.; Peters, D.; Esteve, R.; Bergner, W.; Kück, D.; Aichinger, T.; Basler, T.; Zippelius, B. A SiC trench MOSFET concept offering improved channel mobility and high reliability. In Proceedings of the 2017 19th European Conference on Power Electronics and Applications (EPE'17 ECCE Europe), Warsaw, Poland, 11–14 September 2017; p. P-1.
8. Marek, J.; Chvála, A.; Donoval, D.; Príbytný, P.; Molnár, M.; Mikolášek, M. Compact model of power MOSFET with temperature dependent Cauer RC network for more accurate thermal simulations. *Solid-State Electron.* **2014**, *94*, 44–50. [CrossRef]
9. McDonald, T.; Soldano, M.; Murray, A.; Avram, T. *Power MOSFET Avalanche Design Guidelines*; Application Note AN-1005; International Rectifier Location: El Segundo, CA, USA, 2000.
10. Laska, T.; Hille, F.; Pfirsch, F.; Jereb, R.; Bassler, M. Long term stability and drift phenomena of different trench IGBT structures under repetitive switching tests. In Proceedings of the 19th International Symposium on Power Semiconductor Devices and IC's (ISPSD 2007), Jeju, Korea, 27–31 May 2007; pp. 1–4.
11. Weiss, S.; Kassing, R. Deep Level Transient Fourier Spectroscopy (DLTFS)—A technique for the analysis of deep level properties. *Solid-State Electron.* **1988**, *31*, 1733–1742. [CrossRef]
12. Janzén, E.; Gali, A.; Henry, A.; Ivanov, I.G.; Magnusson, B.; Son, N.T. Defects in SIC. In *Defects in Microelectronic Materials and Devices*; Fleetwood, D.M., Schrimpf, R.D., Eds.; CRC Press: Boca Raton, FL, USA, 2008; pp. 615–671.
13. Achtziger, N.; Witthuhn, W. Selected Aspects of Radiotracer Deep Level Transient Spectroscopy. In *Silicon Carbide*; Choyke, W.J., Matsunami, H., Pensl, G., Eds.; Springer: Berlin/Heidelberg, Germany, 2004; pp. 537–562.
14. Achtziger, N.; Witthuhn, W. Band-gap states of Ti, V, and Cr in 4 H-SiC: Identification and characterization by elemental transmutation of radioactive isotopes. *Phys. Rev. B* **1998**, *57*, 12181. [CrossRef]
15. Achtziger, N.; Pasold, G.; Sielemann, R.; Hülsen, C.; Grillenberger, J.; Witthuhn, W. Tungsten in silicon carbide: Band-gap states and their polytype dependence. *Phys. Rev. B* **2000**, *62*, 12888. [CrossRef]
16. Hemmingsson, C.G.; Son, N.T.; Ellison, A.; Zhang, J.; Janzén, E. Negative-U centers in 4 H silicon carbide. *Phys. Rev. B* **1998**, *58*, R10119–R10122. [CrossRef]
17. Hemmingsson, C.; Son, N.T.; Kordina, O.; Bergman, J.P.; Janzén, E.; Lindström, J.L.; Savage, S.; Nordell, N. Deep level defects in electron-irradiated 4H SiC epitaxial layers. *J. Appl. Phys.* **1997**, *81*, 6155–6159. [CrossRef]
18. Capan, I.; Brodar, T.; Coutinho, J.; Ohshima, T.; Markevich, V.P.; Peaker, A.R. Acceptor levels of the carbon vacancy in 4 H-SiC: Combining Laplace deep level transient spectroscopy with density functional modeling. *J. Appl. Phys.* **2018**, *124*, 245701. [CrossRef]

19. Gelczuk, Ł.; Dąbrowska-Szata, M.; Kolkovsky, V.; Sochacki, M.; Szmidt, J.; Gotszalk, T. Origin and anomalous behavior of dominant defects in 4H-SiC studied by conventional and Laplace deep level transient spectroscopy. *J. Appl. Phys.* **2020**, *127*, 064503. [CrossRef]
20. Capan, I.; Brodar, T.; Pastuovi'c, Z.; Siegele, R.; Ohshima, T.; Sato, S.I.; Makino, T.; Snoj, L.; Radulovi´c, V.; Coutinho, J.; et al. Double negatively charged carbon vacancy at the h- and k-sites in 4H-SiC: Combined Laplace-DLTS and DFT study. *J. Appl. Phys.* **2018**, *123*, 161597. [CrossRef]

materials

MDPI

Review

Silicon and Silicon Carbide Recrystallization by Laser Annealing: A Review

Daniele Arduino [1,2], Stefano Stassi [1,*], Chiara Spano [1], Luciano Scaltrito [1], Sergio Ferrero [1] and Valentina Bertana [1]

[1] Department of Applied Science and Technology, Politecnico di Torino, Corso Duca degli Abruzzi 24, 10129 Torino, Italy; daniele_arduino@polito.it (D.A.); chiara.spano@polito.it (C.S.); luciano.scaltrito@polito.it (L.S.); sergio.ferrero@polito.it (S.F.); valentina.bertana@polito.it (V.B.)

[2] Department of Information Engineering, Electrical Engineering and Applied Mathematics, University of Salerno, Via Giovanni Paolo II 132, 84084 Fisciano, Italy

* Correspondence: stefano.stassi@polito.it

Abstract: Modifying material properties within a specific spatial region is a pivotal stage in the fabrication of microelectronic devices. Laser annealing emerges as a compelling technology, offering precise control over the crystalline structure of semiconductor materials and facilitating the activation of doping ions in localized regions. This obviates the necessity for annealing the entire wafer or device. The objective of this review is to comprehensively investigate laser annealing processes specifically targeting the crystallization of amorphous silicon (Si) and silicon carbide (SiC) samples. Silicon finds extensive use in diverse applications, including microelectronics and solar cells, while SiC serves as a crucial material for developing components designed to operate in challenging environments or high-power integrated devices. The review commences with an exploration of the underlying theory and fundamentals of laser annealing techniques. It then delves into an analysis of the most pertinent studies focused on the crystallization of these two semiconductor materials.

Keywords: laser annealing; silicon; silicon carbide; crystallization; integrated devices

check for updates

Citation: Arduino, D.; Stassi, S.; Spano, C.; Scaltrito, L.; Ferrero, S.; Bertana, V. Silicon and Silicon Carbide Recrystallization by Laser Annealing: A Review. *Materials* **2023**, *16*, 7674. https://doi.org/10.3390/ma16247674

Academic Editors: Michael Lorenz and Fabrizio Roccaforte

Received: 24 October 2023
Revised: 13 December 2023
Accepted: 14 December 2023
Published: 16 December 2023

1. Introduction

The rapid diffusion of microelectronics and their advancement toward improved performance and reduced size has necessitated the development of new technological processes for selectively modifying material properties in specific regions without affecting others. For instance, there is a need to locally crystallize amorphous materials to achieve new properties or activate dopant ions that have been previously implanted in a semiconductor region. This can be accomplished through semiconductor annealing. Traditional annealing methods involve heating samples to high temperatures (above 1000 °C) using furnaces to repair lattice damage and electrically activate dopants.

Thermal annealing is a critical step in semiconductor device fabrication that is typically performed after other processes. Subjecting the entire sample to high temperatures can have undesirable side effects on device functionality, including impurity redistribution and damage to implanted ion disposition [1]. Moreover, it is incompatible with prior metal deposition, as the elevated temperature can cause metal connections to melt and redistribute.

In recent years, various alternatives to thermal annealing in furnaces have been explored, with laser annealing emerging as one of the most effective technologies [2]. Depending on the light frequency, laser annealing offers the advantage of intense absorption on a thin surface layer (a few nanometers deep). This capability allows for the generation of the extremely high temperatures necessary for annealing lattice damage or crystallizing amorphous films in a precisely localized region. Importantly, this occurs without affecting the rest of the sample, mitigating the risk of unwanted alterations to the device [3].

During the laser annealing process, the material phase (crystallinity), optical properties (refractive index, light absorption), and electrical properties (conductivity, energy bandgap, direct or indirect bandgap) of the samples are altered through light absorption [4]. The incident laser energy is absorbed through electronic excitations and rapidly transferred to the lattice, resulting in the melting of the crystal to a certain depth and inducing lattice damage. Subsequently, liquid phase epitaxial regrowth occurs from the undamaged substrate underneath, leading to recrystallization of the melted region into nearly perfect single-crystal material, with dopants occupying substitutional sites in the lattice, thereby activating the implanted ions [5].

Laser annealing is a versatile technique applicable to various types of semiconductors, including group IV semiconductors like silicon, germanium [6], and silicon carbide, as well as III-V compound semiconductors like gallium arsenide [7,8].

Laser thermal annealing is an ultrafast and low thermal budget process solution for the passivation of backside illuminated sensors and power devices. Laser annealing can be a solution for the backside contact of those chips with a vertical flow of electrical current, where an ohmic contact and/or collector on the wafer backside are required [9,10], including SiC power Metal-Oxide-Semiconductor Field Effect Transistors (MOSFET) [11], Insulated Gate Bipolar Transistor (IGBT) [12], and high voltage diodes [13,14]. Moreover, this process can also be useful for the ohmic contact formation process in SiC Schottky diodes, causing a negligible impact on the device's front side [15,16]. The laser annealing process has also been used to activate doping ions in the source and drain region of MOSFET to avoid damage to the channel region induced by global device heating processes such as rapid thermal annealing (RTA) [17–19]. A scheme of laser annealing processes in semiconductor devices is reported in Figure 1. The high-temperature annealing region is restricted to thin layers while keeping underlying layers at low temperatures. An ultrafast annealing time and proper laser parameters may achieve high performance and high yields, locking in the surface properties without damaging buried device layers [20].

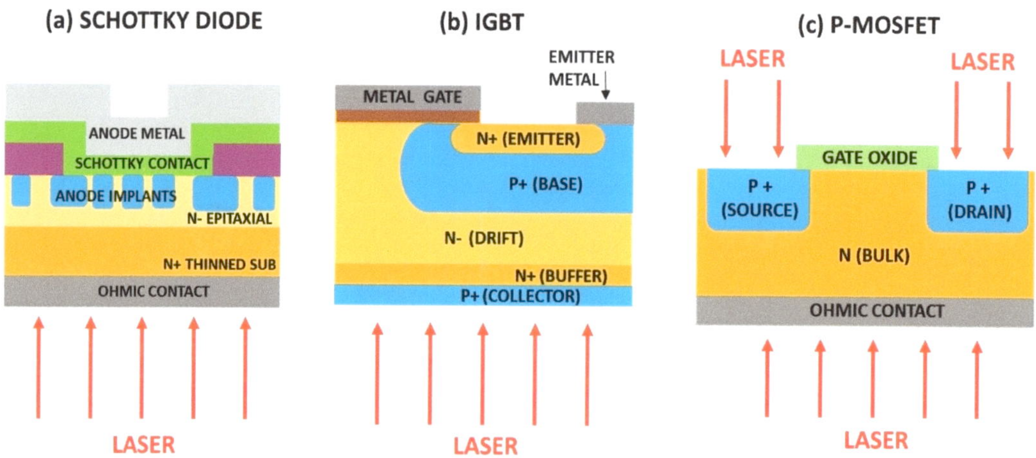

Figure 1. Scheme of the laser annealing process exploited to (**a**) form ohmic contact on the Schottky diode, (**b**) activate doping on the collector layer of IGBT, and (**c**) activate doping in the source and drain region and form ohmic contact in a MOSFET (in this case a pMOSFET, although the process is the same for nMOSFET).

This study aims to provide a comprehensive review of laser annealing processes specifically focused on the crystallization of amorphous silicon (Si) and silicon carbide (SiC) samples. Silicon is widely used in microelectronics and solar cell applications, while SiC is vital for the development of high-power integrated components and devices operating in harsh environments and at high temperatures [21,22]. SiC has also garnered attention for optical applications due to its nonlinear optical properties [23]. The review includes a brief overview of the underlying theory and principles of laser annealing, followed by a presentation of the most relevant studies that focus on the crystallization of the amorphous layer of these two semiconductor materials. This study provides a critical analysis of various processes reported in the literature. The studies are categorized based on the laser wavelength employed, and a detailed examination of the process parameters of the systems involved is conducted. Additionally, the main crystallization results obtained are outlined, accompanied by a discussion of the outcomes of the primary characterization methods employed in these studies.

2. Laser Annealing Theory

Laser annealing is a technique that involves using light absorption to deliver energy to a material. Typically, laser annealing is limited to thin surface layers because the intensity of the incident light decreases as it penetrates the material, based on the material's absorption coefficient, α. The Beer–Lambert law describes this decay of intensity with depth z using the equation:

$$I(z) = I_0 e^{-\alpha z} \tag{1}$$

where I_0 is the intensity on the surface.

The optical penetration or absorption depth, defined as the depth at which the transmitted light's intensity drops to $1/e$ (about 37%) of its initial value on the surface, is denoted as $\delta = 1/\alpha$. Both α and δ depend on the semiconductor type, laser wavelength, and temperature.

Since energy absorption is primarily confined to the absorption depth, this parameter is applicable to all beam profiles, even though it was originally developed for plane waves. Consequently, it is possible to locally modify surface properties without altering the bulk material by using laser wavelengths with short absorption depths [24].

In continuous wave (CW) or nanosecond laser pulses, it is generally assumed that single-photon interactions account for most of the absorption. However, in the case of picosecond (ps) and femtosecond (fs) lasers, the absorption depths can be reduced due to phenomena such as optical breakdown and multiphoton absorption, which result from the extremely high instantaneous intensity of these lasers [25].

The absorption of laser light in insulators and semiconductors is typically achieved through resonant excitations, such as transitions of valence band electrons to the conduction band (interband transitions, Figure 2) or within bands (intersubband transitions) [26].

Typically, photons interact with the electronic or vibrational states present in a material based on their energy. Laser photons of energy greater than that of the band gap generate electron-hole pairs promoted to states of higher kinetic energy in the conduction and valence band. These excited electronic states can subsequently transfer energy to lattice phonons, becoming heat. If there are no other factors such as impurity or defect states or multiphoton absorption, photons with energy lower than the material's band gap will not be absorbed. This generally corresponds to light wavelengths within the infrared to visible spectrum for semiconductors and within the vacuum ultraviolet range (below 200 nm) for insulators.

The laser field to the electronic system is quickly transferred to phonons in less than 1 picosecond, leading to the melting of the near-surface region [27]. The near-surface region of a sample can melt and stay molten for a thermalization time depending on the material treated, during which dopant diffusion in the liquid state and nonequilibrium segregation occur together with ultrarapid recrystallization [28]. The crystallization process can be described in terms of models based on macroscopic diffusion equations for heat and mass transport. The mechanism can be explained in terms of a molten layer that extends all over the amorphous thickness and whose subsequent solidification occurs on a crystalline seed, like a liquid phase epitaxy [27]. In this way, the laser treatment allows the annealing and treatment of the lattice damage caused by ion implantation, diffuse surface-deposited dopant films, and recrystallized doped amorphous films deposited on the substrate [28].

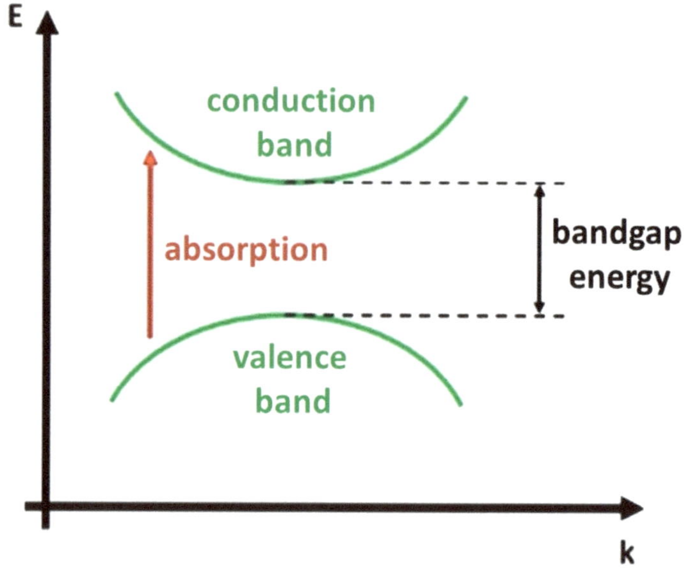

Figure 2. Scheme of the material interband photon absorption.

The specific material and its mechanisms determine the time required for excited electronic states to transfer energy to phonons and thermalize. For non-metals, the thermalization time can be from 10^{-8} s to 10^{-6} s, while it is around 10^{-12}–10^{-10} s for metals [25]. If the laser-induced excitation rate is lower than the thermalization rate, the transient electronically excited states are not significant and the absorbed laser energy can be considered directly converted into heat. This process is known as photothermal (pyrolytic) processing and is commonly observed in semiconductor laser processing with long pulse times (>ns). During this process, material response can be analyzed purely in thermal terms [24].

However, when the laser-induced excitation rate exceeds the thermalization rate, significant excitations can accumulate in the intermediate states. This can lead to direct bond-breaking due to the excitation energies, resulting in non-thermal material modifications. This phenomenon is referred to as photochemical (photolytic) processing, where there are no changes in the system's temperature. Ultrafast femtosecond laser pulses with short-wavelength light, where the photon energy is comparable to the chemical bond energy, can trigger photochemical processing [29].

The laser annealing process is significantly influenced by the technology of the implemented system, which determines the wavelength and the shortest pulse duration of the laser. Several key parameters of the laser system can be adjusted to control the effects of the annealing process on the material surface. A list of the most crucial system specifications and parameters involved in laser annealing processes is provided below. These descriptions aim to enhance understanding of the main results in silicon and silicon carbide annealing processes described in the following sections.

Laser technology: The laser relies on a certain physical mechanism depending on the laser material, which affects the other laser parameters (such as its power, pulse duration, etc.). Possible laser technologies range from gas lasers (such as CO_2 laser) to excimer lasers (based on a combination of a noble gas and a reactive gas, such as KrF and XeCl lasers) and solid-state lasers (based on doped crystal, such as Nd:YAG lasers).

Wavelength: This parameter plays a significant role in the annealing process as it affects light absorption efficiency and the depth of laser effects in the material. Generally, the penetration depth is proportional to $\lambda/4\pi k$ with k, which is the absorption coefficient. The light wavelength (which is directly related to photon energy by the Planck equation) must fit with the material bandgap to be absorbed, otherwise, photons pass through the material. Laser wavelengths used for annealing range from infrared to visible and ultraviolet (UV) regions.

Power and spot size: A combination of these two parameters defines the power density, which indicates the energy quantity delivered by the laser beam to the desired target per unit of area and time. The energy density or fluence represents the energy transferred per unit area by a single pulse. The use of a magnifying lens can increase the given number of photons directed to a specific area, raising the laser fluence and thus the target temperature faster.

Pulse duration. This is the time between the beginning and end edges of an energy pulse, often measured at full-width half maximum (FWHM). Lasers treated in this review are characterized by nanosecond or femtosecond pulses. Generally, considering a fixed frequency, short pulses allow samples to cool down between bursts of light and, hence protect illuminated samples from overheating. However, in general, short pulses with high peak powers (as in the case of femtosecond lasers) may ablate the surface material, avoiding heating the surrounding area. Obviously, this parameter is absent in the case of continuous wave lasers.

Scanning speed: This parameter represents the velocity of the relative motion between the laser and the material. Slower scanning speeds result in the laser remaining in a particular spatial position on the material for a longer time, leading to higher annealing effects.

Beam profile: The laser beam can have various profiles, such as Gaussian, multimode, or rectangular. The energy distribution of the beam determines the spatial effect of the annealing process, yielding different results. Gaussian beams, for example, cause stronger effects near the beam center and weaker effects near the edges, while rectangular beams provide a more uniform spatial distribution.

Environment: Since laser annealing increases the local temperature of the material, environmental conditions can impact process outcomes. The presence of air can cause reactions between the heated material and atmospheric gases, which is not observed in a vacuum or an argon (Ar) atmosphere.

Manipulating these main parameters is possible by taking advantage of the main components of the laser annealing equipment, which consists of a laser source composed of a laser head (which determines the emitted wavelength, frequency, and pulse duration), lenses, and filters for shaping the laser beam profile.

A laser stage motor is normally responsible for the scan speed. Indeed, the laser head can be based on the technology called plotter laser, where the laser light is driven by some fixed internal mirrors, and its movement on the sample surface is controlled by a stage motor that determines the scan speed and the resolution on the X and Y axes. An alternative technology is the galvo head, where the laser ray is driven by rotating dynamic internal mirrors, and a stage motor is used only for large displacement.

3. Silicon Laser Annealing

Silicon is a crucial material for creating microelectronic devices, and various fabrication processes have been developed to manipulate its properties based on specific applications. One important technique being studied is its transformation from amorphous silicon to polycrystalline silicon, which combines the advantages of both single-crystal silicon and amorphous silicon [30]. While amorphous silicon has lower carrier mobility, it is cost-effective to manufacture. On the other hand, polycrystalline silicon, though inferior in mechanical and electrical properties compared to single-crystal silicon, finds extensive use in the field of optoelectronics.

Currently, laser annealing is the most commonly employed technology for preparing polysilicon materials. This method involves the use of a high-power pulsed laser on an amorphous silicon sample. The surface of the amorphous silicon absorbs the laser energy, causing the temperature to rise to the phase transition point. After cooling and solidification, the amorphous silicon is transformed into polysilicon [31,32].

This process exhibits short annealing time and high crystallization efficiency. Additionally, laser annealing keeps the substrate at a low temperature, reducing the requirements on the substrate material and thus lowering manufacturing costs.

The crystallization of silicon grains is influenced by energy density, pulse duration, and laser shape; however, laser wavelength is the primary parameter governing the crystallization phenomenon due to its role in optical absorption [33].

Specifically, crystallization occurs when the photon energy (associated with the photon wavelength) matches the bandgap of the material. The light induces a transition between the ground and excited states, resulting in the loss of a photon and the production of an excited state [34]. In the case of silicon, which has a bandgap energy of 1.1 eV, the minimum wavelength required for electron transition from the valence to the conduction band is around 1100 nm, corresponding to the infrared (IR) region.

Several experiments have been conducted on silicon laser annealing using IR laser sources. Salman et al. [35] successfully crystallized a silicon sample using an IR-pulsed laser with a wavelength of 1064 nm (photon energy of 1.165 eV) and a pulse duration of 200 ns. The experiment was performed under ambient conditions using a pulsed fiber laser built on Ytterbium-doped active fiber, with an average optical power of 20 W and a maximum pulse energy of 0.50 mJ. The repetition rate was set to 20 kHz. Successful crystallization was confirmed through Raman spectroscopy analysis, showing a main scattering peak at 520 cm^{-1}, which is typical of a crystalline silicon structure (Figure 3a). The crystallization process was also verified using Fourier Transform Infrared (FTIR) spectroscopy, Scanning Electron Microscopy (SEM), Transmission Electron Microscopy (TEM), and Atomic Force Microscopy (AFM), which are powerful analysis tools for investigating semiconductor crystalline structures.

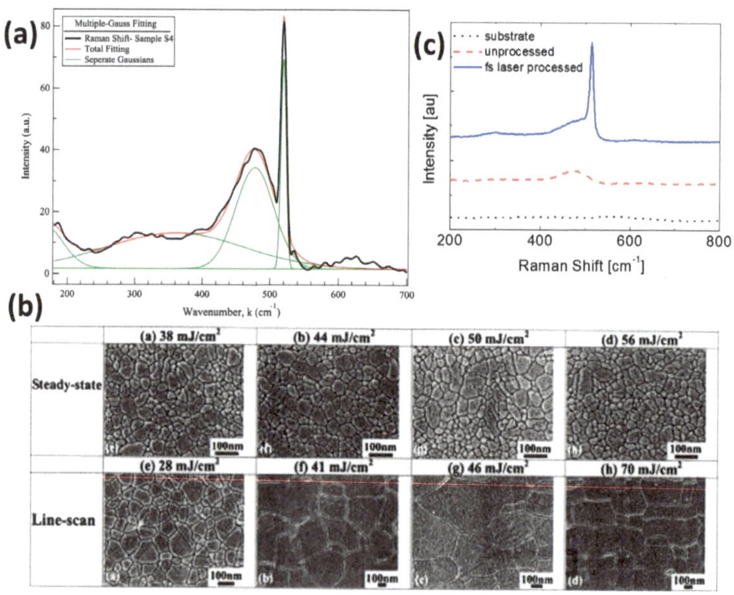

Figure 3. (a) Raman spectrum of annealed Silicon sample (black). The green curves are separated Gaussian fits for different scattering modes, while the red curve is the sum of the fits. Reprinted

from [35], Copyright (2019), with permission from Elsevier. (**b**) SEM pictures of crystalline grains in Silicon samples after femtosecond laser annealing processes with different configurations and fluences. Reprinted from [36], with the permission of AIP Publishing. (**c**) Raman spectra of bare glass substrate and amorphous silicon film without (dash) and with (solid) femtosecond laser processing. Reprinted from [37], Copyright (2019), with permission from Elsevier.

Previous studies [36,38] have utilized infrared laser radiation with a wavelength of 800 nm, generated by a Ti:Sapphire laser system with femtosecond pulse duration. Despite this difference, the presence of crystalline grains can be observed from AFM, SEM (Figure 3b), and TEM measurements, indicating the successful attainment of polysilicon through the annealing process.

Other groups [37,39] have demonstrated the effectiveness of IR radiation (λ = 800 nm by Zhan et al. [37] and λ = 1026 nm by Bronnikov et al. [39]) for silicon crystallization using laser systems based on different technologies (Yb:KGW laser system). Before the laser annealing process, the Raman spectra of the as-grown amorphous film display a broad peak, ranging from 420 to 530 cm^{-1}, associated with the optical vibration modes of amorphous silicon [40,41]. However, after the annealing step, a distinct crystalline peak at 513 cm^{-1} becomes evident (Figure 3c). The sharp peak exhibits a blue shift of approximately 7 cm^{-1} compared to the peak of single crystalline silicon due to phonon confinement in nanocrystalline silicon and mechanical stress induced in the lattice by femtosecond laser treatment [40].

Laser annealing experiments have also explored more energetic wavelengths in the near UV range (200–400 nm) [42] and visible range (400–780 nm) [43,44]. For instance, Pan et al. utilized near ultraviolet ($\lambda \approx$ 400 nm) femtosecond laser annealing in a scanning mode to crystallize amorphous silicon (a-Si) films at room temperature. They investigated the impact of laser fluence, beam overlap, and number of laser shots on the average grain size of the resulting polycrystalline silicon. The experiment revealed that increasing either the beam overlap at a fixed fluence or the fluence for a fixed number of shots generally leads to larger grain sizes [45]. Additionally, they compared the crystallization degree of the polysilicon obtained through UV annealing with that produced by IR annealing ($\lambda \approx$ 800 nm) using Raman spectroscopy. Figure 4a demonstrates that the Raman peak associated with λ = 400 nm is sharper than the peak associated with λ = 800 nm, indicating a higher crystallization degree at λ = 400 nm (near UV) compared to λ = 800 nm (near-infrared).

In contrast, cross-sectional SEM images showed that the 100 nm thick a-Si film is not fully crystallized by UV annealing, unlike in the case of IR annealing (see Figure 4b). This can be attributed to the much shorter penetration depth of 400 nm light in amorphous silicon compared to 800 nm light.

UV radiation with nanosecond pulse durations has also been tested [46–49], as demonstrated by Garcia et al. [33]. They employed solid-state laser systems for annealing and compared the effect of UV radiation with that of visible light. The Nd:YVO_4 systems, with pulse widths of 15 ns and 12 ns at a repetition rate of 50 kHz, were used at different wavelengths: fundamental frequency in the IR (1064 nm), doubled to green (532 nm), and tripled to UV (355 nm). Two fluence thresholds (F1 and F2) governing the melting process were identified, with their values depending on the wavelength employed. These thresholds represent the range of fluence within which the annealing process should take place. When the fluence is lower than F1, the silicon surface is not melted as the energy of the laser pulse is insufficient for a phase change. On the other hand, exceeding the F2 threshold results in ablation and material damage. For a UV wavelength of 355 nm, the F1 and F2 values are 70 mJ/cm^2 and 374 mJ/cm^2, respectively, while for a visible wavelength of 532 nm, F1 and F2 values are 110 mJ/cm^2 and 304 mJ/cm^2, respectively [33]. The authors were not able to crystallize a-Si with their system using a 1064 nm wavelength, because they had to use a longer laser pulse and higher fluence due to lower absorption in the IR range, causing material ejection.

Finally, laser annealing and crystallization of amorphous Si have also been explored using visible light generated by nanosecond pulsed and continuous wave lasers [50–52]. Son et al. [53] employed a Nd:YVO$_4$ continuous wave (CW) laser with a wavelength of 532 nm (green light), an output power of 7.5 W, and a scanning speed of 270 mm/s to crystallize the samples. The laser beam had dimensions of 20 μm (short axis, scan direction) × 800 μm (long axis, transverse direction), with a Gaussian shape in the short axis and a top-hat shape in the long axis. This conventional laser system was compared with a system that utilized a cylindrical microlens array. In the latter system, the laser beam was split into two components: one traveled directly while the other was refracted through the cylindrical microlens. The two components met and interference occurred due to the superposition of the split beams, resulting in enhanced intensity and incident power per unit area on the a-Si due to constructive interference.

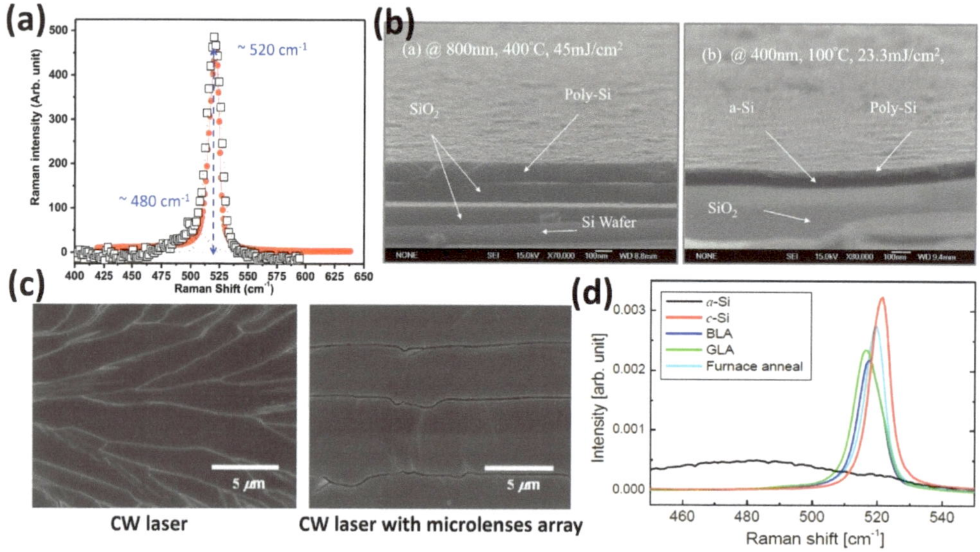

Figure 4. (**a**) Raman spectra of silicon dies showing the peak at 520 cm^{-1} (typical of crystalline Si) for the sample annealed using a 400 nm laser wavelength (sharper, red circles and red curve) and for the sample annealed using an 800 nm laser (smaller, black open squares) [45]. (**b**) Cross-sectional SEM images of samples annealed by femtosecond laser annealing with IR (800 nm, 400 °C, 45 mJ/cm^2, left) and UV (400 nm, 100 °C, 23.3 mJ/cm^2, right) radiation, showing the formation of poly-Si. The scale bars correspond to 100 nm [45]. (**c**) SEM images of crystallized poly-Si film treated with (left) conventional CW laser and (right) CW laser equipped with a cylindrical microlens array. The scale bars correspond to 5 μm. Copyright 2012, Wiley. Reproduced with permission from Ref. [53]. (**d**) Raman spectra of a-Si, c-Si, and polysilicon samples treated with BLA, GLA, and furnace annealing [54].

SEM analysis (Figure 4c) depicted the poly-Si film annealed using the conventional CW laser and the CW laser equipped with a microlens array. The grains of the poly-Si film produced by conventional CW laser crystallization were large but had irregular shapes and orientations. In contrast, the grains obtained by CW laser crystallization using a cylindrical microlens array were larger and more regular.

In order to provide a comprehensive analysis of annealing in the visible range, Pyo et al. [54] investigated the polycrystalline silicon obtained through green laser annealing (GLA) using a 532 nm nanosecond pulsed laser and blue laser annealing (BLA) using a 450 nm continuous wave laser. For GLA, the second harmonic of a Q-switched Nd:YAG laser was utilized with a scanning speed of 4 cm/s, a pulse repetition rate of

30 kHz, and a pulse width of 30 ns. The beam profile was approximately Gaussian with a beam waist of approximately 6 μm, and the peak power was 2.7 W, corresponding to a pulse energy of 86 nJ and an energy density of 153 mJ/cm^2. The CW laser used for BLA consisted of two GaN-based blue laser diodes emitting at 450 nm. The BLA beam shape was rectangular, with a spot size of 6 μm × 2 μm and a power of 530 mW. The laser annealing experiments were conducted at room temperature (25 °C) under a nitrogen environment, with nitrogen pressure maintained at 1.5 times the atmospheric pressure.

After the annealing processes, the crystallinity of the annealed poly-Si was examined using Raman spectroscopy, as shown in Figure 4d, and compared with a standard furnace heating process. The Raman spectrum of amorphous Si exhibited a broad distribution centered at approximately 480 cm^{-1} in the transverse optic (TO) phonon mode, while crystalline Si displayed a sharp peak at around 520 cm^{-1} associated with the TO phonons. The shift and full width at half maximum (FWHM) of the a-Si and c-Si peaks were used to analyze the crystallinity of the annealed poly-Si film. Improved crystallinity is indicated by a shift in the TO phonon peak of poly-Si towards that of c-Si. The Raman shift values for furnace annealing, BLA, and GLA were 519 cm^{-1}, 517.5 cm^{-1}, and 516.5 cm^{-1}, respectively. The FWHM values for furnace annealing, BLA, and GLA were 7.5 cm^{-1}, 9.0 cm^{-1}, and 9.5 cm^{-1}, respectively.

By using the relative values of FWHM and the intensity of TO phonon Raman peaks, it is possible to calculate the crystal volume fraction (f_c) of poly-Si [55]. The f_c of poly-Si can be quantitatively evaluated using the following formula:

$$f_c = \frac{I_c}{I_c + I_a} \tag{2}$$

where f_c is the crystalline volume fraction. I_c and I_a stand for the integrated Raman scattering intensity of crystalline and amorphous sections, respectively. I_c can be obtained by the deconvolution of each Raman spectrum into Gaussian components corresponding to the crystalline phases. The same can be done for I_a, but with the amorphous phases. Looking at Figure 4d, when a-Si becomes poly-Si, the TO phonon peak of the crystalline phase is slowly shifted toward the c-Si peak and FWHM is decreased, resulting in an increase in f_c. The fc values were 90.6% for BLA and 88.2% for GLA, indicating a slightly higher crystal quality for BLA. This finding is consistent with the longer Raman peak shift of GLA observed in Figure 4d. As expected, the furnace-annealed poly-Si sample showed the best crystallinity, while BLA and GLA showed lower crystallinity conversion. On the other hand, BLA and GLA processes have the advantage of heating only the superficial part of the sample, with respect to furnace processes, obtaining a successful degree of crystalline conversion.

A summary of the main parameters used during Si laser annealing processes described in the literature, which successfully convert amorphous silicon into crystalline or polycrystalline silicon, is presented in Table 1.

Research has revealed the potential for transforming amorphous silicon (a-Si) into poly-Si using a range of wavelengths spanning from infrared (IR) to ultraviolet (UV) across different laser systems. Silicon exhibits reduced absorption of IR light compared to visible and UV light, necessitating the use of higher power and shorter pulse durations in these procedures. Despite the challenges associated with crystallization using IR light, it is feasible even on thicker amorphous layers due to its greater penetration depth; conversely, laser annealing processes facilitate crystallization more readily with visible and UV lasers, proving more effective on thin amorphous layers due to the higher absorption rate of silicon.

Table 1. Main parameters of silicon laser annealing processes presented in the literature.

Ref.	Laser Type	Wavelength	Pulse Duration	Beam Profile	Environment	Energy/Power Density *
[35]	Yitterbium doped fiber	IR (1064 nm)	200 ns	/	ambient conditions	/
[36]	Ti-Sapphire	IR (800 nm)	50–125 fs	/	/	38–63 mJ/cm^2
[38]	Ti-Sapphire	IR (800 nm)	/	/	/	160–305 mJ/cm^2
[37]	Yb:KGW	IR(800 nm)	40–200 fs	Gaussian	/	49–69 mJ/cm^2
[39]	Yb:KGW	IR(1026 nm)	230 fs	Gaussian	/	150 mJ/cm^2
[45]	Ti-Sapphire	IR (800 nm)	50 fs	/	/	/
[34]	Ti-Sapphire	UV (400 nm)	50 fs	/	/	20–30 mJ/cm^2
[33]	Nd:YVO_4	UV (355 nm)	12–15 ns	Gaussian	/	240 mJ/cm^2
[33]	Nd:YVO_4	Green (532 nm)	12–15 ns	Gaussian	/	478 mJ/cm^2
[53]	Nd:YVO_4	Green (532 nm)	CW	Gaussian in short axis, top-hat in long axis	/	/
[44]	Solid-state diode	Blue (440 nm)	CW	Elliptically	/	4.61 W
[54]	GaN-based diode	Blue (450 nm)	CW	Gaussian	room T, N_2 atmosphere	/
[54]	Nd:YAG	Green (532 nm)	30 ns	rectangular	room T, N_2 atmosphere	153 mJ/cm^2

* Since the CW (continuous wave) lasers have no pulses, a power density is defined instead of an energy density.

4. Silicon Carbide Laser Annealing

Silicon carbide (SiC) is a valuable alternative to silicon for applications that involve high temperatures and harsh, corrosive environments that would damage standard silicon and polymer electronics [56–61]. The literature includes various examples of high-temperature pressure sensors [62–68], accelerometers [69], and micromotors [70,71] fabricated using SiC.

Furthermore, the bandgap of both amorphous and crystalline forms of silicon carbide (SiC), which is wider than the Si bandgap, makes it a promising material for optoelectronic applications [72,73], as well as high-power and RF/microwave electronics [74–77].

The crystalline form of SiC (c-SiC) is particularly suitable for high-power applications due to its large bandgap (3–3.3 eV), high thermal conductivity (4.9 W cm^{-1} K^{-1}), high breakdown electric field strength (2.2 $\times$ 10^6 V cm^{-1}), and high saturated electron drift velocity (2.0 $\times$ 10^7 cm s^{-1}) [72].

Different arrangements of Si-C bilayers in the [0001] direction give rise to over 200 polytypes of SiC crystal structures [78]. These polytypes, named with the designations C, H, or R to represent cubic, hexagonal, or rhombohedral symmetry, exhibit significant variations in properties such as carrier mobility and electronic gap. Among them, 4H-SiC is often favored for microelectronics applications due to its high carrier mobility.

Lasers have been investigated as tools for both additive (annealing, deposition, surface alteration, and doping) and subtractive (ablation) SiC processing since the early 1980s; however, they have not been widely adopted for microelectronics and MEMS applications. Various types of lasers have been tested for microfabrication on silicon carbide, including traditional excimer, Nd:yttrium aluminum (Nd:YAG), and CO_2 lasers [79–84], as well as more recent lasers such as N_2 [85], Ar^+ [86], Cu vapor [87], and promising picoseconds [88] and femtosecond lasers [89,90].

For MEMS applications, annealing of amorphous and polycrystalline SiC can be useful for surface recovery after ion implantation damage [75] or to transform amorphous silicon carbide into a crystalline phase [91,92].

A subject of ongoing discussion is whether laser-induced recrystallization occurs through solid-state [93–95] or liquid-phase recrystallization [93,94,96–100]. Different effects have been reported by different research groups using the same type of laser. For example, Hishida et al. [95] reported solid-phase recrystallization while Ahmed et al. [99] reported liquid-phase recrystallization even though both groups used XeCl lasers to anneal Al^+ ion-implanted 6H-SiC.

In general, various types of lasers, capable of generating pulsed or continuous wave radiation at different wavelengths, can be used for SiC laser annealing. However, nanosecond-pulsed UV lasers such as excimer and frequency tripled and quadrupled Nd:YAG lasers (base emission at 1064 nm) are the most commonly used due to their prevalence and the high optical absorption of crystalline SiC at UV wavelengths [91,92,101–103].

Basa et al. [104] presented evidence of successful SiC crystallization through laser annealing. They used a KrF excimer laser emitting nanosecond UV pulses with a wavelength of 248 nm and a pulse duration of 30 ns to crystallize a SiC film previously deposited using the PECVD technique in air and at room temperature. The energy density delivered ranged from 123 to 242 mJ/cm^2. X-ray diffraction (XRD) measurements confirmed the crystallization process. The spectra (Figure 5a) displayed peaks associated with crystalline silicon at $2\theta = 28.3°$ (corresponding to the (111) reflection plane of Si), $2\theta = 47.2°$ (for the (220) reflection plane of Si), and $2\theta = 56°$ (for the (311) reflection plane of Si). At energy densities of 188 mJ/cm^2 or higher, a new peak emerged at $2\theta = 35.6°$, corresponding to the (111) reflection plane of cubic silicon carbide 3C-SiC. Moreover, as the laser energy density increased, the intensities of the Si and SiC peaks also increased, providing clear evidence of improved crystallinity.

Hedler et al. [94] also utilized a nanosecond-pulsed KrF excimer laser (30 ns pulse duration) to achieve SiC structure recovery. They employed a wavelength of 248 nm to enable SiC band-to-band absorption and facilitate the crystallization of a previously amorphized SiC film by ion implantation. The experiment was conducted in air and at room temperature using up to 50,000 pulses at a repetition rate of 50 pulses/s and laser fluences ranging from 150 to 900 mJ/cm^2. Subsequent characterization measurements confirmed the crystallization and revealed that higher laser fluences led to a deeper crystallization within the film and an oxidation effect due to the laser irradiation in the air. Additionally, transmission electron microscopy images (Figure 5b) demonstrated that an amorphous layer remained between the upper annealed polycrystalline 3C-SiC layer and the crystalline 4H-SiC substrate, indicating that no epitaxial growth occurred during the annealing process.

To achieve successful recrystallization of previously amorphized silicon carbide (SiC) through Al^+ ion implantation, Mazzamuto et al. [105] utilized an excimer laser system with a different emitted wavelength (308 nm) in the UV range and a short pulse duration of about 160 ns. A higher irradiated fluence allowed for deeper columnar recrystallization of the SiC film after the melting phase induced by laser treatment. The effectiveness of the laser annealing process was confirmed through X-ray diffraction (XRD) measurements. The XRD signal exhibited two prominent peaks at $2\theta = 33.5°$, corresponding to the (100) orientation of 4H-SiC, and $2\theta = 71°$, corresponding to the (201) orientation of 4H-SiC. Following the annealing process, the peak at $2\theta = 33.5°$ experienced a slight shift due to Al doping activation caused by laser treatment. Furthermore, electron energy loss spectroscopy (EELS) analysis clarified the potential for epitaxial regrowth of crystalline SiC. Three main regions resulting from the phenomenon of 4H-SiC epitaxial regrowth were identified: a region consisting of carbon crystallized in thin graphite layers (multi-layer graphene), a region with crystal silicon, and a region of strained 4H-SiC.

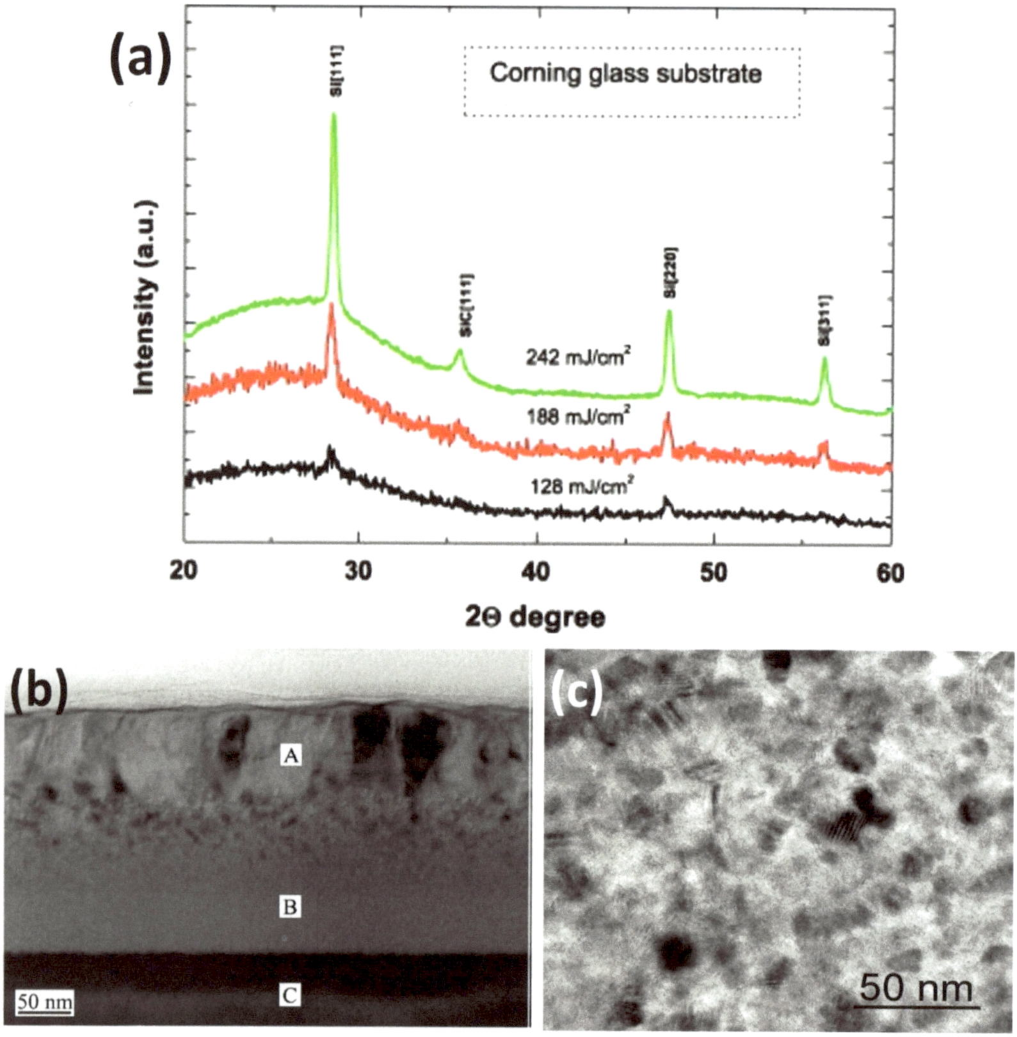

Figure 5. (**a**) XRD spectra of SiC samples annealed at different laser energy densities showing peaks associated with crystalline phases. Reprinted from [104], Copyright (2009), with permission from Elsevier. (**b**) TEM image indicating a polycrystalline 3C-SiC surface layer (A), a remaining amorphous layer (B), and a monocrystalline 4H-SiC substrate (C) after laser annealing treatment. The scale bar corresponds to 50 nm. Reprinted from [94], Copyright (2003), with permission from Elsevier. (**c**) TEM image of a laser-crystallized SiC film. The scale bar corresponds to 50 nm. Reprinted from [96], Copyright (2001), with permission from Elsevier.

Laser annealing experiments were also conducted under various environmental conditions, including a heated sample [96], an argon (Ar) atmosphere [106,107], and under vacuum [95] to facilitate crystallization and remove oxidizing agents. Urban et al. [96] specifically employed a KrF excimer laser with a 25 ns pulse duration and a UV wavelength of 248 nm to anneal a sample placed on a heating stage at 400 °C. A single laser shot with fluences ranging from 100 to 1000 mJ/cm^2 was applied to the material. Optical microscopy, transmission electron microscopy (TEM), and Raman spectroscopy were used for characterization, revealing the presence of a fluence threshold of 250 mJ/cm^2. Below this threshold, no modifications occurred in the SiC samples, leaving the material in its amorphous phase. Conversely, when the SiC film was irradiated with a laser fluence higher than 250 mJ/cm^2, the annealing process successfully crystallized the SiC film, resulting in SiC grains with diameters of 10–20 nm and cubic polycrystalline structures, as depicted by the TEM images (Figure 5c).

Fluences significantly exceeding the crystallization threshold resulted in the segregation of SiC, as shown in Figure 6a. This was indicated by the appearance of Raman peaks at 1500 cm^{-1} (graphite) and 510 cm^{-1} (crystalline Si), while the typical peaks of crystalline SiC around 790 or 960 cm^{-1} were absent. The remaining amorphous SiC near the substrate contributed to the high background signal. It is worth noting that, considering the laser parameters used by Urban et al. [96], such as a 25 ns pulse duration, solid-phase crystallization seems unlikely, suggesting that the annealing process involves a mechanism associated with a metastable liquid phase of SiC.

There have been attempts to anneal SiC films using infrared (IR) wavelength, as demonstrated in the study by Goyes et al. [108]. In their experiment, a continuous wave CO_2 laser with a wavelength of 1060 nm and Gaussian beam distribution was employed to anneal an amorphous SiC film deposited on a silicon wafer under vacuum conditions. The irradiation time ranged from 10 to 30 min, while the laser power was set to 8 W. X-ray diffraction (XRD) measurements were taken to compare sample structures before and after laser treatment. The XRD spectra revealed that the SiC film was initially amorphous, as evidenced by the absence of peaks. However, laser annealing led to crystallization and the formation of α-SiC and β-SiC phases. Specifically, after 10 min of annealing, the β-SiC phase with a (200) orientation was predominant, as it is the most stable phase at high temperatures. After 30 min of annealing, α-SiC became the dominant crystalline phase, although β-SiC with a (111) orientation was still present. Silicon carbide absorption of IR light is negligible, thus crystallization of a-SIC was probably induced by heating the silicon layers underneath, which absorbed the laser radiation, and by thermal conduction of the silicon carbide thin film.

The implementation of laser radiation in the visible range has shown some promising results, allowing it to be considered a possible alternative to UV annealing, as demonstrated by several studies [93,109,110]. Baeri et al. [93] observed that annealing an amorphous SiC sample with a Q-switched ruby laser emitting at a wavelength of 694 nm (red) and fluences ranging from 100 to 1000 mJ/cm^2, along with a 25 ns pulse duration, caused a solid-phase transformation of amorphous SiC into the crystal phase. This process involved the generation of heat through light absorption near the sample surface, followed by cooling through heat transport to the colder substrate. The amorphous SiC reached its equilibrium temperature with the liquid SiC and underwent solidification, resulting in a polycrystalline structure with an average grain size of approximately 30 nm.

Ambrosone et al. [109] investigated the effects of SiC laser annealing using a pulsed frequency Nd:YAG laser with a green wavelength (532 nm) and pulse duration and energy of 10 ns and 2.9 mJ, respectively. The results were compared with those obtained using a pulsed KrF excimer laser with a UV wavelength (248 nm) and pulse duration and energy of 30 ns and 290 mJ, respectively. Both laser annealing experiments were conducted in air at room temperature, with a 60% spatial overlap between successive pulses delivering a fluence of 242 mJ/cm^2. The goal was to crystallize amorphous silicon films with carbon content (χ) ranging from 0.08 to 0.28. The annealed samples were subsequently characterized using XRD spectroscopy to assess how different parameters influence the structure of the final material. Figure 6b illustrates the XRD spectra of films annealed with green laser light, displaying three sharp peaks indicative of the presence of crystalline Si, while peaks associated with crystalline SiC grains were absent. Additionally, it can be observed that the amplitudes of the crystalline Si peaks decrease as the carbon content increases, resulting in a decrease in the crystalline volume fraction (evaluated by the integrated area of the peaks [111]). The film with the highest carbon content (χ = 0.28) exhibited only a small peak at 28.3°.

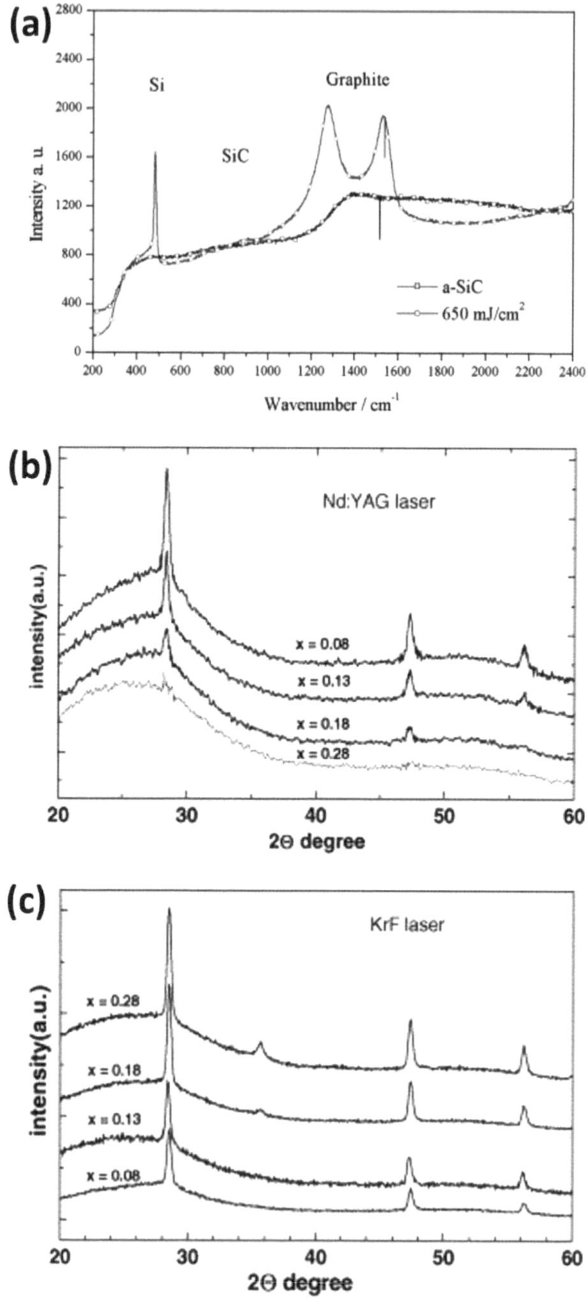

Figure 6. (**a**) Raman spectra of silicon carbide films before (a-SiC) and after high fluence irradiation, showing segregation into c-Si and graphite. Reprinted from [96], Copyright (2001), with permission from Elsevier. (**b**) XRD spectra of SiC films, with different carbon content x, irradiated with Nd:YAG green laser. Reprinted from [109], Copyright (2005), with permission from Elsevier. (**c**) XRD spectra of SiC films, with different carbon content x, irradiated with KrF UV laser. Reprinted from [109], Copyright (2005), with permission from Elsevier.

On the contrary, the XRD spectra in Figure 6c reveals that the peaks associated with crystalline Si become sharper and their intensities increase after UV laser annealing, indicating increasing crystalline volume fractions with higher χ values. Interestingly, crystalline SiC is also present after treatment. The spectra of films with higher χ (i.e., 0.18 and 0.28) show a peak at 35.6° corresponding to the (111) plane of 3C-SiC [109]. Similar to the peaks associated with crystalline Si, a higher carbon content (χ) results in a greater SiC crystalline volume fraction. The average size of the crystalline SiC grains is approximately 30 nm, calculated using the Debye-Sherrer formula.

This difference between the SiC absorption of green and UV lights can be theoretically explained; in fact, the energy associated with green light (532 nm) is about 2.3 eV, which is very similar to the amorphous SiC bandgap (in the range of 2–2.5 eV). This means that the energy of the green photons is just high enough to allow some electrons on the top of the valence band to be promoted to the very bottom of the conductive band. Therefore, by considering that the states in the regions of conductive and valence bands involved in this process are very few, it is possible to assert that the probability is quite low and, hence, the absorption is inefficient. Instead, the energy associated with UV photons (248 nm) is about 4.9 eV, which is much higher than the amorphous SiC bandgap; thence, there are many states involved in the promotion of electrons due to the energy released by UV photons. The absorption process is therefore very efficient in this case.

In another study by Palma et al. [110], a green laser was used for SiC annealing; however, instead of nanosecond pulses, continuous-wave radiation was emitted. The samples were irradiated in the air at room temperature using a beam with a transverse Gaussian intensity distribution delivered by an Argon laser with power densities ranging from 5×10^3 to 5×10^6 W/cm^2 and a wavelength of 514.5 nm. SEM images clearly showed that different regions of the sample experienced varying degrees of SiC crystallization due to the radial intervals of the incident Gaussian energy distribution of the beam. The central region, where the temperature was higher, had polycrystalline carbon while the outer region, where the temperature was lower, had poly-Si.

After the annealing process, Raman spectroscopy was used to study the crystallization outcomes of the SiC film with a carbon content of $\chi = 0.3$. When the sample was irradiated with a low power density (10^4 W/cm^2), only the Si TO phonon line at around 520 cm^{-1} was observed in the Raman spectrum [112]. Conversely, irradiating the SiC film with very high-power density (approximately 10^6 W/cm^2) resulted in the formation of graphite-related lines in the Raman spectrum, with broad bands at 1350 cm^{-1} and 1600 cm^{-1} and a relatively absent crystalline silicon peak.

However, using laser power density in the range of $2–6 \times 10^5$ W/cm^2, the Raman spectrum exhibited both a crystalline silicon peak at 520 cm^{-1} and broad bands at 1350 cm^{-1} and 1600 cm^{-1}, characteristic of polycrystalline graphite. This indicated the simultaneous crystallization of both species, known as the phase segregation effect, highlighting the difficulty of achieving the crystallization of compound semiconductors and the stringent experimental conditions required to achieve this goal [8]. The mechanism of phase segregation in binary systems is not completely understood, but recent studies on silicon carbide demonstrate that this phenomenon can occur in solid-state binary materials that include one element that has the lowest surface energy from the liquid state of the binary system, and this element should have a larger melting temperature than the other element and the binary compound. Therefore, the presence of carbon in silicon carbide facilitates the formation of phase separation.

Lastly, it is important to note that with a slightly higher laser power density (8×10^5 W/cm^2), an amorphous SiC film with a specific carbon content of $\chi = 0.48$ can undergo the crystallization process without the segregation effect. Consequently, the crystalline phase of SiC can be obtained after laser annealing. The Raman spectrum displayed a sharp peak at 790 cm^{-1}, typical of crystalline SiC, and weaker and broader features at 900–1000 cm^{-1}. This analysis confirmed the conversion of amorphous SiC into crystalline SiC. Specifically, since the hexagonal phase α-SiC would have exhibited a peak at 970 cm^{-1}, the Raman analysis demonstrated that the produced polytype is cubic SiC (β-SiC). The mechanism of phase segregation in a binary system is not fully understood; however, recent studies on silicon carbide indicate that this phenomenon can occur in solid-state binary materials containing an element with the lowest surface energy from the liquid state of the binary system. Additionally, this element should have a higher melting temperature than both the other element and the binary compound. The presence of carbon in silicon carbide promotes the formation of phase separation [113]. Stoichiometric SiC (Si:C ratio around 1:1) preserves phase separation along with precise control of power density to more uniformly melt the amorphous silicon carbide layer.

A summary of the main parameters used during SiC laser annealing processes reported in the literature that attempt to crystallize amorphous silicon carbide is presented in Table 2.

Table 2. Main parameters of silicon carbide laser annealing processes presented in the literature.

Ref.	Laser Type	Wavelength	Pulse Duration	Beam Profile	Environment	Energy/Power Density *
[104]	KrF excimer	UV (248 nm)	Nanosecond (30 ns)	/	Room T, air	123–242 mJ/cm^2
[94]	KrF excimer	UV (248 nm)	Nanosecond (30 ns)	/	Room T, air	150–900 mJ/cm^2
[105]	LASSE excimer	UV (308 nm)	Nanosecond (160 ns)	/	/	3200 mJ/cm^2
[96]	KrF excimer	UV (248 nm)	Nanosecond (25 ns)	/	T = 400 °C, air	100–1000 mJ/cm^2
[106]	KrF excimer	UV (248 nm)	Nanosecond (20 ns)	/	Ar atmosphere	200 mJ/cm^2
[95]	XeCl excimer	UV (308 nm)	/	/	Room T, vacuum	/
[82]	XeCl excimer	UV (308 nm)	Nanosecond (160 ns)	/	/	1000–2800 mJ/cm^2
[83]	XeCl excimer	UV (308 nm)	Nanosecond (30 ns)	/	/	500–600 mJ/cm^2
[107]	Nd:YAG	UV (355 nm)	Nanosecond (10 ns)	Gaussian	Ar atmosphere	100–1200 J/cm^2
[109]	KrF excimer	UV (248 nm)	Nanosecond (30 ns)	/	Room T, air	242 mJ/cm^2
[109]	Nd:YAG	Green (532 nm)	Nanosecond (10 ns)	/	Room T, air	242 mJ/cm^2
[93]	q-switched Ruby	Red (694 nm)	Nanosecond (25 ns)	/	/	100–1000 mJ/cm^2
[110]	Argon laser	Green (514 nm)	Continuous wave	Gaussian	Room T, air	8×10^5 W/cm^2
[5]	Nd:YLF	Green (527 nm)	Nanosecond (200 ns)	/	N_2 atmosphere	1170–2500 J/cm^2
[108]	CO_2	DIR (1060 nm)	Continuous wave	Gaussian	Vacuum	5.7 W/cm^2

* Since the CW (continuous wave) lasers have no pulses, a power density is defined instead of an energy density.

Several attempts to crystallize amorphous silicon carbide through laser annealing have indicated that the most effective method involves the use of UV wavelengths. This is because UV wavelengths possess energies well above the bandgap of a-SiC. Conversely, lasers in the visible range are less effective due to inefficient absorption, as the photon energy closely aligns with the a-SiC bandgap. However, by carefully tuning process parameters, crystallization can still be achieved with visible-range lasers. Only one example of an IR laser process has been presented, but the crystallization was inefficient and linked to heating the underlying silicon bulk.

A notable distinction from silicon crystallization is that irradiating amorphous SiC alloys can yield different crystalline results. This outcome depends on the laser energy delivered and the carbon concentration in the film. The controlled formation of the SiC crystalline phase is achievable when the alloy composition is within the quasi-stoichiometric range ($x \approx 0.4$–0.5). However, lower carbon content can result in phase segregation, leading to the formation of crystalline silicon and graphite.

5. Conclusions

This review underscores the pivotal role of laser annealing in semiconductor technology, particularly in transforming amorphous phases into localized crystalline structures, thereby offering precise control over material properties. Studies in the field highlight the numerous advantages of laser annealing over conventional furnace processes. Notably, it allows for localized annealing, preventing unintended damage to delicate regions and enabling spatial control of crystallization. Additionally, laser annealing facilitates rapid local temperature increases and operates under thermodynamic non-equilibrium conditions. Conversely, furnace annealing boasts advantages such as a simpler setup and higher scalability for industrial processes, treating multiple wafers simultaneously. However, it comes with drawbacks, including slower heating ramps and a lack of spatial selectivity in treating specific areas of the device.

In the realm of silicon—the most widely utilized semiconductor—a multitude of approaches has been explored. Various laser technologies, encompassing continuous wave, nanosecond-pulsed, and femtosecond-pulsed lasers, have been applied, producing radiation wavelengths spanning the infrared (IR), visible, and ultraviolet (UV) spectrums, yielding diverse but predominantly positive outcomes. Silicon exhibits diminished absorption of infrared (IR) light compared to visible and ultraviolet (UV) light, necessitating the use of higher power and shorter pulse durations in procedures involving IR light. Despite the challenges associated with crystallization using IR light, it remains viable, even on thicker amorphous layers, owing to its greater penetration depth. Conversely, laser annealing processes prove more efficacious in fostering crystallization with visible and UV lasers, especially on thin amorphous layers, due to silicon's heightened absorption rate in this spectral range. A review of the literature reveals that laser annealing of silicon layers has evolved into a well-established technique for crystallization and doping activation. However, achieving the desired results demands meticulous fine-tuning. Despite this prerequisite, there is considerable potential for the successful implementation of this process, transitioning from research endeavors to an industrial context.

The growing significance of silicon carbide (SiC) in harsh conditions and optical applications has spurred research on laser annealing as a means to achieve a crystalline phase transition. However, in comparison to silicon, the exploration of laser annealing routes for SiC is relatively limited. Excimer lasers, generating nanosecond pulses in the UV range, have been the predominant technology for SiC annealing, demonstrating effective crystallization of amorphous SiC. Only a few prior experiments have delved into the use of continuous wave lasers based on non-excimer technologies or utilized wavelengths in the visible range. Further exploration of these alternative solutions, along with the adoption of new laser technologies, can provide detailed insights into achieving efficient laser annealing of SiC. This advancement would enable precise control over the transition from amorphous SiC to the desired crystalline polytype. Hence, laser annealing emerges as a compelling technique for exploring the crystallization of amorphous SiC. However, its current application is primarily confined to the research fields. The impediments to its industrial implementation stem from various limitations and complexities. These challenges encompass the need for higher-energy photons to effectively transfer energy to the material and the intricacies associated with treating alloy materials. There is a risk of inducing phase segregation with different crystalline phases of the involved atoms, necessitating fine-tuning of the laser process to avoid these phenomena.

Author Contributions: Conceptualization, S.S., L.S., S.F. and V.B; literature investigation, D.A. and C.S.; formal analysis and writing D.A., S.S., L.S., S.F. and V.B.; review and supervision, S.S., L.S., S.F. and V.B.; project administration, funding acquisition, L.S. and S.F. All authors have read and agreed to the published version of the manuscript.

Funding: This work was partly supported by the European Commission through EU H2020 FET Open "SiCOMB" Grant No. 899679.

Institutional Review Board Statement: Not applicable.

Informed Consent Statement: Not applicable.

Data Availability Statement: Not applicable.

Conflicts of Interest: The authors declare no conflict of interest.

References

1. Portavoce, A.; Simola, R.; Mangelinck, D.; Bernardini, J.; Fornara, P. Dopant diffusion during amorphous silicon crystallization. In *Defect and Diffusion Forum*; Trans Tech Publications Ltd.: Stafa-Zurich, Switzerland, 2007; pp. 33–38.
2. Roccaforte, F.; Fiorenza, P.; Vivona, M.; Greco, G.; Giannazzo, F. Selective Doping in Silicon Carbide Power Devices. *Materials* **2021**, *14*, 3923. [CrossRef] [PubMed]
3. Torregrosa, F.; Canino, M.; Li, F.; Tamarri, F.; Roux, B.; Morata, S.; La Via, F.; Zielinski, M.; Nipoti, R. Ion implantation and activation of aluminum in bulk 3C-SiC and 3C-SiC on Si. *MRS Adv.* **2022**, *7*, 1347–1352. [CrossRef]
4. Shtyrkov, E.; Khaibullin, I.; Zaripov, M.; Galyatudinov, M.; Bayazitov, R. Local laser annealing of implantation doped semiconductor layers. *Sov. Phys. Semicond.* **1975**, *9*, 1309–1310.
5. Wu, J.; He, Z.; Guo, Z.; Tian, R.; Wang, F.; Liu, M.; Yang, X.; Fan, Z.; Yang, F. Pulsed Laser Annealing of Phosphorous-Implanted 4H-SiC: Electrical and Structural Characteristics. *J. Electron. Mater.* **2022**, *51*, 172–178. [CrossRef]
6. Wang, C.; Li, C.; Huang, S.; Lu, W.; Yan, G.; Zhang, M.; Wu, H.; Lin, G.; Wei, J.; Huang, W.; et al. Phosphorus diffusion in germanium following implantation and excimer laser annealing. *Appl. Surf. Sci.* **2014**, *300*, 208–212. [CrossRef]
7. Vitali, G.; Rossi, M.; Karpuzov, D.; Budinov, H.; Kalitzova, M. Low-power pulsed-laser annealing of implanted GaAs. *J. Appl. Phys.* **1991**, *69*, 3882–3885. [CrossRef]
8. Vitali, G.V.G. Experimental conditions required to achieve low-power pulsed-laser annealing of implanted GaAs. *Jpn. J. Appl. Phys.* **1992**, *31*, 2049. [CrossRef]
9. Shima, A. Laser Annealing Technology and Device Integration Challenges. In Proceedings of the 2006 14th IEEE International Conference on Advanced Thermal Processing of Semiconductors, Kyoto, Japan, 10–13 October 2006; pp. 11–14.
10. Li, G.; Xu, M.; Zou, D.; Cui, Y.; Zhong, Y.; Cui, P.; Cheong, K.Y.; Xia, J.; Nie, H.; Li, S.; et al. Fabrication of Ohmic Contact on N-Type SiC by Laser Annealed Process: A Review. *Crystals* **2023**, *13*, 1106. [CrossRef]
11. Chen, Y.; Okada, T.; Noguchi, T.; Mazzamuto, F.; Huet, K. Excimer laser annealing for low-voltage power MOSFET. *Jpn. J. Appl. Phys.* **2016**, *55*, 086503. [CrossRef]
12. Rahimo, M.; Corvasce, C.; Vobecky, J.; Otani, Y.; Huet, K. Thin-Wafer Silicon IGBT With Advanced Laser Annealing and Sintering Process. *IEEE Electron Device Lett.* **2012**, *33*, 1601–1603. [CrossRef]
13. Baliga, B.J. Trends in power semiconductor devices. *IEEE Trans. Electron Devices* **1996**, *43*, 1717–1731. [CrossRef]
14. Kim, K.; Kang, Y.; Yun, S.; Yang, C.; Jung, E.; Hong, J.; Kim, K. Reduced On-Resistance and Improved 4H-SiC Junction Barrier Schottky Diodes Performance by Laser Annealing on C-Face Ohmic Regions in Thin Structures. *Coatings* **2022**, *12*, 777. [CrossRef]
15. Langpoklakpam, C.; Liu, A.-C.; Chu, K.-H.; Hsu, L.-H.; Lee, W.-C.; Chen, S.-C.; Sun, C.-W.; Shih, M.-H.; Lee, K.-Y.; Kuo, H.-C. Review of Silicon Carbide Processing for Power MOSFET. *Crystals* **2022**, *12*, 245. [CrossRef]
16. Rascunà, S.; Badalà, P.; Tringali, C.; Bongiorno, C.; Smecca, E.; Alberti, A.; Di Franco, S.; Giannazzo, F.; Greco, G.; Roccaforte, F.; et al. Morphological and electrical properties of Nickel based Ohmic contacts formed by laser annealing process on n-type 4H-SiC. *Mater. Sci. Semicond. Process.* **2019**, *97*, 62–66. [CrossRef]
17. Zhang, Q.; Huang, J.; Wu, N.; Chen, G.; Hong, M.; Bera, L.K.; Zhu, C. Drive-Current Enhancement in Ge n-Channel MOSFET Using Laser Annealing for Source/Drain Activation. *IEEE Electron Device Lett.* **2006**, *27*, 728–730. [CrossRef]
18. Hou, P.-C.; Huang, W.-H.; Kao, M.-H.; Wang, H.-H.; Shieh, J.-M.; Shen, C.-H.; Pan, F.-M.; Chang, L. Source/Drain Activation for Flexible Poly-Si Nanoscale pFETs with a Laser-Buffer Layer by CO_2 laser Annealing. *ECS J. Solid State Sci. Technol.* **2022**, *11*, 065007. [CrossRef]
19. Calabretta, C.; Agati, M.; Zimbone, M.; Boninelli, S.; Castiello, A.; Pecora, A.; Fortunato, G.; Calcagno, L.; Torrisi, L.; La Via, F. 4H-SiC MOSFET Source and Body Laser Annealing Process. *Mater. Sci. Forum* **2020**, *1004*, 705–711. [CrossRef]
20. Huet, K.; Toqué-Tresonne, I.; Mazzamuto, F.; Emeraud, T.; Besaucèle, H. Laser Thermal Annealing: A low thermal budget solution for advanced structures and new materials. In Proceedings of the 2014 International Workshop on Junction Technology (IWJT), Shanghai, China, 18–20 May 2014; pp. 1–6.
21. Pecholt, B.; Gupta, S.; Molian, P. Review of laser microscale processing of silicon carbide. *J. Laser Appl.* **2011**, *23*, 012008. [CrossRef]
22. Scaltrito, L.; Fanchini, G.; Porro, S.; Cocuzza, M.; Giorgis, F.; Pirri, C.F.; Mandracci, P.; Ricciardi, C.; Ferrero, S.; Sgorlon, C.; et al. Surface analysis and defect characterization of 4H–SiC wafers for power electronic device applications. *Diam. Relat. Mater.* **2003**, *12*, 1224–1226. [CrossRef]
23. Ou, H.; Shi, X.; Lu, Y.; Kollmuss, M.; Steiner, J.; Tabouret, V.; Syväjärvi, M.; Wellmann, P.; Chaussende, D. Novel Photonic Applications of Silicon Carbide. *Materials* **2023**, *16*, 1014. [CrossRef]

24. Brown, M.S.; Arnold, C.B. Fundamentals of laser-material interaction and application to multiscale surface modification. In *Laser Precision Microfabrication*; Springer: Berlin/Heidelberg, Germany, 2010; pp. 91–120.
25. Bäuerle, D. *Laser Processing and Chemistry*; Springer Science & Business Media: Berlin/Heidelberg, Germany, 2013.
26. Staudt, W.; Borneis, S.; Pippert, K.D. TFT annealing with excimer laser. Technology and market outlook. *Phys. Status Solidi* **1998**, *166*, 743–749.
27. Baeri, P.; Rimini, E. Laser annealing of silicon. *Mater. Chem. Phys.* **1996**, *46*, 169–177. [CrossRef]
28. Wood, R.F.; Giles, G.E. Macroscopic theory of pulsed-laser annealing. I. Thermal transport and melting. *Phys. Rev. B* **1981**, *23*, 2923–2942. [CrossRef]
29. Chichkov, B.N.; Momma, C.; Nolte, S.; Von Alvensleben, F.; Tünnermann, A. Femtosecond, picosecond and nanosecond laser ablation of solids. *Appl. Phys. A* **1996**, *63*, 109–115. [CrossRef]
30. Ghosh, A.K.; Fishman, C.; Feng, T. Theory of the electrical and photovoltaic properties of polycrystalline silicon. *J. Appl. Phys.* **1980**, *51*, 446–454. [CrossRef]
31. Marqués, L.A.; Pelaz, L.; Aboy, M.a.; Barbolla, J. The laser annealing induced phase transition in silicon: A molecular dynamics study. *Nucl. Instrum. Methods Phys. Res. Sect. B Beam Interact. Mater. At.* **2004**, *216*, 57–61. [CrossRef]
32. Goto, T.; Saito, K.; Imaizumi, F.; Hatanaka, M.; Takimoto, M.; Mizumura, M.; Gotoh, J.; Ikenoue, H.; Sugawa, S. LTPS Thin-Film Transistors Fabricated Using New Selective Laser Annealing System. *IEEE Trans. Electron Devices* **2018**, *65*, 3250–3256. [CrossRef]
33. García, O.; García-Ballesteros, J.; Munoz-Martin, D.; Núnez-Sánchez, S.; Morales, M.; Carabe, J.; Torres, I.; Gandía, J.; Molpeceres, C. Analysis of wavelength influence on a-Si crystallization processes with nanosecond laser sources. *Appl. Surf. Sci.* **2013**, *278*, 214–218. [CrossRef]
34. Demas, J.N.; Demas, S.E. Luminescence. In *Encyclopedia of Physical Science and Technology*, 3rd ed.; Meyers, R.A., Ed.; Academic Press: New York, NY, USA, 2003; pp. 799–823.
35. Salman, H.; Bacıoğlu, A.; Eken, S. Crystallization of hydrogenated amorphous silicon thin film on glass by using ns-pulsed fiber laser operating at 1064 nm. *Mater. Sci. Semicond. Process.* **2019**, *95*, 20–27. [CrossRef]
36. Shieh, J.-M.; Chen, Z.-H.; Dai, B.-T.; Wang, Y.-C.; Zaitsev, A.; Pan, C.-L. Near-infrared femtosecond laser-induced crystallization of amorphous silicon. *Appl. Phys. Lett.* **2004**, *85*, 1232–1234. [CrossRef]
37. Zhan, X.-P.; Hou, M.-Y.; Ma, F.-S.; Su, Y.; Chen, J.-Z.; Xu, H.-L. Room temperature crystallization of amorphous silicon film by ultrashort femtosecond laser pulses. *Opt. Laser Technol.* **2019**, *112*, 363–367. [CrossRef]
38. Carluccio, R.; Cina, S.; Fortunato, G.; Friligkos, S.; Stoemenos, J. Structure of poly-Si films obtained by laser annealing. *Thin Solid Film.* **1997**, *296*, 57–60. [CrossRef]
39. Bronnikov, K.; Dostovalov, A.; Cherepakhin, A.; Mitsai, E.; Nepomniaschiy, A.; Kulinich, S.A.; Zhizhchenko, A.; Kuchmizhak, A. Large-scale and localized laser crystallization of optically thick amorphous silicon films by near-IR femtosecond pulses. *Materials* **2020**, *13*, 5296. [CrossRef] [PubMed]
40. Nalwa, H.S. *Encyclopedia of Nanoscience and Nanotechnology Vol 8. Ne-P*; American Scientific Publishers: Valencia, CA, USA, 2004.
41. Smit, C.; Van Swaaij, R.; Donker, H.; Petit, A.; Kessels, W.; Van de Sanden, M. Determining the material structure of microcrystalline silicon from Raman spectra. *J. Appl. Phys.* **2003**, *94*, 3582–3588. [CrossRef]
42. Koo, K.; Kim, S.; Choi, P.; Kim, J.; Jang, K.; Choi, B. Electrical evaluation of the crystallization characteristics of excimer laser annealed polycrystalline silicon active layer. *Jpn. J. Appl. Phys.* **2018**, *57*, 106503. [CrossRef]
43. Choi, Y.-H.; Ryu, H.-Y. Formation of a Polycrystalline Silicon Thin Film by Using Blue Laser Diode Annealing. *J. Korean Phys. Soc.* **2018**, *72*, 939–942. [CrossRef]
44. Park, M.; Vangelatos, Z.; Rho, Y.; Park, H.K.; Jang, J.; Grigoropoulos, C.P. Comprehensive analysis of blue diode laser-annealing of amorphous silicon films. *Thin Solid Film.* **2020**, *696*, 137779. [CrossRef]
45. Pan, C.-L.; Chen, K.-W.; Wang, Y.-C.; Kao, S.-H.; Wu, P. Room-temperature crystallization of amorphous silicon by near-UV femtosecond pulses. *AIP Adv.* **2020**, *10*, 055321. [CrossRef]
46. Kuriyama, H.; Nohda, T.; Ishida, S.; Kuwahara, T.; Noguchi, S.; Kiyama, S.; Tsuda, S.; Nakano, S. Lateral grain growth of poly-Si films with a specific orientation by an excimer laser annealing method. *Jpn. J. Appl. Phys.* **1993**, *32*, 6190. [CrossRef]
47. Morita, Y.; Noguchi, T. UV pulsed laser annealing of Si+ implanted silicon film and low-temperature super-thin film transistors. *Jpn. J. Appl. Phys.* **1989**, *28*, L309. [CrossRef]
48. Peng, Y.; Fu, G.; Yu, W.; Li, S.; Wang, Y. Crystallization of amorphous Si films by pulsed laser annealing and their structural characteristics. *Semicond. Sci. Technol.* **2004**, *19*, 759. [CrossRef]
49. Sameshima, T.; Watakabe, H.; Andoh, N.; Higashi, S. Pulsed laser annealing of thin silicon films. *Jpn. J. Appl. Phys.* **2006**, *45*, 2437. [CrossRef]
50. Okada, T.; de Dieu Mugiraneza, J.; Shirai, K.; Suzuki, T.; Noguchi, T.; Matsushima, H.; Hashimoto, T.; Ogino, Y.; Sahota, E. Crystallization of Si thin film on flexible plastic substrate by Blue Multi-Laser Diode Annealing. *Jpn. J. Appl. Phys.* **2012**, *51*, 03CA02. [CrossRef]
51. Ratzke, M.; Mchedlidze, T.; Arguirov, T.; Acharya, N.; Kittler, M.; Reif, J. Scanning probe studies of amorphous silicon subjected to laser annealing. *Phys. Status Solidi C* **2011**, *8*, 1351–1355. [CrossRef]
52. Nguyen, T.T.; Hiraiwa, M.; Koganezawa, T.; Yasuno, S.; Kuroki, S.-I. Formation of (100)-oriented large polycrystalline silicon thin films with multiline beam continuous-wave laser lateral crystallization. *Jpn. J. Appl. Phys.* **2018**, *57*, 031302. [CrossRef]

53. Son, Y.D.; Son, N.K.; Kim, K.H.; Kim, E.H.; Oh, J.H.; Jang, J. 7.3: Giant-grain Poly-Si by CW Laser Annealing of a-Si with Cylindrical Microlens Array. In *SID Symposium Digest of Technical Papers*; Blackwell Publishing Ltd.: Oxford, UK, 2007; pp. 76–79.
54. Pyo, J.; Lee, B.; Ryu, H.-Y. Evaluation of Crystalline Volume Fraction of Laser-Annealed Polysilicon Thin Films Using Raman Spectroscopy and Spectroscopic Ellipsometry. *Micromachines* **2021**, *12*, 999. [CrossRef]
55. Shrestha, M.; Wang, K.; Zheng, B.; Mokrzycki, L.; Fan, Q.H. Comparative Study of Furnace and Flash Lamp Annealed Silicon Thin Films Grown by Plasma Enhanced Chemical Vapor Deposition. *Coatings* **2018**, *8*, 97. [CrossRef]
56. Wijesundara, M.; Azevedo, R. *Silicon Carbide Microsystems for Harsh Environments*; Springer Science & Business Media: Berlin/Heidelberg, Germany, 2011; Volume 22.
57. Mehregany, M.; Zorman, C.A. SiC MEMS: Opportunities and challenges for applications in harsh environments. *Thin Solid Film.* **1999**, *355*, 518–524. [CrossRef]
58. Sarro, P.M. Silicon carbide as a new MEMS technology. *Sens. Actuators A Phys.* **2000**, *82*, 210–218. [CrossRef]
59. Cheung, R. *Silicon Carbide Microelectromechanical Systems for Harsh Environments*; World Scientific: Singapore, 2006.
60. Ben Karoui, M.; Gharbi, R.; Alzaied, N.; Fathallah, M.; Tresso, E.; Scaltrito, L.; Ferrero, S. Effect of defects on electrical properties of 4H-SiC Schottky diodes. *Mater. Sci. Eng. C* **2008**, *28*, 799–804. [CrossRef]
61. Pirri, C.F.; Ferrero, S.; Scaltrito, L.; Perrone, D.; Guastella, S.; Furno, M.; Richieri, G.; Merlin, L. Intrinsic 4H-SiC parameters study by temperature behaviour analysis of Schottky diodes. *Microelectron. Eng.* **2006**, *83*, 86–88. [CrossRef]
62. Von Berg, J.; Ziermann, R.; Reichert, W.; Obermeier, E.; Eickhoff, M.; Krötz, G.; Thoma, U.; Boltshauser, T.; Cavalloni, C.; Nendza, J. High temperature piezoresistive β-SiC-on-SOI pressure sensor for combustion engines. In *Materials Science Forum*; U.S. Department of Energy: Oak Ridge, TN, USA, 1998.
63. Pakula, L.; Yang, H.; Pham, H.; French, P.; Sarro, P. Fabrication of a CMOS compatible pressure sensor for harsh environments. *J. Micromechanics Microengineering* **2004**, *14*, 1478. [CrossRef]
64. Young, D.J.; Du, J.; Zorman, C.A.; Ko, W.H. High-temperature single-crystal 3C-SiC capacitive pressure sensor. *IEEE Sens. J.* **2004**, *4*, 464–470. [CrossRef]
65. Kuo, H.-I.; Zorman, C.A.; Mehregany, M. Fabrication and testing of single crystalline 3C-SiC devices using a novel SiC-on-insulator substrate. In Proceedings of the TRANSDUCERS'03. 12th International Conference on Solid-State Sensors, Actuators and Microsystems, Boston, MA, USA, 8–12 June 2003; Digest of Technical Papers. pp. 742–745.
66. Okojie, R.S.; Ned, A.A.; Kurtz, A.D.; Carr, W.N. α(6H)-SiC pressure sensors at 350 °C. In Proceedings of the International Electron Devices Meeting, San Francisco, CA, USA, 8–11 December 1996; Technical Digest. pp. 525–528.
67. Okojie, R.; Nguyen, P.; Nguyen, V.; Savrun, E.; Lukco, D.; Buehler, J.; McCue, T. Failure mechanisms in MEMS based silicon carbide high temperature pressure sensors. In Proceedings of the IEEE International Reliability Physics Symposium Proceedings 45th Annual, Phoenix, AZ, USA, 15–19 April 2007; pp. 429–432.
68. Okojie, R.S.; Ned, A.A.; Kurtz, A.D. Operation of α (6H)-SiC pressure sensor at 500 °C. *Sens. Actuators A Phys.* **1998**, *66*, 200–204. [CrossRef]
69. Atwell, A.R.; Okojie, R.S.; Kornegay, K.T.; Roberson, S.L.; Beliveau, A. Simulation, fabrication and testing of bulk micromachined 6H-SiC high-g piezoresistive accelerometers. *Sens. Actuators A Phys.* **2003**, *104*, 11–18. [CrossRef]
70. Lohner, K.A.; Chen, K.-S.; Ayon, A.A.; Mark Spearing, S. Microfabricated silicon carbide microengine structures. *MRS Online Proc. Libr.* **1998**, *546*, 85–90. [CrossRef]
71. Yasseen, A.A.; Wu, C.H.; Zorman, C.A.; Mehregany, M. Fabrication and testing of surface micromachined polycrystalline SiC micromotors. *IEEE Electron Device Lett.* **2000**, *21*, 164–166. [CrossRef]
72. Foti, G. Silicon carbide: From amorphous to crystalline material. *Appl. Surf. Sci.* **2001**, *184*, 20–26. [CrossRef]
73. Gharbi, R.; Abdelkrim, M.; Fathallah, M.; Tresso, E.; Ferrero, S.; Pirri, C.F.; Brahim, T.M. Observation of negative capacitance in a-SiC:H/a-Si:H UV photodetectors. *Solid-State Electron.* **2006**, *50*, 367–371. [CrossRef]
74. Hornberger, J.; Lostetter, A.B.; Olejniczak, K.; McNutt, T.; Lal, S.M.; Mantooth, A. Silicon-carbide (SiC) semiconductor power electronics for extreme high-temperature environments. In Proceedings of the IEEE Aerospace Conference Proceedings, Big Sky, MT, USA, 6–13 March 2004; pp. 2538–2555.
75. Melzak, J.M. Silicon carbide for RF MEMS. In Proceedings of the IEEE MTT-S International Microwave Symposium Digest, 2003, Philadelphia, PA, USA, 8–13 June 2003; pp. 1629–1632.
76. Scaltrito, L.; Porro, S.; Cocuzza, M.; Giorgis, F.; Pirri, C.F.; Mandracci, P.; Ricciardi, C.; Ferrero, S.; Sgorlon, C.; Richieri, G.; et al. Structural and electrical characterization of epitaxial 4H–SiC layers for power electronic device applications. *Mater. Sci. Eng. B* **2003**, *102*, 298–303. [CrossRef]
77. Ferrero, S.; Porro, S.; Giorgis, F.; Pirri, C.F.; Mandracci, P.; Ricciardi, C.; Scaltrito, L.; Sgorlon, C.; Richieri, G.; Merlin, L. Defect characterization of 4H-SiC wafers for power electronic device applications. *J. Phys. Condens. Matter* **2002**, *14*, 13397. [CrossRef]
78. Li, X.-B.; Shi, E.-W.; Chen, Z.-Z.; Xiao, B. Polytype formation in silicon carbide single crystals. *Diam. Relat. Mater.* **2007**, *16*, 654–657. [CrossRef]
79. Shigematsu, I.; Kanayama, K.; Tsuge, A.; Nakamura, M. Analysis of Constituents Generated With Laser Machining of Si3N4 and SiC. *J. Mater. Sci. Lett.* **1998**, *17*, 737–739. [CrossRef]
80. Sciti, D.; Bellosi, A. Laser micromachining of silicon carbide. *Key Eng. Mater.* **2002**, *206–213*, 305–308. [CrossRef]
81. Lowndes, D.; Wood, R. Studies of pulsed laser melting and rapid solidification using amorphous silicon. *J. Lumin.* **1985**, *30*, 395–408. [CrossRef]

82. Vivona, M.; Giannazzo, F.; Bellocchi, G.; Panasci, S.E.; Agnello, S.; Badalà, P.; Bassi, A.; Bongiorno, C.; Di Franco, S.; Rascunà, S.; et al. Effects of Excimer Laser Irradiation on the Morphological, Structural, and Electrical Properties of Aluminum-Implanted Silicon Carbide (4H-SiC). *ACS Appl. Electron. Mater.* **2022**, *4*, 4514–4520. [CrossRef]
83. Calabretta, C.; Agati, M.; Zimbone, M.; Boninelli, S.; Castiello, A.; Pecora, A.; Fortunato, G.; Calcagno, L.; Torrisi, L.; La Via, F. Laser Annealing of P and Al Implanted 4H-SiC Epitaxial Layers. *Materials* **2019**, *12*, 3362. [CrossRef] [PubMed]
84. Vivona, M.; Giannazzo, F.; Bellocchi, G.; Panasci, S.; Agnello, S.; Badalà, P.; Bassi, A.; Bongiorno, C.; Di Franco, S.; Rascunà, S.; et al. Exploring UV-Laser Effects on Al-Implanted 4H-SiC. *Solid State Phenom.* **2023**, *342*, 85–89. [CrossRef]
85. Medvid, A.; Lytvyn, P.M. Dynamic of laser ablation in SiC. *Mater. Sci. Forum* **2004**, *457–460*, 411–414. [CrossRef]
86. Palma, C.; Rossi, M.C.; Sapia, C. CW-laser crystallisation of amorphous SiC alloys. *Electron. Lett.* **1998**, *34*, 1430–1431. [CrossRef]
87. Dolgaev, S.; Voronov, V.; Shafeev, G.; Ammar, C.F.-B.; Themlin, J.-M.; Cros, A.; Marine, W. Laser-induced fast etching and metallization of SiC ceramics. *Appl. Surf. Sci.* **1997**, *109–110*, 559–562. [CrossRef]
88. Jandeleit, J.; Horn, A.; Weichenhain, R.; Kreutz, E.; Poprawe, R. Fundamental investigations of micromachining by nano-and picosecond laser radiation. *Appl. Surf. Sci.* **1998**, *127–129*, 885–891. [CrossRef]
89. Farsari, M.; Filippidis, G.; Zoppel, S.; Reider, G.A.; Fotakis, C. Efficient femtosecond laser micromachining of bulk 3C-SiC. *J. Micromechanics Microengineering* **2005**, *15*, 1786. [CrossRef]
90. Dong, Y.; Molian, P. In-situ formed nanoparticles on 3C–SiC film under femtosecond pulsed laser irradiation. *Phys. Status Solidi A* **2005**, *202*, 1066–1072. [CrossRef]
91. Kühnapfel, S.; Amkreutz, D.; Klimm, C.; Nickel, N. Excimer laser crystallization of a-SiC x on glass. *Can. J. Phys.* **2014**, *92*, 709–712. [CrossRef]
92. Coscia, U.; Ambrosone, G.; Gesuele, F.; Grossi, V.; Parisi, V.; Schutzmann, S.; Basa, D. Laser annealing study of PECVD deposited hydrogenated amorphous silicon carbon alloy films. *Appl. Surf. Sci.* **2007**, *254*, 984–988. [CrossRef]
93. Baeri, P.; Spinella, C.; Reitano, R. Fast melting of amorphous silicon carbide induced by nanosecond laser pulse. *Int. J. Thermophys.* **1999**, *20*, 1211–1221. [CrossRef]
94. Hedler, A.; Urban, S.; Falk, F.; Hobert, H.; Wesch, W. Excimer laser crystallization of amorphous silicon carbide produced by ion implantation. *Appl. Surf. Sci.* **2003**, *205*, 240–248. [CrossRef]
95. Hishida, Y.; Watanabe, M.; Nakashima, K.; Eryu, O. Excimer laser annealing of ion-implanted 6H-silicon carbide. *Mater. Sci. Forum* **2000**, *338–342*, 873–876. [CrossRef]
96. Urban, S.; Falk, F. Laser crystallization of amorphous SiC thin films on glass. *Appl. Surf. Sci.* **2001**, *184*, 356–361. [CrossRef]
97. Boutopoulos, C.; Terzis, P.; Zergioti, I.; Kontos, A.; Zekentes, K.; Giannakopoulos, K.; Raptis, Y. Laser annealing of Al implanted silicon carbide: Structural and optical characterization. *Appl. Surf. Sci.* **2007**, *253*, 7912–7916. [CrossRef]
98. Chou, S.; Chang, Y.; Weiner, K.; Sigmon, T.; Parsons, J. Annealing of implantation damage and redistribution of impurities in SiC using a pulsed excimer laser. *Appl. Phys. Lett.* **1990**, *56*, 530–532. [CrossRef]
99. Ahmed, S.; Barbero, C.; Sigmon, T. Activation of ion implanted dopants in α-SiC. *Appl. Phys. Lett.* **1995**, *66*, 712–714. [CrossRef]
100. Key, P.; Sands, D.; Schlaf, M.; Walton, C.; Anthony, C.; Brunson, K.; Uren, M. Infra-red reflectivity of ion-implanted and pulsed excimer laser irradiated 4H-SiC. *Thin Solid Film.* **2000**, *364*, 200–203. [CrossRef]
101. Zoppel, S.; Farsari, M.; Merz, R.; Zehetner, J.; Stangl, G.; Reider, G.A.; Fotakis, C. Laser micro machining of 3C–SiC single crystals. *Microelectron. Eng.* **2006**, *83*, 1400–1402. [CrossRef]
102. Sugioka, K.; Midorikawa, K. Novel technology for laser precision microfabrication of hard materials. In Proceedings of the First International Symposium on Laser Precision Microfabrication, Saitama, Japan, 14–16 November 2000; pp. 110–117.
103. Lv, P.; Chen, Z.; Zhang, A.H. Effect of laser annealing on amorphous silicon carbide films prepared by PECVD. In Proceedings of the 4th IEEE International Conference on Nano/Micro Engineered and Molecular Systems, Shenzhen, China, 5–8 January 2009; pp. 743–746.
104. Basa, D.; Ambrosone, G.; Coscia, U.; Setaro, A. Crystallization of hydrogenated amorphous silicon carbon films with laser and thermal annealing. *Appl. Surf. Sci.* **2009**, *255*, 5528–5531. [CrossRef]
105. Mazzamuto, F.; Halty, S.; Mori, Y. Silicon Carbide recrystallization mechanism by non-equilibrium melting laser anneal. *Mater. Sci. Forum* **2016**, *858*, 540–543. [CrossRef]
106. De Cesare, G.; La Monica, S.; Maiello, G.; Masini, G.; Proverbio, E.; Ferrari, A.; Chitica, N.; Dinescu, M.; Alexandrescu, R.; Morjan, I.; et al. Crystallization of silicon carbide thin films by pulsed laser irradiation. *Appl. Surf. Sci.* **1996**, *106*, 193–197. [CrossRef]
107. Paneerselvam, E.; Vasa, N.J.; Nakamura, D.; Palani, I.A.; Higashihata, M.; Ramachandra Rao, M.S.; Thomas, T. Pulsed laser deposition of SiC thin films and influence of laser-assisted annealing. *Mater. Today Proc.* **2021**, *35*, 312–317. [CrossRef]
108. Goyes, C.; Sequeda, F.; Devia, A.; Ferrari, M.; Chiasera, A.; Fonthal, F.; Solarte, E. Effect of CO2 Laser Annealing on Microstructure of Sputtered Silicon Carbide Nano-scale Films. In Proceedings of the Society of Vacuum Coaters 50th Annual Technical Conference, Louisville, KY, USA, 3 May–28 April 2007.
109. Ambrosone, G.; Coscia, U.; Lettieri, S.; Maddalena, P.; Minarini, C.; Parisi, V.; Schutzmann, S. Crystallization of hydrogenated amorphous silicon–carbon films by means of laser treatments. *Appl. Surf. Sci.* **2005**, *247*, 471–476. [CrossRef]
110. Palma, C.; Sapia, C. Laser pattern-write crystallization of amorphous SiC alloys. *J. Electron. Mater.* **2000**, *29*, 607–610. [CrossRef]
111. Houben, L.; Luysberg, M.; Hapke, P.; Carius, R.; Finger, F.; Wagner, H. Structural properties of microcrystalline silicon in the transition from highly crystalline to amorphous growth. *Philos. Mag. A* **1998**, *77*, 1447–1460. [CrossRef]

112. Lannin, J.; Pilione, L.; Kshirsagar, S.; Messier, R.; Ross, R. Variable structural order in amorphous silicon. *Phys. Rev. B Condens. Matter Mater. Phys.* **1982**, *26*, 3506. [CrossRef]
113. Choi, I.; Jeong, H.Y.; Shin, H.; Kang, G.; Byun, M.; Kim, H.; Chitu, A.M.; Im, J.S.; Ruoff, R.S.; Choi, S.-Y.; et al. Laser-induced phase separation of silicon carbide. *Nat. Commun.* **2016**, *7*, 13562. [CrossRef]

materials

MDPI

Communication

Three-Dimensional Epitaxy of Low-Defect 3C-SiC on a Geometrically Modified Silicon Substrate

Gerard Colston, Kelly Turner, Arne Renz, Kushani Perera, Peter M. Gammon, Marina Antoniou and Vishal A. Shah *

School of Engineering, The University of Warwick, Coventry CV4 7AL, UK; g.colston.1@warwick.ac.uk (G.C.); kelly.turner@warwick.ac.uk (K.T.); arne.renz@warwick.ac.uk (A.R.); kushani-hapuhinna.perera@warwick.ac.uk (K.P.); p.m.gammon@warwick.ac.uk (P.M.G.); marina.antoniou@warwick.ac.uk (M.A.)

* Correspondence: vishal.shah@warwick.ac.uk

Abstract: We demonstrate the growth of 3C-SiC with reduced planar defects on a micro-scale compliant substrate. Heteroepitaxial growth of 3C-SiC on trenches with a width and separation of 2 μm, etched into a Si(001) substrate, is found to suppress defect propagation through the epilayer. Stacking faults and other planar defects are channeled away from the center of the patterned structures, which are rounded through the use of H_2 annealing at 1100 °C. Void formation between the columns of 3C-SiC growth acts as a termination point for defects, and coalescence of these columns into a continuous epilayer is promoted through the addition of HCl in the growth phase. The process of fabricating these compliant substrates utilizes standard processing techniques found within the semiconductor industry and is independent of the substrate orientation and offcut.

Keywords: 3C-SiC; epitaxy; compliant substrate; stacking faults; conformal epitaxy

Citation: Colston, G.; Turner, K.; Renz, A.; Perera, K.; Gammon, P.M.; Antoniou, M.; Shah, V.A. Three-Dimensional Epitaxy of Low-Defect 3C-SiC on a Geometrically Modified Silicon Substrate. *Materials* **2024**, *17*, 1587. https://doi.org/10.3390/ma17071587

Academic Editor: Fabrizio Roccaforte

Received: 27 February 2024
Revised: 19 March 2024
Accepted: 25 March 2024
Published: 30 March 2024

1. Introduction

Silicon carbide (SiC) is a wide-bandgap compound semiconductor with high electrical field breakdown and thermal conductivity, which makes it ideal for high-voltage power electronics and other applications in harsh environments [1]. SiC can exist in a number of different crystalline forms due to the stacking sequence of the C-Si bilayers; these are known as polytypes. The hexagonal structured 4H-SiC is highly mature and is available as single crystal substrates, which when paired with high-temperature homoepitaxial growth, form the basis for commercially available power electronic devices. The cost of this material, however, both in terms of the starting substrate and epitaxy, is extremely high and limits its uptake in mass markets [2].

The cubic form of SiC (3C-SiC) has a zinc-blende (diamond) structure and stabilizes at lower temperatures than other polytypes, allowing it to be grown on silicon (Si). The heteroepitaxial growth of 3C-SiC offers many advantages such as low-cost, large-diameter and high-quality starting Si substrates as well as low epitaxial growth temperatures [3,4]. 3C-SiC is of particular importance to power electronics in the 600–1200 V range, as it has a low built-in voltage [5], which accounts for the low specific on-resistance ($R_{on\cdot sp}$) losses and large 3C-SiC/SiO_2 barrier, which potentially support devices' reliability [6].

However, the heteroepitaxial growth of 3C-SiC on Si is plagued by two main difficulties. First, the large lattice mismatch (19.7%) between 3C-SiC and Si introduces a high density of misfit dislocations at the interface and stacking faults that propagate up throughout the epilayer [7]. Stacking faults and other planar defects including micro-twins and anti-phase domains are electrically active and cause high levels of leakage currents in rectifying devices [8,9]. The second issue with heteroepitaxial growth is the high thermal mismatch (8–20%) between 3C-SiC and the Si substrate, which can lead to high levels of thermal stress and wafer bow, which increase with epilayer thickness and can leave epi wafers unsuitable for device fabrication processes [10].

Defect densities can be reduced in 3C-SiC through the epitaxial growth of thicker films, due to the self-annihilation of planar defects; however, this increases epitaxy costs and often increases wafer bow. In addition, this method only reduces the defect density of the 3C-SiC at the surface and does not improve the crystallinity of the epilayer near the interface with the substrate.

More novel processes have been explored to reduce the defectiveness of 3C-SiC, such as the production of substrates suitable for homoepitaxy [11]. Other options include modifying the structure of the starting Si substrate to suppress defect formation and promote the annihilation process. Additionally, there are techniques such as patterning the growth on undulant silicon using diamond slurry [12], growing 3C-SiC on fine-structured nm-scale hexagonal pillars [13] and trapping defects within inverted Si pyramids [14]. The growth of 3C-SiC on porous Si has also been explored to improve adhesion and reduce lattice mismatch between the thin film and substrate [15]. The use of compliant substrates such as these offers a reduction in defects and also the potential to suppress wafer bow; however, several of these methods rely on precise fabrication techniques that are dependent on the substrate's crystal orientation. Any growth on isolated structures also suffers from issues with coalescence of the islands into a continuous film and the introduction of further defects such as anti-phase boundaries [13].

The surface profile of the structures on compliant substrates has been found to have an important impact on the control of defect propagation, as evidenced by growth on pyramids or an undulant surface. Flat surfaces and vertical sidewalls may not offer the ideal template for 3C-SiC growth; however, even with extremely fine patterning processes such as electron-beam lithography, it is often impossible to eliminate the flat regions of the substrate in compliant substrates. Additional processing steps may be necessary to further modify the structure of the patterned silicon prior to epitaxy. One method for achieving this is through annealing processes. The thermal annealing of Si structures in a H_2 atmosphere causes the surface diffusion of Si atoms and can be used to smooth and round Si patterns. The extent of this rounding can be controlled with both the annealing temperature and time [16]. This study investigates the impact that such surface modification has on the heteroepitaxial growth of 3C-SiC on a patterned Si substrate.

2. Materials and Methods

Trenches of 2 μm width and 2 μm separation (defined as 4 μm pitch) were etched to a depth of approximately 4 μm into on-axis 100 mm diameter Si(001) substrates using UV photolithography and reactive ion etch (RIE) processing, with an SF_6 and O_2 plasma. The trenches were aligned to the <110> crystal plane directions. Photoresist was removed by solvent cleaning in acetone before being subjected to an RCA clean. The wafers were then loaded into an epitaxial reactor, ensuring no contamination was introduced into the growth system, and the substrate surface was primed for epitaxy.

Epitaxial growth and thermal annealing were performed within an LPE ACIS-M8 reduced-pressure chemical vapor deposition (RP-CVD) reactor at the University of Warwick. 3C-SiC films were grown at a temperature of 1325 °C using trichlorosilane (TCS, $SiHCl_3$) and ethylene (C_2H_4) with a C/Si ratio of 1.4, within a H_2 carrier gas at a growth rate of ~6 μm/h. A carbonization process was used to initiate the growth of the 3C-SiC, which involved the deposition of a thin seed layer using C_2H_4 at a temperature range starting at 900 °C and ramping up to 1140 °C. On selected samples, the surface profile of the trench walls was modified using H_2 annealing at 1100 °C for 10 min prior to a carbonization process.

The grown 3C-SiC epilayers were characterized by scanning electron microscopy (SEM) and cross-sectional transmission electron microscopy (TEM), with transparent electron cross-sections extracted and polished using a focused ion bean scanning electron microscope

(FIB-SEM). The FIB-SEM process was necessary to extract cross-sections from specific locations on the epi wafers.

3. Results

3.1. Annealing of Si Trenches

Dry etching of silicon can lead to unwanted features such as striations along trench walls and micro-trenches in the base of a trench due to limitations in UV photolithography and imperfect dry etching processes [17]. Annealing the trenches in H_2 modifies the imperfect profile of the mesa walls, giving them a smooth surface, and thereby eliminates these features. Annealing at a sufficiently high temperature can facet and round the surfaces of the Si mesas. Samples not exposed to this baking retain their flat (001)-orientation surface and base and {110} sidewalls (see Figure 1). In this study, the H_2 annealing process was found to have a minimal effect on the trench depth, with the sample depths measured at 4.2 μm and 4.1 μm before and after annealing, respectively.

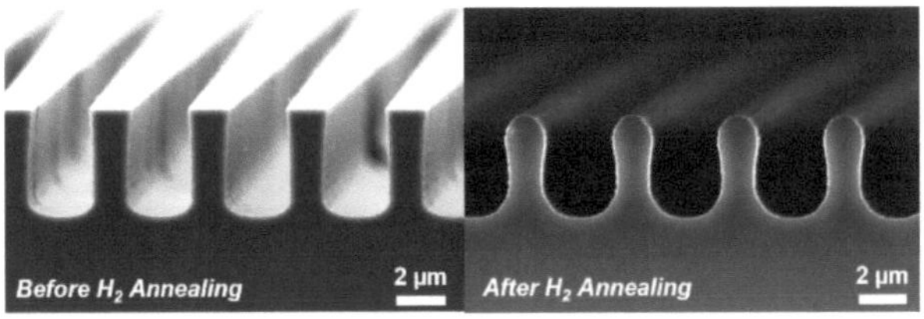

Figure 1. Cross-sectional SEM micrographs highlighting the rounding of the Si trench profile before and after being subject to 1100 °C H_2 annealing.

3.2. 3C-SiC Epitaxy

Conformal epitaxial growth of thin layers of 3C-SiC was found in the Si trenches and on the sidewalls for thin epilayers; however, as the epilayer thickness increased, the sidewall growth was suppressed and voids formed within and above the trenches (see Figure 2). We speculate that this was due to the shadowing effect of the epilayers growing on top of the structures, preventing the flow of precursors into the underlying trenches. Growth on the Si structures that were rounded through thermal annealing appeared more disordered than on the unannealed structures and, in either case, no fusion of the separated epilayers was observed at a growth thickness of 3 μm.

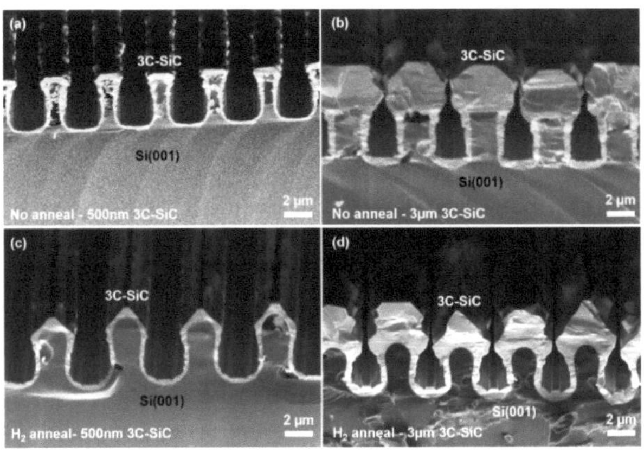

Figure 2. Cross-sectional SEM micrographs of 500 nm and 3 μm 3C-SiC heteroepitaxially grown on Si trench structures without annealing (**a**,**b**) and after 1100 °C H_2 annealing (**c**,**d**).

Cross-sectional TEM showed that the 3C-SiC grown on both rounded and non-rounded structures was crystalline; however, a clear difference in defect density could be observed between the samples (see Figure 3). The rounded surface of the annealed structure was observed to direct stacking faults away from the center of the peak, resulting in an almost defect-free region above the Si structure (see Figure 3e). The selective area electron diffraction (SAED) patterns showed that while the material close to the Si structure had elements of hexagonality caused by the high density of stacking faults within the epilayer, the 3C-SiC growing above the Si structure was monocrystalline in both cases. A clear interface could be observed between adjacent structures, showing that the layers had not coalesced, and, in both cases, the sides of the 3C-SiC growth columns were highly disordered.

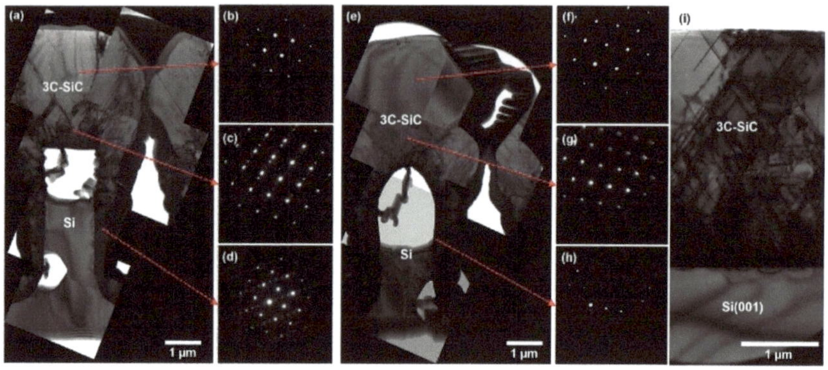

Figure 3. Cross-sectional TEM micrographs of 3 µm 3C-SiC grown on Si trench sidewalls without (**a**) and with 1100 °C H_2 annealing (**e**). SAED patterns of 3C-SiC grown on both structures confirm that the material above the sidewall profile is monocrystalline (**b**,**f**) while 3C-SiC on the surfaces and sidewalls of the structures shows elements of polycrystallinity and defectiveness (**c**,**d**,**g**,**h**). 3C-SiC grown directly on a Si(001) substrate is shown in (**i**) for comparison.

The stacking fault density in the region above the pillars of each sample was estimated from TEM images and compared to that of the growth on non-patterned Si (see Figure 4). The density of stacking faults, based on these local measurements, was found to decrease for the patterned structures, with a further significant reduction after H_2 annealing through the channeling of planar defects away from the center of the mesa.

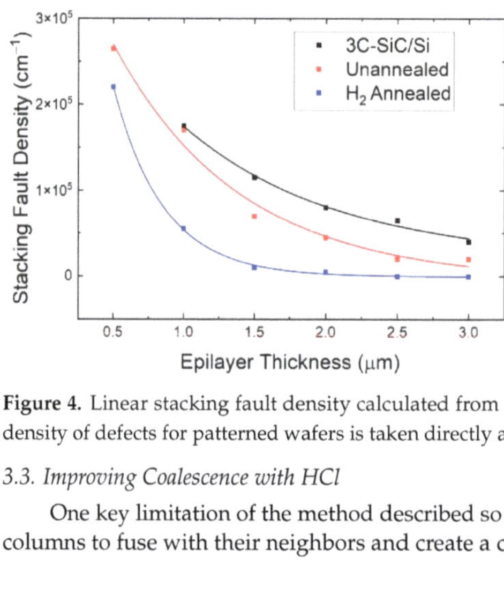

Figure 4. Linear stacking fault density calculated from TEM micrographs at varying depths. The density of defects for patterned wafers is taken directly above the Si structures.

3.3. *Improving Coalescence with HCl*

One key limitation of the method described so far is the inability of individual 3C-SiC columns to fuse with their neighbors and create a coherent epilayer. It has been proposed that the lack of coalescence of these 3C-SiC films may be due to their thinness; however, increasing the growth of 3C-SiC to 10 μm did not trigger any fusion. Instead, separate individual columns of 3C-SiC grew upon each of the Si structures (see Figure 5a). In another idea, the disordered structure of 3C-SiC is thought to cause the lack of coalescence, and so 500 sccm of HCl was introduced into the growth process of 3C-SiC to assess the impact of this etching agent. At typical growth temperatures of 3C-SiC, HCl is found to etch highly dislocated regions of material faster than ordered crystal areas and, hence, can be used to promote the faceted growth of 3C-SiC [18]. In the case of growth on Si(001), the addition of sufficient HCl increased the prevalence of the 3C(001) and 3C(111) surfaces. When these regions met, they were observed to coalesce much more effectively (see Figure 5b). Al doping markers were grown in situ during the epitaxial growth of the 10 μm 3C-SiC epilayer with additional HCl at every 2 μm of growth to highlight the growth evolution of the film when observed by SEM. Even with the H_2 annealing process, the growth fronts were found to be the (001) and (111) crystal planes, and the length of (001) was found to increase as the epilayer increased in thickness, indicating the fusion of the (111) planes. Were this epilayer to be grown up to >10 μm, a flat epilayer surface may be achieved, as the (111) planes may fuse together entirely.

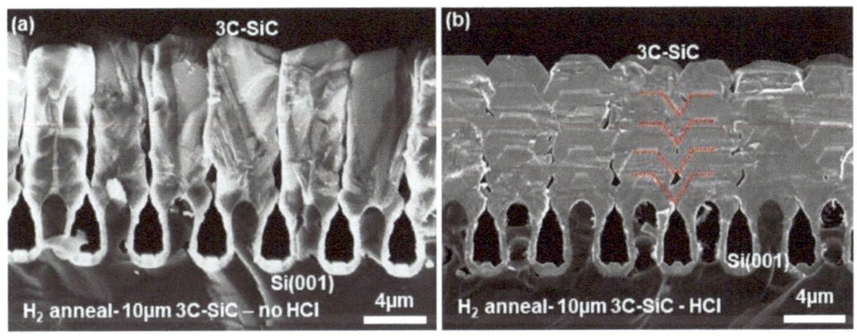

Figure 5. Cross-sectional SEM micrograph of 10 μm 3C-SiC grown on Si trench structures subject to 1100 °C H_2 annealing without HCl (**a**) and with additional HCl (**b**) during epitaxy. The growth fronts of one column are highlighted in red to show the growth evolution.

Further optimization of this growth process is required to improve the coalescence of these columns and to improve the surface morphology of the layer. Increasing the flow of HCl will impact the prevalence on different 3C-SiC growth fronts; however, improvement could also be achieved by modifying the aspect ratio of the mesas and trench separation. Chemical mechanical polishing (CMP) could be employed to give a smooth top surface, making the material suitable for further epitaxy, wafer bonding or device fabrication, as well as producing a high density of voids at the 3C-SiC/Si interfaces on the patterned structures due to the increased Si surface area.

4. Discussion

The growth of 3C-SiC on patterned substrates leads to a reduction in the stacking fault density by trapping defects at the sidewalls of 3C-SiC columns. A further reduction in defect density is obtained by modifying the geometry of the Si structures, which can be achieved through thermal annealing in H_2. Rounding the starting Si structure is shown to generate planar defects, which propagate along the {111} crystal planes that direct away from the [001] growth direction. Conformal epitaxial growth of 3C-SiC on such microstructures begins to form on Si; however, once a thickness comparable to the separation between structures is obtained, growth within the trenches is suppressed and voids are formed. These voids act as natural defect termination points and may help reduce thermally induced wafer bow by breaking up the continuous interface of 3C-SiC/Si.

The challenge with the growth of epilayers on separate structures such as this is promoting the coalescence of the individual columns into a continuous film; however, this has been achieved through the introduction of sufficiently high levels of HCl into RP-CVD during epitaxy, which results in more ordered growth and enables the fusion of layers grown on the Si structures. A similar technique is used in the process of 4H-SiC trench refill epitaxy, whereby additional HCl is added to the epitaxial process to modify the growth and etch rates of 4H-SiC on different faces of the patterned structure [19,20].

Using a patterned substrate for the epitaxial growth of 3C-SiC has a clear impact on the formation of planar defects such as stacking faults and micro-twins; however, its impact on the generation of point defects is unknown. The formation of point defects in SiC is mainly controlled by the growth rate and temperature as well as the precursor composition and, as such, compliant substrates have little impact on point defect generation directly [14]; however, the change in stacking sequence caused by planar defects can lead to the formation of interstitials, which may be reduced by this technique [21].

5. Conclusions

The density of planar defects in 3C-SiC can be reduced through the use of the demonstrated micron-scale compliant substrate fabricated by the formation of trenches in a Si(001) substrate followed by thermal annealing in a H_2 atmosphere. The rounded shape of the Si structure, formed by this annealing process, assists in channeling stacking faults away from the center of each pillar, thus resulting in an area of lower defect density. Such areas of 3C-SiC must then fuse with adjacent growth columns to produce a coalesced thin film, which can be enabled through the selective etching of additional HCl added into the growth phase. The pattern used in this study included linear trenches, which limited the investigation to the propagation of defects in one dimension; however, the technique could be expanded to arrays of pillars, which would enable defect trapping in both in-plane directions. The method presented offers a low-cost and highly scalable process for the fabrication of micron-scale compliant substrates for 3C-SiC as it relies solely on standard semiconductor processing techniques including photolithography and dry etching. The process does not rely on the crystal orientation and hence could be applied to other substrates beyond the (001) orientation and with various levels of offcut, which would help suppress the formation of anti-phase domains.

Author Contributions: Conceptualization, G.C., K.T. and V.A.S.; methodology, G.C., K.T. and V.A.S.; validation, P.M.G. and M.A.; formal analysis, G.C.; investigation, G.C. and K.T.; resources, V.A.S.; data curation, G.C., K.T., A.R. and K.P.; writing—original draft preparation, G.C.; writing—review and editing, all; supervision, V.A.S.; project administration, V.A.S.; funding acquisition, V.A.S. All authors have read and agreed to the published version of the manuscript.

Funding: This research was funded by the EPSRC, with grant number EP/W004291/1, and by the Henry Royce Institute (MCAP065) funded by the EPSRC, with grant number EP/X527257/1.

Institutional Review Board Statement: Not applicable.

Informed Consent Statement: Not applicable.

Data Availability Statement: The research data will be made available upon a reasonable request to the corresponding author.

Conflicts of Interest: The authors declare no conflicts of interest.

References

1. Östling, M.; Ghandi, R.; Zetterling, C.M. SiC power devices—Present status, applications and future perspective. In Proceedings of the 2011 IEEE 23rd International Symposium on Power Semiconductor Devices and Ics, San Diego, CA, USA, 23–26 May 2011; p. 10.
2. Burk, A.A., Jr.; O'Loughlin, M.J.; Tsvetkov, D.; Ustin, S. Industrial Perspective of SiC Epitaxy. In *Wide Bandgap Semiconductors for Power Electronics*; Wiley Online Library: Hoboken, NJ, USA, 2021; p. 75.
3. Ferro, G. 3C-SiC Heteroepitaxial Growth on Silicon: The Quest for Holy Grail. *Crit. Rev. Solid State Mater. Sci.* **2014**, *40*, 56. [CrossRef]

4. Li, F.; Roccaforte, F.; Greco, G.; Fiorenza, P.; La Via, F.; Pérez-Tomas, A.; Evans, J.E.; Fisher, C.A.; Monaghan, F.A.; Mawby, P.A.; et al. Status and Prospects of Cubic Silicon Carbide Power Electronics Device Technology. *Materials* **2021**, *14*, 5831. [CrossRef] [PubMed]
5. Li, F.; Renz, A.B.; Pérez-Tomás, A.; Shah, V.; Gammon, P.; Via, F.L.; Jennings, M.; Mawby, P. A study on free-standing 3C-SiC bipolar power diodes. *Appl. Phys. Lett.* **2021**, *118*, 242101. [CrossRef]
6. Li, F.; Song, Q.; Perez-Tomas, A.; Shah, V.; Sharma, Y.; Hamilton, D.; Fisher, C.; Gammon, P.; Jennings, M.; Mawby, P. A First Evaluation of Thick Oxide 3C-SiC MOS Capacitors Reliability. *IEEE Trans. Electron Devices* **2020**, *67*, 237. [CrossRef]
7. Yamasaki, J.; Inamoto, S.; Nomura, Y.; Tamaki, H.; Tanaka, N. Atomic structure analysis of stacking faults and misfit dislocations at 3C-SiC/Si(0 0 1) interfaces by aberration-corrected transmission electron microscopy. *J. Phys. D Appl. Phys.* **2012**, *45*, 494002. [CrossRef]
8. Chung, G.S.; Kim, K.S.; Yakuphanoglu, F. Electrical characterization of Au/3C-SiC/n-Si/Al Schottky junction. *J. Alloys Compd.* **2010**, *507*, 508. [CrossRef]
9. Eriksson, J.; Weng, M.H.; Roccaforte, F.; Giannazzo, F.; Di Franco, S.; Leone, S.; Raineri, V. On the Viability of Au/3C-SiC Schottky Barrier Diodes. *Mater. Sci. Forum* **2010**, *645–648*, 677–680. [CrossRef]
10. Anzalone, R.; Litrico, G.; Piluso, N.; Reitano, R.; Alberti, A.; Fiorenza, P.; Coffa, S.; La Via, F. Carbonization and transition layer effects on 3C-SiC film residual stress. *J. Cryst. Growth* **2017**, *473*, 11. [CrossRef]
11. La Via, F.; Severino, A.; Anzalone, R.; Bongiorno, C.; Litrico, G.; Mauceri, M.; Schoeler, M.; Schuh, P.; Wellmann, P. From thin film to bulk 3C-SiC growth: Understanding the mechanism of defects reduction. *Mater. Sci. Semicond. Process.* **2018**, *78*, 57. [CrossRef]
12. Nagasawa, H.; Yagi, K.; Kawahara, T.; Hatta, N. Reducing Planar Defects in 3C–SiC. *Chem. Vap. Depos.* **2006**, *12*, 502. [CrossRef]
13. von Känel, H.; Isa, F.; Falub, C.V.; Barthazy, E.J.; Gubler, E.M.; Chrastina, D.; Isella, G.; Kreiliger, T.; Taboada, A.G.; Meduna, M.; et al. Three-Dimensional Epitaxial Si1-XGex, Ge and SiC Crystals on Deeply Patterned Si Substrates. *ECS Trans.* **2014**, *64*, 631. [CrossRef]
14. Via, F.L.; Zimbone, M.; Bongiorno, C.; La Magna, A.; Fisicaro, G.; Deretzis, I.; Scuderi, V.; Calabretta, C.; Giannazzo, F.; Zielinski, M.; et al. New Approaches and Understandings in the Growth of Cubic Silicon Carbide. *Materials* **2021**, *14*, 5348. [CrossRef] [PubMed]
15. Suchikova, Y.; Kovachov, S.; Bohdanov, I.; Kozlovskiy, A.L.; Zdorovets, M.V.; Popov, A.I. Improvement of β-SiC Synthesis Technology on Silicon Substrate. *Technologies* **2023**, *11*, 152. [CrossRef]
16. Lee, M.-C.; Wu, M. Thermal Annealing in Hydrogen for 3-D Profile Transformation on Silicon-on-Insulator and Sidewall Roughness Reduction. *J. Microelectromech. Syst.* **2006**, *15*, 338–343. [CrossRef]
17. Mohammadi, Z.; Shah, V.; Jennings, M.R.; Fisher, C.A.; Mawby, P. Elimination of Microtrenching in Trenches in 4H-Silicon Carbide Using Shadow Masking. *Mater. Sci. Forum* **2015**, *821–823*, 533. [CrossRef]
18. Pedersen, H.; Leone, S.; Kordina, O.; Henry, A.; Nishizawa, S.I.; Koshka, Y.; Janzén, E. Chloride-Based CVD Growth of Silicon Carbide for Electronic Applications. *Chem. Rev.* **2012**, *112*, 2434. [CrossRef] [PubMed]
19. Ji, S.; Kosugi, R.; Kojima, K.; Adachi, K.; Kawada, Y.; Mochizuki, K.; Yonezawa, Y.; Yoshida, S.; Okumura, H. Fast-filling of 4H-SiC trenches at 10 μm/h by enhancing partial pressures of source species in chemical vapor deposition processes. *J. Cryst. Growth* **2020**, *546*, 125809. [CrossRef]
20. Ji, S.; Kojima, K.; Kosugi, R.; Saito, S.; Sakuma, Y.; Tanaka, Y.; Yoshida, S.; Himi, H.; Okumura, H. Filling 4H-SiC trench towards selective epitaxial growth by adding HCl to CVD process. *Appl. Phys. Express* **2015**, *8*, 065502. [CrossRef]
21. Xi, J.; Liu, B.; Zhang, Y.; Weber, W.J. Ab initio study of point defects near stacking faults in 3C-SiC. *Comput. Mater. Sci.* **2016**, *123*, 131. [CrossRef]

materials MDPI

Review

Amorphous SiC Thin Films Deposited by Plasma-Enhanced Chemical Vapor Deposition for Passivation in Biomedical Devices

Scott Greenhorn [1,2,3,*], Edwige Bano [3], Valérie Stambouli [2] and Konstantinos Zekentes [1,3,*]

[1] The Institute of Electronic Structure and Laser of the Foundation for Research and Technology-Hellas (MRG-IESL/FORTH), GR-70013 Heraklion, Greece
[2] Laboratoire des Matériaux et de la Génie Physique, Université Grenoble Alpes, Centre National de la Recherche Scientifique, Institut Polytechnique de Grenoble, 38016 Grenoble, France
[3] Centre de Radiofréquences, Optique et Micro-nanoélectronique des Alpes, Université Grenoble Alpes, Centre National de la Recherche Scientifique, Institut Polytechnique de Grenoble, 38016 Grenoble, France
* Correspondence: scott.greenhorn@grenoble-inp.fr (S.G.); trifili@physics.uoc.gr (K.Z.)

Abstract: Amorphous silicon carbide (a-SiC) is a wide-bandgap semiconductor with high robustness and biocompatibility, making it a promising material for applications in biomedical device passivation. a-SiC thin film deposition has been a subject of research for several decades with a variety of approaches investigated to achieve optimal properties for multiple applications, with an emphasis on properties relevant to biomedical devices in the past decade. This review summarizes the results of many optimization studies, identifying strategies that have been used to achieve desirable film properties and discussing the proposed physical interpretations. In addition, divergent results from studies are contrasted, with attempts to reconcile the results, while areas of uncertainty are highlighted.

Keywords: a-SiC; amorphous; PECVD; biomedical; passivation

Citation: Greenhorn, S.; Bano, E.; Stambouli, V.; Zekentes, K. Amorphous SiC Thin Films Deposited by Plasma-Enhanced Chemical Vapor Deposition for Passivation in Biomedical Devices. *Materials* **2024**, *17*, 1135. https://doi.org/10.3390/ma17051135

Academic Editor: Fabrizio Roccaforte

Received: 30 January 2024
Revised: 22 February 2024
Accepted: 26 February 2024
Published: 29 February 2024

1. Introduction

Amorphous silicon carbide (a-SiC or in hydrogenated form, a-SiC:H) is a promising material for various applications in opto- and micro-electronic devices. This thermally and chemically stable wide-bandgap dielectric material has superior mechanical strength, high radiation resistance, and high optical transmittance [1]. Amorphous SiC appears to be suitable as protective coating against corrosion, moisture, etching, and abrasion [2,3], and has shown promise in optoelectronics [4,5], electronic device insulation [6,7], as well as for biomedical applications [8–17]. Amorphous SiC is specifically considered as an advantageous material for the isolation of neural implantable devices [11,18–20] as it is inherently insulating; has high-K dielectric properties, high wear resistance, and direct binding to Si; and works well as a durable, biocompatible, and hemocompatible coating. Each application has a tendency to focus on different a-SiC composition ranges, where optoelectronic applications prefer more conductive Si-rich films, and electronic device insulation tends to optimize for highly resistive C- and H-rich films, while mechanical and biomedical applications aim for stoichiometric, robust, and chemically inert films.

The passivation layer of an implantable device [21,22] is responsible for isolating adjacent channels electrically, while also protecting the device from the immune response after implantation [23]. In many cases, the thickness of the passivation layer is also non-negligible (up to several microns [24]), and thus its mechanical properties (stress, Young's modulus, hardness) have a significant effect on the device behavior as well [25]. An optimized passivation layer displays high resistivity, high chemical resistance, low stress, excellent biocompatibility, and negligible toxicity [26]. The mechanical properties must strike a balance between minimizing mechanical mismatch with the brain [27], ensuring high flexibility to minimize micromovement-induced issues [28], and ensuring sufficient rigidity and strength for precise insertion into the brain [29,30].

a-SiC compares favorably to typical passivation materials, particularly in terms of the robustness and lifetime. In particular, the dissolution rate in phosphate-buffered saline (PBS) is an order of magnitude slower than that of more conventional passivation materials like Si_3N_4 or SiO_2 [18,31–33], and significantly longer than parylene or polyimide passivation [34,35] due to a-SiC's reduced permeability to water. This results in longer implant lifetimes and reduces material leeching into surrounding tissue. Implants coated in a-SiC showed reduced etching in a buffer solution compared to fluorapatite (dental glass) ceramic, particular for low pH solutions [17]. While the rigidity of a-SiC makes matching the mechanical properties of tissue more difficult [36,37], its strength allows for very thin devices (~6 μm) and thus sufficiently flexible behavior to minimize the biological response and at the same time reduce insertion trauma [30,38,39].

a-SiC was first used in biomedical applications as a biocompatible coating for stainless steel stents starting in 1996 [11,14,40], where it demonstrated significantly reduced protein activation compared to the uncoated implants. Further studies have cemented a-SiC as a biocompatible and hemocompatible material for a variety of applications including bone implant scaffolds and biosensors (2003-present) [12,18,41–45]. In particular, a-SiC has been used as a passivation material for SiC [24,46,47], polymeric [38], and metallic [39,48] implantable microelectrode arrays, starting in 2003, as well as a coating for prefabricated Utah arrays [18,19], showing its potential for implantable electrical devices. Note that the US-FDA (Food and Drug Administration) approved in 2005 under PMA #P030037 an intravascular stent coated with a-SiC. Finally, a-SiC has also been used as the structural support material as well as passivation for implantable electrodes [39,48].

A previous general review of a-SiC is given by Bullot and Schmidt [49], reviewing results in a-SiC based on data published between 1968 and 1987. Another review on a-SiC film developments was presented by Choi [50], based on publications between 1987 and 2000. Fraga et al. [51] provide a more recent synthesis of results across a wide variety of deposition methods, but do not cover film optimization for any deposition method. A review of a-SiC biocompatibility and suitability for biomedical applications was performed by Mahmoodi and Ghazanfari [52] in 2011 without including details on the deposition and physical characterization of the material. Similar topics have also been addressed in sections of a chapter by Locke et al. [33] in 2012 and an article by Saddow [10] in 2022, both of which focus on crystalline SiC material more generally and only briefly discussing the a-SiC material aspects. a-SiC is also briefly addressed in a more general review by Kaloyeros and Arkles [53] in 2023. While their review provides a comprehensive look at deposition methods and applications of SiC, it does not investigate in detail the relation between process conditions and material properties.

With ongoing research and improved deposition processes [6,7,31,54–58] as well as increasing opportunities for applications of a-SiC, there is now a need for a comprehensive introduction to a-SiC film optimization. To maintain a reasonable scope for the review, the main focus will be on PECVD deposition optimization as it has the most extensive literature and is most frequently used in studies on biomedical applications. The result is a synthesis of the current state of knowledge of the optimization of PECVD a-SiC, providing a starting point for designing optimization studies of a-SiC to new researchers, outlining demonstrated processes to obtain films with desirable properties for various biomedical applications, and reviewing recent results for those already working in a-SiC technology. Our proper results are incorporated in many cases to strengthen our conclusions on a-SiC optimization.

2. Deposition of a-SiC Films

A survey of papers on a-SiC films published over the past 30 years shows that the majority of such films were deposited by PECVD, and the second-most common technique is sputtering.

2.1. Sputtering Method Employed for a-SiC Film Deposition

A summary of deposition techniques given in [59] describes sputtered SiC films as having "fair" uniformity, "very good" substrate versatility, "good" stress control, and "fair" throughput. The same study also reports that, due to the chamber geometry with the sputtering target far from the substrate, the sidewall coverage is often poor, which results in poor conformal coatings for non-planar devices.

a-SiC films can be deposited via multiple sputtering procedures, including a single SiC target, a single target (Si) in a C-containing atmosphere, or co-sputtering with a Si target and a C target [49,60]. A SiC target is preferred as the composition of the sputtered film usually replicates that of the target, allowing great control of the resulting film's composition. Unhydrogenated films are usually prepared by sputtering of a polycrystalline SiC target using argon ions (Ar^+) [49]. This also leads to a very low level of impurities in the resulting sputtered film. As the sputtering process removes Si–C molecules rather than individual C or Si atoms, the method results in perfect stoichiometry for the deposited film. These films exhibit higher chemical resistance due to the lack of H bonds.

It is noted in one study that sputtered a-SiC films deposited at room temperature have been observed to display high pinhole density and low resistivity [60], which is said to make them unsuitable for biomedical applications. However, optimized resistivity values for sputtered films (often deposited at temperatures above 150 °C) range from 10^4 $\Omega \cdot cm$ [61] up to 10^{11} $\Omega \cdot cm$ [62] for non-hydrogenated sputtered films and as high as 10^{14} $\Omega \cdot cm$ for hydrogenated films [63], and do not report the presence of pinholes. Stress values range from highly compressive (−1400 Mpa) to slightly tensile (100 MPa) depending on the deposition conditions [64].

The optical constants (refractive index, extinction coefficient, bandgap) of a-SiC films obtained by sputtering have been reported to show a variability of up to 10%, including by ageing or annealing, that is not explained purely by the deposition process or film composition [65]. This suggests that as-deposited sputtered a-SiC films do not result in time-stable films, a factor which must be considered when determining deposition recipes.

2.2. PECVD Method Employed for a-SiC Film Deposition

PECVD a-SiC films have consistently been shown to have the best electrical and mechanical properties of the various growth methods [60]. The previously-referenced summary of deposition techniques [59] describes PECVD as having "fair" uniformity, "very good" substrate versatility, "very poor" stress control, and "very good" throughput. In particular, PECVD has been shown to be vulnerable to "edge effects" in small samples, with non-uniform edges at the extreme edges of the samples [66]. This can be reduced by moving to Induction-Coupled Plasma (ICP) CVD [67], although the literature available for this method is substantially smaller than for conventional PECVD.

PECVD [68] is a variation on CVD techniques using gas-phase precursors in an inert carrier gas to grow thin films, which allows low-temperature growth due to capacitive coupling with a plasma. Ionizing the gases in the chamber results in free electrons, which transfer energy very efficiently to the gas while the neutral atoms remain near ambient temperature. The energetic electrons can induce high-energy processes (precursor dissociation, creation of free radicals) desired for film deposition without the drawbacks of high-temperature deposition for the substrate (crystallization, damage to heat-sensitive substrates, etc.).

The plasma tends to be more positively charged than the chamber (due to the high mobility of electrons); in other words, there is a consistent flow of electrons from the plasma to the chamber walls and substrate. All these surfaces receive an energetic ion bombardment due to the induced electrostatic force. This bombardment results in a high film density and few contaminants, but has also been noted to increase film stress [6,69]. With high-density plasmas, sputtering can occur simultaneously, which planarizes the film by filling in cavities.

PECVD of SiC can be performed using a single gas precursor providing the silicon and carbon atoms or can be performed by using separate precursors for the silicon and the carbon atoms individually. Unlike sputtering, however, the preferred method here is to use separate precursors, primarily using a mixture of silane and methane (SiH_4 + CH_4) gases. Precursors of ethylene (C_2H_4) [70–72] and methyltrichlorosilane (CH_3SiCl_3) [73] have also been reported. Inert (such as Ar or He) and active (such as H_2) carrier gases are also frequently used.

In addition to the precursor gases and their flow rates, the chamber temperature and pressure as well as the radio frequency (RF) power and, occasionally, frequency can be independently controlled in most PECVD setups. Each of these parameters has been reported to have distinct influences on deposited film properties, which will be discussed in Section 3.

As PECVD chambers can be used to deposit multiple materials, it is necessary to ensure cleanliness prior to a-SiC deposition. For proper recipe repeatability, it is crucial to guarantee that start conditions of the chamber are always the same. Many chemistries are used to clean the chamber and prevent deposited materials from building up on the chamber surfaces. Most studies [60,74] agree that a plasma cleaning pre-treatment using fluorinated (CF_4, NF_3, ...) and oxygen-containing (O_2, N_2O) gases is useful to remove a buildup of material in the chamber. In addition, a short pre-deposition of the material before introducing the sample can be performed to ensure consistent chamber conditions throughout the full deposition process. A flush [60] or sputter cleaning of the sample [75] with Ar prior to deposition has been reported as necessary to avoid pinhole formation. The chamber is evacuated to pressures below $5{\cdot}10^{-6}$ Torr just prior to process gases starting to flow.

2.3. Optimal Deposition Technique for Biomedical Applications

While sputtering can achieve films with some desirable properties, it has not found use in biomedical applications, which prefer PECVD [8,9,11–18,20,24]. This is mainly due to the relatively poor sidewall coverage and resulting inhomogeneous coating achieved by sputtering. As such, this review will focus primarily on the physics and optimization of PECVD a-SiC in order to obtain desirable film qualities for these applications.

A potential application for sputtered a-SiC is in multilayer, multifunctional a-SiC films [24]. In the referenced study, sputtered a-SiC was used as a thin chemical protection layer on top of a thicker electrical passivation layer of a-SiC:H deposited by PECVD. This approach allows the combination of the high electrical isolation from PECVD (see next section) in addition to the maximal chemical passivation from stoichiometric sputtered a-SiC.

3. Controlling a-SiC Film Properties for PECVD

Most studies of a-SiC film deposition focus on a subset of the film properties and deposition parameters, studying individual interactions between single deposition variables and properties of the resulting films. The most commonly studied deposition parameters include the flow rates of precursor gases, the inclusion of H_2 as a precursor, the deposition temperature, the RF power, and the pressure of the chamber. Post-treatment of the films involving plasma and annealing has also been studied. Various properties are studied to characterize the resulting films: the composition, in terms of Si, C, H, and occasionally O; the bonding structure in terms of Si–C, Si–H, and C–H, as well as Si–Si and C–C; the film organization in terms of an average number of bonds per atom, the surface roughness, the conductivity or resistivity of the films, the optical properties including the bandgap and refractive index, the stress, the hardness, the Young's modulus, and the resistance to a chemical attack. A thorough analysis of the relationship between the deposition conditions and the a-SiC film properties is performed below and a conclusion on optimal deposition conditions for a-SiC for biomedical applications is reported for each material property.

3.1. Stoichiometry of a-SiC Films

The stoichiometry of a-SiC films can be varied from pure Si to pure C. If a layer is not stoichiometric, the following symbolism of a-Si_xC_{1-x} is employed.

In the case of an amorphous binary material such as silicon carbide (a-Si_xC_{1-x}), three main types of bonding are possible: Si–C (heteronuclear), Si–Si, and C–C (homonuclear). Most of the free energy of the amorphous phase is stored in the material as angular distortion in the Si–C bonds [76], which is always present in the a-SiC. Moreover, other types of defects such as chemical disorder (i.e., the presence of homonuclear Si–Si and C–C bonds) and dangling bonds are present, as well as bonds to typical ambient atoms such as H or O, which are present in most films.

For stoichiometric films with 50% Si and 50% C and with perfect chemical ordering, a given Si is expected to be surrounded by four C, and vice versa. As the composition ratio x is varied, one would expect that the relative proportions of the three types of bonds also change (chemical disorder). Since C–C bonds are thermodynamically more stable than Si–Si and Si–C bonds, it is probable to find C–C bonds already in a material with a C/Si fraction slightly higher than 1, starting around C/Si = 1.1 and increasing with the C content [76]. At higher carbon concentrations, clusters may be formed, which are expected to contain graphitic configurations. Annealing a-SiC films in a dry atmosphere is also reported to increase graphitic cluster formation [77]. Also, the consistent Si and C lattice is in most cases broken by the incorporation of hydrogen atoms, which only bond to a single Si or C atom. The organization of the films can be given in terms of an <r> value describing the average number of bonds for a given atom in the network [56,78].

a-SiC film stoichiometry has been shown to affect the conductivity (Section 3.2.1), optical bandgap (Section 3.3.1), refractive index and dielectric constant (Section 3.3.2), stress (Section 3.4.1), and hardness and Young's modulus (Section 3.5.1).

3.1.1. Variation of Stoichiometry with Precursor Gas Ratio

Importantly, the Si/C ratio in the film does not match the ratio of silicon to carbon in the precursor gas flows as SiH_4 is more reactive than CH_4 (and indeed several other C-containing precursors such as C_2H_4) [79]. Studies show that for a given deposition setup, the relation between the precursor gases' ratio and film composition is nearly linear, with different slopes for each experiment (Figure 1) [55,79–81].

The slope of the composition–precursor ratio curve correlates most strongly to the power density used (power/chamber area, Table 1, Figure 2), with high power densities resulting in a less steeply increasing atomic Si/C composition as a function of the precursors (see next section on RF power). The chamber size used for the calculation is approximate, based on the reported maximum wafer size for the PECVD system used.

Table 1. Comparison of PECVD deposition systems using SiH_4 and CH_4 precursors for the studies of Figure 1.

Study	Daves [55]	Janz [79]	Pham [80]	Adithi [82]	Li/Guruvenket [81,83]
PECVD system	Oxford Plasmalab 100 Abingdon, UK	Roth & Rau AK400 Hohenstein-Ernstthal, Germany	Novellus Concept One San Jose, California	Oxford Plasmalab 100 Abingdon, UK	Applied Materials P5000 Santa Clara, CA, USA
Max Diameter (cm)	20	15.6	50	20	20
Power (W)	200	150	100	40	

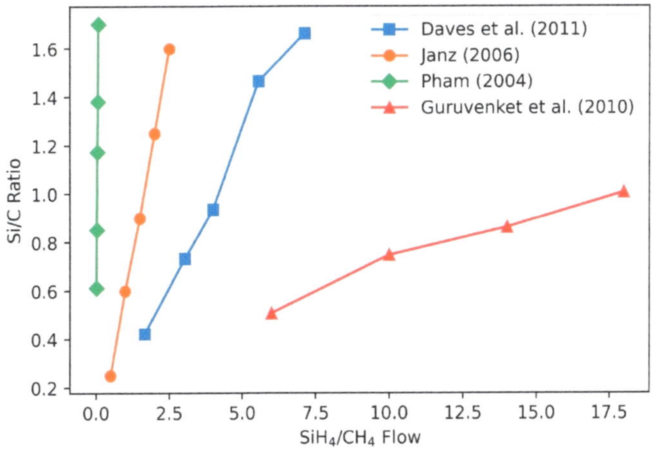

Figure 1. Comparison of measured Si/C atomic ratios to precursor SiH_4/CH_4 ratios for all reproducible studies. Compositional studies were performed using X-ray Photoelectron Spectroscopy (XPS) (Daves et al.), Elastic Recoil Detection (Time-of-Flight mode) (ERD-ToF) (Guruvenket et al.), electron-probe microanalyzer (EPMA) (Pham), and Auger Electron Spectroscopy (AES) (Janz et al.). Daves et al. used 1000 mTorr pressure and high temperature (400 °C) with a SiH_4/CH_4 ratio of 0.25; Janz used a temperature of 350 °C, pressure of 37.5 mTorr, and power density of 250 mW/cm^2; Pham used a temperature of 400 °C, 2250 mTorr pressure, and a power of 100 W; and Guruvenket et al. used temperature of 300 °C and pressure of 100 mTorr. Data from [55,79–81]. Reproduced with permission.

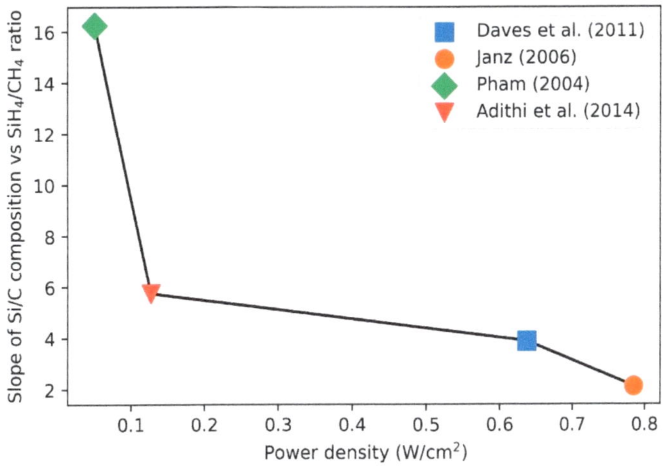

Figure 2. Comparison of the slope of the Si/C curve (Figure 1, right) to the power density for the studies, where available. Values calculated from [55,79,80,82].

In conclusion, the composition of a-SiC films is linear as a function of the ratio of SiH_4 and CH_4 precursors, with the slope determined by the other deposition characteristics, in particular, the power density.

3.1.2. Variation of Stoichiometry with RF Power

Three studies [55,84,85] have found that increasing the RF power decreases the Si/C ratio of the film (Figure 3). This can be explained by the relative difficulty of forming reactive species from CH_4 compared to SiH_4, due to the higher bond energy of C–H (98.8 Kcal/mole) compared to that of Si–H (70.9 Kcal/mole). Higher RF powers allow reactive species of CH_4 to be formed and independently adsorbed/bonded on the substrate [54]. Once C-reactive species are on the growing surface, they form stable C–H much easier than Si–H bonds (see section below on hydrogen incorporation). At low RF powers, mainly reactive species of SiH_4 are formed, and C is mostly incorporated through whole CH_4 molecules bonding to the SiH_4 radicals present on the surface [86–88]. This situation with the presence of SiH_4 radicals but not CH_4 radicals is referred to as a "low-power-density regime" with the most commonly-used case of a "silane-starving plasma" where all silane molecules entering the chamber are radicalized [19,54]. Silane-starving plasma requires both an RF power sufficiently low to radicalize only silane as well as a sufficiently low silane flow such that all the silane can be radicalized. Silane-starving plasma has been reported using either very low power density (8.3 mW/cm^2) with an SiH4/CH4 ratio near 0.3 [54], or using a higher power density (250 mW/cm^2) combined with a lower SiH4/CH4 ratio of 0.17 [19]. This has been reported to result in films with diamond-like properties and high Si–C ordering [86].

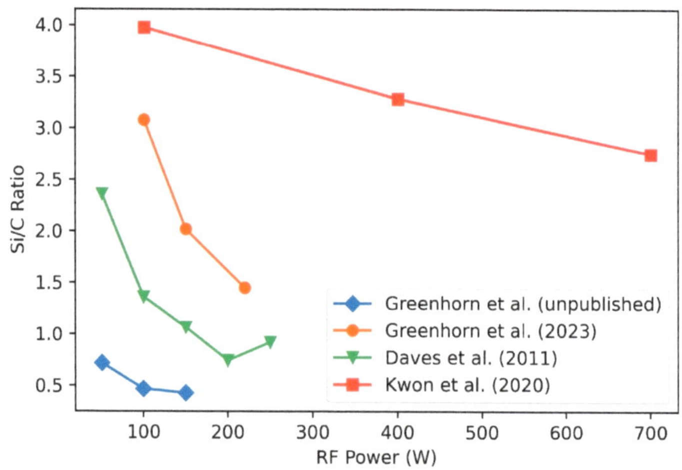

Figure 3. Change in Si/C ratio as a function of RF power. The studies by Greenhorn et al. use C_2H_4 as a C precursor, and the unpublished data use much lower precursor flow rates combined with high Ar flow. Daves et al. used 1000 mTorr pressure and high temperature (400 °C) with a SiH_4/CH_4 ratio of 0.25 while Kwon used a large industrial-type PECVD chamber with high flow rates, a pressure of 2500 mTorr, and a very high temperature of 550 °C and precursor gases of Si_2H_6 and CH_4. Data from [55,84,85] and additional unpublished data from the authors. Reproduced with permission.

Note that one study by Deku et al. [89] does not report a change in the composition as a function of the RF power.

In conclusion, most studies agree that a-SiC films become more Si-rich as the RF power is decreased.

3.1.3. Variation of Stoichiometry with Deposition Temperature

The deposition temperatures used for a-SiC typically range from 150 °C up to 450 °C. Diaz-Botia et al. [60] note more C–H bonding and less Si–C bonding for films deposited at 200 °C than for those deposited at 350 °C using precursor SiH_4/CH_4 ratios from 0.1 to 0.4. Hsu et al. [19] also found that films deposited at 200 °C have less Si–C compared to films deposited at 400 °C. No variation in the Si/C or O content ratio has been observed [19,60,90].

However, Huran et al. [91] found an increase in the Si/C ratio (from 0.625 to 0.857) as the temperature increases from 250 °C to 450 °C. Deku et al. [89] also found a slight increase in the Si/C ratio for deposition temperatures above 300 °C, for films near Si/C = 0.5.

In conclusion, studies do not agree on the influence of deposition temperature on Si/C composition, with some reporting a slight increase in Si with increasing temperature while others report no effect. Nevertheless, most studies agree that the increase in chamber temperature increases the Si–C bonding.

3.1.4. Presence of Hydrogen in a-SiC

PECVD a-SiC films contain hydrogen in a sufficiently important quantity to change the stoichiometry and in this case the film is often referred to as hydrogenated and denoted as a-Si_xC_{1-x}:H_y. Reported hydrogen contents range from 15 to 58% [56,92–96] for films deposited using SiH_4–CH_4 gas mixtures, Ar as carrier gas, and no hydrogen flow. Hydrogen acts as a terminating atom in the crystalline structure, being 1-fold coordinated compared to Si and C, which are 4-fold coordinated. The presence of hydrogen thus disrupts the regular structure of Si and C and functions as a terminating atom.

In principle, hydrogen incorporation is unavoidable due to the PECVD gas precursors containing hydrogen (CH_4, C_2H_4, etc.). C–H bonds are more stable than the Si–H bonds, so carbon atoms are expected to be more hydrogenated than silicon atoms and thus C–H bonds are more prevalent than Si–H bonds. Demichelis et al. [92] found that hydrogen content in the film increases linearly with the amount of CH_4 in the precursor gas, which supports the result that hydrogen incorporation in the film is brought on primarily by CH_4. This is also found by King et al. [56] and by Fujimoto et al. [71].

Using H_2 as the main carrier gas has been shown to decrease CH_n and SiH_2 groups while increasing Si–C bonding [19,60,76]. In [19], the effect is reported to be magnified in the silane-starving case. The same study identifies two potential mechanisms: either that H flow removes methyl radicals as methane before they can be deposited, or that hydrogen plasma preferentially etches CH groups, removing H from the deposited material. The structural result is a decrease in microvoids (and increase in film density) and a reduction in graphitic cluster formation [56,97].

Decreasing the RF power is found to increase the H content of the films [71,84,91]. King et al. [56] also notes that very-low-frequency RF (300–400 kHz) has been used to achieve high H content (>50%) in some films. Yang et al. [90], Lee et al. [70], and Herrera-Celis et al. [54] observed that hydrogen content decreases with increasing deposition temperature.

In conclusion, hydrogen incorporation always occurs in PECVD a-SiC with hydrogen-containing precursors. C–H bonds are more stable than Si–H bonds, so most H is incorporated with the carbon-containing precursor [49], and correspondingly, the hydrogen content tends to increase for higher C content [92]. The H content can be reduced by increasing the temperature [54,70,90] or RF power [56,71,84,91], or by using H_2 as a carrier gas [19].

3.1.5. Presence of Oxygen in a-SiC

Oxygen incorporation is less studied than hydrogen in a-SiC films but is mentioned in several studies. In particular, oxygen content ranging from 0.1 to 23 atomic % is reported [6,20,55,74,98]. The study with the highest reported O content (23%) notes that this was due to chamber contamination [20]. Another study [89] examined more samples in order to observe trends in oxygen content, finding increased oxygen in the bulk in films deposited at lower temperatures (150 °C, 250 °C vs. 325 °C, 350 °C). This is consistent with the expectation of increased porosity for low-temperature films described in Section 3.1.3, which allows O to percolate through the 'porous maze' [99]. XPS measurements allowed the binding structure of the incorporated oxygen to be studied in more detail. For high temperatures, the O atom presence is well-described by O–Si–C bonds, while the larger and broader peak for lower temperatures suggests both O–Si–C and O–Si bonding. The O–Si–C peak was observed to increase over time with the sample exposed to the air. Yang [90] found an increase in C–O–H bonds in films with very low C/Si precursor ratios.

Bulk measurements report significantly lower O content compared to the surface due to the formation of the surface oxide [70]. A surface oxide is also reported by Tomastik et al. [100] for doped a-SiC, where the oxide layer increases when the sample is annealed at higher temperatures. The oxide has a thickness of approximately 70 nm and is reported to be mechanically softer than the bulk a-SiC.

3.1.6. Stoichiometry of a-SiC Films for Biomedical Applications

Most demonstrations of implantable devices employ stoichiometric or near-stoichiometric a-SiC films when optimizing for biomedical applications [18,19,44]. Coatings for bone implants have used slightly more C-rich films (Si/C = 0.745) [15]. Stoichiometric composition is associated with high Si–C bonding, which results in robust, long-lasting films. For the same purpose, the H_2 content must be minimized, which can be achieved by using silane-starving plasma and adding H_2 flow [19,54]. These deposition conditions allow a wider range of compositions to be explored in order to optimize other material properties.

3.2. *Resistivity of a-SiC Films*

a-SiC resistivity has been measured using both in-plane [101,102] and vertical (using metal-insulator–metal capacitor-like structures) [18] measurements, and indirectly through leakage current measurements (see Section 3.6.1). The two main methods have not been compared with a-SiC deposited using the same parameters. Unlike single-crystal SiC, the high resistivity of a-SiC films [6,76,102–105] comes about from the amorphous microstructure and relatively low deposition temperatures used compared to crystalline SiC. The lower deposition temperature prevents the activation of dopants from the deposition, thereby resulting in an electrical insulator. The mechanisms of electrical conduction in a-SiC films are not trivial [106–109] and are not pursued in detail here in favor of focusing on the methods of controlling the resistivity during deposition. The conductivity of a-SiC is known to increase in a power relationship as the measurement's AC frequency is increased for frequencies up to 1 MHz [96,109], resulting in resistivity decreases of up to 3 orders of magnitude from DC to 1 MHz. This reduces the passivation capability of a-SiC for high-frequency applications.

3.2.1. Variation of Resistivity with Stoichiometry

According to Du et al. [110], the resistivity decreases exponentially from $1 \cdot 10^{11}$ to $1 \cdot 10^{6}$ $\Omega \cdot$cm as the atomic Si/C ratio increases, showing that a higher fraction of C leads to higher resistivity. This is explained in terms of nanocrystallinity, as crystalline Si regions reduce the resistivity. High Si content is also associated with 4-coordinate sp^3-like σ bonding, which does not allow complete relaxation and thus the elastic strain is relieved by dangling bonds, which contribute to conduction [109]. Increasing the number of C atoms breaks the Si crystalline pattern, forming C–C diamond-like bonds [105], and thus increasing the resistivity. Loulou et al.'s [102] and Donercark et al.'s [58] results are consistent with Du et al. [110] across their respective ranges (see Figure 4).

Desalvo [76] found that H incorporation in PECVD-grown a-SiC:H does not influence the dark conductivity. Alternatively, King [6] found very high resistivities for highly porous, H-rich films, finding resistivities above 10^{16} Ohm.cm for films with H% greater than 40%. However, this study increased both H and C content simultaneously, making the influence of H alone uncertain. Dutta et al. [111] compared sputtered hydrogenated and unhydrogenated a-SiC, and found a significant increase in resistivity from a-SiC to a-SiC:H along with smaller increases as the partial pressure of H_2 was increased.

In conclusion, resistivity increases with the amount of carbon in the film according to studies using PECVD deposition. Hydrogen content may also play a role by increasing the nanoporosity but is difficult to distinguish from the variation of C content in the case of PECVD. Si-rich films, by contrast, are reported to have regions of nanocrystalline Si, which cause a lower resistivity. The combination of the two effects describes the reported results well: as Si content decreases, nanocrystalline Si decreases and nanoporous C–H increases, resulting overall in an increase in resistivity.

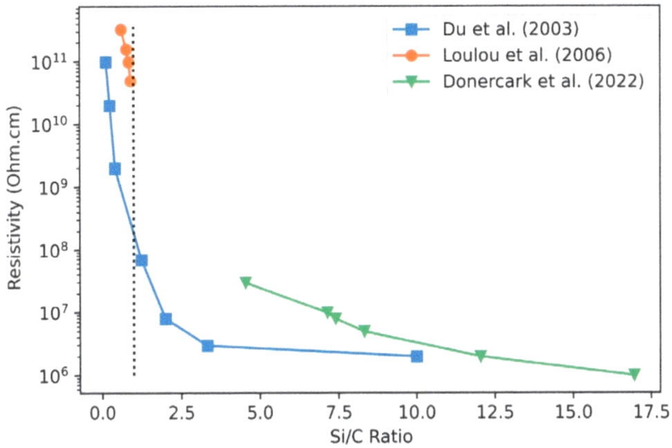

Figure 4. Resistivity as a function of precursor ratio. Stoichiometric (1:1) composition is marked with a vertical dotted black line. Du et al. used C_2H_4 as a precursor gas with H_2 as a carrier, while Loulou et al. used CH_4 and He carrier. Donercark et al. used SiH_4, CH_4, and H_2 gases. From [58,102,110]. Reproduced with permission.

3.2.2. Variation of Resistivity with Deposition Temperature

One study [84] reports a resistivity dependence on deposition temperature under consistent conditions; lower temperatures correspond to higher resistivity (Figure 5). Du et al. [110] suggest that higher temperatures are more likely to form nanocrystalline Si regions in the film, which exhibit increased conductivity compared to a-SiC with more Si–C bonding. Harder et al. [11], alternately, attribute the increase in conductivity to the lower quantity of hydrogen in films deposited at high temperature, which increases the density of states in the bandgap.

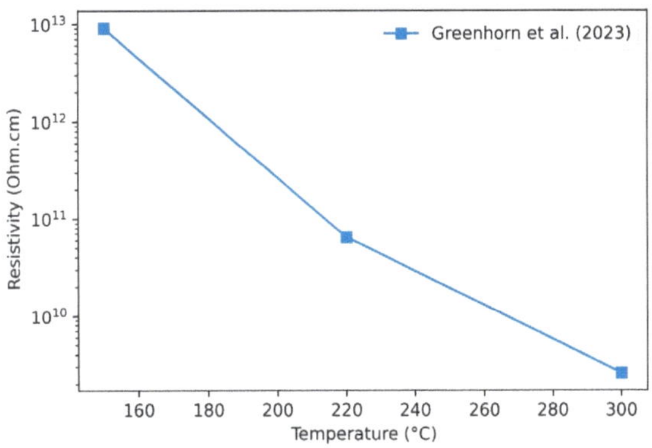

Figure 5. Variation of resistivity with deposition temperature. The films were deposited at 300 °C with 220 W RF power and 1200 mTorr pressure with a SiH_4/C_2H_4 ratio of 0.57. From [84] and reproduced with permission.

Prolonged studies of electrical isolation over time find that the use of lower deposition temperatures (200 °C vs. 350 °C) gives higher resistivity when measured immediately, but exhibit a higher rate of eventual failure in thin films due to leakage currents through micropores in the film [60]. This is consistent with the model of porous material, leading to high initial resistivity but also allowing chemical attacks to easily form channels.

3.2.3. Varying a-SiC Resistivity by Doping and/or Post-Deposition Annealing

Studies in optoelectronics have sought to achieve lower-resistivity a-SiC by doping. Meanwhile, amorphous materials are often relatively difficult to dope, as the large existing variations in local structure accommodate impurities without strongly affecting the electronic properties. However, studies have shown that a combination of doping and annealing can result in an appreciable decrease in resistivity.

Common dopants for a-SiC films are nitrogen (from NH_3) [103], phosphorus (from PH_3), or boron (from B_2H_6) [74,112]. By adding a flow of a dopant precursor with proportions of up to 10% of the combined silane and methane precursor flow rates, resistivity decreases by approximately two orders of magnitude. An additional annealing step, at 650 °C for two hours, was reported to further lower the resistivity by an additional four orders of magnitude [74].

Vetter et al. [101] further studied the effect of low-temperature annealing in the case of low-RF-power films deposited at 400 °C. Transitions in the resistivity are observed at T = 80 °C, where a decrease in resistivity is observed, and at T = 170 °C, where a permanent increase in resistivity is observed. The first decrease is attributed to the breaking of weak Si–H and C–H bonds, which are expected to be plentiful due to the low RF power used during deposition, leaving dangling bonds that result in higher conductivity. The permanent increase in resistivity is then attributed to the reformation of covalent bonds in the a-SiC network, reorganizing to reduce dangling bonds in a more stable arrangement that is not destroyed by further heating/cooling. High conductivity has been achieved in a-SiC using a combination of doping and annealing [113].

3.2.4. Resistivity of a-SiC Films for Biomedical Applications

High-resistivity a-SiC is desirable for the passivation of neural implants, with values of the order of 10^{12}–10^{13} Ω·cm reported in [11,18] for near-stoichiometric films. Although increasing the C content could further increase the resistivity, this typically results in a decrease in chemical inertness (see Section 3.6.1). Therefore, near-stochiometric films are employed for this class of biomedical applications where increasing the C content is undesirable.

3.3. *Optoelectronic Properties of a-SiC Films*

3.3.1. Optical Bandgap

The Urbach rule [114]—in which optical absorption at the band edge is found to be exponential with respect to the photon energy—is observed in an extraordinarily large number of materials systems. This rule is used for the determination of the optical bandgap in amorphous semiconductors including a-SiC. Various measured values have been reported for the a-SiC optical bandgap, showing the dependence of this parameter with growth conditions.

Three studies [58,79,115] have examined the variation of the optical bandgap with stoichiometry (Figure 6), finding a decrease in the bandgap as the Si content increases. Mastelaro et al. [116] attribute the previous trend to the formation of diamond-like C–C bonds at high C content rather than to voids due to hydrogen inclusion and the formation of microvoids. The high-C behavior of the bandgap is reported to depend on the bonding structure: graphite-like bonds cause a reduction in the optical bandgap for C concentrations above 50%, while C–C diamond-like bonds (achieved by silane-starving conditions) allow increases in the bandgap to continue for C concentrations up to 70% [105].

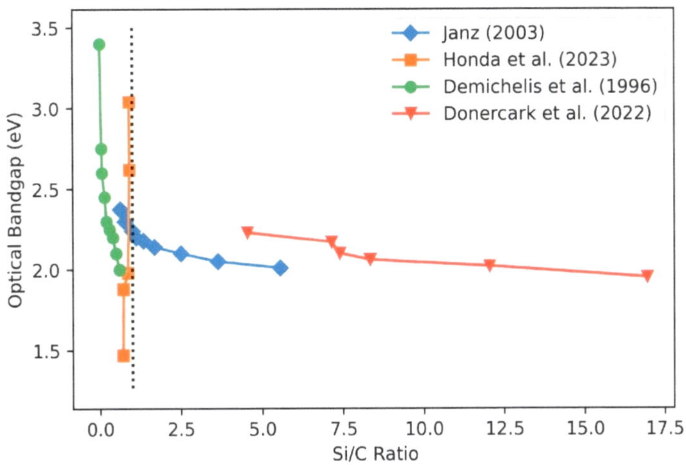

Figure 6. Bandgap dependence on C/Si stoichiometric ratio. A dotted black line marks stoichiometric composition. Janz used SiH_4 and CH_4 gases, a temperature of 350 °C, pressure of 37.5 mTorr, and power density of 250 mW/cm^2. Honda et al. used N as a dopant and tetramethylsilane, 1,1,3,3-tetramethyldisilane, and n-hexane precursors. Demichelis et al. used SiH_4 and CH_4 gases, deposition temperature of 195 °C, and pressure of 640 mTorr. Donercark et al. used SiH_4, CH_4, and H_2 gases at 200 °C and 1200 mTorr. Data from [57,58,79,115]. Reproduced with permission.

Honda et al. [57,117] observed a significant increase in the bandgap of N-doped C-rich a-SiC by decreasing the RF power during deposition including achieving optical bandgaps of 1.25 eV using 26 W RF power, opposite the expected trend from the change in composition observed in Section 3.1.2. Their deposition was optimized to reduce the formation of C–C and Si–Si clusters, which, combined with the presence of dopants, may explain the difference in their results.

In conclusion, the optical bandgap of a-SiC is reported to vary from 1.25 eV [117] to 4.0 eV [105], most strongly correlated to the composition, with higher C content being associated with a higher bandgap. The bandgap is reported to increase significantly by decreasing the RF power. Systematic studies on the effect of deposition temperature on the bandgap values are missing.

3.3.2. Refractive Index

a-SiC's wide bandgap makes it an almost ideal optoelectronic material, transparent for all visible wavelengths above 0.5 μm and thus highly suitable for guiding light in the visible and infrared optical spectrum. A refractive index typically greater than 2.5 (significantly larger than that of SiO_2 and even than that of Si_3N_4) also makes α-SiC an excellent candidate for optical waveguides [74]. The refractive index has been used as a measure of the repeatability of a deposition recipe [6] and has been seen to be consistent across multiple depositions while using the same parameters. As a function of wavelength, the refractive index increases rapidly with the wavelength for small wavelengths and reaches a maximum near 400 nm and decreasing exponentially to a stable value in the near-IR range [16,66]. Refractive indices reported in the literature are typically measured at a wavelength between 500 nm and 677 nm.

Refractive index values reported in the literature are typically measured at a wavelength between 500 nm and 677 nm and they are between 1.66 and 3.4 [6,16,20,60,82,118]. The refractive index is mostly determined from the film Si and C composition [19,55,60,82], although the H content and porosity play a role [6]. The refractive index increases with increasing Si content according to all studies (Figure 7).

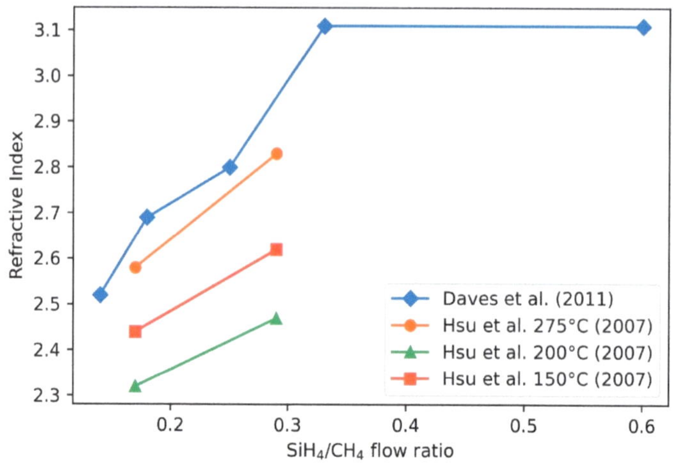

Figure 7. Variation of refractive index with SiH_4/CH_4 precursor flow ratio. Daves et al. used a temperature of 400 °C, 200 W RF power, and 1000 mTorr pressure, and measured the refractive index at 630 nm. Hsu et al. used three different temperatures, 300 W RF power, and 400 mTorr pressure, and measured the refractive index at 500 nm. Data from [19,55]. Reproduced with permission. Data from [60,82] are not available for reproduction.

Increasing the hydrogen content decreases the refractive index for all Si/C compositions and temperatures studied [6,19]. This is associated with a decrease in film density and an increase in nanoporosity, which explain the changes in optical properties. Post-deposition annealing, which densifies the film and removes hydrogen, is found to moderately increase the refractive index, for example, an increase of 0.5 when annealed at 1000 °C [119].

No relation was observed between the refractive index and the RF power in [74], and there was only a small increase in the refractive index with increasing pressure (from 2.7 to 2.9 as the pressure increases from 500 mTorr to 2000 mTorr).

3.3.3. Optical Properties of a-SiC Films for Biomedical Applications

Refractive index values for biomedical a-SiC are often not considered a critical parameter and thus are not always reported in the literature. Values of 1.7–2.4 are reported by Diaz-Botia [60], Hsu et al. [19] obtain values ranging from 2.3 to 2.6, and Chen et al. [16] obtain values from 1.93 to 2.53, which span most of the range obtained in a-SiC. Chen et al. note that higher refractive indices (2.5 and above) are preferred for waveguide applications, but the precise value used for devices is not mentioned in most biomedical studies. While a-SiC has seen relatively little use in biophotonic, optogenetic, or imaging applications, its transparency and the ability to widely vary the refractive index could allow for such applications in the future in conjunction with crystalline SiC [120,121].

3.4. Stress in a-SiC Films

Controlling stress in a-SiC:H films is one of the most significant challenges for PECVD deposition, as it depends on the film composition and bonding structure, the substrate, and the chamber conditions. El Khakani et al. propose a model for the stress in which the film stress is given by the bonding states, thermal stress from the substrate, and ion bombardment [69].

$$\sigma_{total} = -AN_{C-H} - BN_{Si-H} + CN_{Si-C} + \sigma_{th} + \sigma_b \quad (1)$$

A, B, and C are experimentally determined positive constants; $N_{C\text{–}H}$, $N_{Si\text{–}H}$, and $N_{Si\text{–}C}$ are the bond densities of C–H, Si–H, and Si–C bonds, respectively. The σ_{th} represents thermal stress from non-equal thermal expansion coefficients for the a-SiC film and the substrate, and σ_b is the stress induced by ion bombardment during deposition. The—sign represents contributions to compressive stress.

Alternately, Daves et al. [55] propose that the stress occurs due to the coalescence of separately growing islands during deposition.

3.4.1. Variation of Stress with Film Composition

Several studies report the importance of the Si/C composition for the stress in a-SiC films (Figure 8). Studies in which the C content is increased while keeping all other parameters constant find an increase in the compressive stress [93,122], which matches the expected result from (1) given that H content tends to follow C content [95]. On the contrary, Register et al. [20] observed a consistent increase in tensile stress for increasing C content due to high contamination with oxygen during deposition that is not accounted for in (1).

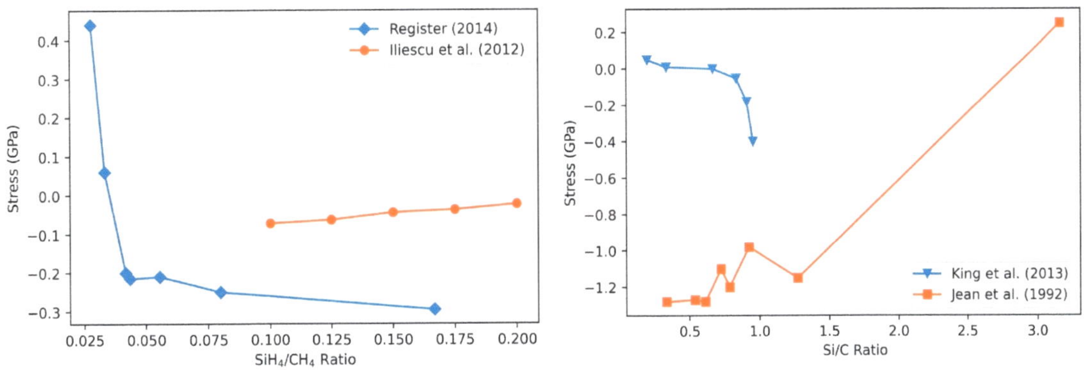

Figure 8. Stress as a function of stoichiometry or gas-flow composition. Both precursor gas ratio and measured composition are shown on separate axes. Compositional studies were performed using RBS (King et al.) and XPS/ERD (Jean et al.). Register used 250 °C, 50 W, and 900 mTorr, while Iliescu et al. used 300 °C, 150 W, and 1100 mTorr. King et al. varied most deposition parameters in order to obtain H-rich films. Jean et al. used a temperature of 250 °C and a pressure of 200 mTorr. Data from [6,20,93,122].

Hydrogen content also influences the film stress as both Si–H and C–H bonds are expected to contribute to compressive stress according to model (1). The results of Windischmann et al. [123] and Herrera-Celis et al. [54] are in agreement with this model. Herrera-Celis et al. systematically study the bond densities of C–H, C–H_2, and C–H_3 as a function of the stress for a series of films, observing the expected trend of increasing compressive stress for increasing C–H bonding; however, they do not provide the stress values as a function of absolute H or C content.

King et al. [6] obtained a conflicting result, observing an increased tensile stress for increased C content and (correspondingly) H content (certified by the high porosity of their C- and H-rich films as well as by RBS measurements), in contradiction to the above results and model (1). The use of high-energy bombardment was proposed as a possible explanation for this behavior. Hsu et al. [19] also found that the compressive stress increases with decreasing hydrogen content, as certified by the corresponding decrease in Si–H and C–H bonds observed in the infrared spectra. The authors do not propose any physical explanation for these results.

In addition, Pham [80] found that the compressive stress is maximal (−600 GPa) for low SiH_4 precursor flow (50 standard cubic centimeters per second, sccm), and becomes less compressive (−150 GPa) for an increased Si flow rate (250 sccm). If the C precursor is held constant, this result is in agreement with [93,122] and model (1) as described above.

Other studies do not report a specific effect from the flow rate as opposed to the ratio of precursors.

In conclusion, while simple descriptive model (1) accurately predicts film stress variation with the film composition, changing multiple parameters simultaneously leads to significantly more complicated effects.

3.4.2. Variation of Stress with RF Power

Decreasing the RF power makes the residual stress more tensile according to [12,54,55,118] or has only a small influence on the stress [84,85,89] (Figure 9). Recall that decreasing RF power results in increased H content (see Section 3.1.1) while also decreasing the bombardment energy. Referring again to (1), this suggests that the contribution from the decreased bombardment energy outweighs or equilibrates that of the increase in hydrogen content. It is noted that in the Greenhorns' unpublished data and the study of Daves et al. [55], low precursor flow rates are used, which may explain the significantly larger influence of the RF power.

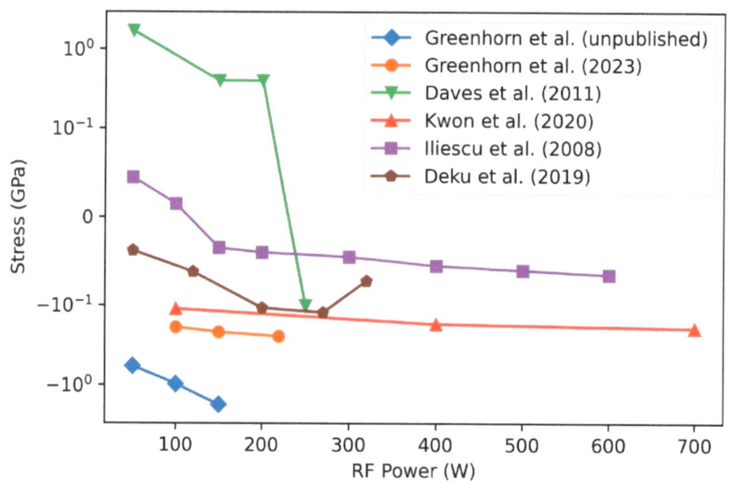

Figure 9. Residual stress as a function of plasma power. The studies by Greenhorn et al. use C_2H_4 as a C precursor, and the unpublished data use much lower precursor flow rates (23 sccm in total). Daves et al. used 1000 mTorr pressure and high temperature (400 °C) with a SiH_4/CH_4 ratio of 0.25, as well as a total flow rate of 200 sccm (90% Ar) while Kwon used a large industrial-type PECVD chamber with high flow rates, a pressure of 2500 mTorr, and a very high temperature of 550 °C and precursor gases of Si_2H_6 and CH_4. Iliescu used a temperature of 300 °C, power of 150 W, and pressure of 1100 mTorr, and precursor/carrier flow rates of 70 sccm/500 sccm. Deku used a power density of 0.27 W:cm^2 and a pressure of 1000 mTorr and fixed the total gas flow at 800 sccm. Data from [55,84,85,89,118]. Reproduced with permission.

3.4.3. Variation of Stress with Deposition Pressure

Daves et al. [55] found a weak compressive stress increase with chamber pressure for pressures below 1500 mTorr. Above 1500 mTorr, a dramatic increase in compressive stress is observed. This is in agreement with Equation (1) as high pressure results in a decreased bombardment energy, σ_b. Another study finds a nonlinear relation between stress and pressure, with a minimum compressive stress at 800 mTorr [54]. Pham, by contrast, finds that very high pressures (2500 mTorr as compared to 1500 mTorr) result in more tensile stress [80], contrary to model (1).

Note, however, that all studies observe compressive stress in all cases, indicating that the other deposition conditions favored compressive stress.

3.4.4. Variation of Stress with Deposition Temperature

According to Iliescu et al. [118], the deposition temperature has no effect on residual stress in the case of the low-frequency (380 kHz) generator while it has a dramatic effect in the case of the typical high-frequency (13.56 MHz) RF generator, with the stress becoming more tensile as the deposition temperature increases while the stoichiometry is not modified (Figure 10). Since an increase in temperature is expected to result in increased Si–C bonding (see Section 3.1.1), this result is consistent with the above relation, relation (1).

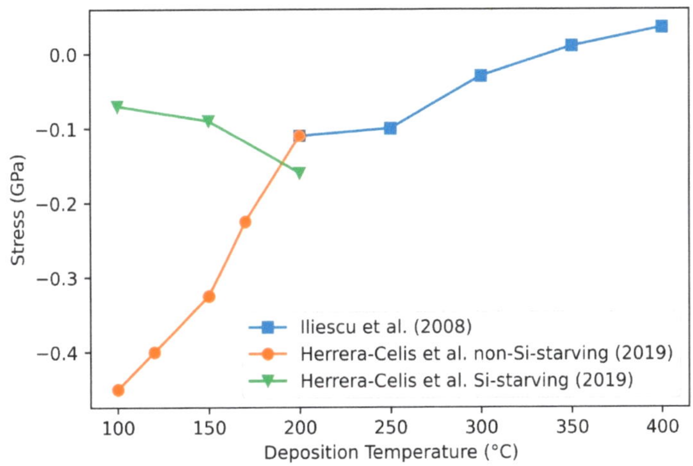

Figure 10. Iliescu et al. [118] show increasing residual stress with temperature, while Herrera-Celis [54] found two modes depending on the type of plasma: SiH_4-starving and non-SiH_4-starving. Iliescu et al. used a power of 150 W, pressure of 1100 mTorr, and reported films near stoichiometric composition. Herrera-Celis used power density of 16.7 mW/cm^2 for the non-Si-starving case, 8.3 mW/cm^2 for the Si-starving case, and pressure of 1100 Torr for both cases. The composition of the films was not reported in either study. Reproduced with permission.

A more complex effect is observed by Herrera-Celis et al. [54], who found that the relationship between stress and temperature has two modes, depending on the species of radicals in the plasma chamber. In the "silane-starving regime" [116], the stress becomes more tensile as the temperature increases, in agreement with other studies. However, increasing the silane fraction ($SiH_4/(SiH_4 + CH_4)$) to 0.26 inverses the trend, with the stress becoming increasingly compressive as the temperature increases in agreement with Hsu et al. [19] under similar conditions. This behavior agrees with (1) as, in the silane-starving regime, one expects a low probability of the reaction between silane radicals and methane, thus reducing the C–H bonds and increasing the formation of Si–C bonds.

3.4.5. Modifying Film Stress with Post-Deposition Annealing

Post-PECVD-deposition annealing modifies the stress of the a-SiC layers [55], where the stress typically becomes more tensile after annealing due to the reduction in hydrogen content and C–H bonds. Typically, films with as-deposited compressive stress of several hundred MPa can be made to have tensile stress (of up to, again, several hundred MPa) after annealing in air at 600 °C [55,69,80]. The physics behind this effect was studied in detail by Frischmuth et al. [119] using mass effusion measurements, and they found that the change in stress is initially caused by the removal of H_2, CH_4, and Ar starting at 350 °C–400 °C and continuing to 900 °C, along with a corresponding decrease in film thickness and an increase in density. At 1000 °C, a second rapid increase in tensile stress occurs, also attributed to H_2. Guivarc'h et al. [124] report similar results, with the evacuation of H beginning at 400 °C, and confirm the decrease in film thickness and density during annealing.

King et al. [6] also notes a small increase of approx. 80 MPa toward tensile stress after heating up to 400 °C in a vacuum, a change which slowly reverses during exposure to air. It is suggested that this smaller effect is due to the evacuation of moisture during heating, which is then reincorporated after the experiment, returning the stress to a normal value.

3.4.6. Evolution of Film Stress over Time

Several studies [6,84,89,125] observe that films deposited at low temperature or at low RF power exhibit increased compressive stress over time. Samples maintained in an N_2 atmosphere do not exhibit such a change. The change is attributed to the incorporation of O into the bulk of films by replacing Si–H bonds after deposition, which increase the stress by deforming the lattice structure. This conclusion is supported by XPS measurements showing an increase in oxygen in the material. The films deposited at low power have the lowest stress, but the stress increases significantly as a function of time, approaching that of films deposited at higher power. Another study notes that water vapor is particularly effective at oxidizing a-SiC [77].

Jousseaume et al. [125] reported decreasing the change in stress by using plasma treatments after deposition using He, O_2, and H_2. All treatments result in a more consistent and lower stress over time. Based on compositional measurements, different mechanisms are provided for each of the treatments. The He plasma increases the density of the film, the O_2 plasma forms a dense oxide at the surface, and the H_2 plasma passivates the dangling bonds.

The contribution from incorporated oxygen to the stress is not included in Formula (1). While it is reported to be associated with the more porous, H/C-rich films, it suggests that the actual dynamic stress measured in a-SiC films exposed to typical environments is indeed more complicated than what (1) initially suggests.

In conclusion, a-SiC films deposited at both low RF power and low temperature (so, more porous structure with high H content) are the most vulnerable to a stress increase, likely due to oxygen passing through the porous structure [99]. An additional plasma treatment reduces the surface reactivity and improves the moisture barrier properties of the films, providing further evidence that the evolution in stress over time is caused by the incorporation of moisture.

3.4.7. Stress in a-SiC Films for Biomedical Applications

Achieving minimal stress is a critical point for biomedical devices, with most studies achieving values below 200 MPa for their optimized films [12,18,20,39,60], compressive in all cases. High stress can lead to small cracks in the passivation [11] or, for thin implantable devices, an unwanted curvature that complicates insertion and placement [39]. Optimizing the stress has been one of the major focuses for biomedical a-SiC studies, and thus several different strategies for achieving low stress have been demonstrated. A careful balancing of the process parameters (deposition temperature, RF power, and pressure depending on the study) is necessary as there are multiple contributors to both tensile and compressive stress. Unfortunately, most studies do not account for changes in stress over time, which can vary significantly particularly in humid (biological) environments, which could significantly affect in vivo performance. The use of plasma post-treatment and/or thermal annealing has not been tested for biomedical applications, but has shown good results in minimizing changes in stress over time.

3.5. *Mechanical Properties of a-SiC Films*

a-SiC is noted for its desirable mechanical properties, in particular, a higher hardness/elastic modulus ratio than that of single-crystal SiC, suggesting its potential in wear-resistant applications [51], as well as extremely high tensile strength [126]. The maximum Young's modulus is consistently achieved with 1:1 stoichiometric a-SiC films. While hardness is less studied, it typically follows the similar trends to the Young's modulus.

3.5.1. Variation of Film Mechanical Properties with Stoichiometry

Guruvenket et al. [81] found that the Young's modulus and hardness decrease for increasing atomic Si/C ratios from 0.9 to 2, which correlate to a decrease in Si–C bonding (Figure 11). Janz et al. [79] also report increasing hardness with increasing C content for Si-rich films, in agreement with the other results (data not available for reproduction). On the contrary, Adithi et al. [82] determined that the Young's modulus ranges from 124 to 167 GPa, roughly aligned to an increasing Si fraction, with Si/C ratios ranging from 0.85 to 1.2 (data not available for reproduction).

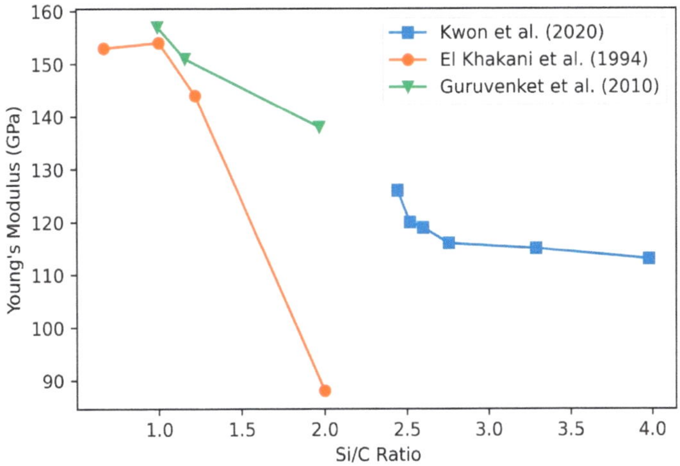

Figure 11. Comparison of results obtained by different studies comparing Young's modulus to composition. Compositional studies were performed using XPS (Kwon et al.), ERD (El Khakani et al.), and ERD-ToF (Guruvenket et al.). Kwon et al. used a large industrial-type PECVD chamber with high flow rates, a pressure of 2500 mTorr, and a very high temperature of 550 °C and precursor gases of Si_2H_6 and CH_4. Guruvenket et al. used temperature of 300 °C and pressure of 100 mTorr. El Khakani et al. used a temperature of 250 °C and pressure of 200 mTorr with a power density of 0.3 W/cm^2. Data from [36,81,85]. Reproduced with permission.

El Khakani et al. [36] studied hydrogenated (27%) and H-free SiC thin films fabricated with varying techniques including PECVD on undoped (100) Si substrates. They show that the nearly stoichiometric a-SiC:H films present higher hardness and Young's modulus values than the Si-rich a-Si_xC_{1-x}:H films. Hydrogen-free a-SiC films present both hardness and Young's modulus values (about 30 GPa and 240 GPa, respectively) higher by about 50% than those of hydrogenated a-SiC:H PECVD films. This study related the Young's modulus linearly to the level of Si–C bonding in the material, finding maximal Young's modulus with the maximal Si–C bonding at stoichiometric 1:1 Si:C. King et al.'s [6] results agreed with this conclusion.

Matsuda et al. [127] found that the hardness and Young's modulus are inversely proportional to the porosity, which is associated with high C and H content. For increasing porosity/H content beyond a composition-dependent threshold, the films are found to become flexible. The importance of chemical ordering for the increased rigidity of Si-rich films is also emphasized in [109].

A study of a-SiC film doped with N at up to 40% [100] found improvements in the scratch resistance, nanoindentation resistance, and fracture resistance. Several of the improved surface properties are attributed to the formation of a thin layer of SiO_x on the surface, which is softer than a-SiC and prevents crack formation on the surface. The use of an oxide layer to improve the mechanical properties of a-SiC has been further studied by Bae et al. [128] for a-SiC fibers with similar results.

3.5.2. Variation of Mechanical Properties with Deposition Temperature

Ivashchenko et al. [73] also found that the hardness and Young's modulus increase on average with increasing temperature, although not linearly. This is likely caused by the effusion of H molecules at higher temperatures, which also typically results in increased Si–C bonding, in alignment with the previous conclusions.

3.5.3. Variation of Mechanical Properties with Post-Deposition Annealing

The hardness and Young's modulus of films annealed at 600 °C or higher have been observed to increase compared to their as-deposited values [119]. The effect is most likely explained by the reformulation of Si–C bonds after the effusion of hydrogen from the film.

3.5.4. Mechanical Properties of a-SiC Films for Biomedical Applications

Most of the effort for optimizing the mechanical properties of a-SiC has focused on non-biological applications, requiring dedicated microresonator fabrication that is not pursued by biomedical studies. As such, hardness and Young's moduli are not reported in these studies. For biomedical implants, an important consideration is the tissue–implant interface and mechanical mismatch. Due to the high hardness and Young's modulus of a-SiC, it is impossible to directly match the properties of soft tissue with a-SiC. However, a SiC probes can be thinner and more compliant/flexible than the current Si-based implantable devices, leading to a reduced biotic response. The idea of employing stiff but thin (<10 μm) a-SiC for a minimal biotic response was the basis of flexible MEAs' development with excellent results from Deku et al. [39]. It is really the device stiffness, which includes cross-sectional area, rather than just the device modulus, that seems to matter the most for a reduced biotic response [129].

3.6. *Passivation Properties*

The following sections address additional aspects of a-SiC film performance specific to biomedical device applications.

3.6.1. Dissolution Rate in Chemical Solutions

The dissolution rate of a-SiC:H films in chemical solutions is film-thickness-dependent. Indeed, Diaz-Botia et al. [60] tested the dissolution rate of their films in PBS in the case of thin films (380 nm), and many rectangle defects appeared after two weeks while a 650-nm-thick film remained intact and defect-free after a 6-week soaking test. The etch rate of a 110-nm-thick film was 0.05–2 nm/h at 90 °C in PBS whilst a 450-nm-thick film had no measurable etching. Likewise, Cogan et al. [18] observed a dissolution rate of 0.1 nm/h at 90 °C in a PBS solution, and no appreciable dissolution at 37 °C after 40 weeks in PBS for 1-μm-thick films. So, thick a-SiC films > 1 μm have a higher dissolution resistance than thin ones.

The deposition temperature having a strong impact on the film organization also plays an important role in dissolution resistance. a-SiC layers deposited at lower than 200 °C suffered from large quantities of rectangle defects after a 1-week PBS soaking test [19]. More dissolution-resistant films were deposited at 200 °C in silane-starving conditions. These results suggest that a low Si–C bond density and porous film may provide little protection in a saline solution. Diaz-Botia [60] also found that films deposited at higher temperatures have a reduced long-term leakage current through pinholes under accelerated aging in PBS, as samples deposited at 200 °C show a dramatic increase in the leakage current (from ~0 to 1250 nA/cm^2) after 18 h whereas samples deposited at 350 °C do not show such an increase even after 600 h. The effect is most strongly visible for thin films (180 nm), whereas films above 500 nm do not show early breakdown. Contrary to the above studies, high chemical resistance for low-temperature depositions has been achieved [130], using a deposition temperature of 180 °C as compared to the more typical 300 °C–500 °C. This was achieved using He as the carrier gas and by increasing the silane precursor flow, avoiding the porosity associated with higher C contents.

Chemical resistance tests using other chemicals like HF and KOH have also been performed [131], demonstrating low etch rates. Avram et al. [132] tested in 30% KOH at 80 degrees and observed etching at a rate of 13 A/min, and no significant etching (below 10 A/h) in 49% HF. A clear trend between residual stress and chemical resistance was confirmed in an acid resistance test by using 40% HF [75] with no appreciable etching after the test for the lowest tensile stress (+41 MPa). It is noted that such chemical treatments may affect other film properties beyond just the removal of material; Iliescu et al. [13] found that KOH-treated a-SiC films had a significant reduction in biocompatibility when used for cell culturing.

In conclusion, the general trends in terms of chemical resistance are (i) that porous H-rich films with low Si–C bonding are more vulnerable to chemical etching while films with high Si–C bonding and low H content exhibit higher chemical resistance, and (ii) that thick films are more chemically resistant.

3.6.2. Conformal Coating and Uniformity

a-SiC films deposited by PECVD are quite uniform [60,118] with minimal defects [74] according to optical inspection and impedance tests in liquid, and are therefore acceptable for conformal coatings. Pham reports pinholes in Si-rich a-SiC but not in C-rich films [80]. In the case of Utah or metal-wire electrode arrays of needles with heights of the order of 2 mm, a ratio of 5 [18,19] to 10 [75] has been found between the thickness at the base and the tip of the probe, with a significantly thicker film at the wider base compared to the thin tip of the probes. When coated on implantable metallic stents, small cracks (<1 μm) were visible at junctions in the structure, which were noted as regions of high stress [11]. After torquing tests, an a-SiC coating on a Ti implant (with 20 nm SiO_2 as an interfacial adhesion layer) remained unchanged, suggesting good adhesion [15].

3.6.3. Surface Roughness

While the influence of the deposition conditions on a-SiC film roughness is unclear, the dependence on thickness is strongly supported by all available studies. Indeed, from AFM and SEM measurements, Deku et al. [39] observed a roughness of 4 nm RMS on 2-μm-thick a-SiC films, while Register et al. [20] found a surface roughness of 0.55 nm on 230-nm-thick a-SiC films. Herrera-Celis et al. [54] also report a roughness below 1 nm for all sub-micron films in their study. It seems that for sub-micron films, a roughness below 1 nm is observed while the latter increases to some nm for film thickness higher than 1 μm. Other studies using thicker films also find higher roughness, ranging from 2.4 to 6.1 nm [83,133]. In particular, the addition of NH_2 gas as a precursor is reported to significantly increase the roughness, from 4 nm to 12 nm [134]. A 200 nm coating of a-SiC on Ti implants with roughness of 9.8 nm was found to reduce the roughness to 8.3 nm [15].

Gelvez-Lizarazo et al. [12] studied the variation of surface roughness systematically for low RF powers and low deposition temperatures and were able to obtain significant variation in the roughness (although they do not note the film thickness used). They found that the root-mean-square (RMS) roughness increases from 1.69 nm to 5.12 nm as the RF power increases from 15 W to 30 W, and increases from 0.88 nm to 5.12 nm as the temperature increases from 100 °C to 200 °C. They relate the surface roughness to the size of the nuclei attached to the surface in the early stages of deposition. At low pressure, there is increased inelastic scattering, leading to an increased dissociation of the silane and methane precursors, and thus the deposited nuclei are smaller.

4. Perspectives

Interest in a-SiC is increasing with new techniques and applications frequently emerging. Newer deposition techniques, including ICP CVD, are showing promise with high reliability and uniformity [67], but lack the large body of literature to perform a reliable comparative study of film property optimization. While ICP CVD also allows the use of lower temperatures during depositions, it remains unclear whether similar trends to those of PECVD will occur.

Application areas for a-SiC are also expected to increase, with recent results in ultra-hard materials [126], as a platform for functionalization in biosensors [135], and as a platform for photoelectrochemical CO_2 conversion [136]. Furthermore, new characterization techniques for C-based materials and amorphous materials could open up new possibilities for optimization, such as thermal-optic effects studied using two-wave mixing [137], bandgap measurements using Kelvin force microscopy [138], structural determination using atomic resolution electron tomography [139], and material simulations using molecular dynamics [140,141]. As a-SiC becomes easier to be reliably produced with a wider variety of optimized properties, it will see increasing use in novel devices.

5. Conclusions

Passivation with high conformability, biocompatibility, long lifetimes, high resistance, and low stress is critical for developing next-generation biomedical devices. While there has been significant work to optimize the deposition parameters and control the properties of PECVD a-SiC films, work remains to reliably obtain films with simultaneously optimal electrical resistivity, stress, and chemical resistance. Many properties of a-SiC film deposition are well understood and strongly supported by the literature, while others require additional study. With the innate high resistivity and excellent chemical resistance, a-SiC is a promising material for neural interface passivation if these properties can be preserved alongside low stress and desirable mechanical properties.

The most common parameter for tuning the film properties is the Si and C composition, which is primarily determined by the precursor flow rates of the Si- and C-containing species. Increasing the C content in a-SiC:H is known to increase the film electrical resistivity, decrease the refractive index, and increase the optical bandgap (and vice versa for increasing Si). Moving away from a 1:1 stoichiometry is also expected to decrease the hardness and Young's modulus.

However, the Si and C composition alone does not fix all of the relevant properties. Another significant factor is the bonding and organization of the film, which is largely determined by H content. The hydrogen content can be increased by using a more C-rich precursor mixture, by decreasing the RF power, or by lowering the temperature. The resulting films have lower organization and are nanoporous, with low refractive indices and high resistivity due to the passivation of dangling bonds with H, and are mechanically more flexible, and tend to have low compressive stress. Their properties are also reported to be less stable in time, showing increased vulnerability to a chemical attack and increased oxygen intake when exposed to air, resulting in an increase in compressive stress. Alternately, using high RF power, high deposition temperature, or H_2 flow during deposition has been shown to decrease H content and correspondingly increase Si–C bonding. These films are reported to be more stable with excellent chemical resistance and more stable properties over time.

While low-stress films are frequently reported, the effect of this optimization on other film properties, such as resistivity or chemical resistance, is not often considered. Stress control is one of the most important issues in PECVD a-SiC, with a wide range of stresses achievable from tensile to compressive. The most well-understood method to control the film stress is through the bonding structure, where Si–H and C–H bonds contribute to compressive stress and Si–C bonds contribute to tensile stress. However, ion bombardment during deposition can also lead to an increase in tensile stress that outweighs the contribution from bonding and is relatively understudied. Additionally, thermal mismatch with the substrate can contribute another source of stress, which must be considered when designing a recipe. Compressive stress can be reduced or made tensile by post-deposition annealing, at the expense of a decrease in resistivity.

The typical strategy for obtaining reliable a-SiC films for biomedical devices is to target a 1:1 stoichiometry using "silane-starving flow", which results in dense, diamond-like films with high Si–C ordering. The addition of H_2 gas flow and the use of high deposition temperatures further reduce disorder from terminating H bonds and increase Si–C bonding. Choices of RF power and gas flow rates will depend on the size of the reactor being used in order to achieve silane-starving flow. Note, however, that very low precursor gas flow rates (below 25 sccm) have resulted in films with high stress even in typical research-scale PECVD reactors.

It is clear that PECVD of a-SiC films is complicated to optimize given the significant interactions between process parameters that can influence all of the properties of the films. Moving forward, it is critical for the a-SiC research community to document all aspects of their processes and report a broad variety of characterization results in order to gain a more holistic understanding of the film growth.

Author Contributions: Conceptualization, K.Z. and S.G.; investigation, K.Z. and S.G.; data curation, S.G.; writing—original draft preparation, K.Z. and S.G.; writing—review and editing, K.Z., S.G., E.B. and V.S.; visualization, S.G.; supervision, K.Z., E.B. and V.S.; project administration, E.B.; funding acquisition, K.Z., E.B. and V.S. All authors have read and agreed to the published version of the manuscript.

Funding: This research was funded by the Agence Nationale de la Recherche: Interface neuronale à longue durée de vie en Carbure de Silicium–SiCNeural.

Data Availability Statement: Restrictions apply to the availability of these data. Data were obtained from the respective copyright holders and are available in full from the original publications with the permission of the copyright holders.

Conflicts of Interest: The authors declare no conflicts of interest.

References

1. Wijesundara, M.B.J.; Azevedo, R. *Silicon Carbide Microsystems for Harsh Environments*; MEMS Reference Shelf; Springer: New York, NY, USA, 2011; Volume 22. [CrossRef]
2. Ivashchenko, V.I.; Porada, O.K.; Ivashchenko, L.A.; Rusakov, G.V.; Dub, S.M.; Popov, V.M. Hydrogenated Amorphous Silicon Carbide Films as Perspective Tribological Coatings and Semiconductor Layers. In *Hydrogen Materials Science and Chemistry of Carbon Nanomaterials*; Veziroglu, T.N., Zaginaichenko, Y.S., Schur, D.V., Baranowski, B., Shpak, A.P., Skorokhod, V.V., Eds.; NATO Science Series II: Mathematics, Physics and Chemistry; Springer: Dordrecht, The Netherlands, 2004; Volume 172, pp. 339–346. [CrossRef]
3. Porada, O.K.; Ivashchenko, V.I.; Ivashchenko, L.A.; Rusakov, G.V.; Dub, S.N.; Stegnij, A.I. a-SiC:H films as perspective wear-resistant coatings. *Surf. Coat. Technol.* **2004**, *180–181*, 122–126. [CrossRef]
4. Barbouche, M.; Benabderrahmane Zaghouani, R.; Ben Ammar, N.E.; Aglieri, V.; Nasser, H.; Turan, R.; Ezzaouia, H. Effect of amorphous SiC layer on electrical and optical properties of Al/a-SiC/c-Si Schottky diode for optoelectronic applications. *J. Mater. Sci. Mater. Electron.* **2021**, *32*, 20598–20611. [CrossRef]
5. Cocorullo, G.; Terzini, E.; della Corte, F.G.; de Rosa, R.; Rendina, I.; Rubino, A. a-Si:Hra-SiC:H waveguides and modulators for low-cost silicon-integrated optoelectronics. *J. Non-Cryst. Solids* **1998**, *227–230*, 1118–1122. [CrossRef]
6. King, S.W.; Bielefeld, J.; Xu, G.; Lanford, W.A.; Matsuda, Y.; Dauskardt, R.H.; Kim, N.; Hondongwa, D.; Olasov, L.; Daly, B.; et al. Influence of network bond percolation on the thermal, mechanical, electrical and optical properties of high and low-k a-SiC:H thin films. *J. Non-Cryst. Solids* **2013**, *379*, 67–79. [CrossRef]
7. Chen, J.; Calvin, J.J.; King, S.W.; Woodfield, B.F.; Navrotsky, A. Energetics of porous amorphous low-k SiOCH dielectric films. *J. Chem. Thermodyn.* **2019**, *139*, 105885. [CrossRef]
8. Knaack, G.L.; McHail, D.G.; Borda, G.; Koo, B.; Peixoto, N.; Cogan, S.F.; Dumas, T.C.; Pancrazio, J.J. In vivo Characterization of Amorphous Silicon Carbide As a Biomaterial for Chronic Neural Interfaces. *Front. Neurosci.* **2016**, *10*, 301. [CrossRef]
9. Yakimova, R.; Petoral, R.M.; Yazdi, G.R.; Vahlberg, C.; Lloyd Spetz, A.; Uvdal, K. Surface functionalization and biomedical applications based on SiC. *J. Phys. Appl. Phys.* **2007**, *40*, 6435–6442. [CrossRef]
10. Saddow, S. Silicon Carbide Technology for Advanced Human Healthcare Applications. *Micromachines* **2022**, *13*, 346. [CrossRef]
11. Harder, C.; Rzany, A.; Schaldach, M. Coating of Vascular Stents with Antithrombogenic Amorphous Sil. *Prog. Biomed. Res.* **1999**, *4*, 71–77.
12. Gelvez- Lizarazo, O.; Reyes-Betanzo, C. Characterization of Residual Stress in a-SiC:H Deposited by RF-PECVD for Manufacturing of Membranes for Cell Culture. *MRS Proc.* **2016**, *1812*, 117–122. [CrossRef]
13. Iliescu, C.; Chen, B.; Poenar, D.P.; Lee, Y.Y. PECVD amorphous silicon carbide membranes for cell culturing. *Sens. Actuators B Chem.* **2008**, *129*, 404–411. [CrossRef]

14. Amon, M.; Bolz, A.; Schaldach, M. Improvement of stenting therapy with a silicon carbide coated tantalum stent. *J. Mater. Sci. Mater. Med.* **1996**, *7*, 273–278. [CrossRef]
15. Fares, C.; Hsu, S.-M.; Xian, M.; Xia, X.; Ren, F.; Mecholsky, J.J.; Gonzaga, L.; Esquivel-Upshaw, J. Demonstration of a SiC Protective Coating for Titanium Implants. *Materials* **2020**, *13*, 3321. [CrossRef]
16. Chen, Z.; Fares, C.; Elhassani, R.; Ren, F.; Kim, M.; Hsu, S.-M.; Clark, A.E.; Esquivel-Upshaw, J.F. Demonstration of SiO_2/SiC-based protective coating for dental ceramic prostheses. *J. Am. Ceram. Soc.* **2019**, *102*, 6591–6599. [CrossRef] [PubMed]
17. Hsu, S.-M.; Ren, F.; Chen, Z.; Kim, M.; Fares, C.; Clark, A.E.; Neal, D.; Esquivel-Upshaw, J.F. Novel Coating to Minimize Corrosion of Glass-Ceramics for Dental Applications. *Materials* **2020**, *13*, 1215. [CrossRef] [PubMed]
18. Cogan, S.F.; Edell, D.J.; Guzelian, A.A.; Ping Liu, Y.; Edell, R. Plasma-enhanced chemical vapor deposited silicon carbide as an implantable dielectric coating. *J. Biomed. Mater. Res.* **2003**, *67A*, 856–867. [CrossRef] [PubMed]
19. Hsu, J.-M.; Tathireddy, P.; Rieth, L.; Normann, A.R.; Solzbacher, F. Characterization of a-SiCx:H thin films as an encapsulation material for integrated silicon based neural interface devices. *Thin Solid Film.* **2007**, *516*, 34–41. [CrossRef] [PubMed]
20. Register, J. SiC for Advanced Biological Applications. Ph.D. Dissertation, University of South Florida, Tampa, FL, USA, 2014. Available online: http://scholarcommons.usf.edu/etd/5113 (accessed on 10 January 2023).
21. Schmitt, G. Passivation and corrosion of microelectrode arrays. *Electrochim. Acta* **1999**, *44*, 3865–3883. [CrossRef]
22. Faßbender, F.; Schmitt, G.; Schöning, M.J.; Lüth, H.; Buß, G.; Schultze, J.-W. Optimization of passivation layers for corrosion protection of silicon-based microelectrode arrays. *Sens. Actuators B Chem.* **2000**, *68*, 128–133. [CrossRef]
23. Lotti, F.; Ranieri, F.; Vadalà, G.; Zollo, L.; Di Pino, G. Invasive Intraneural Interfaces: Foreign Body Reaction Issues. *Front. Neurosci.* **2017**, *11*, 497. [CrossRef]
24. Diaz-Botia, C.A.; Luna, L.E.; Neely, R.M.; Chamanzar, M.; Carraro, C.; Carmena, J.M.; Sabes, P.N.; Maboudian, R.; Maharbiz, M.M. A silicon carbide array for electrocorticography and peripheral nerve recording. *J. Neural Eng.* **2017**, *14*, 056006. [CrossRef] [PubMed]
25. Sridharan, A.; Muthuswamy, J. Quantitative Assessment of the Mechanical Properties of the Neural Interface. In *Handbook of Neuroengineering*; Thakor, N.V., Ed.; Springer Nature: Singapore, 2023; pp. 213–259. [CrossRef]
26. Szostak, K.M.; Grand, L.; Constandinou, T.G. Neural Interfaces for Intracortical Recording: Requirements, Fabrication Methods, and Characteristics. *Front. Neurosci.* **2017**, *11*, 665. [CrossRef] [PubMed]
27. Axpe, E.; Orive, G.; Franze, K.; Appel, E.A. Towards brain-tissue-like biomaterials. *Nat. Commun.* **2020**, *11*, 3423. [CrossRef]
28. Sharafkhani, N.; Kouzani, A.Z.; Adams, S.D.; Long, J.M.; Lissorgues, G.; Rousseau, L.; Orwa, J.O. Neural tissue-microelectrode interaction: Brain micromotion, electrical impedance, and flexible microelectrode insertion. *J. Neurosci. Methods* **2022**, *365*, 109388. [CrossRef]
29. Thielen, B.; Meng, E. A comparison of insertion methods for surgical placement of penetrating neural interfaces. *J. Neural Eng.* **2021**, *18*, 041003. [CrossRef] [PubMed]
30. Geramifard, N.; Dousti, B.; Nguyen, C.; Abbott, J.; Cogan, S.F.; Varner, V.D. Insertion mechanics of amorphous SiC ultra-micro scale neural probes. *J. Neural Eng.* **2022**, *19*, 026033. [CrossRef]
31. Nguyen, C.K.; Abbott, J.R.; Negi, S.; Cogan, S.F. Evaluation of Amorphous Silicon Carbide on Utah Electrode Arrays by Thermal Accelerated Aging. In Proceedings of the 2021 43rd Annual International Conference of the IEEE Engineering in Medicine & Biology Society (EMBC), Virtual, 1–5 November 2021; IEEE: Piscataway, NJ, USA, 2021; pp. 6623–6626. [CrossRef]
32. Fu, X.-A.; Jezeski, R.; Zorman, C.A.; Mehregany, M. Use of deposition pressure to control residual stress in polycrystalline SiC films. *Appl. Phys. Lett.* **2004**, *84*, 341–343. [CrossRef]
33. Locke, C.W.; Severino, A.; La Via, F.; Reyes, M.; Register, J.; Saddow, S.E. Chapter 2—SiC Films and Coatings: Amorphous, Polycrystalline, and Single Crystal Forms. In *Silicon Carbide Biotechnology*; Elsevier: Amsterdam, The Netherlands, 2012; pp. 17–61. Available online: https://www.sciencedirect.com/science/article/pii/B9780123859068000027 (accessed on 10 January 2023).
34. Knaack, G.L.; Charkhkar, H.; Cogan, S.F.; Pancrazio, J.J. Chapter 8—Amorphous Silicon Carbide for Neural Interface Applications. In *Silicon Carbide Biotechnology*, 2nd ed.; Elsevier: Amsterdam, The Netherlands, 2016; pp. 249–260.
35. Prasad, A.; Xue, Q.-S.; Sankar, V.; Nishida, T.; Shaw, G.; Streit, W.J.; Sanchez, J.C. Comprehensive characterization and failure modes of tungsten microwire arrays in chronic neural implants. *J. Neural Eng.* **2012**, *9*, 056015. [CrossRef]
36. El Khakani, M.A.; Chaker, M.; Jean, A.; Boily, S.; Kieffer, J.C.; O'Hern, M.E.; Ravet, M.F.; Rousseaux, F. Hardness and Young's modulus of amorphous *a*-SiC thin films determined by nanoindentation and bulge tests. *J. Mater. Res.* **1994**, *9*, 96–103. [CrossRef]
37. Sommakia, S.; Lee, H.C.; Gaire, J.; Otto, K.J. Materials approaches for modulating neural tissue responses to implanted microelectrodes through mechanical and biochemical means. *Curr. Opin. Solid State Mater. Sci.* **2014**, *18*, 319–328. [CrossRef]
38. Feng, C.; Frewin, C.L.; Tanjil, M.R.-E.; Everly, R.; Bieber, J.; Kumar, A.; Wang, M.C.; Saddow, S.E. A Flexible a-SiC-Based Neural Interface Utilizing Pyrolyzed-Photoresist Film (C) Active Sites. *Micromachines* **2021**, *12*, 821. [CrossRef]
39. Deku, F.; Cohen, Y.; Joshi-Imre, A.; Kanneganti, A.; Gardner, T.J.; Cogan, S.F. Amorphous silicon carbide ultramicroelectrode arrays for neural stimulation and recording. *J. Neural Eng.* **2018**, *15*, 016007. [CrossRef]
40. Rzany, A.; Schaldach, M. Smart Material Silicon Carbide: Reduced Activation of Cells and Proteins on a-SiC:H-coated Stainless Steel. *Prog. Biomed. Res.* **2001**, *6*, 182–194.
41. Frewin, C.L. The Neuron-Silicon Carbide Interface: Biocompatibility Study and BMI Device Development. Ph.D. Dissertation, University of South Florida, Tampa, FL, USA, 2009. Available online: https://digitalcommons.usf.edu/etd/1973/ (accessed on 10 January 2023).

42. Knaack, G.L. In Vitro and In Vivo Biocompatibility Testing of Silicon Carbide for Neural Interfaces. Ph.D. Dissertation, George Mason University, Fairfax, VA, USA, 2011. Available online: https://mars.gmu.edu/server/api/core/bitstreams/489502f5-0c11-45b9-823b-b1520ab8036f/content (accessed on 10 January 2023).
43. Rizal, U.; Swain, B.S.; Rameshbabu, N.; Swain, B.P. Biocompatibility of Hydrogen-Diluted Amorphous Silicon Carbide Thin Films for Artificial Heart Valve Coating. *J. Mater. Eng. Perform.* **2018**, *27*, 2679–2686. [CrossRef]
44. Nezafati, M. Biomaterial Testing Methodology for Long-Term in vivo Applications: Silicon Carbide Corrosion Resistance, Biocompatibility and Hemocompatibility. Ph.D. Dissertation, University of South Florida, Tampa, FL, USA, 2014. Available online: https://digitalcommons.usf.edu/cgi/viewcontent.cgi?article=6479&context=etd (accessed on 10 January 2023).
45. Oliveros, A.; Guiseppi-Elie, A.; Saddow, S.E. Silicon carbide: A versatile material for biosensor applications. *Biomed. Microdevices* **2013**, *15*, 353–368. [CrossRef] [PubMed]
46. Beygi, M.; Bentley, J.T.; Frewin, C.L.; Kuliasha, C.A.; Takshi, A.; Bernardin, E.K.; La Via, F.; Saddow, S.E. Fabrication of a Monolithic Implantable Neural Interface from Cubic Silicon Carbide. *Micromachines* **2019**, *10*, 430. [CrossRef]
47. Frewin, C.; Beygi, M.; Bernardin, E.; Feng, C.Y.; La Via, F.; Dominguez-Viqueria, W.; Saddow, S.E. The Development of Monolithic Silicon Carbide Intracortical Neural Interfaces for Long-Term Human Implantation. *Mater. Sci. Forum* **2022**, *1062*, 195–203. [CrossRef]
48. Jeakle, E.N.; Abbott, J.R.; Usoro, J.O.; Wu, Y.; Haghighi, P.; Radhakrishna, R.; Sturgill, B.S.; Nakajima, S.; Thai, T.T.D.; Pancrazio, J.J.; et al. Chronic Stability of Local Field Potentials Using Amorphous Silicon Carbide Microelectrode Arrays Implanted in the Rat Motor Cortex. *Micromachines* **2023**, *14*, 680. [CrossRef] [PubMed]
49. Bullot, J.; Schmidt, M.P. Physics of Amorphous Silicon–Carbon Alloys. *Phys. Status Solidi B* **1987**, *143*, 345–418. [CrossRef]
50. Choi, W.K.; Lee, L.P.; Foo, S.L.; Gangadharan, S.; Chong, N.B.; Tan, L.S. Oxidation study of plasma-enhanced chemical vapor deposited and rf sputtered hydrogenated amorphous silicon carbide films. *J. Appl. Phys.* **2001**, *89*, 1942. [CrossRef]
51. Fraga, M.; Pessoa, R. Progresses in Synthesis and Application of SiC Films: From CVD to ALD and from MEMS to NEMS. *Micromachines* **2020**, *11*, 799. [CrossRef]
52. Mahmoodi, M.; Ghazanfari, L. Fundamentals of Biomedical Applications of Biomorphic SiC. In *Properties and Applications of Silicon Carbide*; Gerhardt, R., Ed.; InTech: Horwich, UK, 2011. [CrossRef]
53. Kaloyeros, A.E.; Arkles, B. Silicon Carbide Thin Film Technologies: Recent Advances in Processing, Properties, and Applications—Part I Thermal and Plasma CVD. *ECS J. Solid State Sci. Technol.* **2023**, *12*, 103001. [CrossRef]
54. Herrera-Celis, J.; Reyes-Betanzo, C.; Gelvez-Lizarazo, O.; Arriaga, L.G.; Itzmoyotl-Toxqui, A. Low residual stress in hydrogenated amorphous silicon-carbon films deposited by low-temperature PECVD. *J. Mater. Res. Technol.* **2019**, *8*, 5581–5590. [CrossRef]
55. Daves, W.; Krauss, A.; Behnel, N.; Häublein, V.; Bauer, A.; Frey, L. Amorphous silicon carbide thin films (a-SiC:H) deposited by plasma-enhanced chemical vapor deposition as protective coatings for harsh environment applications. *Thin Solid Film.* **2011**, *519*, 5892–5898. [CrossRef]
56. King, S.W.; Ross, L.; Li, H.; Xu, G.; Bielefeld, J.; Atkins, R.E.; Henneghan, P.D.; Davis, K.; Johnson, D.C.; Lanford, W.A. Influence of hydrogen content and network connectivity on the coefficient of thermal expansion and thermal stability for a-SiC:H thin films. *J. Non-Cryst. Solids* **2014**, *389*, 78–85. [CrossRef]
57. Honda, K.; Matsumoto, A.; Kondo, B.; Shimizu, Y. High-performance carbon-rich amorphous silicon–carbon alloy semiconductors with low optical gaps. *Phys. E Low-Dimens. Syst. Nanostruct.* **2023**, *148*, 115652. [CrossRef]
58. Donercark, E.; Sedani, S.H.; Kabaçelik, I.; Salimi, A.; Turan, R. Interface and material properties of wide band gap a-SiCx:H thin films for solar cell applications. *Renew. Energy* **2022**, *183*, 781–790. [CrossRef]
59. Amorim, M.; Savio, R.; Massi, M.; Santiago, H. Applications of SiC-Based Thin Films in Electronic and MEMS Devices. In *Physics and Technology of Silicon Carbide Devices*; Hijikata, Y., Ed.; InTech: Horwich, UK, 2012. [CrossRef]
60. Diaz-Botia, C.A. Silicon Carbide Technologies for Interfacing with the Nervous System. Ph.D. Dissertation, University of California, Berkeley, CA, USA, 2017. Available online: https://escholarship.org/uc/item/35w3d01b (accessed on 10 January 2023).
61. Matsunami, H.; Masahiro, H.; Tanaka, T. Structures and physical properties of sputtered amorphous SiC films. *J. Electron. Mater.* **1979**, *8*, 249–260. [CrossRef]
62. Choi, W.K.; Loo, F.L.; Ling, C.H.; Loh, F.C.; Tan, K.L. Structural and electrical studies of radio frequency sputtered hydrogenated amorphous silicon carbide films. *J. Appl. Phys.* **1995**, *78*, 7289–7294. [CrossRef]
63. Choi, W.K.; Loo, F.L.; Loh, F.C.; Tan, K.L. Effects of hydrogen and rf power on the structural and electrical properties of rf sputtered hydrogenated amorphous silicon carbide films. *J. Appl. Phys.* **1996**, *80*, 1611–1616. [CrossRef]
64. Ledermann, N.; Baborowski, J.; Muralt, P.; Xantopoulos, N.; Tellenbach, J.-M. Sputtered silicon carbide thin films as protective coating for MEMS applications. *Surf. Coat. Technol.* **2000**, *125*, 246–250. [CrossRef]
65. Favaro, G.; Amato, A.; Arciprete, F.; Bazzan, M.; Cesarini, E.; De Matteis, F.; Dao, T.H.; Granata, M.; Honrado-Benítez, C.; Gutiérrez-Luna, N.; et al. Measurement and simulation of mechanical and optical properties of sputtered amorphous SiC coatings. *arXiv* **2022**, arXiv:2202.04458. Available online: http://arxiv.org/abs/2202.04458 (accessed on 22 August 2023). [CrossRef]
66. Lopez-Rodriguez, B.; Van Der Kolk, R.; Aggarwal, S.; Sharma, N.; Li, Z.; Van Der Plaats, D.; Scholte, T.; Chang, J.; Pereira, S.F.; Groeblacher, S.; et al. High-quality amorphous Silicon Carbide for hybrid photonic integration at low temperature. *arXiv* **2023**, arXiv:2306.04491. Available online: http://arxiv.org/abs/2306.04491 (accessed on 3 February 2024). [CrossRef]

67. Frischmuth, T.; Schneider, M.; Maurer, D.; Grille, T.; Schmid, U. Inductively-coupled plasma-enhanced chemical vapour deposition of hydrogenated amorphous silicon carbide thin films for MEMS. *Sens. Actuators Phys.* **2016**, *247*, 647–655. [CrossRef]
68. Sterling, H.F.; Swann, R.C.G. Chemical vapour deposition promoted by r.f. discharge. *Solid-State Electron.* **1965**, *8*, 653–654. [CrossRef]
69. El Khakani, M.A.; Chaker, M.; Jean, A.; Boily, S.; Pépin, H.; Kieffer, J.C.; Gujrathi, S.C. Effect of rapid thermal annealing on both the stress and the bonding states of *a* -SiC:H films. *J. Appl. Phys.* **1993**, *74*, 2834–2840. [CrossRef]
70. Lee, W. X-ray photoelectron spectroscopy and Auger electron spectroscopy studies of glow discharge $Si_{1-x}C_x$:H films. *J. Appl. Phys.* **1980**, *51*, 3365–3372. [CrossRef]
71. Fujimoto, F.; Ootuka, A.; Komaki, K.; Iwata, Y.; Yamane, I.; Yamashita, H.; Hashimoto, Y.; Tawada, Y.; Nishimura, K.; Okamoto, H.; et al. Hydrogen Content in a-SiC:H Films Prepared by Plasma Decomposition of Silane and Methane or Ethylene. *Jpn. J. Appl. Phys.* **1984**, *23*, 810. [CrossRef]
72. Myong, S.Y.; Lee, H.K.; Yoon, E.; Lim, K.S. High quality microcrystalline silicon-carbide films prepared by photo-CVD method using ethylene gas as a carbon source. *MRS Online Proceeding Libr. OPL* **1999**, *557*, 603. [CrossRef]
73. Ivashchenko, V.I.; Dub, S.N.; Porada, O.K.; Ivashchenko, L.A.; Skrynskyy, P.L.; Stegniy, A.I. Mechanical properties of PECVD a-SiC:H thin films prepared from methyltrichlorosilane. *Surf. Coat. Technol.* **2006**, *200*, 6533–6537. [CrossRef]
74. Janz, S.; Reber, S.; Glunz, S.W. Amorphous SiC: Applications for Silicon Solar Cells. In Proceedings of the 21st European Photovoltaic Solar Energy Conference, Dresden, Germany, 4–8 September 2006; pp. 660–663.
75. Joshi-Imre, A.; Black, B.J.; Abbott, J.; Kanneganti, A.; Rihani, R.; Chakraborty, B.; Danda, V.R.; Maeng, J.; Sharma, R.; Rieth, L.; et al. Chronic recording and electrochemical performance of amorphous silicon carbide-coated Utah electrode arrays implanted in rat motor cortex. *J. Neural Eng.* **2019**, *16*, 046006. [CrossRef] [PubMed]
76. Desalvo, A.; Giorgis, F.; Pirri, C.F.; Tresso, E.; Rava, P.; Galloni, R.; Rizzoli, R.; Summonte, C. Optoelectronic properties, structure and composition of a-SiC:H films grown in undiluted and H2 diluted silane-methane plasma. *J. Appl. Phys.* **1997**, *81*, 7973–7980. [CrossRef]
77. Vasin, A.V.; Muto, S.; Ishikawa, Y.; Rusavsky, A.V.; Kimura, T.; Lysenko, V.S.; Nazarov, A.N. Comparative study of annealing and oxidation effects in a-SiC:H and a-SiC thin films deposited by radio-frequency magnetron sputtering. *Thin Solid Film.* **2011**, *519*, 2218–2224. [CrossRef]
78. Thakur, S.; Dionne, C.J.; Karna, P.; King, S.W.; Lanford, W.; Li, H.; Banerjee, S.; Merrill, D.; Hopkins, P.E.; Giri, A. Density and atomic coordination dictate vibrational characteristics and thermal conductivity of amorphous silicon carbide. *Phys. Rev. Mater.* **2022**, *6*, 094601. [CrossRef]
79. Janz, S. Amorphous Silicon Carbide for Photovoltaic Applications. Ph.D. Dissertation, Fraunhofer Institute for Solar Energy Systems, Freiburg im Breisgau, Germany, 2006.
80. Pham, H.T.M. PECVD Silicon Carbide—A Structural Material for Surface Micromachined Devices. Ph.D. Dissertation, TU Delft, Delft, The Netherlands, 2004.
81. Guruvenket, S.; Azzi, M.; Li, D.; Szpunar, J.A.; Martinu, L.; Klemberg-Sapieha, J.E. Structural, mechanical, tribological, and corrosion properties of a-SiC:H coatings prepared by PECVD. *Surf. Coat. Technol.* **2010**, *204*, 3358–3365. [CrossRef]
82. Adithi, U.; Deshpande, S.; Bhat, K.N. Material and mechanical characterization of PECVD deposited a-SiC:H with H2 dilution. In Proceedings of the 2014 IEEE 2nd International Conference on Emerging Electronics (ICEE), Bengaluru, India, 4–6 December 2014; IEEE: Bengaluru, India, 2014; pp. 1–4. [CrossRef]
83. Li, D.; Guruvenket, S.; Szpunar, J.A.; Klemberg-Sapieha, J.E. Effect of C/Si Ratio on the Electrochemical Behavior of a-SiC_x:H Coatings on SS301 Substrate Deposited by PECVD. *Int. J. Corros.* **2014**, *2014*, 565109. [CrossRef]
84. Greenhorn, S.; Zekentes, K.; Bano, E.; Stambouli, V.; Uvarov, A. Optimizing PECVD a-SiC:H Films for Neural Interface Passivation. *Key Eng. Mater.* **2023**, *947*, 83–88. [CrossRef]
85. Kwon, S.; Park, Y.; Ban, W.; Youn, C.; Lee, S.; Yang, J.; Jung, D.; Choi, T. Effect of plasma power on properties of hydrogenated amorphous silicon carbide hardmask films deposited by PECVD. *Vacuum* **2020**, *174*, 109187. [CrossRef]
86. Pereyra, I.; Carreño, M.N.P.; Tabacniks, M.H.; Prado, R.J.; Fantini, M.C.A. The influence of "starving plasma" regime on carbon content and bonds in a-Si1-xCx:H thin films. *J. Appl. Phys.* **1998**, *84*, 2371–2379. [CrossRef]
87. Pereyra, I.; Carrefio, M.N.P. Wide gap a-Sil_xC x:H thin films obtained under starving plasma deposition conditions. *J. Non-Cryst. Solids* **1996**, *201*, 110–118. [CrossRef]
88. Solomon, I.; Schmidt, M.P.; Tran-Quoc, H. Selective low-power plasma decomposition of silane-methane mixtures for the preparation of methylated amorphous silicon. *Phys. Rev. B* **1988**, *38*, 9895–9901. [CrossRef] [PubMed]
89. Deku, F.; Mohammed, S.; Joshi-Imre, A.; Maeng, J.; Danda, V.; Gardner, T.J.; Cogan, S.F. Effect of oxidation on intrinsic residual stress in amorphous silicon carbide films. *J. Biomed. Mater. Res. B Appl. Biomater.* **2019**, *107*, 1654–1661. [CrossRef]
90. Yang, C.-C. Hydrogenated Amorphous Silicon Carbide Prepared using DC Saddle Field PECVD for Photovoltaic Applications. Master's Dissertation, University of Toronto, Toronto, ON, Canada, 2011. Available online: https://tspace.library.utoronto.ca/bitstream/1807/31649/3/Yang_ChengChieh_201111_MASc_Thesis.pdf (accessed on 10 January 2023).
91. Huran, J.; Hotovy, I.; Kobzev, A.P.; Balalykin, N.I. RBS study of amorphous silicon carbide films deposited by PECVD. *Czechoslov. J. Phys.* **2004**, *54*, C1006–C1010. [CrossRef]
92. Demichelis, F.; Pirri, C.F.; Tresso, E.; Rigato, V.; DellaMea, G. Hydrogen diffusion and related defects in hydrogenated amorphous silicon carbide. *J. Non-Cryst. Solids* **1991**, *128*, 133–138. [CrossRef]

93. Jean, A.; Chaker, M.; Diawara, Y.; Leung, P.K.; Gat, E.; Mercier, P.P.; Pépin, H.; Gujrathi, S.; Ross, G.G.; Kieffer, J.C. Characterization of *a* -SiC:H films produced in a standard plasma enhanced chemical vapor deposition system for x-ray mask application. *J. Appl. Phys.* **1992**, *72*, 3110–3115. [CrossRef]
94. Windischmann, H.; Collins, R.W.; Cavese, J.M. Effect of Hydrogen on The Intrinsic Stress in Ion Beam Sputtered Amorphous Silicon Films. *J. Non-Cryst. Solids* **1986**, *85*, 261–272. [CrossRef]
95. Demichelis, F.; Pirri, C.F.; Tresso, E. Influence of doping on the structural and optoelectronic properties of amorphous and microcrystalline silicon carbide. *J. Appl. Phys.* **1992**, *72*, 1327–1333. [CrossRef]
96. Flannery, A.F.; Mourlas, N.J.; Storment, C.W.; Tsai, S.; Tan, S.H.; Heck, J.; Monk, D.; Kim, T.; Gogoi, B.; Kovacs, G.T.A. PECVD silicon carbide as a chemically resistant material for micromachined transducers. *Sens. Actuators Phys.* **1998**, *70*, 48–55. [CrossRef]
97. Demichelis, F.; Pirri, C.F.; Tresso, E.; Stapinski, T. Differences in physical properties of hydrogenated and fluorinated amorphous silicon carbide prepared by reactive sputtering. *J. Appl. Phys.* **1992**, *71*, 5641–5645. [CrossRef]
98. Li, X.; Wong, T.K.S.; Rusli; Yang, D. Structural and electronic properties of low dielectric constant carbon rich amorphous silicon carbide. *Diam. Relat. Mater.* **2003**, *12*, 963–967. [CrossRef]
99. Arce, R.; Koropecki, R.R.; Buitrago, R.H.; Alvarez, F.; Chambouleyron, I. Direct evidence of porosity in carbon-rich hydrogenated amorphous silicon carbide films. *J. Appl. Phys.* **1989**, *66*, 4544–4546. [CrossRef]
100. Tomastik, J.; Ctvrtlik, R.; Ingr, T.; Manak, J.; Opletalova, A. Effect of Nitrogen Doping and Temperature on Mechanical Durability of Silicon Carbide Thin Films. *Sci. Rep.* **2018**, *8*, 10428. [CrossRef] [PubMed]
101. Vetter, M.; Voz, C.; Ferre, R.; Martín, I.; Orpella, A.; Puigdollers, J.; Andreu, J.; Alcubilla, R. Electronic properties of intrinsic and doped amorphous silicon carbide films. *Thin Solid Film.* **2006**, *511–512*, 290–294. [CrossRef]
102. Loulou, M.; Gharbi, R.; Fathallah, M.A.; Ambrosone, G.; Coscia, U.; Abbate, G.; Marino, A.; Ferrero, S.; Tresso, E. Structural, optical and electrical properties of helium diluted a-Si1−xCx:H films deposited by PECVD. *J. Non-Cryst. Solids* **2006**, *352*, 1388–1391. [CrossRef]
103. Huran, J.; Hotový, I.; Kobzev, A.P.; Balalykin, N.I. Influence of nitrogen concentration on conductivity of N-doped a-SiC:H films deposited by PECVD. *Vacuum* **2002**, *67*, 567–570. [CrossRef]
104. Geramifard, N.; Joshi-Imre, A.; Cogan, S.F. Electrical Characterizations of Amorphous Silicon Carbide. *ECS Meet. Abstr.* **2019**, *MA2019-01*, 2160. [CrossRef]
105. Carreño, M.N.P.; Pereyra, I.; Fantini, M.C.A.; Takahashi, H.; Landers, R. Microvoids in diamond-like amorphous silicon carbide. *J. Appl. Phys.* **1994**, *75*, 538–542. [CrossRef]
106. Leitl, B.; Schmidt, G.; Pobegen, G.; Knoll, P.; Krenn, H. Conduction mechanisms in hydrogenated amorphous silicon carbide. *J. Non-Cryst. Solids* **2020**, *528*, 119750. [CrossRef]
107. Sel, K.; Akaoğlu, B.; Atilgan, İ.; Katircioğlu, B. Effects of tail states on the conduction mechanisms in silicon carbide thin films with high carbon content. *Solid-State Electron.* **2011**, *57*, 1–8. [CrossRef]
108. Le Comber, P.G. Electrical conduction in amorphous semiconductors. *Sci. Prog.* **1979**, *66*, 105–118.
109. Ozdemir, O.; Atilgan, I.; Akaoglu, B.; Sel, K.; Katircioglu, B. Frequency dependence of conductivity in intrinsic amorphous silicon carbide film, assessed through admittance measurement of metal insulator semiconductor structure. *Thin Solid Film.* **2006**, *497*, 149–156. [CrossRef]
110. Du, P.; Song, C.; Weng, W.; Han, G.; Shen, G. Effects of carbon additions on crystallinity and resistivity in Si–C–H thin films deposited by CVDs. *J. Phys. Chem. Solids* **2003**, *64*, 777–784. [CrossRef]
111. Dutta, R.; Banerjee, P.K.; Mitra, S.S. Effect of hydrogenation on the electrical conductivity of amorphous silicon carbide. *Solid State Commun.* **1982**, *42*, 219–222. [CrossRef]
112. Huran, J.; Boháček, P.; Sasinková, V.; Kleinová, A.; Mikolášek, M.; Kobzev, A.P. Amorphous silicon carbide thin films doped with P or B for the photoelectrochemical water splitting devices. *Curr. Appl. Phys.* **2022**, *34*, 101–106. [CrossRef]
113. Shan, D.; Sun, D.; Tang, M.; Yang, R.; Kang, G.; Tao, T.; Cao, Y. Structures, Electronic Properties and Carrier Transport Mechanisms of Si Nano-Crystalline Embedded in the Amorphous SiC Films with Various Si/C Ratios. *Nanomaterials* **2021**, *11*, 2678. [CrossRef]
114. Urbach, F. The long-wavelength edge of photographic sensitivity and of the electronic absorption of solids. *Phys. Rev.* **1953**, *92*, 1324. [CrossRef]
115. Demichelis, F.; Crovini, G.; Giorgis, F.; Pirri, C.F.; Tresso, E. Effect of Power Density and Molecular Dwell Time on Compositional and Optoelectronic Properties of a-SiC:H. *Solid State Commun.* **1996**, *98*, 617–622. [CrossRef]
116. Mastelaro, V.; Flank, A.M.; Fantini, M.C.A.; Bittencourt, D.R.S.; Carreño, M.N.P.; Pereyra, I. On the structural properties of *a* -$Si_{1-x}C_x$:H thin films. *J. Appl. Phys.* **1996**, *79*, 1324–1329. [CrossRef]
117. Honda, K.; Yoshinaga, K.; Nagata, Y. Amorphous Carbon-Based Semiconductor Capable of Controlling Its Optical Gap and Conductivity by Incorporating Silicon and Nitrogen Atoms. *ECS J. Solid State Sci. Technol.* **2016**, *5*, P590. [CrossRef]
118. Iliescu, C.; Chen, B.; Wei, J.; Pang, A.J. Characterisation of silicon carbide films deposited by plasma-enhanced chemical vapour deposition. *Thin Solid Film.* **2008**, *516*, 5189–5193. [CrossRef]
119. Frischmuth, T.; Schneider, M.; Maurer, D.; Grille, T.; Schmid, U. High temperature annealing effects on the chemical and mechanical properties of inductively-coupled plasma-enhanced chemical vapor deposited a-SiC:H thin films. *Thin Solid Film.* **2016**, *611*, 6–11. [CrossRef]

120. Nguyen, T.-K.; Yadav, S.; Truong, T.-A.; Han, M.; Barton, M.; Leitch, M.; Guzman, P.; Dinh, T.; Ashok, A.; Vu, H.; et al. Integrated, Transparent Silicon Carbide Electronics and Sensors for Radio Frequency Biomedical Therapy. *ACS Nano* **2022**, *16*, 10890–10903. [CrossRef]
121. Register, J.; Muller, A.; King, J.; Weeber, E.; Frewin, C.L.; Saddow, S.E. Silicon Carbide Waveguides for Optogenetic Neural Stimulation. *MRS Online Proc. Libr.* **2012**, *1433*, 19–24. [CrossRef]
122. Iliescu, C.; Poenar, D.P. PECVD Amorphous Silicon Carbide (α-SiC) Layers for MEMS Applications. In *Physics and Technology of Silicon Carbide Devices*; Hijikata, Y., Ed.; InTech: Horwich, UK, 2012. [CrossRef]
123. Windischmann, H. Intrinsic Stress in Sputter-Deposited Thin Films. *Crit. Rev. Solid State Mater. Sci.* **1992**, *17*, 547–596. [CrossRef]
124. Guivarc'h, A.; Richard, J.; LeContellec, M.; Ligeon, E.; Fontenille, J. Hydrogen content of amorphous silicon carbide prepared by reactive sputtering: Effects on films properties. *J. Appl. Phys.* **1980**, *51*, 2167. [CrossRef]
125. Jousseaume, V.; Rochat, N.; Favennec, L.; Renault, O.; Passemard, G. Mechanical stress in PECVD a-SiC:H: Aging and plasma treatments effects. *Mater. Sci. Semicond. Process.* **2004**, *7*, 301–305. [CrossRef]
126. Xu, M.; Shin, D.; Sberna, P.M.; van der Kolk, R.; Cupertino, A.; Bessa, M.A.; Norte, R.A. High-Strength Amorphous Silicon Carbide for Nanomechanics. *Adv. Mater.* **2023**, *36*, 2306513. [CrossRef] [PubMed]
127. Matsuda, Y.; Kim, N.; King, S.W.; Bielefeld, J.; Stebbins, J.F.; Dauskardt, R.H. Tunable Plasticity in Amorphous Silicon Carbide Films. *ACS Appl. Mater. Interfaces* **2013**, *5*, 7950–7955. [CrossRef] [PubMed]
128. Bae, S.-G.; Kim, S.; Lee, J.-A.; Jeong, Y.-G.; Shin, D.-G. Improving the mechanical properties of amorphous silicon carbide fibers by forming a protective silicon dioxide layer. *Ceram. Int.* **2022**, *48*, 30745–30753. [CrossRef]
129. Shoffstall, A.J.; Srinivasan, S.; Willis, M.; Stiller, A.M.; Ecker, M.; Voit, W.E.; Pancrazio, J.J.; Capadona, J.R. A Mosquito Inspired Strategy to Implant Microprobes into the Brain. *Sci. Rep.* **2018**, *8*, 122. [CrossRef] [PubMed]
130. Jiang, L.; Chen, X.; Wang, X.; Xu, L.; Stubhan, F.; Merkel, K.-H. a-SiCx:H ®lms deposited by plasma-enhanced chemical vapor deposition at low temperature used for moisture and corrosion resistant applications. *Thin Solid Film.* **1999**, *352*, 97–101. [CrossRef]
131. Flannery, A.F.; Mourlas, N.J.; Storment, C.W.; Tsai, S.; Tan, S.H.; Kovacs, G.T.A. PECVD silicon carbide for micromachined transducers. In Proceedings of the International Solid State Sensors and Actuators Conference (Transducers'97), Chicago, IL, USA, 6–10 June 1997; IEEE: Chicago, IL, USA, 1997; Volume 1, pp. 217–220. [CrossRef]
132. Avram, M.; Avram, A.; Bragaru, A.; Chen, B.; Poenar, D.P.; Iliescu, C. Low stress PECVD amorphous silicon carbide for MEMS applications. In Proceedings of the CAS 2010 Proceedings (International Semiconductor Conference), Sinaia, Romania, 11–13 October 2010; IEEE: Sinaia, Romania, 2010; pp. 239–242. [CrossRef]
133. Peri, B.; Borah, B.; Dash, R.K. Effect of RF power and gas flow ratio on the growth and morphology of the PECVD SiC thin film s for MEMS applications. *Bull. Mater. Sci.* **2015**, *38*, 1105–1112. [CrossRef]
134. Awad, Y.O. Characterization of Amorphous Silicon Carbide and Silicon Carbonitride Thin Films Synthesized by Polymer-Source Chemical Vapor Deposition: Mechanical Structural and Metal-Interface Properties. Ph.D. Dissertation, University of Sherbrooke, Sherbrooke, CA, USA, 2006. Available online: https://savoirs.usherbrooke.ca/handle/11143/1821 (accessed on 12 August 2023).
135. Retana-González, R.A.; Reyes-Betanzo, C.; Gómez-Montaño, F.J.; Orduña-Díaz, A. Development of a hydrogenated amorphous silicon carbide-based biosensor for E. coli detection. *MRS Adv.* **2023**, *8*, 1438–1444. [CrossRef]
136. Boussaa, S.A.; Benfadel, K.; Khodja, A.T.; Ayachi, M.; Boulil, R.; Bekhedda, K.; Talbi, L.; Boukezzata, A.; Ouadah, Y.; Allam, D.; et al. Elaboration and Characterization of Amorphous Silicon Carbide Thin Films (a-SiC) by Sputerring Magnetron Technique for Photoelectrochemical CO_2 Conversion. *Silicon* **2023**, *15*, 1145–1157. [CrossRef]
137. Hernández-Acosta, M.A.; Martines-Arano, H.; Soto-Ruvalcaba, L.; Martínez-González, C.L.; Martínez-Gutiérrez, H.; Torres-Torres, C. Fractional thermal transport and twisted light induced by an optical two-wave mixing in single-wall carbon nanotubes. *Int. J. Therm. Sci.* **2020**, *147*, 106136. [CrossRef]
138. Fabbri, L.; Bordoni, C.; Barquinha, P.; Crocco, J.; Fraboni, B.; Cramer, T. Accurate determination of band tail properties in amorphous semiconductor thin film with Kelvin probe force microscopy. *APL Mater.* **2023**, *11*, 061123. [CrossRef]
139. Xie, Z.; Zhang, Y.; Huang, S.; Li, Z.; Cheng, Q.; Zhou, J. Towards quantitative determination of atomic structures of amorphous materials in three dimensions. *Natl. Sci. Open* **2023**, *2*, 20220048. [CrossRef]
140. Ivashchenko, V.I.; Turchi, P.E.A.; Shevchenko, R.V.; Gorb, L.; Leszczynski, J.; Kozak, A.O. An effect of nitrogen incorporation on the structure and properties of amorphous SiC: First-principles molecular dynamics simulations. *Thin Solid Film.* **2022**, *756*, 139349. [CrossRef]
141. Yan, Z.; Liu, R.; Liu, B.; Shao, Y.; Liu, M. Molecular Dynamics Simulation Studies of Properties, Preparation, and Performance of Silicon Carbide Materials: A Review. *Energies* **2023**, *16*, 1176. [CrossRef]

materials

MDPI

Article

Single-Ion Counting with an Ultra-Thin-Membrane Silicon Carbide Sensor

Enrico Sangregorio [1,2], Lucia Calcagno [1,*], Elisabetta Medina [2,3,4], Andreo Crnjac [5,*], Milko Jakšic [5], Anna Vignati [3,4], Francesco Romano [6], Giuliana Milluzzo [6], Marzio De Napoli [6] and Massimo Camarda [2,7]

1 Department of Physics and Astronomy "Ettore Majorana", University of Catania (Italy), Via Santa Sofia 64, 95123 Catania, Italy; enrico.sangregorio@phd.unict.it
2 STLab srl, Via Anapo 53, 95126 Catania, Italy; elisabetta.medina@unito.it (E.M.); massimo.camarda@stlab.eu (M.C.)
3 Physics Department, Università degli Studi di Torino, Via Pietro Giuria 1, 10125 Turin, Italy; anna.vignati@unito.it
4 INFN—National Institute for Nuclear Physics, Turin Division, Via Pietro Giuria 1, 10125 Turin, Italy
5 Division of Experimental Physics, Ruđer Bošković Institute, 10000 Zagreb, Croatia; jaksic@irb.hr
6 INFN—National Institute for Nuclear Physics, Catania Division, Via S. Sofia 64, 95123 Catania, Italy; francesco.romano@ct.infn.it (F.R.); giuliana.milluzzo@ct.infn.it (G.M.); marzio.denapoli@ct.infn.it (M.D.N.)
7 SenSiC GmbH, DeliveryLAB, 5234 Villigen, Switzerland
* Correspondence: lucia.calcagno@ct.infn.it (L.C.); andreo.crnjac@irb.hr (A.C.)

Abstract: In recent times, ion implantation has received increasing interest for novel applications related to deterministic material doping on the nanoscale, primarily for the fabrication of solid-state quantum devices. For such applications, precise information concerning the number of implanted ions and their final position within the implanted sample is crucial. In this work, we present an innovative method for the detection of single ions of MeV energy by using a sub-micrometer ultra-thin silicon carbide sensor operated as an in-beam counter of transmitted ions. The SiC sensor signals, when compared to a Passivated Implanted Planar Silicon detector signal, exhibited a 96.5% ion-detection confidence, demonstrating that the membrane sensors can be utilized for high-fidelity ion counting. Furthermore, we assessed the angular straggling of transmitted ions due to the interaction with the SiC sensor, employing the scanning knife-edge method of a focused ion microbeam. The lateral dimension of the ion beam with and without the membrane sensor was compared to the SRIM calculations. The results were used to discuss the potential of such experimental geometry in deterministic ion-implantation schemes as well as other applications.

Keywords: silicon carbide; membrane sensor; deterministic ion implantation; counting efficiency; spatial resolution

Citation: Sangregorio, E.; Calcagno, L.; Medina, E.; Crnjac, A.; Jakšic, M.; Vignati, A.; Romano, F.; Milluzzo, G.; De Napoli, M.; Camarda, M. Single-Ion Counting with an Ultra-Thin-Membrane Silicon Carbide Sensor. *Materials* **2023**, *16*, 7692. https://doi.org/10.3390/ma16247692

Academic Editor: Alexander A. Lebedev

Received: 6 November 2023
Revised: 12 December 2023
Accepted: 14 December 2023
Published: 18 December 2023

1. Introduction

Ion implantation has been widely applied in the semiconductor industry, as introducing dopants is an easy and fast way to engineer the electrical and optical properties of semiconductors. Over the last decades, deterministic single-ion implantation has attracted wide interest in the semiconductor field because of its application in solid-state quantum technology across various material systems, including silicon and diamond. Some non-exhaustive examples of Si-based single-dopant devices include donors coupled to quantum dots [1] for charge [2], electron [3,4], and nuclear spin [5,6] qubits (quantum bits). Alternatively, single-color centers in diamond substrates, including nitrogen-vacancy (NV) centers [7] and coupling silicon-vacancy (SiV) centers [7,8], are studied as quantum electrodynamics (QED) devices based on diamond technology. Motivated by the proposed quantum applications, the demand for deterministically placing single dopants into nanostructured devices has prompted the development of various techniques related to silicon and diamond material systems [9,10]. Single-ion implantation is achieved through accurate control over the ion's final position and the number of implanted ions. Single-atom lithographic techniques based on scanning probes have successfully achieved the positioning of single dopants with nanometer-scale precision [11–13]. However, this technique is currently limited to a small number of species and is relatively slow. In contrast, direct ion implantation offers less precision in terms of dopant atom positioning but offers more flexibility in the choice of ion species, potentially allowing for faster and more scalable processes [9,14]. The primary challenges in direct ion-implantation methods lie in accurately counting individual ions as they approach the substrate and precisely predicting their final position within the sample. Several techniques have been developed to monitor the number of ions reaching the sample during single-ion-implantation processes. One of the most commonly used methods involves the detection of secondary electrons emitted upon the impact of ions on the sample [15,16]. This approach requires the presence of a secondary electron detector and can be applied to nearly all samples (provided that the ion energy is sufficiently high to yield a detectable secondary electron emission). Alternatively, other frequently used techniques exploit integrated structures within the implanted sample to generate detectable signals during the process. Integrated structures such as PiN diodes, which utilize electron–hole pairs generated by ion–matter interactions, can produce a detectable signal upon free-charge capture [17,18]. Additionally, integrated field-effect transistors (FETs) are employed, where ion implantation modulates the drain current [18,19]. However, these methods are exclusively applicable to samples featuring PiN or FET structures, thereby limiting their utility in various areas.

In the present work, we present an innovative single-ion detection method using an independent silicon carbide sensor to be placed ahead of the sample to be implanted. The sensor is a sub-micrometer SiC membrane realized through a state-of-the-art, doping-selective electrochemical etching process [20,21]. The described sensor geometry was employed here for detecting ion beams in the MeV energy range. In this setup, the ions lose only a portion of their energy (ΔE) in the device and are transmitted further without significant influence on the impact trajectory and total energy (E) of the ions. The electron/hole pairs generated through the ion–sensor interaction are amplified and collected, resulting in a distinct signal corresponding to each ion passing through the SiC membrane. The results were compared with measurements obtained using a reference silicon detector. Moreover, a study of the alteration in the ion beam after crossing the membrane was conducted, and the lateral ion straggling, a crucial parameter for single-ion-implantation applications, was calculated and compared with simulations.

2. Materials and Methods

The device tested is an advanced silicon carbide ultra-thin radiation sensor engineered as a free-standing membrane with a parallel-plate electrode configuration. This device is a semiconductor Schottky barrier diode consisting of an ultra-thin n^- silicon carbide active layer characterized by a low doping concentration of 10^{14} cm^{-3} on top of an inert n^+ highly doped silicon carbide substrate approximately 370 μm thick with a doping concentration of 10^{18} cm^{-3}. The fabrication of the free-standing membrane at the core of the sensor was accomplished through a state-of-the-art, doping-selective electrochemical etching (ECE) technique, which allowed for precision material removal down to sub-micrometer thickness levels. In more detail, the electrochemical etching of silicon carbide in hydrogen fluoride (HF)-based solutions consists of two steps: the first step is the oxidation of SiC driven by holes, and the second step is the dissolution of the formed SiO_{2-x} in HF [22]. In the case of the highly doped 370 μm substrate, holes were generated by tunneling effects. In contrast, for the low-doped n-type SiC epitaxial layer, tunneling was negligible. Therefore, the thin epitaxial layer acted as a stopping layer for the etching process, hence resulting in the formation of the free-standing membrane [20]. The total sensor area (5×5 mm^2) was divided into four independent pads, and the ECE process was carried out within a 2 mm diameter circular region in the central area of the device. In this study, only one of the four pads was connected to the data acquisition system and analyzed.

The metal contact, needed for both the ECE process and for the subsequent sensor operation, was established by depositing a 30 nm aluminum layer on the front surface of the device to create a Schottky contact. In contrast, the back contact was applied after the ECE process and involved a 100 nm aluminum layer. The metal depositions were conducted using an electron-beam (E-beam) evaporation system. A schematic structure of the SiC membrane sensor is presented in the inset of Figure 1b.

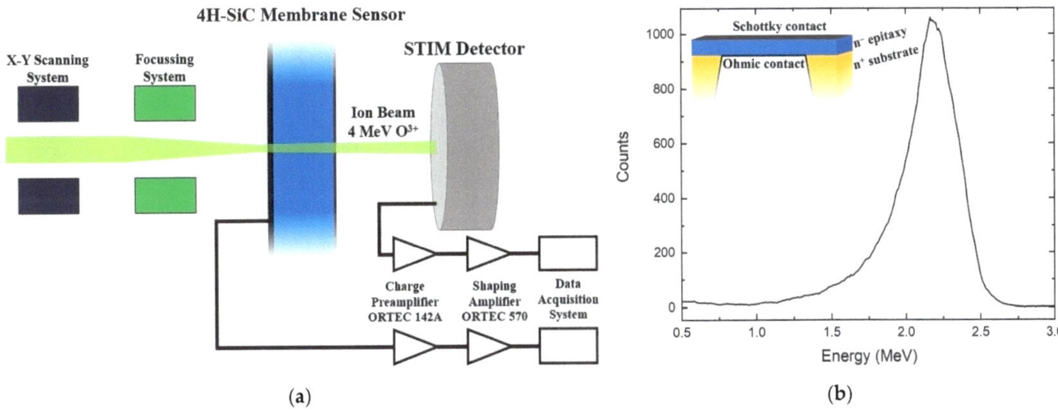

Figure 1. (**a**) Schematic of the experimental setup for the calculation of the sensor thickness and counting measurements; (**b**) energy spectrum obtained by the STIM detector used for the determination of the SiC sensor thickness. In the insert, a schematic cross-section of the SiC sensor is shown.

The main characterization of the sensor was performed by exposing the device to accelerated ions of different masses and energies in the MeV range using the ion beam facility of the Ruđer Bošković Institute [23]. Techniques based on the interaction of MeV ion beams with materials offer a powerful analytical framework for semiconductor detector characterization [24,25]. The device was mounted in a vacuum irradiation chamber attached to the 6 MV Tandem Van de Graaff electrostatic accelerator. The accelerator is equipped with a sputtering ion source used for the production of a wide range of ion species, from light ions such as H or Li to very heavy ions (up to Au). Ions are accelerated and transmitted through a range of ion beam optics elements downstream to the experimental end station, where the samples are positioned. In our setup, an ion microprobe end station was employed, allowing the focusing of ion beams to a micrometer-sized spot and enabling the scanning of the beam across the sample surface. These experimental conditions were utilized to acquire spatially resolved information about the ion–sample interaction as determined by the beam spot size. This setup was crucial for testing the device for single-ion detection, as the ion beam current could be reduced to ~ Hz rates and positioned in different spatial regions of interest of our device.

In our experimental scenarios, the ion beam was transmitted through the SiC membrane portion of the sensor, leaving only a portion of the energy inside, and was stopped 6 cm downstream on the in-beam-positioned Si detector, which was a Passivated Implanted Planar Silicon (PIPS) detector with very thin top dead layer. The PIPS detector will be referred to as a Scanning Transmission Ion Microscopy (STIM) detector (Canberra Semiconductor, Olen, Belgium), as it was used to detect transmitted ions. Both devices were connected to the same low-noise signal-processing chain, which was based on a charge-sensitive preamplifier (ORTEC 142A) and a shaping amplifier (ORTEC 570), both provided by ORTEC, Oak Ridge, TN, USA. This geometry enabled independent detection of ions by the device under test (the SiC membrane) and a well-characterized solid-state Si detector (the STIM detector). When an ion interacts with a semiconductor detector, it produces electron–hole pairs that are collected by the electric field applied through the electrodes. This technique is often referred to as the ion beam induced charge (IBIC) technique [24]. Signals collected from the sensor electrode can be used to quantify different parameters, such as the deposited energy, transient collection behavior, timing properties, and so on. When combined with a scanning microbeam setup, the IBIC technique can be seen as a 3D-like microscopic technique for the investigation of charge transport properties in semiconductor detectors.

A 4 MeV O^{3+} ion beam was employed to precisely determine the thickness of the silicon carbide membrane. A schematic representation of the experimental setup used during this investigation is presented in Figure 1a. The incident ions that passed through the SiC sensor deposited a portion of their energy (ΔE, approximately 45%) within the free-standing membrane sensor. Subsequently, these ions were collected by the STIM detector, and the acquired signal was represented as a count-versus-energy plot (Figure 1b). After subtracting the energy deposited in the STIM detector, which is represented by the peak position in Figure 1b, from the initial ion beam energy, the energy deposited in the SiC membrane sensor (ΔE) was determined. Furthermore, the full-width half maximum of the peak in Figure 1b was used to calculate the uncertainty in the membrane thickness. The Stopping and Range of Ions in Matter (SRIM) Monte Carlo simulation tool [26] was used to estimate the energy loss $\Delta E/\Delta x$ (eV/nm) of the beam in the sensor, enabling the calculation of the membrane sensor thickness T_{SiC}. Using this method, a total sensor thickness of $T_{SiC} = 727.3 \pm 57.6$ nm was calculated. Considering the thickness of the aluminum electrodes, the membrane active layer thickness resulted in about 597 nm. The relatively high 8% error associated with this measurement can be primarily attributed to the surface roughness of the membrane resulting from the doping-selective ECE process used for the formation of the SiC membrane [20,21]. The ion-counting fidelity of the SiC membrane was determined using the same oxygen beam and experimental setup while employing the IBIC technique.

The electron–hole pairs generated as a result of energy deposition during ion–membrane interactions were collected, and 2D-IBIC maps were generated using the homemade software SPECTOR v2.0 [27]. During the acquisition of IBIC signals, a reverse bias of -5 V was applied to the SiC Schottky diode. After this first interaction, the ions had enough energy to reach the STIM detector, thereby allowing for a simultaneous generation of a second IBIC map corresponding to transmitted ions. A comparison between the two acquired maps was performed to determine the number of recorded events in the two devices while assuming a 100% collection efficiency in the STIM detector. This comparison enabled the evaluation of the single-ion detection efficiency of the SiC membrane sensor.

The decrease in ion energy was not the sole effect of the interaction between the ion beam and the SiC membrane sensor. As the ions collide with the atomic electrons of the solid sensor, the trajectory angle of the ions in the material can be altered [25]. This phenomenon, commonly denoted as "ion lateral straggling", increases the uncertainty in the final position of ions within the implanted sample. In our experiment, we quantified ion lateral straggling resulting from the interaction between ions and the SiC sensor by using a finely machined metal grid with a defined pitch dimension. The grid was positioned between the membrane and the STIM detector, allowing the scanning transmitted ion beam to form a projection image of the grid. Using this experimental setup, a 10 MeV C^{4+} ion beam was scanned across the grid to acquire 2D-IBIC maps both with and without the presence of the membrane. This enabled the determination of the beam spot dimension in the two cases using the knife-edge analysis technique based on the grid projection.

3. Results

3.1. Counting Measurement

Figure 2a shows an IBIC map obtained with a 4 MeV O^{3+} beam on the SiC membrane. Since IBIC signals are proportional to the amount of energy deposited by ions, three different regions are visible on the map. Region A corresponds to an area where the 370 µm bulk beneath the membrane remained intact after the etching process. In this region, the elevated IBIC signal can be explained by the contribution of two types of free charges to the overall signal: charges generated in the membrane that were driven to the sensor electrodes by the electric field (drift current) and charges produced in a shallow region of the bulk below that reached the electrodes through diffusion processes (diffusion current). In region B, the membrane signal was acquired. In this region, only the charges generated in the epitaxial membrane were collected on the sensor electrodes, and a lower IBIC signal was generated. Finally, region C represents a non-bonded sensor pad, therefore this region is dominated by background noise. Figure 2c shows the IBIC map acquired by the STIM detector simultaneously with the events mapped in Figure 2a. Here, region A exhibits no STIM signal due to the thick silicon carbide substrate, in which the ions are fully absorbed and not transmitted to the PIPS detector. Region B and part of region C represent the membrane section of the SiC sensor, which shows a STIM signal, as the ion of the beam had enough energy to cross the membrane and reach the silicon reference detector.

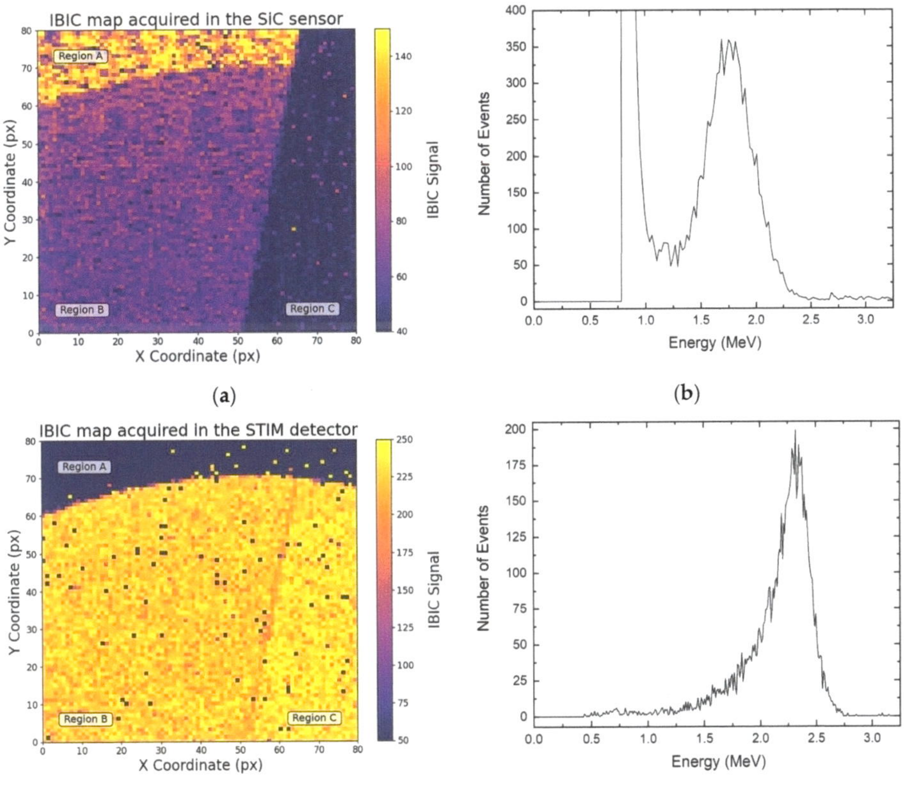

Figure 2. (**a**) IBIC map acquired by the SiC sensor and the corresponding energy spectrum (**b**); (**c**) IBIC map acquired by the STIM detector and its corresponding energy spectrum (**d**). Both maps were acquired simultaneously using a 4 MeV O^{3+} ion beam.

To assess the SiC sensor's capability to detect individual ions, a comparative analysis was performed between the counts recorded independently in the two detection systems. Assuming a 100% detection efficiency for the STIM detector, the relative SiC detection efficiency was determined accordingly. Identical regions in both maps were selected, and the number of event histograms (corresponding only to the ions recorded within those regions of interest) were extracted both for the SiC sensor (Figure 2b) and the STIM detector (Figure 2d).

The count plot in Figure 2b shows a high-count, low-energy (0.8 MeV) peak attributed to electronic noise. In the process of quantifying the number of ions detected by the membrane sensor, these noise events were removed from the total count, resulting in the exclusive contribution of the 1.7 MeV signal due to the overall membrane sensor. The extended low-energy tail observed in the histogram of transmitted ion energies (Figure 2d) can be attributed to the aforementioned energy-straggling effects, leading to the generation of an asymmetric 2.3 MeV peak. The difference in the peak position in Figure 2b,d derived from the different energy that the 4 MeV ions deposited in the two sensors. This procedure was repeated for various areas within region B, and the comparison of events recorded simultaneously by the two detection systems resulted in a $96.5 \pm 0.9\%$ ion-detection efficiency for the SiC sensor. The approximately 3% difference in the recorded events within the membrane sensor can be related to the complexities involved in subtracting electronic noise from the signal, which was primarily due to the proximity of the two peaks and the asymmetry of the membrane signal. While further improvements can be made by reducing background noise, this result underscores the excellent charge-collection efficiency of the tested ultra-thin membrane, highlighting its strong potential for high-fidelity ion-counting applications.

3.2. *Lateral Straggling*

In single-ion-implantation applications, careful control is applied over the implantation process to guarantee the precise localization of each ion within the target material. However, the occurrence of ion straggling resulting from the interactions between the ion beam and the independent ion sensor has the potential to compromise the determinism of the implantation procedure. This phenomenon introduces uncertainty regarding the final position of the implanted ions. Furthermore, the lateral straggling discussed earlier should be considered in conjunction with other non-improvable phenomena such as lateral straggling arising from ion–sample interaction and diffusion processes occurring during post-implantation annealing, which is essential for the activation of implanted species.

The ion lateral straggling was calculated by measuring the spatial profile of the ion beam before and after interaction with the silicon carbide free-standing membrane sensor. To achieve this, an electroformed metal grid with a well-known pitch dimension of 25.4 μm was mounted inside the experimental chamber between the SiC membrane and the downstream STIM detector (Figure 3a). The function of the grid was to act as an obstacle for the ions as they were transmitted toward the downstream detector, resulting in the generation of IBIC maps with the projected image of the grid formed by ions passing through the open regions of the grid. This allowed us to determine the ion beam spot size at the plane of the grid.

With this experimental setup, a 10 MeV C^{4+} ion microbeam was focused on the PIPS detector, and STIM data were acquired under two different configurations. In the first configuration, the carbon ion beam traversed the grid and reached the PIPS detector without any interaction with the silicon carbide free-standing membrane. In this case, the grid edges of the resulting STIM signal exhibited a high-definition level due to the unaltered ion beam's convergence (Figure 3b). In the second scenario, the SiC membrane was mounted before the metal grid, causing the ion beam to interact with the SiC membrane before encountering the grid. As previously mentioned, this interaction between the SiC sensor and the carbon ions introduced straggling effects. Consequently, the divergence of the ion beam led to a lower edge definition of the grid in the final STIM map (Figure 3c).

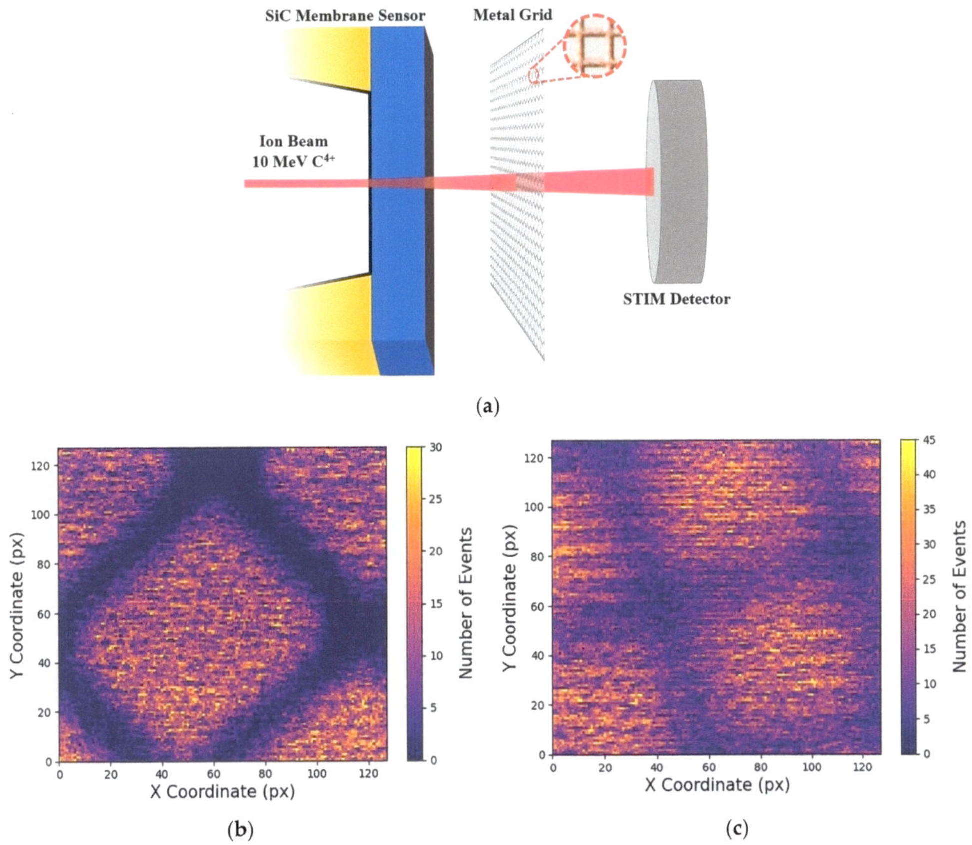

Figure 3. (**a**) Schematic representation of the experimental setup used for the lateral straggling determination; (**b**) STIM map acquired without the SiC membrane sensor; (**c**) STIM map acquired with ions passing through the SiC membrane sensor. Both maps were acquired using a 10 MeV C^{4+} ion beam.

The ion beam's spatial profile in the two configurations was calculated using the knife-edge calculation procedure. Data from the regions near the grid edge shadow in the STIM map have been reported as the number of events versus position (Figure 4). The sigmoidal profile of the recorded events corresponds to the lateral profile of the scanning beam spot and was analyzed using a Boltzmann sigmoid function (the dashed red curves in Figure 4). The beam's spatial profile was determined in the two scenarios (with and without the SiC free-standing membrane). This determination was carried out by calculating the number of pixels (Δx) falling in the range defined by the two points marked in red in Figure 4, which corresponded to the position of the 10% and the 90% values of the upper plateau of the function. Using this procedure, different measurements were carried out, and a main value of $\Delta x = 11.84 \pm 1.85$ px was determined for the without-membrane configuration. The beam spot size could be calculated by multiplying the Δx main value and the pixel-to-micron conversion factor for this configuration ($F_1 = 0.29\ \mu m \times px^{-1}$), resulting in a beam spot size of $r_{Beam} = 3.43 \pm 0.54\ \mu m$. The same procedure was employed for the STIM data acquired with the beam passing through the SiC membrane sensor. With this configuration, a main value of $\Delta x = 21.46 \pm 1.74$ px was obtained. The beam size after membrane interaction was calculated by considering the conversion factor $F_2 = 0.38\ \mu m \times px^{-1}$, leading to a beam dimension of $D_{Beam} = 8.15 \pm 0.66\ \mu m$. The high 8% error is ascribable to the roughness of the SiC membrane. The difference in the conversion factors F_1 and F_2 derive from the different magnifications at which the two STIM data were acquired. This difference is visible by comparing Figure 3b,c.

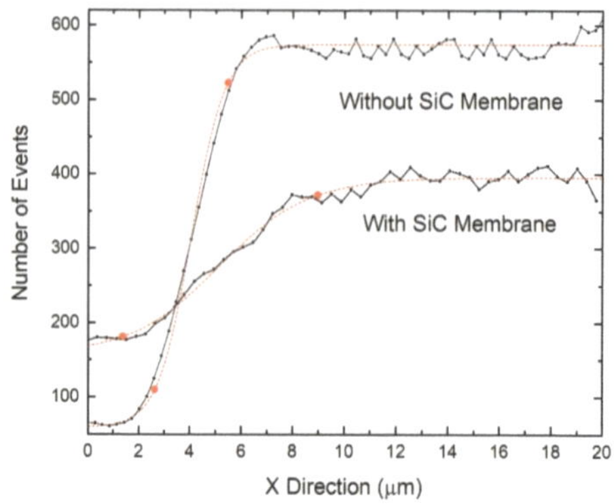

Figure 4. Count event profile close to the grid projection edges measured by the STIM detector without and with the SiC membrane sensor. The black dotted line represents the number of events measured as a function of position. The dashed red curves represent the Boltzmann sigmoidal function fitted to the experimental data. The points marked in red enclose the data interval between 10% and 90% of the upper plateau value.

To verify the reliability of the experimental results, a simulation was conducted using the SRIM simulation tool to determine ion beam straggling under ideal conditions. Considering the experimental setup as well as the beam and sample characteristics, SRIM simulations offer precise calculations for various parameters, including but not limited to the ion range, implanted ion spreading (both transversal and longitudinal), and lateral straggling of transmitted ions. The latter parameter was employed to calculate the beam's spatial profile, enabling the comparison with the experimental results. The lateral straggling parameter, as determined by SRIM simulations, characterized the beam's lateral profile measured immediately after ions passed through the membrane, resulting in $r_{SRIM} = 2.42 \pm 0.30$ nm. In order to obtain a parameter to be compared with the experimental result, the scheme of the setup, shown in Figure 5, had to be taken into account.

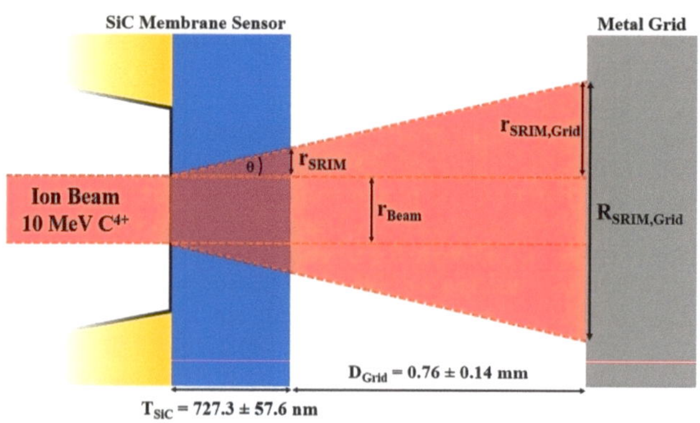

Figure 5. Geometrical representation of the ion beam path for theoretical lateral straggling calculation.

The metal grid, which shadowed the ion beam for the experimental measurement, was positioned at a distance of $D_{Grid} = 0.76 \pm 0.14$ mm from the SiC membrane. It was essential to account for the beam divergence in this region. The SRIM spreading parameter r_{SRIM} was used to calculate the straggling angle θ_{SRIM} through trigonometric considerations. Using this methodology, we calculated the straggling angle relative to the incident direction of the ions, which resulted in $\theta_{SRIM} = 0.19 \pm 0.02°$. By knowing θ_{SRIM} and the distance $D_T = T_{SiC} + D_{Grid}$, where $T_{SIC} = 727.3 \pm 57.6$ nm is the thickness of the SiC sensor calculated previously, the lateral straggling $r_{SRIM,Grid}$ on the grid could be calculated. With these considerations, a lateral beam straggling of $r_{SRIM,Grid} = 2.53 \pm 0.60$ μm was obtained. To determine the total ion beam lateral profile, the initial beam dimension ($r_{BEAM} = 3.43 \pm 0.54$ μm) had to be added to the spreading already calculated, resulting in a final beam dimension of $R_{SRIM} = 8.49 \pm 0.81$ μm.

The comparison between the theoretical ion beam's final dimension calculated using SRIM simulations (R_{SRIM}) and the experimental beam dimension (D_{Beam}) shows that the calculation method used in this work is quite reliable and allows the prediction of the ion lateral straggling in various scenarios. It also demonstrates the importance of mounting the sample to be implanted as close as possible to the membrane sensor to minimize the adverse effects of ion straggling. In scenarios involving heavy MeV ions, these distances should be less than 100 μm. Based on these results, further upgrades to the setup are planned. These upgrades aim to enable the mounting of the targets as close as 10 μm behind the membrane sensor, significantly enhancing the accuracy of the impact position of transmitted ions with the target.

4. Discussion

In deterministic ion implantation, the exact counting of ions as well as its spatial precision represent an ongoing challenge. The device presented here utilizes a membrane solid-state sensor and a low-noise, charge-sensitive electronic chain. This system collects a signal generated by ions transmitted through the sensitive membrane volume. In our experiments, the energy loss in the membrane active layer was about 1.5 MeV, and the number of generated pairs was on the order of 10^5 pairs per ion. However, typical ion-implantation energies are in the range of a few hundred keV, which requires low energy loss inside the membrane and the detection of a signal derived by $10^3 \div 10^4$ electron–hole pairs. These limits impose the use of a nanometric-thin membrane and a very low noise generated both by the detector and by the stage electronics.

Concerning the thickness of the sensor, a 100 nm SiC free-standing membrane can be produced using the ECE process described earlier. Thin membranes of this nature have already demonstrated favorable mechanical properties, including a high fracture strength and deformation [28]. The energy loss in ionization, i.e., the energy that ions lose in collisions with atomic electrons generating free charges in the solid material, depends on the mass and energy of the implanted ion. Using a SiC sensor with a 100 nm SiC epitaxy sandwiched between 20 nm and 70 nm aluminum electrodes, a typical dopant such as P at 250 keV loses a total energy of 190 keV in the sensor and generates approximately 9×10^3 electron–hole pairs in the sensor active layer. With low noise, this system will allow the implantation of a deterministic number of 60 keV P ions. The detector noise is mainly determined by the leakage current and by the detector capacitance. The leakage current in our detector was sufficiently low (a few pA) thanks to the wide bandgap of the silicon carbide semiconductor. The capacitance, on the other hand, considerably influenced the noise level due to the large sensor area (~1.6 mm^2) and the low sensor thickness (~730 nm). Although a 100-nanometer thickness may have a negative impact on the sensor capacitance, this effect can be substantially alleviated by reducing the surface area of the sensor, resulting in an enhanced signal-to-noise ratio. To minimize electronic noise, a custom charge-sensitive amplifier with effective capacitance matching of the input stage to the detector capacitance can be employed. Further experiments are in the planning stage to utilize an even thinner device along with upgrades to the signal-processing electronics.

Concerning the measured lateral ion straggling, the obtained value of 8.15 μm seems to be relatively high for single-ion-implantation applications. This high value introduces a significant level of uncertainty in determining the final position of the ions, potentially compromising the deterministic nature of the implantation process. However, it is important to note that the primary contribution to the final ion beam size is attributable to the divergence of the beam in the region between the SiC membrane sensor and the sample. In this work, this distance is represented by the distance between the SiC sensor and the metal grid (D_{Grid} = 0.76 ± 0.14 mm). The membrane straggling contribution was very low (r_{SRIM} = 2.42 ± 0.30 nm calculated with SRIM) compared to the final beam dimension. Therefore, by reducing the distance between the SiC sensor and the implanted sample, a higher determination of the ion's final position can be achieved. For example, by reducing the D_{Grid} distance to a few micrometers (5 ÷ 10 μm), the membrane straggling contribution on the sample will be 19.1 ÷ 35.7 nm (calculated with the same θ_{SRIM} angle). In this case, the initial beam dimension r_{Beam} will strongly affect the final beam lateral profile (in this experiment, r_{Beam} was 3.43 ± 0.54 μm).

Hence, through the reduction in the initial beam size to a few tens of nanometers and the detector thickness to 100 nm, it becomes possible to attain a final beam size of approximately 100 nanometers. This reduction significantly mitigates the uncertainty associated with the final position of the implanted atom.

5. Conclusions

In this study, we introduced an innovative approach to single-ion detection during the ion-implantation process by utilizing an advanced SiC ultra-thin solid-state sensor. The sub-micrometer membrane enables ion detection through the creation of electron–hole pairs due to interactions between ions and the solid sensor without causing substantial interference with the ion trajectory, thereby enabling simultaneous control of the implantation process. The ion-beam-induced charge signal was collected for ions transmitted through the membrane and fully stopped in the PIPS detector positioned behind, resulting in a 96.5 ± 0.9% calculated ion counting confidence for the membrane. While this result can be improved by minimizing the signal background, it demonstrates the potential of utilizing a thin sub-micrometer membrane as a high-fidelity in-beam ion detector. Such a detector could be useful for novel true maskless deterministic implantation schemes needed for the fabrication of novel solid-state technologies and devices. However, the presence of the membrane interfered with the ion beam trajectory, introducing ion straggling effects that resulted in an increase in the uncertainty of the final position of the ion in the target. To quantify the ion straggling caused by the SiC membrane, a knife-edge measurement technique was employed, and the results were compared to SRIM simulations, yielding a final beam size of 8.15 μm. The observed high straggling value can be reduced through improvements in experimental conditions. These optimizations include minimizing the separation distance between the SiC sensor and the implanted sample to a few micrometers while simultaneously reducing the initial dimension of the ion beam to the nanometer scale and the membrane thickness to 100 nm. Implementing these conditions makes it possible to achieve a final beam dimension on the order of 100 nanometers, effectively minimizing the uncertainty associated with the final position of the implanted atom.

Author Contributions: Conceptualization, A.C., M.J. and M.C.; methodology, E.S., L.C., E.M., A.C. and M.C.; validation, E.S., L.C., A.C. and M.C.; formal analysis, E.S., L.C. and A.C.; investigation, E.S., E.M., A.C., M.J. and M.C.; resources, A.C., M.J. and M.C.; data curation, E.S., L.C. and A.C.; writing—original draft preparation, E.S.; writing—review and editing, E.S., L.C., E.M., A.C., M.J., A.V., M.D.N., F.R., G.M. and M.C.; visualization, E.S. and M.D.N.; supervision, L.C., M.J. and M.C.; project administration, M.J. and M.C.; funding acquisition, M.J. and M.C. All authors have read and agreed to the published version of the manuscript.

Funding: This project received funding from the European Union's Horizon Europe Research and Innovation program under grant agreement No. 101057511 (EURO-LABS). This project was partially funded by the SAMOTHRACE project (Avviso 3277-"SAMOTHRACE" ECS00000022) and partially funded by the RADIATE project (European Union's Horizon 2020 Research and Innovation program under grant agreement No. 824096).

Data Availability Statement: Data are contained within the article.

Acknowledgments: The authors are grateful to Karla Ivanković and Matija Matijević for their technical support during the experimental measurements.

Conflicts of Interest: Author Massimo Camarda was employed by the company SenSiC GmbH. The remaining authors declare that the research was conducted in the absence of any commercial or financial relationships that could be construed as a potential conflict of interest.

References

1. Kane, B.E. A silicon-based nuclear spin quantum computer. *Nature* **1998**, *393*, 133–137. [CrossRef]
2. Hollenberg, L.C.L.; Dzurak, A.S.; Wellard, C.; Hamilton, A.R.; Reilly, D.J.; Milburn, G.J.; Clark, R.G. Charge-based quantum computing using single donors in semiconductors. *Phys. Rev. B* **2004**, *69*, 113301. [CrossRef]
3. Morello, A.; Pla, J.J.; Zwanenburg, F.A.; Chan, K.W.; Tan, K.Y.; Huebl, H.; Möttönen, M.; Nugroho, C.D.; Yang, C.; van Donkelaar, J.A.; et al. Single-shot readout of an electron spin in silicon. *Nature* **2010**, *467*, 687–691. [CrossRef] [PubMed]
4. Tyryshkin, A.M.; Lyon, S.A.; Astashkin, A.V.; Raitsimring, A.M. Electron spin relaxation times of phosphorus donors in silicon. *Phys. Rev. B* **2003**, *68*, 193207. [CrossRef]
5. Kalra, R.; Laucht, A.; Hill, C.D.; Morello, A. Robust Two-Qubit Gates for Donors in Silicon Controlled by Hyperfine Interactions. *Phys. Rev. X* **2014**, *4*, 021044. [CrossRef]
6. Pla, J.J.; Tan, K.Y.; Dehollain, J.P.; Lim, W.H.; Morton, J.J.L.; Zwanenburg, F.A.; Jamieson, D.N.; Dzurak, A.S.; Morello, A. High-fidelity readout and control of a nuclear spin qubit in silicon. *Nature* **2013**, *496*, 334–338. [CrossRef] [PubMed]
7. Shields, B.J.; Unterreithmeier, Q.P.; de Leon, N.P.; Park, H.; Lukin, M.D. Efficient Readout of a Single Spin State in Diamond via Spin-to-Charge Conversion. *Phys. Rev. Lett.* **2015**, *114*, 136402. [CrossRef] [PubMed]
8. Englund, D.; Shields, B.; Rivoire, K.; Hatami, F.; Vučković, J.; Park, H.; Lukin, M.D. Deterministic Coupling of a Single Nitrogen Vacancy Center to a Photonic Crystal Cavity. *Nano Lett.* **2010**, *10*, 3922–3926. [CrossRef]
9. van Donkelaar, J.; Yang, C.; Alves, A.D.C.; McCallum, J.C.; Hougaard, C.; Johnson, B.C.; Hudson, F.E.; Dzurak, A.S.; Morello, A.; Spemann, D.; et al. Single atom devices by ion implantation. *J. Phys. Condens. Matter* **2015**, *27*, 154204. [CrossRef]
10. Rueß, F.J.; Pok, W.; Reusch, T.C.G.; Butcher, M.J.; Goh, K.E.J.; Oberbeck, L.; Scappucci, G.; Hamilton, A.R.; Simmons, M.Y. Realization of Atomically Controlled Dopant Devices in Silicon. *Small* **2007**, *3*, 563–567. [CrossRef]
11. Fuechsle, M.; Miwa, J.A.; Mahapatra, S.; Ryu, H.; Lee, S.; Warschkow, O.; Hollenberg, L.C.L.; Klimeck, G.; Simmons, M.Y. A single-atom transistor. *Nat. Nanotechnol.* **2012**, *7*, 242–246. [CrossRef] [PubMed]
12. Stock, T.J.Z.; Warschkow, O.; Constantinou, P.C.; Li, J.; Fearn, S.; Crane, E.; Hofmann, E.V.S.; Kölker, A.; McKenzie, D.R.; Schofield, S.R.; et al. Atomic-Scale Patterning of Arsenic in Silicon by Scanning Tunneling Microscopy. *ACS Nano* **2020**, *14*, 3316–3327. [CrossRef] [PubMed]
13. Koenraad, P.M.; Flatté, M.E. Single dopants in semiconductors. *Nat. Mater.* **2011**, *10*, 91–100. [CrossRef] [PubMed]
14. Yang, C.; Jamieson, D.N.; Pakes, C.; Prawer, S.; Dzurak, A.; Stanley, F.; Spizziri, P.; Macks, L.; Gauja, E.; Clark, R.G. Single Phosphorus Ion Implantation into Prefabricated Nanometre Cells of Silicon Devices for Quantum Bit Fabrication. *Jpn. J. Appl. Phys.* **2003**, *42*, 4124–4128. [CrossRef]
15. Shinada, T.; Okamoto, S.; Kobayashi, T.; Ohdomari, I. Enhancing semiconductor device performance using ordered dopant arrays. *Nature* **2005**, *437*, 1128–1131. [CrossRef] [PubMed]
16. Schenkel, T.; Persaud, A.; Park, S.J.; Nilsson, J.; Bokor, J.; Liddle, J.A.; Keller, R.; Schneider, D.H.; Cheng, D.W.; Humphries, D.E. Solid state quantum computer development in silicon with single ion implantation. *J. Appl. Phys.* **2003**, *94*, 7017–7024. [CrossRef]
17. Jamieson, D.N.; Yang, C.; Hopf, T.; Hearne, S.M.; Pakes, C.I.; Prawer, S.; Mitic, M.; Gauja, E.; Andresen, S.E.; Hudson, F.E.; et al. Controlled shallow single-ion implantation in silicon using an active substrate for sub-20-keV ions. *Appl. Phys. Lett.* **2005**, *86*, 202101. [CrossRef]
18. Johnson, B.C.; Tettamanzi, G.; Yang, C.; Alves, A.; Van Donkelaar, J.; Thompson, S.; Verduijn, A.; Mol, J.A.; Wacquez, R.; Vinet, M.; et al. Single Ion Implantation into Si-Based Devices. *ECS Trans.* **2010**, *33*, 179–189. [CrossRef]
19. Batra, A.; Weis, C.D.; Reijonen, J.; Persaud, A.; Schenkel, T.; Cabrini, S.; Lo, C.C.; Bokor, J. Detection of low energy single ion impacts in micron scale transistors at room temperature. *Appl. Phys. Lett.* **2007**, *91*, 193502. [CrossRef]
20. Nida, S.; Tsibizov, A.; Ziemann, T.; Woerle, J.; Moesch, A.; Schulze-Briese, C.; Pradervand, C.; Tudisco, S.; Sigg, H.; Bunk, O.; et al. Silicon carbide X-ray beam position monitors for synchrotron applications. *J. Synchrotron Radiat.* **2019**, *26*, 28–35. [CrossRef]
21. Mokhtarzadeh, M.; Carulla, M.; Kozak, R.; David, C. Optimization of etching processes for the fabrication of smooth silicon carbide membranes for applications in quantum technology. *Micro Nano Eng.* **2022**, *16*, 100155. [CrossRef]
22. Gautier, G.; Defforge, T.; Gommé, G.; Valente, D.; Alquier, D. Electrochemical Formation of Porous Silicon Carbide for Micro-Device Applications. *Mater. Sci. Forum* **2018**, *924*, 943–946. [CrossRef]
23. Accelerators—Ruđer Bošković Institute. Available online: https://www.irb.hr/eng/Divisions/Division-of-Experimental-Physics/Laboratory-for-ion-beam-interactions/Articles/Accelerators (accessed on 20 October 2023).
24. Breese, M.; Vittone, E.; Vizkelethy, G.; Sellin, P. A review of ion beam induced charge microscopy. *Nucl. Instrum. Methods Phys. Res. B* **2007**, *264*, 345–360. [CrossRef]

25. Breese, M.B.H.; Jamieson, D.N.; King, P.J.C. *Materials Analysis Using a Nuclear Microprobe*; John Wiley: Hoboken, NJ, USA, 1996; ISBN 0471106089.
26. James Ziegler—SRIM & TRIM. Available online: http://www.srim.org/ (accessed on 20 October 2023).
27. Cosic, D.; Bogovac, M.; Jakšić, M. Data acquisition and control system for an evolving nuclear microprobe. *Nucl. Instruments Methods Phys. Res. Sect. B Beam Interactions Mater. Atoms* **2019**, *451*, 122–126. [CrossRef]
28. Nguyen, T.-K.; Phan, H.-P.; Kamble, H.; Vadivelu, R.; Dinh, T.; Iacopi, A.; Walker, G.; Hold, L.; Nguyen, N.-T.; Dao, D.V. Superior Robust Ultrathin Single-Crystalline Silicon Carbide Membrane as a Versatile Platform for Biological Applications. *ACS Appl. Mater. Interfaces* **2017**, *9*, 41641–41647. [CrossRef]

materials

Article

Application of Grazing-Incidence X-ray Methods to Study Terrace-Stepped SiC Surface for Graphene Growth

Boris S. Roschin [1], Tatiana S. Argunova [2,*], Sergey P. Lebedev [2], Victor E. Asadchikov [1], Alexander A. Lebedev [2], Yuri O. Volkov [1] and Alexander D. Nuzhdin [1]

1 Federal Research Center "Crystallography and Photonics", Russian Academy of Sciences, Leninsky ave. 59, 119333 Moscow, Russia
2 Ioffe Institute, Russian Academy of Sciences, Polytekhnicheskaya st. 26, 194021 St. Petersburg, Russia
* Correspondence: argunova@mail.ioffe.ru; Tel.: +7-921-387-1057

Abstract: The synthesis of graphene by the graphitization of SiC surface has been driven by a need to develop a way to produce graphene in large quantities. With the increased use of thermal treatments of commercial SiC substrates, a comprehension of the surface restructuring due to the formation of a terrace-stepped nanorelief is becoming a pressing challenge. The aim of this paper is to evaluate the utility of X-ray reflectometry and grazing-incidence off-specular scattering for a non-destructive estimate of depth-graded and lateral inhomogeneities on SiC wafers annealed in a vacuum at a temperature of 1400–1500 °C. It is shown that the grazing-incidence X-ray method is a powerful tool for the assessment of statistical parameters, such as effective roughness height, average terrace period and dispersion. Moreover, these methods are advantageous to local probe techniques because a broad range of spatial frequencies allows for faster inspection of the whole surface area. We have found that power spectral density functions and in-depth density profiles manifest themselves differently between the probing directions along and across a terrace edge. Finally, the X-ray scattering data demonstrate quantitative agreement with the results of atomic force microscopy.

Keywords: graphene; epitaxial; SiC; terrace-stepped relief; X-ray scattering; atomic force microscopy

Citation: Roschin, B.S.; Argunova, T.S.; Lebedev, S.P.; Asadchikov, V.E.; Lebedev, A.A.; Volkov, Y.O.; Nuzhdin, A.D. Application of Grazing-Incidence X-ray Methods to Study Terrace-Stepped SiC Surface for Graphene Growth. *Materials* **2022**, *15*, 7669. https://doi.org/10.3390/ma15217669

Academic Editors: Fabrizio Roccaforte and Antonio Di Bartolomeo

Received: 23 September 2022
Accepted: 28 October 2022
Published: 31 October 2022

1. Introduction

A solid surface with an ordered nanorelief has been the subject of scientific and practical interest due to its potential use in nanoelectronic devices [1]. In addition, the nanorelief on a substrate surface affects the epitaxial growth. Sapphire (α-Al_2O_3) and silicon carbide (SiC) are examples of commonly used substrates whose polished vicinal surface is modified from a flat to a terrace-stepped structure by thermal or chemical treatment.

Techniques to enable probing surface properties typically use a probe located within a specific area of the sample. Atomic force microscopy (AFM) is indispensable for detecting the evolution of surface morphology. In particular, islet coarsening or terrace formation on sapphire substrates annealed in oxygen were studied in [2,3]. The authors of [4,5] illustrated the utility of AFM by analyzing nanosteps on a SiC surface annealed in hydrogen (H_2) or hydrochloric (HCl) vapor. AFM provides reliable measurements in areas up to several tens of square microns. The need to go beyond a micrometer-scale mapping can only be satisfied if the sample moves sequentially in very small increments of motion, which is not practical. In contrast to AFM, grazing-incidence X-ray scattering (XRS) methods solve the problem of obtaining information about the entire surface of large samples.

The practical use of a nanostructured surface is moving forward rapidly in connection with the production of epitaxial graphene on SiC substrates (see, e.g., a review [6]). At a high temperature, the sublimation of Si atoms at a rate faster than C eventually results in the formation of epitaxial graphite film. A number of experiments have proved that the film is characterized by low mosaicity and good homogeneity [7–10]. However, some basic aspects involved in the graphitization process are still poorly understood.

The quality of graphene depends on the substrate preparation. Difficulties in understanding and predicting the evolution of morphology are, in particular, associated with a vicinal surface. High-vacuum annealing or thermal etching of SiC in H_2 and HCl vapor leads to a terrace-stepped surface structure. The growth of graphene on a stepped surface of vicinal SiC can give reliefs that are more complex than terraces. The increased effectiveness of preparation procedures has raised the standard of graphene technology [11–14]. For instance, graphene layers with improved morphology were obtained under an argon atmosphere [9,15].

A common method of evaluation of roughness height, terrace period and thickness is AFM [4,5,11–15]. Unlike AFM, which is insensitive to a buried interface (for example, between the graphene layer and the substrate), XRS methods provide information at a depth of up to tens of nanometers. This depth value is at least an order of magnitude greater than AFM can give. XRS is preferable to AFM for the evaluation of stochastically distributed roughness [16]. In addition, the authors of [17–19] used highly sensitive XRS methods to study sapphire substrates with a terrace-stepped nanorelief. However, processed SiC surface is still a subject of investigation with this approach. Furthermore, there is still a lack of understanding of the interplay between values obtained from AFM and XRS data.

Probing a larger surface area is advantageous because it allows for a full set of parameters per sample, which results in improved characterization. In particular, one can estimate the period of terraces over the measurement area, the period dispersion, irregularities in terrace widths and material density variation across the depth. In this paper, we test the feasibility of using XRS under grazing incidence geometry and X-ray specular reflectometry to determine the surface structure and roughness parameters resulting from the vacuum annealing of SiC substrates. We compare effective roughness height, average terrace periods and deviations calculated according to XRS and AFM data and analyze the observed trends.

2. Materials and Methods

2.1. Sample Preparation

SiC (0001) wafers were purchased from different commercial channels. Semi-insulating crystals had a charge-unbalanced donor concentration of ~10^{15}–10^{16} cm^{-3}. Epipolished substrates with low miscut angles ($\leq 2^\circ$ off-axis) were certified as flat within commercial standards. After chemical-mechanical polishing a damaged layer remains on the surface of the substrate. To improve this situation we used ultra-high-vacuum (UHV) annealing in a closed tantalum cell. Before an experiment, a 4-inch diameter wafer was cut in equal samples 1 cm^2 in area. The samples were cleaned and washed in distilled water and organic solvents.

Based on previous work in this area, we established the optimal temperature range that enables etching without changing the stoichiometric composition of the substrate surface. Namely, well-defined terraces emerge on a vicinal SiC surface at 1200 °C, when step-bunching is avoided, or it may be observed with higher offcut angles. A severe degradation takes place at 1800 °C [20,21]. Between surface restructuring and degradation, there are the most favorable conditions for obtaining atomically smooth terraces [13,22,23]. The annealing was carried out in the temperature range 1300–1500 °C at a residual pressure of 10^{-4} Pa.

A Si polar face of 4H-SiC was used to prepare the substrate. The growth of graphene on the Si face is well studied. Early experiments performed by grazing incidence X-ray diffraction [7], STM, LEED and spectroscopy [24,25] revealed the precursor phase for graphene growth on 6H- and 4H-SiC. Later, a schematized representation of the layer-by-layer graphene growth via the solid-state decomposition of the precursor [7] was improved. A new idea of the buffer layer between the graphene and Si-terminated surface of 4H-SiC resulted from X-ray reflectivity measurements [26]. The fact that the graphene on the Si-face is partly isolated from interactions with the substrate means that the sheet may grow perfectly continuously all around the surface. Meanwhile, the bi-layer graphene can replace the buffer layer in favor of better electronic properties [27]. Despite two decades of research, there are still issues to be clarified in relation to advanced processes or traditional treatments of a substrate surface.

2.2. Experimental Methods and Techniques

The AFM experiments were carried out using a P47 scanning probe microscope (NT-MDT Spectrum Instruments, Moscow, Russia). The operation of a hard probe was performed in the tapping mode. In addition, we used an independent method for determining the step heights in terrace-stepped nanostructures: high-resolution transmission electron microscopy (HRTEM). A Tecnai Osiris electron microscope (FEI Technologies Inc., Hillsboro, OR, USA) with 200 kV accelerating voltage was employed to investigate the cross-section samples. To prepare the samples we used a focused Ga^+ ion beam provided by the FIB station. Before etching, a platinum film was deposited onto the surface of SiC.

The X-ray measurements were performed on a laboratory diffractometer. The setup was equipped with a mobile X-ray source-detector system in a "butterfly" geometry [28]. A conventional X-ray tube with a wide-aperture Cu-anode served as a source. Probing radiation was tuned to the wavelength of λ = 0.154 nm by the use of a Si(111) single-reflection monochromator providing $\Delta\lambda/\lambda$~10^{-5}. The experimental configuration included a three-slit collimator with vacuum paths to reduce the radiation absorption and scattering by air molecules. The collimator delivered the beam of width $d \approx 0.55$ mm and divergence $\Delta\theta \approx 10^{-4}$ rad at total radiation intensity of $I \approx 3 \times 10^6$ s^{-1}.

The sample was placed on an adjustment table so that its surface was parallel to the propagation direction of the incident beam. During further measurements, the sample remained stationary while the source and detector rotated around the sample at specified angles. The angular position error of the source and detector rotation stages was kept within 2 arcsec. For data collection, a scintillation detector SCSD-4 (Radicon Ltd., St. Petersburg, Russia) was used in conjunction with a position-sensitive strip Dectris MYTHEN 2R (Dectris AG, Baden-Dättwil, Switzerland). It allowed us to measure the intensity of both X-ray beams reflected and scattered by the sample surface. We note that the scintillation detector provided better signal-to-noise ratio. It is important for measuring super-smooth surfaces since the scattering intensity declines within 8 orders of magnitude. The overall angle scanning range was 0–3°, while the angle of incidence for scattering experiments was set to 0.236°. The corresponding scattering vectors q_z and spatial frequencies ν covered a range of q_z = 0–0.42 Å^{-1} and ν = 0.05–10 μm^{-1}, respectively.

To analyze the surface structure and roughness parameters with X-rays, we applied methods of specular reflectometry and grazing-incidence off-specular scattering. These methods are widely used for studying thin films and buried interfaces due to their angstrom-scale precision and high sensitivity to depth-graded and planar density distributions. Reflectivity and scattering data were acquired along each sample using the same setup.

To represent the kinematics of X-ray scattering from a macroscopically flat horizontal surface under grazing incidence, we used the following coordinate system (Figure 1). The origin point O is the center of illuminated area. The x,y plane coincides with the air-sample interface. The z-axis is normal to the sample surface. The distribution of scattered intensity $\Phi(\theta, \varphi)$ of a beam falling on the surface at a grazing angle of θ_0 is expressed in terms of scattering amplitude $A(\boldsymbol{q})$ by the following:

$$\Phi(\theta,\varphi) = \frac{1}{Q_{inc}} \cdot \frac{dQ_{sc}}{d\Omega} = \frac{\langle |A(q)|^2 \rangle}{sin\theta_0 \int d^2p} \quad (1)$$

where Q_{inc} is the incident radiation power, dQ_{sc} is the fraction of power scattered by the rough surface within the solid angle $d\Omega$, $\boldsymbol{q} = k\{\cos\theta\cos\varphi;\cos\theta\sin\varphi\}$ is the projection of scattering wave vector onto the x,y plane, $k = (2\pi/\lambda)$ is the wave vector, and $\boldsymbol{p} = (x,y)$ is the lateral coordinate vector. In turn, we assumed the spatial distribution of roughness along the surface to be independent from the density distribution normal to the surface, resulting in polarizability $\chi(\boldsymbol{r}) = \chi_0\Theta[z-\zeta(\boldsymbol{p})]$, where ζ is the function describing the surface relief and Θ is the Heaviside step function. In that case, the exact expression for the scattering amplitude can be presented as [29]:

$$A(q) = \frac{k^2}{2\pi} \int exp(-iqp)\Delta\chi(r)\psi_0(z,q)\Psi(r)d^3r \quad (2)$$

where ψ_0 (z,q) is the incident wave, and $\Psi(\mathbf{r})$ is the exact solution of the scattered wave equation. Expanding the polarizability parameter $\Delta\chi$ into the series by small delta-like perturbations, the scattering amplitude in the first order can be written as [30]:

$$A_{PT}(q) = \frac{k^2}{4\pi}\chi_b t(q_0)t(q) \int exp[i(q_0 - q)p]\zeta(p)d^2p \quad (3)$$

where χ_b is the bulk polarizability within the sample, and $t(q)$ is the amplitude transmission factor for the perfectly smooth interface, which can be calculated using, e.g., widely known Fresnel formalism.

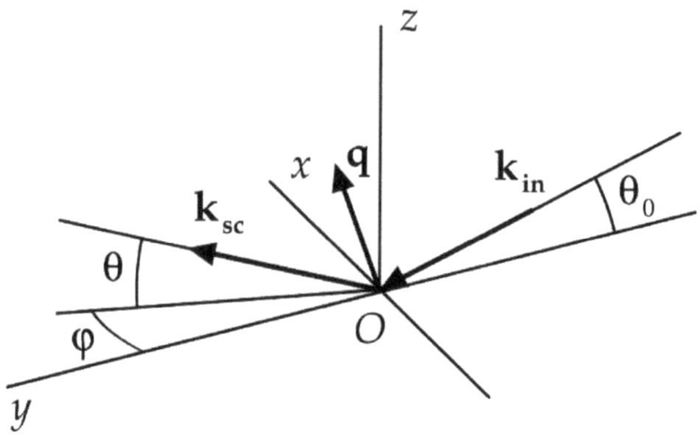

Figure 1. Schema of the X-ray scattering process and associated coordinate system. The origin point O is at the center of the illuminated region, the x,y plane coincides with the sample surface, the z axis is directed along the normal to the surface plane, and the x axis is perpendicular to the incident beam direction. The incident beam vector is $\mathbf{k}_{in}$; the grazing angle is θ_0; the scattered beam vector is $\mathbf{k}_{sc}$; the scattering vector equals $\mathbf{q} = \mathbf{k}_{in} - \mathbf{k}_{sc}$.

It should be noted that at small grazing incident angles ($\theta_0 << 1$) the distribution of scattered intensity is significantly higher along the grazing angle θ than along the azimuthal angle φ. Therefore, it is practical to use the one-dimensional scattering diagram integrated over the azimuth, $\Pi(\theta) = \int \Phi(\theta,\varphi)d\varphi$. As shown in [30,31], the respective scattering distribution is presented as:

$$\Pi(\theta) = \frac{k^3\chi_b^2}{16\pi sin\theta_0\sqrt{cos\theta_0 cos\theta}}|t(\theta_0)t(\theta)|^2 PSD(\nu) \quad (4)$$

Here $\nu = (2\pi)/|q - q_0|$ is the spatial frequency along the surface plane, and PSD is the so-called power spectral density function of surface roughness, which in turn is the real part of the Fourier spectrum of autocorrelation function for height-to-height roughness [32]:

$$PSD(\nu) = 4\int \langle\zeta(0)\zeta(p)\rangle cos(2\pi\nu p)dp \quad (5)$$

As such, power spectral density for the surface roughness can be extracted directly from the diffuse scattering measurements. Note that for a quantitative comparison the effective roughness height σ_{eff} can be defined by the relation:

$$\sigma_{eff} = \sqrt{\int_{v_{min}}^{v_{max}} PSD(v)dv} \tag{6}$$

In turn, the angular distribution of specular reflectivity (for $\theta = \theta_0$) from the surface within kinematical approximation is proportional to the depth-graded distribution of polarizability [33]:

$$R(\theta) \equiv \frac{I(\theta)}{I_0} = R_F \times \left| \frac{1}{\chi_b} \int_{-\infty}^{\infty} \frac{d\chi}{dz} exp(2iqz)dz \right|^2 \tag{7}$$

where R_F is the classical Fresnel reflection from an ideally smooth step-like surface.

For the analysis of the experimental reflectivity data $R(\theta)$ and reconstruction of the electron density distribution $\rho(z)$, we applied a model-independent approach proposed in [34], which is based on the extrapolation of the asymptotic component of reflectivity R into the range of large angles. One of the significant advantages of the said approach is that it does not require any prior assumptions on the sample internal structure in the form of a parametric model, but provides a direct numerical search of the depth-graded distribution of electron density instead. That makes model-independent methods useful for studies of disoriented and amorphous layers, where the choice of appropriate model is often ambiguous.

It should be noted that numerous recent works applied AFM data in conjunction with X-ray experiments to characterize interfaces, for example [35,36]. However, in these publications interfacial roughness has been only considered in terms of root-mean-square (rms) roughness height σ, which has several downsides. In particular, σ by definition is an integral parameter, and as such it does not account properly for neither stochasticity nor anisotropy of interfacial relief. Furthermore, estimations of σ from X-ray reflectivity data treat it as a fitting parameter in either Debye–Waller or Nevot–Croce factors, which *a priori* assume the interface to be fully stochastic, isotropic and having normally-distributed height-to-height roughness correlations. These assumptions, evidently, are not fulfilled for the oriented terrace-stepped relief. On the other hand, it has been previously shown [31,37] that the introduction of a non-Gaussian component into a height-to-height roughness distribution changes the integral specular reflection intensity in the asymptotic region up to one order of magnitude while preserving the same absolute σ value. As such, for non-stochastic non-isotropic interfaces estimation of σ from Debye–Waller or Nevot–Croce factors always contains a substantial error.

On the contrary, analysis of the interface roughness in terms of PSD function provides the spectrum of height-to-height correlations, which unambiguously describes the non-stochasticity of the roughness in question. In addition to that, the application of the grazing incidence XRS approach allows one to extract the PSD function directly from experimental data without any initial assumptions about its statistical properties. Furthermore, the extended self-consistent algorithm described in [38] allows one to account for the stochastic surface roughness during the calculation of $\rho(z)$ by simultaneously processing both reflectivity and scattering data. All numerical X-ray calculations were implemented in the custom software written in Python language by the use of the Scientific Python and PyLab libraries and environment [39].

3. Results and Discussion

3.1. AFM and TEM

In this study we used the AFM/HRTEM complementary analysis to add more information to the previous statements concerning SiC surfaces slightly misoriented from on-axis. The authors of [40] established that the oxide was removed and minor scratches after mechanical polishing were partially healed during the first stage of annealing in vacuum at low temperatures of 900–1000 °C. From a temperature of 1200 °C terraces appeared between the scratches. Finally, flat terraces replaced the damaged layer on the surface of mechanically polished substrates. In the present paper, we investigated chemical-mechanical-polished (CMP) samples whose surfaces were similarly improved after vacuum annealing. Nevertheless, processing results showed clear differences between the smooth and scratched surfaces. Terrace propagation without interruption was observed when scratches were eliminated by CMP.

Attempts to control the terrace widths exhibit unsatisfactory aspects because miscut angles pertaining to different substrates vary slightly. At the same time, a manageable duration of annealing with an optimal fixed temperature allows for obtaining steps with a given height [22]. Annealing for 5 min. at a temperature of 1300 °C ends with the formation of steps with a height of h = 0.75 nm, which corresponds to the half unit cell of 6H-SiC. With the increase in the annealing time from 5 to 10 min. under the same temperature, the step height doubles and becomes equal to the unit cell: h = 1.5 nm. In addition, the controlled annealing experiments showed a gradual destruction of the terrace edges. Furthermore, the beginning of surface graphitization started above 1550–1600 °C.

The goal of current investigations was a direct measurement of step heights by HRTEM. The annealing of Si-face at 1450 °C led to the emergence of terraces and steps. The sample preparation for HRTEM analysis began with finding a low-index crystallographic direction, which is closest to the path of terrace propagation. A thin electron-transparent lamella was prepared by cutting a substrate along this direction. When the electron beam is parallel to a low-index zone axis, one obtains cross-sectional representations along six non-equivalent directions. Sequential alternation of steps was revealed under different viewing angles. However, the terrace width is not measurable in this configuration. The width may be less than or equal to that seen in the AFM maps.

Figure 2a shows the AFM image of a typical step pattern in planar view. Well-defined terraces have kinks protruding in lateral directions. According to AFM data, the step heights may vary from 3 to 5 nm. The single step of a similar pattern in an HREM cross-section view is presented in Figure 2b. The value of the step height is 4 nm. This value corresponds to the step with a height of four layers of close-packed 4H-SiC. The layers themselves are seen between well-defined boundaries separated by a period of c = 1.005 nm. Throughout this substrate, other HRTEM estimates of the step heights are 3 and 5 nm.

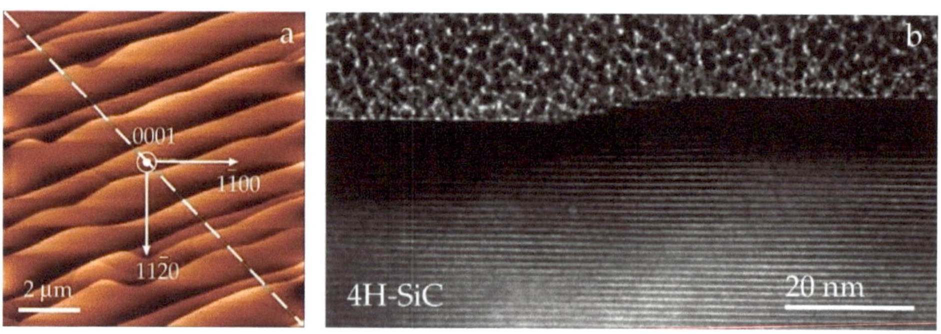

Figure 2. (**a**) AFM image of the terrace-stepped surface morphology of the Si-face of 4H-SiC. The annealing temperature is 1450 °C. The plot for height against distance (not shown) is obtained along the dashed line. (**b**) HRTEM image of a single step on a 4H-SiC surface imaged along a zone axis *[1-210]*. The period of the stripes corresponds to the unit cell parameter.

Note that these steps have a rather large height. Graphene grown at the edges of high terraces and graphene between steps differ in their properties. A graphene device covering the surface step has a much higher resistance than the one on the terrace [41]. Mammadov et al. [42] and Lebedev et al. [43], using Kelvin-probe force microscopy, showed that the position of multilayer graphene, whose potential is higher compared to the monolayer graphene, corresponds to the borders of the terraces, while monolayer graphene covers the terraces.

The combination of minimum step height with maximum terrace width provides optimal conditions for obtaining homogeneous graphene layers [44]. Since these requirements are very tight and difficult to be met within the SiC industry, the problem arises of how to choose optimal thermal conditions for the growth of graphene. In particular, by controlling the heating rate, one can suppress the step-bunching process [45,46]. Lebedev et al. [43], using a heating rate of ~250 °C/min, obtained a homogeneous monolayer of graphene (with a minimum number of double-layer inclusions) on surface steps up to about several nanometers.

At increasing temperatures, the terrace-stepped relief may become more complex than at lower temperatures. The detection of changes requires probing a large area with nanoscale resolution. However, AFM or HRTEM provide information within a local probing range of a few tens of micrometers or a few nanometers, respectively. Meanwhile, the XRS setup described above makes it possible to inspect an area down to 10×10 mm^2 in any location on samples 10 to 200 mm in size and obtain information about the entire surface.

3.2. Grazing-Incidence X-ray Scattering and Reflectivity Measurements

A commercial 4H-SiC substrate was compared with a UHV annealed (1550 °C) specimen cut out from the substrate. Scintillation and a position-sensitive strip detector measured the scattered intensity. Power spectral density (PSD) functions were extracted from the scattering data according to Equation (4). In addition, similar functions have been calculated directly from the AFM information according to Equation (5).

As an illustration, in Figure 3 we show such PSD functions versus the spatial frequency. One can see that the scattering curves of the polished wafer, represented by the green and blue markers, decrease rapidly until they reach the background noise level. A fast descent of the PSD functions indicates that the wafer has isotropic roughness distribution over a well-polished atomically smooth surface. The effective roughness height calculated from Equation (6) equals σ_{eff} = 1.6 Å. Very low roughness values result in the very small signal-to-noise ratio, which makes it hard to obtain reliable data with a strip detector.

Terrace-stepped surfaces were made by annealing samples as described above, following the procedures of Lebedev et al. [22,40]. The annealed surface may demonstrate the presence of different scales of roughness. Their physical justification relates to the appearance of large-scale roughness that is intrinsic to step formation, while low-scale part corresponds to the roughness of the terraces. In particular, the terrace formation made the roughness increase by almost an order in magnitude, yielding σ_{eff} = 9.3 Å.

In Figure 3 the PSD functions of the stepped surface are drawn with red and black markers. In contrast to a polished face, the related curves have a convex form. The PSD functions exhibit characteristic maxima at $\nu_L \approx 0.54$ μm^{-1}. The value $L = 1/\nu_L$ corresponds to the area-averaged period of the terrace-stepped nanostructure. Structural factors, such as the deviation of the terrace period from the average value and uneven terrace edges, broaden these maxima. We evaluated the deviation parameter $\Delta\nu_L$ taking into account the full width at half maximum (FWHM) of the peaks, by fitting them with a standard Gaussian function. Since the AFM scan area is approximately three orders of magnitude smaller than the X-ray irradiation area, it turns out that the peak found from the AFM data is narrower. Because of that, the terrace period deviation is lower. Table 1 presents the resulting parameters.

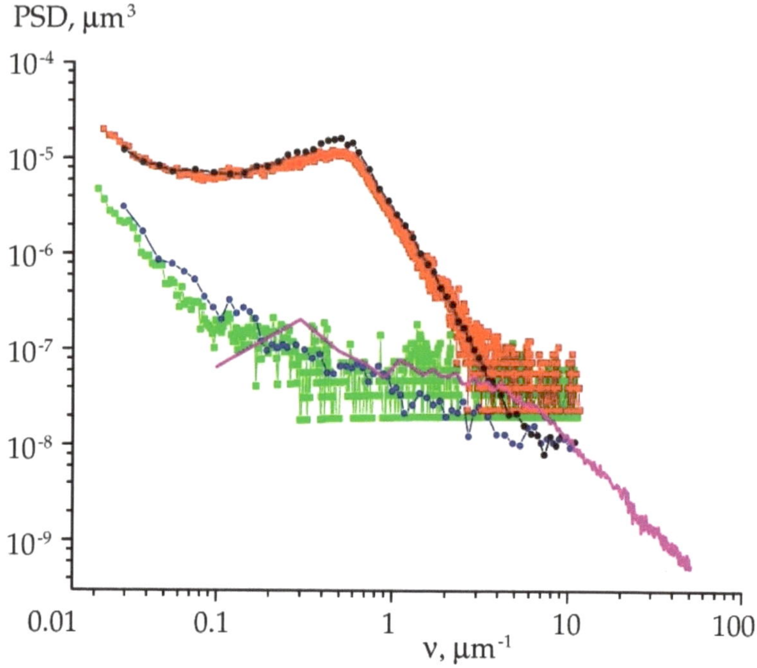

Figure 3. Roughness PSD functions from X-ray scattering measurements plotted against spatial frequency. Colored markers and lines represent the data obtained from the surface of polished (green, blue) and annealed (red, black) specimens with scintillation (circles) and position sensitive (squares) detectors. The solid magenta line defines PSD function calculated from AFM data for the polished surface.

Table 1. Area-averaged period of terrace-stepped relief.

Average Terrace Period	Period Deviation, AFM	Period Deviation, XRS
1.85 μm	2.13 μm	4.2 μm

Terrace-stepped surfaces are highly anisotropic. X-ray scattering curves registered parallel to a terrace propagation direction may have features different to those seen along the orthogonal direction. To account for the anisotropic roughness characteristics of annealed specimens, we followed the methodology of Asadchikov et al. [19]. The scattering curves were measured under different lateral directions across the sample surface. The respective PSD distributions are presented in Figure 4. AFM images (Figure 4a,b) show the surface topography with well-defined terraces and kinks. The topography is not exactly the same for these two areas of the sample. Nevertheless, the kinks take place exclusively at the edges of steps and overall the terrace surface is essentially flat.

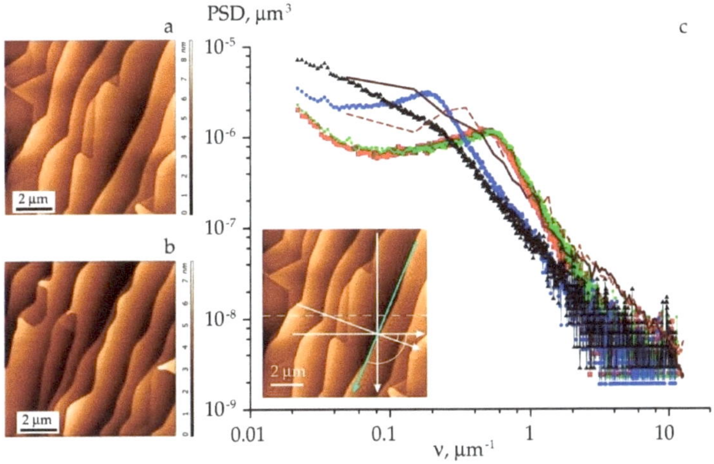

Figure 4. AFM images (**a**,**b**) and roughness PSD functions vs. spatial frequency (**c**) calculated for the surface of the annealed specimen (1450 °C). In the inset figure within (**c**) the dotted line indicates the initial AFM scan. The PSD functions along probing directions, which deviate 0°, −20°, −90° and −110° from the initial scan line, are represented with red, green, blue and black markers, respectively. Arcs mark angles of −20° and −110°. The cyan arrow shows the probing direction along a terrace edge. Brown lines depict the PSD functions extracted from AFM data. In particular, the solid brown line corresponds to the cyan arrow.

Processing of the scattering data showed that the integral value of the root mean square roughness remained in the range $\sigma_{eff} = 9.0 \pm 0.6$ Å. At the same time, the angular position of the peak was shifted according to azimuth rotation of the sample. Figure 4 illustrates the peak motion along the ν-axis. The shift is determined by a characteristic period L between the steps. From the peak position we obtained $L = 1/\nu_L \approx 1.85$ μm. In this case, the terraces extend along the direction marked in the cyan line in Figure 4, while the period is parallel to the orthogonal direction. Finally, we emphasize that the PSD peak is broadened along the ν axis more than 10 times, which indicates the presence of microscopic irregularities in the terrace width and shape across the surface of the sample.

The directions of AFM scans may exert a similar influence as above. These directions had different lateral angles with respect to the edges of the steps. From the scans, we calculated one-dimensional PSD functions. We noticed that the particular PSD functions extracted from the scan along a step edge have no peak. The corresponding distribution drawn with a solid brown line in Figure 4 is smoothed out and no peak is observed. The related rms roughness equals $\sigma_{eff} = 8.3$ Å. Meanwhile, the dashed brown line in Figure 4 exhibits a regular peak that corresponds to the area-averaged period of the terraces.

As we see from the examples above, grazing-incidence off-specular scattering allows one to reconstruct the statistical properties of lateral inhomogeneities, in particular of surface roughness and morphology. In turn, specular reflectometry offers the possibility of studying depth-wise inhomogeneities. By measuring the angular dependence of the reflection coefficient at small glancing angles one can obtain the depth-dependence of density ρ with a resolution of less than a nanometer.

In our present experiments, multiple reflectivity curves were measured along the same lateral directions that we used for the case of scattering. The results are shown in Figure 5a. Coloration represents a sample itself and the direction on the sample surface. In particular, magenta corresponds to the flat SiC substrate not subjected to annealing. Other colors indicate the relevant directions on the stepped surface of the annealed specimen. For example, we used black to show the intensity acquired along the step edges, while green markers represent an orthogonal direction. The caption of Figure 5 explains the variety of colors. Note that the asymptotic decay in intensity obtained for the annealed specimen fundamentally differs from that of the original substrate, which directly indicates the difference in morphology.

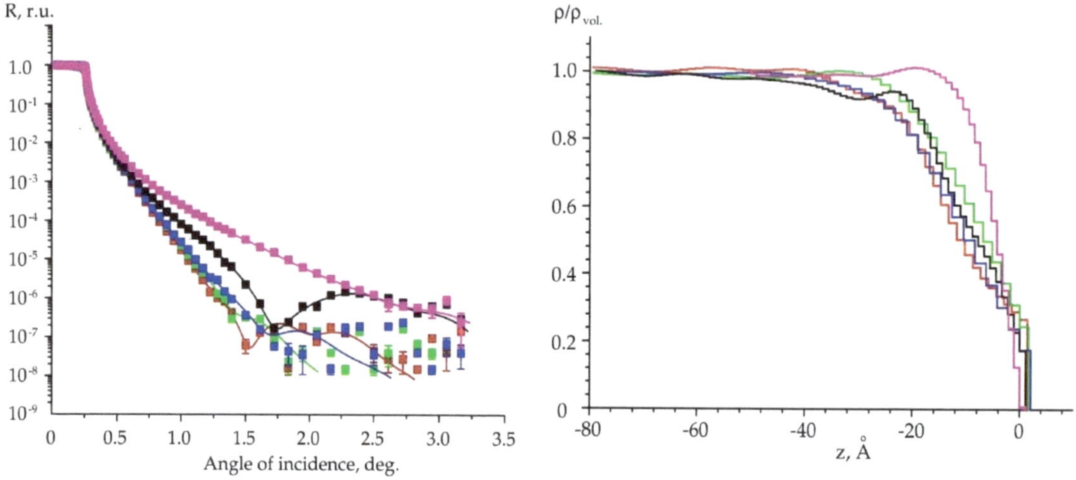

(**a**) (**b**)

Figure 5. (**a**) Reflectivity curves measured for two samples: original substrate and annealed specimen (1450 °C). The magenta color defines the original substrate. The red, green, blue and black markers represent experimental data along the probing directions, which deviate 0°, −20°, −90° and −110°, respectively, from the AFM scan line in inset of Figure 4. The lines indicate fit results. (**b**) Distributions of depth-graded volumetric density, normalized to the density of bulk SiC, versus depth *z*.

Depth-graded distributions of volumetric density are displayed in Figure 5b. Information was acquired from reflectivity data according to [34,38]. We found that the original sample has a disturbed surface layer as deep as 25 Å. The increased layer density may be due to the hardening effect caused by the surface-finishing treatment. In turn, density profiles for the annealed specimen reveal the presence of a so-called 'transition layer.' The layer depths vary from 40 to 60 Å, depending on the probing direction. We determined a pronounced change in density at a depth of 30 Å in the direction normal to the step edges. This value agrees with an average terrace height obtained from AFM scans.

There are, therefore, strong reasons to assume that redistribution of the probing beam intensity between specular (reflected) and off-specular (scattered) components depends on the lateral orientation of the sample. These components are provided by different parts of the terrace-stepped surface areas due to their inclination relative to the beam axis. This, in turn, corresponds to different 'effective density' values for a near-surface region averaged across the irradiated area. It should be noted that the more precise analysis of subsurface structure is constrained by non-uniform and irregular terrace edges. Various publications have studied and discussed the formation of a more regular surface relief under different treatments [4,14,40,47].

4. Conclusions

A detailed comparative analysis has shown that there is a good agreement between data obtained by AFM and grazing-incidence XRS methods. Meanwhile, the X-ray methods allow determination of not only lateral but also depth grade distribution of inhomogeneities. In particular, the oxidized layer with 25 Å thickness and the 'transition layer' were revealed. These near-surface regions are allegedly tied to the molecular adsorption from an ambient air medium. Furthermore, the depth-graded distributions of volumetric density provide an assumption that the oxidized layer manifests itself differently between the probing directions along and across a terrace edge.

X-ray scattering data is always averaged over an irradiation area significantly larger than the area measured in a regular AFM experiment. Due to this fact, we successfully extracted a number of meaningful statistical parameters, including effective roughness height, average terrace period and dispersion. Finally, a general structure of the oxidized and damaged near-surface layers was set out. Such information cannot be obtained by using conventional AFM or TEM methods.

An important aspect of X-ray measurements is the broad range of spatial frequencies associated with surface roughness. We stress that a detailed comparison of AFM and X-ray data would be possible if AFM measurements were conducted on numerous areas of different sizes with different linear scales. This, in turn, is challenging and takes longer than a regular X-ray scattering experiment.

Author Contributions: S.P.L., B.S.R. and A.D.N.: experimental investigations. Y.O.V., B.S.R.: software, computer simulations. T.S.A., B.S.R. and Y.O.V.: paper writing. A.A.L. and V.E.A.: conceptualization, project administration. All authors have read and agreed to the published version of the manuscript.

Funding: Purchase of SiC: sample preparations, graphene growth and characterization received funding from the Ministry of Science and Higher Education of the Russian Federation, project 075-15-2021-1349. X-ray reflectivity and scattering measurements, data processing and software were funded by the Ministry of Science and Higher Education of the Russian Federation, state assignment of Federal Research Center 'Crystallography and Photonics'.

Institutional Review Board Statement: Not applicable.

Informed Consent Statement: Not applicable.

Data Availability Statement: Data is contained within the article.

Acknowledgments: The authors would like to acknowledge M.S. Dunaevskii and A.V. Myasoedov from the Ioffe Institute for conducting the AFM and HRTEM investigation.

Conflicts of Interest: The authors declare no conflict of interest.

References

1. Rosei, F. Nanostructured surfaces: Challenges and frontiers in nanotechnology. *J. Phys. Condens. Matter* **2004**, *16*, S1373–S1436. [CrossRef]
2. Nguyen, T.T.T.; Bonamy, D.; Pham Van, L.; Barbier, L.; Cousty, J. Coarsening of two-dimensional Al_2O_3 islands on vicinal (1,-1,0,2) sapphire surfaces during annealing in air. *Surf. Sci.* **2008**, *602*, 3232–3238. [CrossRef]
3. Wang, J.; Howard, A.; Egdell, R.G.; Pethica, J.B.; Foord, J.S. Arrangement of rotational domains of the (root 31 × root 31) R9 reconstruction of Al_2O_3(0001) revealed by non-contact AFM. *Surf. Sci.* **2002**, *515*, 337–343. [CrossRef]
4. Nakajima, A.; Yokoya, H.; Furukawa, Y.; Yonezu, H. Step control of vicinal 6H-SiC(0001) surface by H_2 etching. *J. Appl. Phys.* **2005**, *97*, 104919. [CrossRef]
5. Nakamura, S.; Kimoto, T.; Matsunami, H.; Tanaka, S.; Teraguchi, N.; Suzuki, A. Formation of periodic steps with a unit-cell height on 6H-SiC (0001) surface by HCl etching. *Appl. Phys. Lett.* **2000**, *76*, 3412–3414. [CrossRef]
6. Yazdi, G.R.; Iakimov, T.; Yakimova, R. Epitaxial graphene on SiC: A review of growth and characterization. *Crystals* **2016**, *6*, 53. [CrossRef]
7. Charrier, A.; Coati, A.; Argunova, T.; Thibaudau, F.; Garreau, Y.; Pinchaux, R.; Forbeaux, I.; Debever, J.M.; Sauvage-Simkin, M.; Themlin, J.M. Solid-state decomposition of silicon carbide for growing ultra-thin heteroepitaxial graphite films. *J. Appl. Phys.* **2002**, *92*, 2479–2481. [CrossRef]
8. Berger, C.; Song, Z.; Li, T.; Li, X.; Ogbazghi, A.Y.; Feng, R.; Dai, Z.; Marchenko, A.N.; Conrad, E.H.; First, P.N.; et al. Ultrathin epitaxial graphite: 2D electron gas properties and a route toward graphene-based nanoelectronics. *J. Phys. Chem. B* **2004**, *108*, 19912–19916. [CrossRef]
9. Emtsev, K.V.; Bostwick, A.; Horn, K.; Jobst, J.; Kellogg, G.L.; Ley, L.; McChesney, J.L.; Ohta, T.; Reshanov, S.A.; Rotenberg, E.; et al. Towards wafer-size graphene layers by atmospheric pressure graphitization of silicon carbide. *Nat. Mater.* **2009**, *8*, 203–207. [CrossRef]
10. Virojanadara, C.; Syväjarvi, M.; Yakimova, R.; Johansson, L.I.; Zakharov, A.A.; Balasubramanian, T. Homogeneous large-area graphene layer growth on 6H-SiC(0001). *Phys. Rev. B* **2009**, *78*, 245403. [CrossRef]

11. Owman, F.; Hallin, C.; Mårtensson, P.; Janzén, E. Removal of polishing-induced damage from 6H-SiC (0001) substrates by hydrogen etching. *J. Cryst. Growth* **1996**, *167*, 391–395. [CrossRef]
12. Robinson, Z.R.; Jernigan, G.G.; Currie, M.; Hite, J.K.; Bussmann, K.M.; Nyakiti, L.O.; Garces, N.Y.; Nath, A.; Rao, M.V.; Wheeler, V.D.; et al. Challenges to graphene growth on SiC(000-1): Substrate effects, hydrogen etching and growth ambient. *Carbon* **2015**, *81*, 73–82. [CrossRef]
13. Lebedev, S.P.; Petrov, V.N.; Kotousova, I.S.; Lavrentev, A.A.; Dementev, P.A.; Lebedev, A.A.; Titkov, N. Formation of periodic steps on 6H-SiC (0001) surface by annealing in a high vacuum. *Mater. Sci. Forum* **2011**, *679*, 437–440.
14. Lebedev, S.P.; Barash, I.S.; Eliseyev, I.A.; Dementev, P.A.; Lebedev, A.A.; Bulat, P. Investigation of the hydrogen etching effect of the SiC surface on the formation of graphene films. *Technol. Phys.* **2019**, *64*, 1843–1849. [CrossRef]
15. Yakimova, R.; Virojanadara, C.; Gogova, D.; Syväjarvi, M.; Siche, D.; Larsson, K.; Johansson, L.J. Analysis of the formation conditions for large area epitaxial graphene on SiC substrates. *Mater. Sci. Forum* **2010**, *645–648*, 565–568.
16. Asadchikov, V.E.; Kozhevnikov, I.V.; Krivonosov, Y.S.; Mercier, R.; Metzger, T.H.; Morawe, C.; Ziegler, E. Application of X-ray scattering technique to the study of supersmooth surfaces. *Nucl. Instrum. Meth. Phys. Res. A* **2004**, *530*, 575–595. [CrossRef]
17. Blagov, A.E.; Prosekov, P.A.; Grischenko, J.V.; Zanaveskin, M.L.; Roschin, B.S.; Butashin, A.V.; Feodorov, V.A.; Kanevskii, V.M.; Asadchikov, V.E. Features of X-ray diffraction on sapphire single crystals with a nanostructured surface. *J. Surf. Invest.* **2009**, *3*, 447–449. [CrossRef]
18. Prokhorov, I.A.; Zakharova, B.G.; Asadchikov, V.E.; Butashin, A.V.; Roshchin, B.S.; Tolstikhina, A.L.; Zanaveskin, M.L.; Grishchenko, Y.V.; Muslimov, A.E.; Yakimchuk, I.V.; et al. Characterization of single-crystal sapphire substrates by X-ray methods and atomic force microscopy. *Cryst. Rep.* **2011**, *56*, 456–462. [CrossRef]
19. Asadchikov, V.E.; Blagov, A.E.; Butashin, A.V.; Volkov, Y.O.; Deryabin, A.N.; Kanevskii, V.M.; Muslimov, A.E.; Protsenko, A.I.; Roshchin, B.S.; Targonskii, A.V.; et al. Lateral inhomogeneities of sapphire plates determined with the aid of X-ray and probe methods. *Technol. Phys.* **2020**, *65*, 400–406. [CrossRef]
20. Nishiguchi, T.; Ohshima, S.; Nishino, S. Thermal etching of 6H-SiC substrate surface. *Jpn. J. Appl. Phys.* **2003**, *42*, 1533–1537. [CrossRef]
21. Van der Berg, N.G.; Malherbe, J.B.; Botha, A.J.; Friedland, E. Thermal etching of SiC. *Appl. Surf. Sci.* **2012**, *258*, 5561–5566. [CrossRef]
22. Lebedev, S.P.; Dement'ev, P.A.; Lebedev, A.A.; Petrov, V.N.; Titkov, A.N. Fabrication and use of atomically smooth steps on 6H-SiC for calibration of z-displacements in scanning probe microscopy. *Mater. Sci. Forum* **2010**, *645–648*, 767–770.
23. Dunaevskii, M.S.; Makarenko, I.V.; Petrov, V.N.; Lebedev, A.A.; Lebedev, S.P.; Titkov, A.N. Using atomic-step-structured 6H-SiC(0001) surfaces for the calibration of nanotranslations in scanning probe microscopy. *Technol. Phys. Lett.* **2009**, *35*, 47–49. [CrossRef]
24. Riedl, C.; Starke, U.; Bernhardt, J.; Franke, M.; Heinz, K. Structural properties of the graphene-SiC(0001) interface as a key for the preparation of homogeneous large-terrace graphene surfaces. *Phys. Rev. B* **2007**, *76*, 245406. [CrossRef]
25. Lauffer, P.; Emtsev, K.V.; Graupner, R.; Seyller, T.; Ley, L.; Reshanov, S.A.; Weber, H.B. Atomic and electronic structure of few-layer graphene on SiC(0001) studied with scanning tunneling microscopy and spectroscopy. *Phys. Rev. B* **2008**, *77*, 155426. [CrossRef]
26. Hass, J.; Millán-Otoya, J.E.; First, P.N.; Conrad, E.H. Interface structure of epitaxial graphene grown on 4H-SiC(0001). *Phys. Rev B* **2008**, *78*, 205424. [CrossRef]
27. Virojanadara, C.; Zakharov, A.A.; Yakimova, R.; Johansson, L.I. Buffer layer free large area bi-layer graphene on SiC (0001). *Surf. Sci.* **2010**, *604*, L4–L7. [CrossRef]
28. Asadchikov, V.E.; Babak, V.G.; Buzmakov, A.V.; Dorokhin, Y.P.; Glagolev, I.P.; Zanevskii, Y.V.; Zryuev, V.N.; Krivonosov, Y.S.; Mamich, V.F.; Moseiko, L.A.; et al. An X-ray diffractometer with a mobile emitter-detector system. *Instrum. Exp. Technol.* **2002**, *48*, 364–372. [CrossRef]
29. Vinogradov, A.V.; Zorev, N.N.; Kozhevnikov, I.V.; Yakushkin, I.G. Phenomenon of total external reflection of x-rays. *Sov. Phys. JETP* **1985**, *62*, 1225–1229.
30. Kozhevnikov, I.V.; Pyatakhin, M.V. Use of DWBA and perturbation theory in X-ray control of the surface roughness. *J. X-Ray Sci. Technol.* **1998**, *8*, 253–275.
31. Kozhevnikov, I.V. General laws of x-ray reflection from rough surfaces: II. Conformal roughness. *Cryst. Rep.* **2012**, *57*, 490–498. [CrossRef]
32. Whitehouse, D.J. *Surfaces and Their Measurements*; Hermes Penton: London, UK, 2002; ISBN 1-56032-969-6.
33. Tolan, M. *X-ray Scattering from Soft-Matter Thin Films: Materials Science and Basic Research*; Springer: Berlin/Heidelberg, Germany, 1999; ISBN 978-3-540-49525-3.
34. Kozhevnikov, I.V. Physical analysis of the inverse problem of X-ray reflectometry. *Nucl. Instrum. Methods Phys. Res. A.* **2003**, *508*, 519–541. [CrossRef]
35. Wang, H.; Zhang, H.; Wang, Y.; Tan, W.; Huo, D. Spin glass feature and exchange bias effect in metallic Pt/antiferromagnetic LaMnO3heterostructure. *J. Phys. Condens. Matter* **2021**, *33*, 285802. [CrossRef] [PubMed]
36. Zhang, H.; Wang, Y.; Wang, H.; Huo, D.; Tan, W. Room-temperature magnetoresistive and magnetocaloric effect in $La_{1-x}Ba_xMnO_3$ compounds: Role of Griffiths phase with ferromagnetic metal cluster above Curie temperature. *J. Appl. Phys.* **2022**, *131*, 043901. [CrossRef]
37. Church, E.; Takasz, P. Light scattering from non-Gaussian surfaces. *Proc. SPIE* **1995**, *2541*, 91–107.

38. Kozhevnikov, I.V.; Peverini, L.; Ziegler, E. Development of a self-consistent free-form approach for studying the three-dimensional morphology of a thin film. *Phys. Rev. B* **2012**, *85*, 125439. [CrossRef]
39. Virtanen, P.; Gommers, R.; Oliphant, T.E.; Haberland, M.; Reddy, T.; Cournapeau, D.; Van Mulbregt, P. SciPy 1.0 Contributors. SciPy 1.0: Fundamental algorithms for scientific computing in Python. *Nat. Methods* **2020**, *17*, 261–272. [CrossRef]
40. Lebedev, A.A.; Kotousova, I.S.; Lavrent'ev, A.A.; Lebedev, S.P.; Makarenko, I.V.; Petrov, V.N.; Titkov, A.N. Formation of nanocarbon films on the SiC surface through sublimation in vacuum. *Phys. Solid State* **2009**, *51*, 829–832. [CrossRef]
41. Lin, Y.M.; Farmer, D.B.; Jenkins, K.A.; Wu, Y.; Tedesco, J.L.; Myers-Ward, R.L.; Eddy, C.R.; Gaskill, D.K.; Dimitrakopoulos, C.; Avouris, P. Enhanced performance in epitaxial graphene FETs with optimized channel morphology. *IEEE Electron Device Lett.* **2011**, *32*, 1343–1345. [CrossRef]
42. Mammadov, S.; Ristein, J.; Krone, J.; Raidel, C.; Wanke, M.; Wiesmann, V.; Speck, F.; Seyller, T. Work function of graphene multilayers on SiC(0001). *2D Mater.* **2017**, *4*, 015043. [CrossRef]
43. Lebedev, S.P.; Amel'chuk, D.G.; Eliseyev, I.A.; Barash, I.S.; Dementev, P.A.; Zubov, A.V.; Lebedev, A.A. Surface morphology control of the SiC (0001) substrate during the graphene growth. *Fuller. Nanotub.* **2020**, *28*, 281–285. [CrossRef]
44. Nyakiti, L.O.; Myers-Ward, R.L.; Wheeler, V.D.; Imhoff, E.A.; Bezares, F.J.; Chun, H.; Caldwell, J.D.; Friedman, A.L.; Matis, B.R.; Baldwin, J.W.; et al. Bilayer Graphene grown on 4H-SiC (0001) step-free mesas. *Nano Lett.* **2012**, *12*, 1749–1756. [CrossRef] [PubMed]
45. Oliveira, M.H., Jr.; Schumann, T.; Ramsteiner, M.; Lopes, J.M.J.; Riechert, H. Influence of the silicon carbide surface morphology on the epitaxial graphene formation. *Appl. Phys. Lett.* **2011**, *99*, 111901. [CrossRef]
46. Bao, J.; Yasui, O.; Norimatsu, W.; Matsuda, K.; Kusunoki, M. Sequential control of step-bunching during graphene growth on SiC (0001). *Appl. Phys. Lett.* **2016**, *109*, 081602. [CrossRef]
47. Ostler, M.; Speck, F.; Gick, M.; Seyller, T. Automated preparation of high-quality epitaxial graphene on 6H-SiC(0001). *Phys. Status Solidi B* **2010**, *247*, 2924–2926. [CrossRef]

materials

MDPI

Article

Fast Joule Heating for the Scalable and Green Production of Graphene with a High Surface Area

Zakhar Ivanovich Evseev *, Aisen Ruslanovich Prokopiev, Petr Stanislavovich Dmitriev, Nikolay Nikolaevich Loskin and Dmitrii Nikolaevich Popov

Institute of Physics and Technologies, North-Eastern Federal University, 677000 Yakutsk, Russia; ar.prokopyev@s-vfu.ru (A.R.P.); dm.ni.popov@s-vfu.ru (D.N.P.)
* Correspondence: zi.evseev@s-vfu.ru

Abstract: The rapid development of electric vehicles, unmanned aerial vehicles, and wearable electronic devices has led to great interest in research related to the synthesis of graphene with a high specific surface area for energy applications. However, the problem of graphene synthesis scalability, as well as the lengthy duration and high energy intensity of the activation processes of carbon materials, are significant disadvantages. In this study, a novel reactor was developed for the green, simple, and scalable electrochemical synthesis of graphene oxide with a low oxygen content of 14.1%. The resulting material was activated using the fast joule heating method. The processing of mildly oxidized graphene with a high-energy short electrical pulse (32 ms) made it possible to obtain a graphene-based porous carbon material with a specific surface area of up to 1984.5 m^2/g. The increase in the specific surface area was attributed to the rupture of the original graphene flakes into smaller particles due to the explosive release of gaseous products. In addition, joule heating was able to instantly reduce the oxidized graphene and decrease its electrical resistance from >10 MΩ/sq to 20 Ω/sq due to sp^2 carbon structure regeneration, as confirmed by Raman spectroscopy. The low energy intensity, simplicity, and use of environment-friendly chemicals rendered the proposed method scalable. The resulting graphene material with a high surface area and conductivity can be used in various energy applications, such as Li-ion batteries and supercapacitors.

Keywords: graphene; mildly oxidized graphene; electrochemically exfoliated graphene; fast joule heating; activated carbon; specific surface area

check for updates

Citation: Evseev, Z.I.; Prokopiev, A.R.; Dmitriev, P.S.; Loskin, N.N.; Popov, D.N. Fast Joule Heating for the Scalable and Green Production of Graphene with a High Surface Area. *Materials* **2024**, *17*, 576. https://doi.org/10.3390/ma17030576

Academic Editor: Fabrizio Roccaforte

Received: 30 December 2023
Revised: 21 January 2024
Accepted: 23 January 2024
Published: 25 January 2024

1. Introduction

Porous carbon materials (PCMs), due to their structural features and chemical properties, have a wide range of applications [1] and have attracted significant attention from the scientific community [2]. The development of methods for fast, environmentally friendly, and inexpensive production of PCMs is especially relevant in connection with the rapid development of the energy industry for electric vehicles, unmanned aerial vehicles, and wearable electronics devices. Recently, great progress has been made in the production of synthetic carbon nanomaterials, such as carbon nanotubes and graphene materials [3]. Due to their high thermal and electrical conductivities, these materials are considered promising for use in power supplies with high power densities [4]. A large amount of research has been carried out on the use of graphene-based materials for the production of PCMs. [2]. In addition, carbon nanomaterials, such as graphene oxide (GO), have been shown to be promising additives for modifying the mechanical properties of various materials, making large-scale production relevant [5,6]. Research related to GO has received much attention, owing to the low cost of its production, as well as the possibility of flexible modification [7]. GO is an oxidized form of graphene that contains a large number of functional groups in various configurations [8]. GO is an electric insulator [6]. The electrical conductivity of GO can be significantly increased by reduction, which removes oxygen functional groups [9].

Various methods for GO reduction have been proposed, such as thermal annealing [10], laser annealing [11], and microwave irradiation [12]. In terms of scalability, chemical methods using reducing agents, such as hydrazine (N_2H_4) [13] and sodium borohydride ($NaBH_4$) [14], have been well established. However, chemical reduction requires lengthy treatment times, leads to the contamination of the GO by the reducing agent, and is often performed using toxic and hazardous agents [15]. In addition, the high content of oxygen groups in the GO makes deep reduction problematic, which negatively affects the electrical conductivity. Methods for the synthesis of mildly oxidized graphene (MOG), such as electrochemical exfoliation [16], are of great interest. This method makes it possible to obtain GO with a low oxygen content of up to 20% [16], compared to 50% for GO obtained using classical methods [8]. In addition, electrochemical exfoliation can significantly reduce the synthesis time and is highly scalable. It can also be performed using non-toxic materials with high water solubility, which contributes to environmental friendliness and improves the purity of the product. A study [17] showed that electrochemical exfoliation of graphite can be carried out using various inorganic salts. The salts containing SO_4^- anions were the most effective. Na_2SO_4, owing to its low cost and toxicity, can be used for large-scale production. However, this method has some major shortcomings. The main disadvantage is the exfoliation of large graphite particles during synthesis [18]. When detached from the graphite electrode, these particles do not participate in the electrochemical process and reduce the yield of the few-layer graphene flakes. Also, during swelling and oxidation of the original electrode, an uneven distribution of the electric field occurs in the volume of the electrode [19]. In addition, the molding of graphite electrodes is a separate technological step in which various impurities in the form of various binders and contaminants can be introduced. To solve these problems, Achee et al. [18] proposed electrochemical exfoliation in a contained volume. The authors used a dialysis bag containing an electrode and a graphite powder. The use of a contained volume made it possible to increase the yield to 65%, and increased the quality of the resulting MOG. In addition, a scalable flow reactor prototype was proposed. However, in the proposed design, it is impossible to stir the graphite powder to ensure the uniform processing of the entire volume.

The next technological step in the PCM production is the activation process [20]. Upon activation, a micro- and nanoporous structure is created. This allows for high specific surface area, helping to increase adsorption or specific energy density in energy applications [21]. The most common methods for activating carbon materials are chemical and physical activations [20]. Both methods involve lengthy thermal treatment of the original carbon precursor with a strong alkali solution, such as KOH [20]. The main disadvantages of these methods are their high energy intensity, process duration, and the use of dangerous chemicals [20,21]. Recently, research on methods for processing carbon materials using fast joule heating has gained much interest [22]. This method can be used to activate oxidized carbon materials [23]. The fast joule heating technique involves the rapid reduction of oxidized material during flash heating, associated with a large discharge current. The intense release of oxygen groups occurs in the form of gaseous products, which leads to the formation of pores when the graphene flakes rupture [23]. For the application of this method, MOG is more suitable, due to its electrical conductivity, in contrast to GO obtained by the Hummers method, which is a dielectric material. In [23], the authors coated carbon fabric with a layer of GO. Electrical pulses with a duration of 50 ms were then applied to the fabric at a voltage of 30 V. After treatment with 20 pulses, the specific surface area of the composite material increased to 166 m^2/g. By mixing the initial GO with KOH, the specific surface area increased to 974 m^2/g. In [24], carbon fibers were activated via fast joule heating. It was shown that when pulses of 4 to 8 V were applied, pores formed on the surface of the fibers. Increased specific surface area had a positive effect on the electrochemical performance of the treated material. However, the specific surface area was not analyzed in this study. To the best of our knowledge, activation by fast joule heating has only been investigated in these studies, making additional research in this area relevant.

In this study, a novel reactor was developed for the electrochemical exfoliation of graphite in a closed, expandable volume, which addressed the issue of large particle exfoliation. The proposed reactor design also provided the possibility of stirring the processed powder to increase the uniformity of exfoliation. To activate the MOG, a fast joule-heating setup was developed, and optimal parameters were selected to maximize the increase in specific surface area. This study presents a cost-effective and green method for the quick production of PCMs with a high specific surface area without the use of chemical processing. Additionally, fast joule heating resulted in the simultaneous reduction of MOG and the increase in electrical conductivity. The BET method was used to study the specific surface area, revealing a substantial increase in the specific area of MOG after processing in the fast joule heating installation.

2. Materials and Methods

The synthesis of MOG was carried out in a reactor with two volumes positioned vertically to one another and separated by a polyethylene terephthalate membrane with 1 μm pores (Figure 1). The upper volume served as a container for the electrolyte and was equipped with flat gold electrodes with a surface area of 5 cm^2. A 0.1 M solution of Na_2SO_4 (Rushim, Moscow, Russia) was used as an electrolyte. The lower volume was equipped with a movable piston with a flat gold electrode with a surface area of 4 cm^2 fixed on its surface.

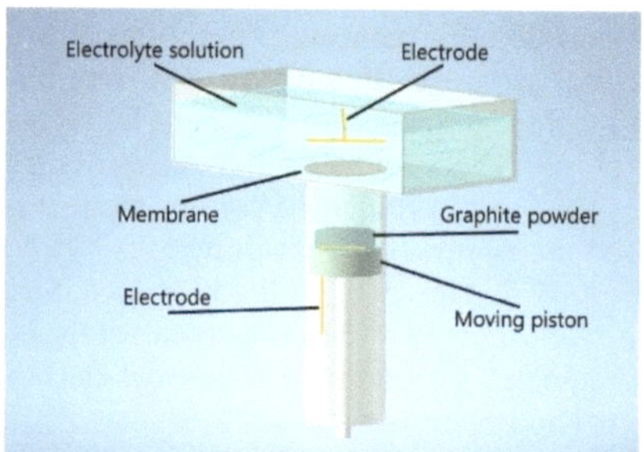

Figure 1. Schematic representation of the electrochemical reactor for MOG synthesis.

The graphite powder was processed through a number of steps. First, 2 g of flake graphite (Sigma Aldrich, St. Louis, MO, USA) was placed in a lower volume, which periodically replenished the electrolyte and mixed graphite powder by moving the piston. A positive voltage of +10 V was applied to the graphite, causing it to oxidize and exfoliate, resulting in a volume increase in the original graphite powder. The exfoliation process took 6 h, after which the small particles of oxidized graphite were collected for further treatment. The obtained product was then rinsed using a vacuum filtration unit with a copious amount of deionized water. The resulting dry residue was mixed with 100 mL of deionized water to form an aqueous suspension of oxidized graphite. The suspension was treated with ultrasound at 60 W for 3 h using an Up 200St installation (Hielscher Ultrasonics, Teltow, Germany), resulting in the peeling of oxidized graphite particles and the formation of a suspension of the few-layer MOG. The MOG suspension was then centrifuged using Eppendorf MiniSpin plus (Eppendorf AG, Hamburg, Germany) at 14,500 rpm for 10 min to separate the unexfoliated particles by decanting it over the sediment. To obtain MOG powder, freeze-drying was performed using the vacuum freeze-dryer Biobase Scientz-10N (Ningbo Scientz Biotechnology Co., Ltd., Zhejiang, China).

To activate the MOG, the methodology outlined in a previous study [25] and the studies discussed in [22] were employed. A 32 mF capacitor bank (Epcos AG, Munich, Germany) was charged to the required value and then discharged into the reaction chamber at voltages ranging from 100 to 220 V. The reaction chamber consisted of two copper electrodes that pressed the MOG powder in a quartz tube (Dominik Co., Moscow, Russia) with a diameter of 14 mm. The discharge occurred in a nitrogen atmosphere at a pressure of 0.3 mbar and took approximately 32 ms. The activated material was labeled activated MOG (aMOG), and the samples corresponding to the processing voltages were aMOG–100, aMOG–140, aMOG–180, aMOG–200, and aMOG–220.

The Raman spectra of the resulting material were examined using the NTegra Spectra installation (NT–MDT, Zelenograd, Russia). The surface morphology of the resulting material was analyzed using scanning electron microscopy (SEM) with a JEOL-7800F microscope (Jeol, Tokyo, Japan). Atomic force microscopy (AFM) was employed to study the individual MOG and aMOG flakes using Solver Next (NT–MDT, Zelenograd, Russia). The elemental composition was determined using X-ray energy-dispersive spectroscopy (EDS) with a NanoAnalysis microanalysis system (Oxford Instruments, Oxford, UK) attachment of the SEM. The functional composition was studied through Fourier-transform infrared spectroscopy (FTIR) with a Spotlight 200i spectrometer (PerkinElmer, Waltham, MA, USA). Measurements of current–voltage characteristics (C–V) were performed using a two-probe method on ASEC-03 (Prokhorov General Physics Institute of the Russian Academy of Sciences, Moscow, Russia) and AMM-3046 (Aktakom, Moscow, Russia) in the voltage range of −1 to +1 V. The specific surface area was determined using the Brunauer–Emmett–Teller method on NOVAtouch LX (Quantachrome Instruments, Inc., Boynton Beach, FL, USA) with a static volumetric method to measure the amount of adsorbed nitrogen. The preparation of the samples was carried out in accordance with the ISO 9277:2022 standard [26]. The samples were degassed at a residual pressure of 10 µT and a degassing temperature of 350 °C for 12 h. The free space ("dead volume") of the cells was initially determined using helium. The purities of N_2 and He were >99.999 %.

3. Results

Figure 2 shows AFM images of the initial MOG (a) and aMOG–200 (b) individual flakes. Thickness profiles are displayed in each image. The average lateral dimensions of individual MOG flakes were in the range between 0.05 and 0.8 µm, with thicknesses varying between 6 and 20 nm. In comparison, the lateral dimensions of the aMOG–200 flakes were within 100 nm, with thicknesses of 0.8 to 1.2 nm. On average, individual flakes of MOG exhibited a greater number of graphene layers than aMOG, as well as larger lateral dimensions. This indicates that the initial MOG flakes underwent rupture during the activation process with fast joule heating.

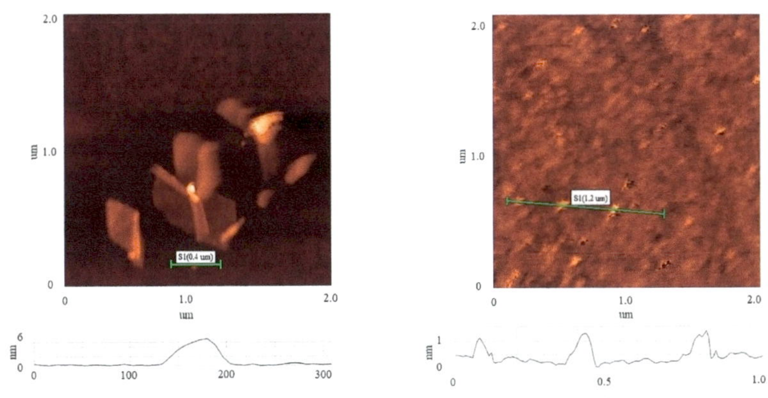

Figure 2. AFM images of individual flakes: (**a**) MOG; (**b**) aMOG–200. The height profiles of the flakes are shown at the bottom.

Images of the MOG and aMOG–200 film surfaces were obtained using the SEM (Figure 3). It is shown that the MOG film consisted of agglomerated particles formed during the drying of the MOG suspension. The surface of aMOG–200 had a micro- and nanoporous structure formed as a result of activation.

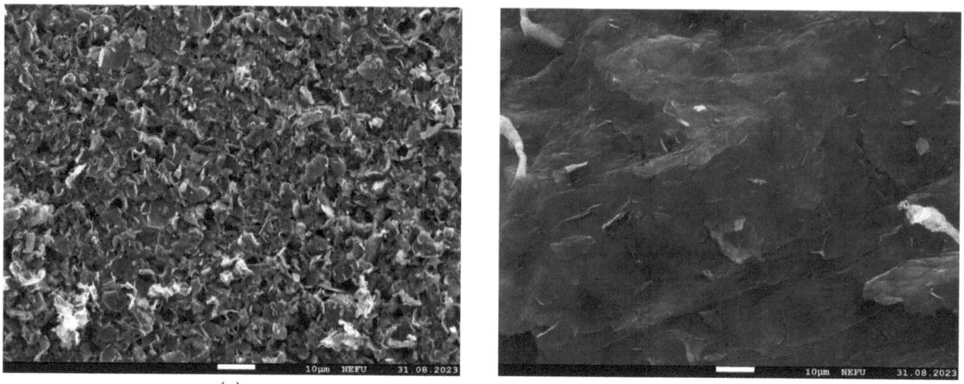

(**a**) (**b**)

Figure 3. SEM images at 1000× magnification: (**a**) MOG; (**b**) aMOG–200.

The elemental analyses of the initial MOG and aMOG–200 are presented in Table 1. It should be noted that the EDS did not allow the identification of hydrogen atoms. The data indicate that the developed reactor successfully synthesized MOG with a low oxygen content of 14.1%. MOG was reduced by rapid joule heating to an oxygen content of 4.2%.

Table 1. Carbon and oxygen contents in MOG and aMOG–200 determined by EDS.

Content	MOG	aMOG–200
C, at. %	85.9	95.8
O, at. %	14.1	4.2

Figure 4a shows the FTIR spectra of the MOG and aMOG–220. The peak localized in the vicinity of 1250 cm^{-1} corresponds to the O–H hydroxyl group [27,28]. The presence of the hydroxyl group was confirmed by the presence of vibrations in the region of 3200–3400 cm^{-1}, also associated with O–H groups and H_2O [28]. The peak associated with stretching vibrations of carboxyl groups (–COOH) was observed in the region of 1725 cm^{-1}. The disturbances localized in the vicinity of 1091 cm^{-1} were associated with the presence of epoxy groups (C–O–C). Stretching vibrations of C–O bonds of alkoxy groups in the region of 1044 cm^{-1} were present [27]. It is worth noting that for aMOG, the C–O bonds (1131 cm^{-1}) associated with carbonyl groups [27,28] were absent. The presence of C=C bonds was also observed (1680–1710 cm^{-1}), which are related to the vibrations of sp^2 crystallites of graphite [27,28].

Figure 4b shows the Raman spectra of MOG and aMOG–220 samples obtained after joule heating. All powders were characterized by the presence of the main peaks, D (1350 cm^{-1}) and G (1580 cm^{-1}), corresponding to graphite- and graphene-containing structures [29]. Raman spectra of MOG samples demonstrated the presence of a wide-band 2D peak in the region of 2700–3100 cm^{-1}. After the fast joule heating, the Raman spectra of aMOG–220 changed significantly. The D peak intensity decreased. A clearly defined peak at 2701 cm^{-1}, corresponding to the second-order 2D peak, was observed [30]. The intensity of the 2D peak shows the graphitization of the initial MOG, which, coupled with the low intensity of the D peak and the sharp G peak, indicates the formation of ordered graphene flakes [31].

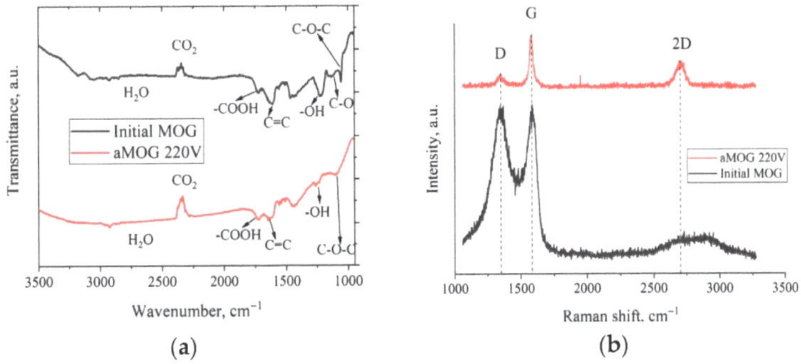

Figure 4. (**a**) FTIR spectra of MOG and aMOG–220; (**b**) Raman spectra of the MOG and aMOG–220.

The Raman spectra of aMOG obtained under various discharge voltages are shown in Figure 5a. Peaks corresponding to graphite- and graphene-containing structures (D, G, 2D, and D + G) were identified [32]. The first-order Raman peaks, called the D and G peaks in the Raman spectra, are associated with the disordering of the lattice and vibrations of sp^2-hybridized carbon bonds [33], respectively. The 2D (2700 cm^{-1}) peak of the Raman spectra corresponds to the overtone of the D peak [34], which represents the presence of graphene layers [29]. The band located at ~2900 cm^{-1} (D + G) is a combination of the D and G peaks and is also associated with defects [34,35]. As can be seen from Figure 5a, the peak at 3150 cm^{-1} disappeared with increasing voltage. This peak corresponds to the hydroxyl and carbon–hydrogen groups [36]. An increase in the energy of discharge contributed to the removal of peaks localized in the frequency range between 2900–3100 cm^{-1}. Figure 5b shows the decomposition of the Raman spectra of aMOG–220 into Lorentzian peaks. The ratio of the integrated peak intensities (I_D/I_G), which is responsible for assessing the disorder of the carbon structure, was 0.45, which corresponds to the defective graphene [37]. From the empirical formula [38] defined by Cancado et al., the lateral dimensions of sp^2 crystallites of nanographite were calculated (Table 2). The ratio of I_{2D}/I_G shows that the aMOG–220 was similar to few-layer graphene. The estimate at the first approximation was up to 7 layers [39], which correlates with the data obtained by the AFM method.

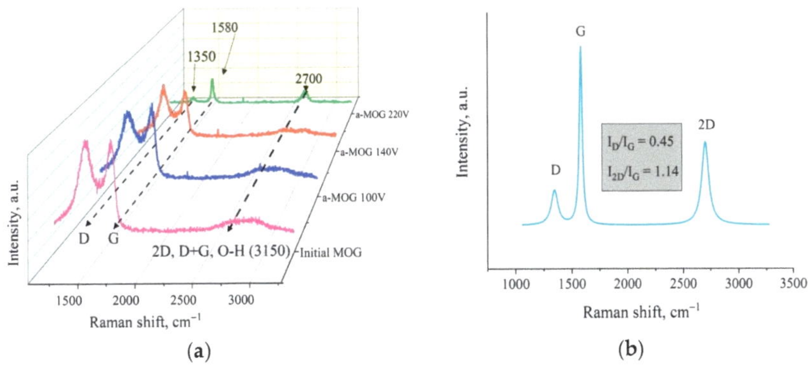

Figure 5. (**a**) Raman spectra of samples obtained after fast joule heating of the MOG at different voltages; (**b**) Lorentzians of the aMOG–220 Raman spectra.

Table 2. Values of the surface resistance (R/sq.), ratio of Raman I_D/I_G peaks, and lateral sizes of sp^2 crystallites (L_a), depending on the activation voltage.

Voltage, V	Resistance, R/sq	I_D/I_G	L_a, nm
Initial MOG	>10 MΩ	>3	–
100	<1 MΩ	>3	–
140	40 kΩ	<2	9.6
180	8 kΩ	1.4	13.7
200	0.7 kΩ	0.9	21.3
220	20 Ω	0.45	42.6

Table 2 shows the values of the surface resistance (R/sq.) of the aMOG, obtained from measurements of C–V characteristics using the two-probe method.

Figure 6 shows the average values of the surface resistance, depending on the treatment voltage. The surface resistance values were plotted as columns, and the estimated lateral sizes of sp^2 crystallites were plotted as a line. It can be seen that with increasing voltage, R decreased from the initial values (from the insulator) to tens of Ω/sq., indicating the reduction of MOG and regeneration of the graphene structure, which is confirmed by the growth of the sp^2 crystallites L_a. Due to the high degree of disorder in the initial MOG and aMOG–100, L_a was not assessed. With an increase in the lateral dimensions of the sp^2 crystallites, the electrical conductivity of aMOG increased by up to four orders of magnitude.

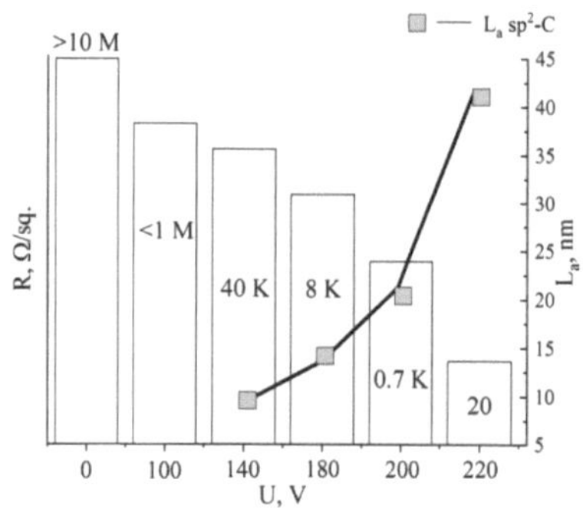

Figure 6. Histogram of R/sq. and L_a distributions depending on the voltage of joule heating. Rectangles—L_a values; text above and inside of the histogram—resistance values.

The specific surface areas measured using the BET method are listed in Table 3. From the calculated values, it can be concluded that with increasing discharge voltage, the specific surface area also increased to a maximum at 200 V—1984.5 m^2/g. However, at the discharge voltage of 220 V, a sharp decrease in the surface area was observed.

Table 3. Dependence of the specific surface area on the discharge voltage.

Voltage, V	S_{BET}, m^2/g
Initial MOG	181.5
100	278.2
140	313.3
180	414.6
200	1984.5
220	260.5

BET studies showed that the adsorption isotherms for all samples, according to the IUPAC classification, belonged to type IV. The hysteresis loops in all isotherms were of the

H1 type, which is characteristic of agglomerates of spherical particles that are uniformly packed and similar in size. The nature of the adsorption branches at low relative pressures indicated the presence of a certain number of micropores. All the samples exhibited a multimodal distribution of pore size in the range of 2–10 nm. The pore size distribution was correlated with the thicknesses of the aMOG samples measured by AFM. The distribution of the pores by surface area of the aMOG–200 with the highest measured specific surface area is presented in Figure 7b. It is shown that the pore distribution had maxima at 2.8 and 3.7 nm.

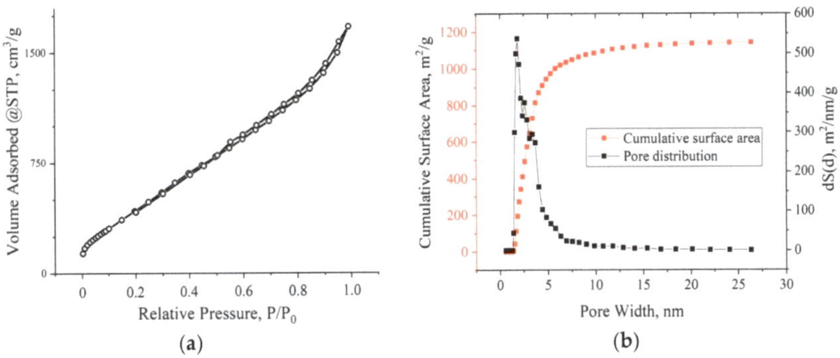

Figure 7. (**a**) Typical isotherm of nitrogen adsorption–desorption on aMOG samples; (**b**) distribution of the pore surface area of aMOG–200.

4. Discussion

The results obtained from the data indicate that the reactor developed in this study can produce multilayer flakes of MOG with an oxygen content up to 14.1%. The composition of the functional groups of the initial MOG correlates with the data obtained in other studies [16]. At the same time, the reactor design has the possibility of significant scaling. The use of the fast joule heating method made it possible to increase the conductivity of the aMOG by four orders of magnitude, which can be associated with the thermal reduction of the initial MOG to an oxygen content of 4.2%. Raman spectra analysis showed an increase in sp^2 crystallite size L_a with the increase in fast joule heating voltage, which is an indication of the reduction of MOG and the regeneration of the sp^2 carbon lattice. Simultaneously, the explosive release of oxygen groups in the form of gaseous products during fast joule heating (32 ms) led to the rupture of the initial MOG individual flakes [23]. AFM studies showed decreases in lateral dimensions from 0.8 μm to 100 nm and thicknesses from 6–20 nm to 0.8–1.2 nm for aMOG, compared to the initial MOG. This explains the significant increase in the specific surface area of aMOG–200 to 1984.5 m^2/g at a processing voltage of 200 V. The hysteresis loops of the adsorption isotherms of the material were of type H1, which is characteristic of agglomerates of spherical particles that are uniformly packed and similar in size. At the same time, the Raman spectra showed a decrease in the D peak associated with structural defects. This could indicate that percolation did not occur in the lateral plane of the flakes during activation by fast joule heating. The mechanism of flake rupture appears to be the delamination of the multilayer graphene into thinner flakes and lateral cracking into smaller particles. The data obtained from BET analysis indicated a significant increase in the specific surface area as the processing voltage was increased to a maximum area at 200 V. At a processing voltage of 220 V, the increase in the specific area was significantly lower. It can be speculated that this effect may be associated with a regeneration of the sp^2 structure, which leads to the restacking of individual flakes via the Van der Waals forces. This is indirectly confirmed by the high electrical conductivity of aMOG–220 (~20 Ω/sq). In addition, the pore size distribution in aMOG–220 shifted towards a multimodal distribution in the range >5 nm, compared to the <5 nm distribution in aMOG–200, which can be attributed to the restacking of the flakes. Additional research is required to determine the reasons for this effect.

In conclusion, an original reactor was developed for the electrochemical synthesis of MOG. The synthesis of MOG was carried out without the use of toxic and dangerous agents and also had low energy intensity, which is the basis for possible scalability. To activate the synthesized MOG and obtain graphene-based PCMs, fast joule heating was used. This method made it possible to obtain graphene-based PCMs with a high specific surface area of up to 1984.5 m^2/g. Due to its speed, simplicity, and low energy consumption, the developed technique can be used for the green production of PCMs for various energy sources, such as Li-ion batteries and supercapacitors.

Author Contributions: Conceptualization, Z.I.E. and A.R.P.; methodology, Z.I.E., A.R.P., N.N.L. and D.N.P.; validation, Z.I.E. and A.R.P.; investigation, Z.I.E., A.R.P., P.S.D., N.N.L. and D.N.P.; writing—original draft preparation, Z.I.E., A.R.P. and P.S.D.; writing—review and editing, Z.I.E., A.R.P. and P.S.D. All authors have read and agreed to the published version of the manuscript.

Funding: This research was funded by the Ministry of Education and Science of the Russian Federation (the state task FSRG–2022–0011).

Institutional Review Board Statement: Not applicable.

Informed Consent Statement: Not applicable.

Data Availability Statement: The data presented in this study are available upon request from the corresponding author. The data are not publicly available due to the data protection policy of the university.

Acknowledgments: We thank Vinokurov P.V. for his help with the AFM and Raman spectra measurements.

Conflicts of Interest: The authors declare no conflicts of interest. The funders had no roles in the design of the study; in the collection, analyses, or interpretation of data; in the writing of the manuscript; or in the decision to publish the results.

References

1. Liu, P.; Chen, G.-F. Applications of Porous Ceramics. In *Porous Materials Processing and Applications*, 1st ed.; Elsevier Inc.: Amsterdam, The Netherlands, 2014; pp. 303–344.
2. Huang, H.; Shi, H.; Das, P.; Qin, J.; Li, Y.; Wang, X.; Su, F.; Wen, P.; Li, S.; Lu, P.; et al. The chemistry and promising applications of graphene and porous graphene materials. *Adv. Funct. Mater.* **2020**, *30*, 1909035. [CrossRef]
3. Wang, Y.; Chen, J.; Ihara, H.; Guan, M.; Qiu, H.D. Preparation of porous carbon nanomaterials and their application in sample preparation: A review. *TrAC Trend. Anal. Chem.* **2021**, *143*, 116421. [CrossRef]
4. Han, J.; Johnson, I.; Chen, M. 3D continuously porous graphene for energy applications. *Adv. Mater.* **2022**, *34*, 2108750. [CrossRef]
5. Das, A.; Deka, J.; Rather, A.M.; Bhunia, B.K.; Saikia, P.P.; Mandal, B.B.; Raidongia, K.; Manna, U. Strategic formulation of graphene oxide sheets for flexible monoliths and robust polymeric coatings embedded with durable bioinspired wettability. *ACS Appl. Mater. Interfaces* **2017**, *9*, 42354–42365. [CrossRef]
6. Das, A.; Deka, J.; Raidongia, K.; Manna, U. Robust and self-healable bulk-superhydrophobic polymeric coating. *Chem. Mater.* **2017**, *29*, 8720–8728. [CrossRef]
7. Hooch Antink, W.; Choi, Y.; Seong, K.D.; Kim, J.M.; Piao, Y. Recent progress in porous graphene and reduced graphene oxide-based nanomaterials for electrochemical energy storage devices. *Adv. Mater. Interfaces* **2018**, *5*, 1701212. [CrossRef]
8. Zhu, Y.; Murali, S.; Cai, W.; Li, X.; Suk, J.W.; Potts, J.R.; Ruoff, R.S. Graphene and graphene oxide: Synthesis, properties, and applications. *Adv. Mater.* **2010**, *22*, 3906–3924. [CrossRef]
9. Pei, S.; Cheng, H.-M. The reduction of graphene oxide. *Carbon* **2012**, *50*, 3210–3228. [CrossRef]
10. Alam, S.N.; Sharma, N.; Kumar, L. Synthesis of graphene oxide (GO) by modified hummers method and its thermal reduction to obtain reduced graphene oxide (rGO). *Graphene* **2017**, *6*, 1–18. [CrossRef]
11. Huang, L.; Liu, Y.; Ji, L.C.; Xie, Y.Q.; Wang, T.; Shi, W.Z. Pulsed laser assisted reduction of graphene oxide. *Carbon* **2011**, *49*, 2431–2436. [CrossRef]
12. Voiry, D.; Yang, J.; Kupferberg, J.; Fullon, R.; Lee, C.; Jeong, H.Y.; Shin, H.S.; Chhowalla, M. High-quality graphene via microwave reduction of solution-exfoliated graphene oxide. *Science* **2016**, *353*, 1413–1416. [CrossRef] [PubMed]
13. Park, S.; An, J.; Potts, J.R.; Velamakanni, A.; Murali, S.; Ruoff, R.S. Hydrazine-reduction of graphite- and graphene oxide. *Carbon* **2011**, *49*, 3019–3023. [CrossRef]

14. Shin, H.J.; Kim, K.K.; Benayad, A.; Yoon, S.; Park, H. Efficient reduction of graphite oxide by sodium borohydride and its effect on electrical conductance. *Adv. Funct. Mater.* **2009**, *19*, 1987–1992. [CrossRef]
15. de Barros, N.G.; Gonzaga Neto, A.C.; Vaccioli, K.B.; Angulo, H.R.V.; de Andrade e Silva, L.G.; Toffoli, S.M.; Valera, T.S. Graphene Oxide: A Comparison of Reduction Methods. *C* **2023**, *9*, 73. [CrossRef]
16. Vasilieva, F.D.; Kapitonov, A.N.; Yakimchuk, E.A.; Smagulova, S.A.; Antonova, I.V.; Kotin, I.A. Mildly oxidized graphene oxide suspension for printing technologies. *Mater. Res. Express* **2018**, *5*, 65608. [CrossRef]
17. Parvez, K.; Wu, Z.S.; Li, R.; Liu, X.; Graf, R.; Feng, X.; Müllen, K. Exfoliation of graphite into graphene in aqueous solutions of inorganic salts. *J. Am. Chem. Soc.* **2014**, *136*, 6083–6091. [CrossRef] [PubMed]
18. Achee, T.C.; Sun, W.; Hope, J.T.; Quitzau, S.G.; Sweeney, C.B.; Shah, S.A.; Habib, T.; Green, M.J. High-Yield Scalable Graphene Nanosheet Production from Compressed Graphite Using Electrochemical Exfoliation. *Sci. Rep.* **2018**, *8*, 14525. [CrossRef] [PubMed]
19. Muhsan, A.A.; Lafdi, K. Numerical study of the electrochemical exfoliation of graphite. *SN Appl. Sci.* **2019**, *1*, 276. [CrossRef]
20. Heidarinejad, Z.; Dehghani, M.H.; Heidari, M.; Javedan, G.; Ali, I.; Sillanpää, M. Methods for preparation and activation of activated carbon: A review. *Environ. Chem. Lett.* **2020**, *18*, 393–415. [CrossRef]
21. Wang, G.; Yu, M.; Feng, X. Carbon materials for ion-intercalation involved rechargeable battery technologies. *Chem. Soc. Rev.* **2021**, *50*, 2388–2443. [CrossRef]
22. Wyss, K.M.; Luong, D.X.; Tour, J.M. Large-scale syntheses of 2D Materials: Flash joule heating and other methods. *Adv. Mater.* **2021**, *34*, 2106970. [CrossRef]
23. Karim, G.M.; Dutta, P.; Majumdar, A.; Patra, A.; Deb, S.K.; Das, S.; Dambhare, N.V.; Rath, A.K.; Maiti, U.N. Ultra-fast electro-reduction and activation of graphene for high energy density wearable supercapacitor asymmetrically designed with MXene. *Carbon* **2023**, *203*, 191–201. [CrossRef]
24. Zhao, Y.; Liu, H.; Li, S.; Chen, P.; Jiang, S.; Liu, J.; Meng, F. Rapid joule-heating activation boosted capacitive performance of carbon fibers. *Compos. Commun.* **2022**, *34*, 101263. [CrossRef]
25. Prokop'ev, A.R.; Neustroev, E.P.; Loskin, N.N.; Vinokurov, P.V.; Evseev, Z.I.; Popov, D.N. Poluchenie turbostratnogo grafena iz plastikovyh othodov. *Vestn. Sev.—Vostochnogo Fed. Univ. MK Ammosova* **2023**, *3*, 33–41. (In Russian)
26. *ISO 9277:2022;* Determination of the Specific Surface Area of Solids by Gas Adsorption. IUPAC: Zurich, Switzerland, 2022.
27. Jamil, A.; Mustafa, F.; Aslam, S.; Arshad, U.; Ahmad, M.A. Structural and optical properties of thermally reduced graphene oxide for energy devices. *Chin. Phys. B* **2017**, *26*, 086501–086508. [CrossRef]
28. Chen, D.; Feng, H.; Li, J. Graphene Oxide: Preparation, Functionalization, and Electrochemical Applications. *Chem. Rev.* **2012**, *112*, 6027–6053. [CrossRef]
29. Malard, L.M.; Pimenta, M.A.; Dresselhaus, G.; Dresselhaus, M.S. Raman spectroscopy in graphene. *Phys. Rep.* **2009**, *473*, 51–87. [CrossRef]
30. Das, A.; Chakraborty, B.; Sood, A.K. Raman spectroscopy of graphene on different substrates and influence of defects. *Bull. Mater. Sci.* **2008**, *31*, 579–584. [CrossRef]
31. Roscher, S.; Hoffmann, R.; Ambacher, O. Determination of the graphene-graphite ratio of graphene powder by Raman 2D band symmetry analysis. *Anal. Methods* **2019**, *11*, 1224–1228. [CrossRef]
32. Ferrari, A.C.; Basko, D.M. Raman spectroscopy as a versatile tool for studying the properties of graphene. *Nat. Nano* **2013**, *8*, 235–246. [CrossRef]
33. Compagnini, G.; Giannazzo, F.; Sonde, S.; Raineri, V.; Rimini, E. Ion irradiation and defect formation in single layer graphene. *Carbon* **2009**, *47*, 3201–3207. [CrossRef]
34. Díez-Betriu, X.; Álvarez-García, S.; Botas, C.; Álvarez, P.; Sánchez-Marcos, J.; Prieto, C.; Menendez, R.; De Andrés, A. Raman spectroscopy for the study of reduction mechanisms and optimization of conductivity in graphene oxide thin films. *J. Mater. Chem. C* **2013**, *1*, 6905. [CrossRef]
35. Giannazzo, F.; Dagher, R.; Schilirò, E.; Panasci, S.E.; Greco, G.; Nicotra, G.; Roccaforte, F.; Agnello, S.; Brault, J.; Cordier, Y.; et al. Nanoscale structural and electrical properties of graphene grown on AlGaN by catalyst-free chemical vapor deposition. *Nat. Nanotechnol.* **2020**, *32*, 015705. [CrossRef]
36. Beams, R.; Gustavo Cançado, L.; Novotny, L. Raman characterization of defects and dopants in graphene. *J. Phys. Condens. Matter* **2015**, *27*, 083002. [CrossRef]
37. Bhatt, M.D.; Kim, H.; Kim, G. Various defects in graphene: A review. *RSC Adv.* **2022**, *12*, 21520–21547. [CrossRef] [PubMed]
38. Cançado, L.G.; Takai, K.; Enoki, T.; Endo, M.; Kim, Y.A.; Mizusaki, H.; Jorio, A.; Coelho, L.N.; Magalhães-Paniago, R.; Pimenta, M.A. General equation for the determination of the crystallite size La of nanographite by Raman spectroscopy. *Appl. Phys. Lett.* **2006**, *88*, 163106. [CrossRef]
39. Graf, D.; Molitor, F.; Ensslin, K.; Stampfer, C.; Jungen, A.; Hierold, C.; Wirtz, L. Spatially resolved Raman spectroscopy of single-and few-layer graphene. *Nano Lett.* **2007**, *7*, 238–242. [CrossRef]

materials

MDPI

Article

Evaluation of Effective Mass in InGaAsN/GaAs Quantum Wells Using Transient Spectroscopy

Lubica Stuchlikova [1], Beata Sciana [2], Arpad Kosa [1], Matej Matus [1], Peter Benko [1], Juraj Marek [1], Martin Donoval [1], Wojciech Dawidowski [2], Damian Radziewicz [2] and Martin Weis [1,*]

[1] Institute of Electronics and Photonics, Slovak University of Technology in Bratislava, Ilkovicova 3, 81219 Bratislava, Slovakia
[2] Faculty of Electronics, Photonics and Microsystems, Wrocław University of Technology, 11/17 Janiszewskiego St., 50-372 Wrocław, Poland
* Correspondence: martin.weis@stuba.sk

Abstract: Transient spectroscopies are sensitive to charge carriers released from trapping centres in semiconducting devices. Even though these spectroscopies are mostly applied to reveal defects causing states that are localised in the energy gap, these methods also sense-charge from quantum wells in heterostructures. However, proper evaluation of material response to external stimuli requires knowledge of material properties such as electron effective mass in complex structures. Here we propose a method for precise evaluation of effective mass in quantum well heterostructures. The infinite well model is successfully applied to the InGaAsN/GaAs quantum well structure and used to evaluate electron effective mass in the conduction and valence bands. The effective mass m/m_0 of charges from the conduction band was 0.093 ± 0.006, while the charges from the valence band exhibited an effective mass of 0.122 ± 0.018.

Keywords: quantum well; electron effective mass; transient spectroscopy

Citation: Stuchlikova, L.; Sciana, B.; Kosa, A.; Matus, M.; Benko, P.; Marek, J.; Donoval, M.; Dawidowski, W.; Radziewicz, D.; Weis, M. Evaluation of Effective Mass in InGaAsN/GaAs Quantum Wells Using Transient Spectroscopy. *Materials* **2022**, *15*, 7621. https://doi.org/10.3390/ma15217621

Academic Editor: Fabrizio Roccaforte

Received: 7 October 2022
Accepted: 28 October 2022
Published: 30 October 2022

1. Introduction

The effort to decarbonise energy production and the recent energy crisis caused by the natural gas shortage are driving forces to implement renewable energy sources such as photovoltaic (PV) systems. However, even though silicon-based PV systems are the most common on the market today, the implementation of advanced semiconducting materials and structures is needed to achieve higher power conversion efficiency. Several semiconducting material families have been investigated for PV applications [1,2]; unfortunately, the single-junction solar cell cannot easily cover the whole solar spectrum. The introduction of quantum wells into PV semiconducting material is one of the envisioned approaches for improving its power conversion efficiency by expanding the absorption spectrum [3,4].

High-performance PV devices require novel materials, advanced device design, and optimised fabrication technology that provides almost defectless semiconducting materials. The semiconductor structural defects can be represented by energy states localised in the energy gap and cause photogenerated charge carrier trapping and recombination. There are two fundamental approaches to studying charge carrier nature. The photogenerated carrier lifetime before trapping and recombination is investigated by lifetime spectroscopy [5], whereas deep-level transient spectroscopy studies the charge released from traps [6]. Here should be noted that the spectroscopies characterise defects only in bulk semiconductors; the study of heterostructures always reveals the material properties of each layer as well as the influence of mutual interfaces. The application of transient spectroscopies to observe charge carriers released from the quantum well was already reported [7–9]; however, the analysis did not go beyond the observation of the charge release phenomenon. A more detailed analysis requires an estimation of the electron's effective mass. Despite great progress in developing novel materials and advanced heterostructure devices, many fundamental material parameters remain not fully understood or unknown. Furthermore, ternary or quaternary semiconductors or complex heterostructures are still a great challenge for experimental methods. There are two fundamental approaches for effective mass estimation in quantum well heterostructures. The first group of methods uses optical phenomena such as angle-resolved photoemission spectroscopy (ARPES) [10], optically detected cyclotron resonance (ODCR) [11], or room-temperature optoelectronic properties (RTOP) such as reflectivity, photoluminescence or photovoltaic properties [12]. The second group of experimental techniques is based on the application of the magnetic field such as de Haas–van Alphen effect (dHvA) [13], Shubnikov-de Haas (SdH) oscillations [14], or optical quantum Hall effect (QHE) [15]. There are also other experimental techniques, such as electron energy-loss spectroscopy (EELS) using scanning electron microscopy [16]; however, their application is rare. The disadvantage of these experimental techniques is that they cannot be applied directly on an electronic device such as a diode or solar cell; hence, the results may not be fully applicable due to differences in investigated materials and/or structures. Interestingly, the pure electrical method for the evaluation of effective mass at electronic devices is still missing. As a result, the lifetime or deep-level transient spectroscopies have no other option but to apply effective mass estimated on different subjects of study.

The GaAs-based heterostructures are used as a benchmark for quantum well devices since they can be fabricated at very high reliability, and the GaAs materials are still very promising in photovoltaic applications [4,17]. Even though there are several reports on effective mass InGaAsN/GaAs or GaAsN/GaAs quantum wells, the analysis is based mostly on computer simulations [18,19] or analysis of optical properties using band anticrossing model [20,21]. As a result, there is a need for effective mass experimental estimation using an electronic device structure.

Here we report the method for evaluating the effective mass in InGaAsN/GaAs quantum wells structures using a deep-level transient spectroscopy technique. The infinite potential well model was applied to analyse the activation energies of charge carriers released from quantum wells, and the effective mass is estimated across a large number of structures.

2. Materials and Methods

The quantum well heterostructures were grown by atmospheric pressure metalorganic vapour phase epitaxy (APMOVPE) using a horizontal reactor (model AIX200 R&D by AIXTRON, Germany) on (100)-oriented semi-insulating GaAs or Si-doped n-type GaAs substrates. Trimethylgallium, trimethylindium, tertiarybutylhydrazine and arsine (10% AsH_3:H_2 mixture) were employed as the precursors, while the high-purity hydrogen was used as a carrier gas. The heterostructures consisted of 450 nm thick GaAs buffer layer, followed by triple $In_yGa_{1-y}As_{1-x}N_x$/GaAs quantum wells region, and capped by 40–50 nm thick GaAs, as depicted in Figure 1 [22]. A family of 15 heterostructures of various InGaAsN thicknesses and indium/nitrogen ratios was fabricated for this study. The InGAsN layer thicknesses d varied from 6 to 19.8 nm. The nitrogen content x ranged from 0 to 1.2%, whereas the indium content y ranged from 0 to 16%. The concentric ring geometry was applied for electrode topology. Here, the voltage pulse was applied at the Pt/Ti/Pt/Au electrode, while the AuGe/Ni/Au provided the ground.

High-resolution X-ray diffractometry (HRXRD) was applied to study the structural properties of the fabricated heterostructures. The diffractograms were recorded by MRD X'Pert diffractometer (Philips, The Netherlands), using a four-crystals Bartels (220) monochromator and CuKα1 radiation (1.5406 Å). The diffractograms of the (004) symmetrical reflection were evaluated using dynamical diffraction analysis. X'Pert Epitaxy v.4.1 (PANalytical B.V., The Netherlands) software tool was employed for diffractogram modelling.

The transient spectroscopy was done by recording the capacity response on external voltage stimulus using DL8000 system (BIO-RAD Micromeasurement, Mountain View, CA, USA). The temperature varied from 80 to 450 K to observe shallow states related to quantum well structures as well as the deep states representing structural defects. The excitation voltage was 0.05 V, while the charge release was observed at the reverse voltage of -0.5 V. The filling pulses of at least 3 ms were long enough to fill all states, while the transients were recorded for the period T_w up to 1 s to observe released charge carriers. The Fourier transform analysis was employed to analyse the multiexponential transients and evaluate the Arrhenius plots [23].

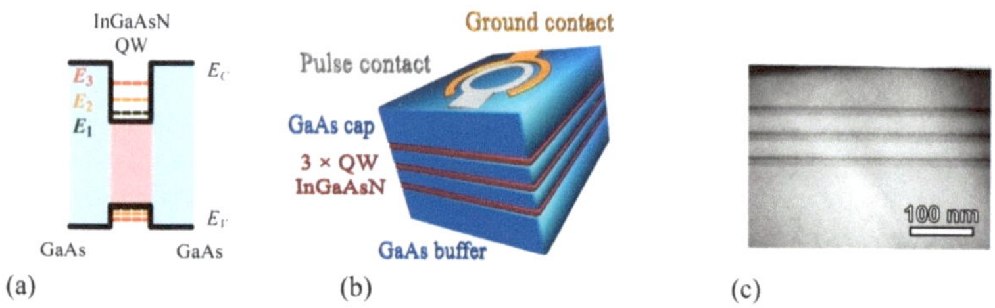

Figure 1. (**a**) Sketch of the energy band diagram of InGaAsN/GaAs quantum well structure. (**b**) A simplified view of the triple quantum well structure and (**c**) micrograph (transmission electron microscopy in the bright field) of fabricated heterostructure.

3. Results and Discussion

All fabricated devices exhibited diode-like rectifying nature with a low leakage current in the negative voltage region, as depicted in Figure 2a, which is required for the capacitance-based transient spectroscopy. Hence, the unoccupied states were filled at 0.05 V, and capacitance transients representing the charge release were investigated at the negative voltage of -0.5 V. Note that the discrete Fourier coefficients a_n, b_n can be estimated from each transient since the time constant τ follows the relation

$$\tau = (n\omega)^{-1} b_n / a_n, \tag{1}$$

where a_n, b_n stand for the cosine and sine coefficients of n-th order, and the angular frequency ω is defined by the measurement period $\omega = 2\pi/T_w$ [23,24]. Figure 2b illustrates the typical spectrum of b_1 coefficient, which denotes the first sine coefficient. The Fourier coefficient spectrum reveals the high concentration of charge released that is usually assigned to deep states of structural defects and the small concentration of charge released from quantum well states or interface charging. The deep insight can provide the Arrhenius relation for emission rate $e_n = \tau^{-1}$ as

$$e_n = N_c \langle v_n \rangle \sigma_n \exp(-(E_c - E_t)/\mathrm{k}T), \tag{2}$$

where is N_c effective density of states at the conduction band edge, $\langle v_n \rangle$ is the average thermal velocity of electrons, σ_n is the capture cross-section, $\mathrm{k}T$ is the thermal energy (k is the Boltzmann constant, and T is the thermodynamic temperature) and $E_c - E_t$ is the energy difference between the conduction band edge and the trap level localised in the energy gap. It should be mentioned here that the density of states and average thermal velocity depend on electron effective mass m_n^* that can be recognised as a material parameter. As a result, the energy difference $E_c - E_t$ (also denoted as the activation energy) can be estimated precisely, whereas the capture cross-section can be evaluated only if the effective mass is known. Figure 2c depicts a typical Arrhenius plot of InGaAsN/GaAs heterostructure. Since the effective mass in InGaAsN/GaAs heterostructure is not known, we assume $m_n^*/m_0 = 1$. Interestingly, the charge released from the trap states can be easily assigned to the structural defects with activation energies of 0.694 ± 0.004 and 0.411 ± 0.001 eV. Both 0.4 and 0.7 eV trap states are well known and belong to the defects denoted as EL2 and EL5, respectively. The trap state with the activation energy of 0.411 eV is assigned to the complex defect involving the arsenic antisite, usually denoted as EL2 [25]. The trap state with the activation energy of 0.694 eV belongs to the Ga vacancy or Ga-As divacancy, known as EL5 [26].

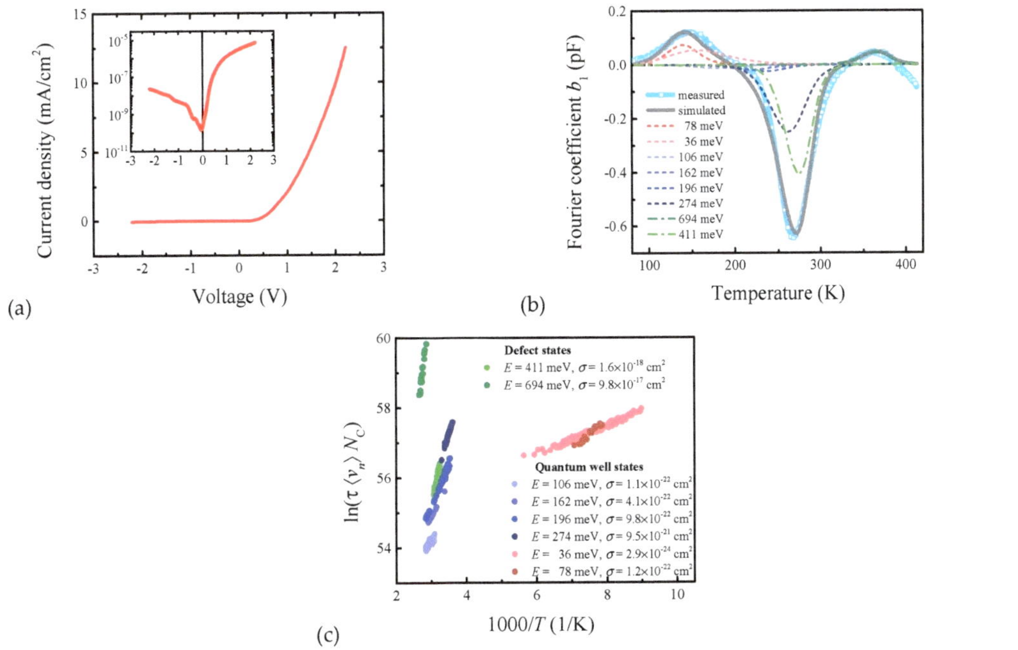

Figure 2. (**a**) Steady-state current-voltage characteristic of heterostructure with $x = 0.32\%$, $y = 13.3\%$, and $d = 16.6$ nm. (**b**) The spectrum of the Fourier coefficient b_1 together with simulations of various energy states. (**c**) The Arrhenius plot.

On the other hand, the charge released from quantum wells exhibits lower activation energies and extremely low capture cross-section that points out different physical phenomena. Since the quantum well in InGaAsN/GaAs heterostructure is due to energy band offsets, as shown in Figure 1a, the slight differences in In/N ratios affect the quantum well depth. Hence, the depth was estimated assuming linear superposition of indium and nitrogen dependences of conduction (valence) band edges as illustrated in Supplementary Material Section S1 [27]. The quality of this estimation can be verified by quantum well states occupancy analysis. For energies higher than 3kT above the Fermi energy E_F (i.e., $E \gg E_F + 3\mathrm{k}T$), the distribution function f can be approximated by the Boltzmann distribution

$$f = 1/\exp((E-\mu)/\mathrm{k}T), \tag{3}$$

where μ is the chemical potential that represents the quantum well depth. Since the energy level separation in quantum well always satisfy this requirement, the approximation should be valid. Furthermore, the capture cross-section estimated by the Arrhenius plot is proportional to the occupancy probability; hence, it can be used for chemical potential analysis. Figure 3 illustrates the heterostructure with $x = 0.32\%$, $y = 13.3\%$, and $d = 16.6$ nm (evaluated in Figure 2). Interestingly, the chemical potential reaches a value of 281 ± 43 meV that is close to the value of 289 meV estimated on reported energy band offsets.

The potential distribution of InGaAsN/GaAs quantum well can be approximated as being of rectangular shape; hence, the infinity potential well can be used as a simple model [28]. Here, the electrons can occupy states with specific energy eigenvalues E_n of the n-th quantum number that follow

$$E_n = \frac{\hbar^2 \pi^2}{2m_n^*}\left(n^2/d^2\right), \tag{4}$$

where is $\hbar$ is reduced Planck constant. It should be mentioned here that the InGaAsN layer thickness d representing the quantum well width was estimated by HRXRD as shown in Supplementary Material Section S2. The GaAs thermal expansion coefficient has thermal dependence; however, its average value in the examined temperature range reaches a value of about 5×10^{-6} K^{-1} [29]. As a result, the quantum well with a width of 14 nm in a temperature difference of 300 K changes by 0.15% (i.e., 0.02 nm). This difference can be neglected since it is in the range of the experimental error. Furthermore, it is interesting to note that the quantum well eigenvalue energies in Equation (4) are linearly proportional to the square of the quantum number and quantum well width ratio, n^2/d^2.

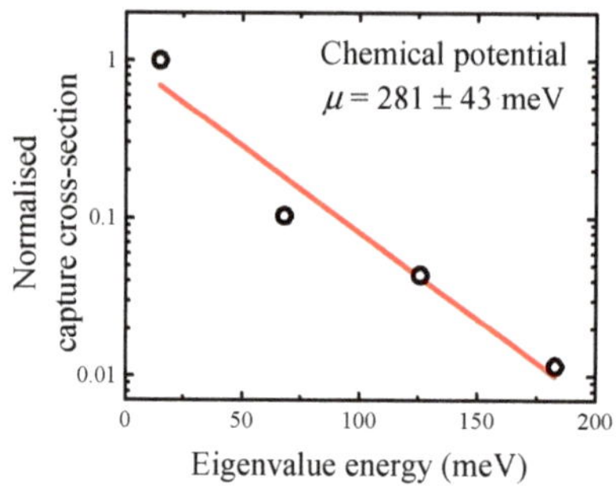

Figure 3. Normalised capture cross-section dependence on energy eigenvalues for heterostructure with $x = 0.32\%$, $y = 13.3\%$, and $d = 16.6$ nm (shown in Figure 2) with the quantum well depth of 289 meV as estimated by energy band offset reports.

As a result, it can be applied to estimate the effective mass of electrons or holes, as depicted in Figure 4. Interestingly, we can identify eigenvalues energies up to the quantum number of 4.

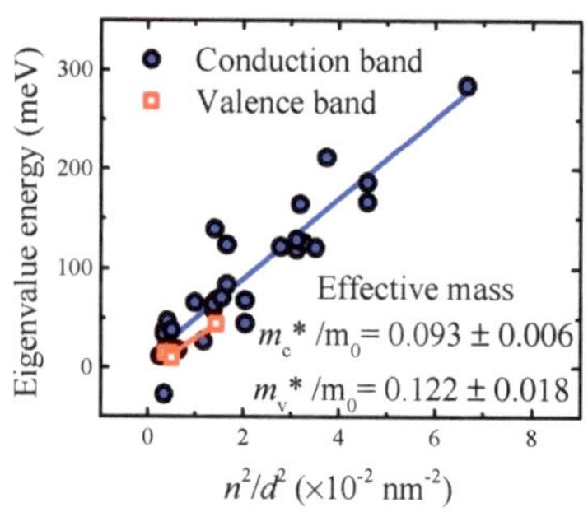

Figure 4. Eigenvalue energy dependence on n^2/d^2 ratio for conduction and valence bands. The effective mass is evaluated according to Equation (3).

The conduction band effective mass of InGaAsN/GaAs quantum well was $m_n^*/m_0 = 0.093 \pm 0.006$, while the valence band effective mass reached a value of $m_p^*/m_0 = 0.122 \pm 0.018$. These values agree with already reported effective masses in InGaAsN/GaAs quantum well heterostructures, as it is summarised in Table 1. Note that the transient capacitance spectroscopy is capable of estimating both electron and hole effective masses in a single device, whereas commonly used experimental techniques are capable of evaluating only effective mass of major charge carriers.

Table 1. Electron effective masses of InGaAsN/GaAs quantum well heterostructures estimated by various techniques.

Effective Mass $m^*_{n,p}/m_0$	Technique	Reference
0.05 ~ 0.09 (n)	RTOP	[12]
0.06 ~ 0.08 (n)	RTOP	[30]
0.077 ~ 0.078 (n)	QHE, SdH	[31]
0.07 ~ 0.12 (p)	SdH	[32]
0.07 (n)	ODCR	[33]
0.093 ± 0.006 (n)	Transient spectroscopy	This work
0.122 ± 0.018 (p)	Transient spectroscopy	This work

RTOP—room-temperature optical properties, QHE—optical quantum Hall effect, SdH—Shubnikov-de Haas oscillations, ODCR—optically detected cyclotron resonance.

4. Conclusions

The InGaAsN/GaAs quantum well structure was investigated using transient spectroscopy. The detailed evaluation of transients revealed the presence of charges released from deep energy states of structural defects as well as shallow states of quantum wells. The capture cross-section analysis confirmed the quantum well depth estimated by the energy band offsets. The infinite potential well model was employed to analyse the activation energies of released charges, and the effective masses of electrons in conduction and valence bands were estimated. The proposed methodology illustrates a novel approach to the evaluation of effective masses in quantum well heterostructures and offers a new roadmap for advanced device analysis.

Supplementary Materials: The following supporting information can be downloaded at: https://www.mdpi.com/article/10.3390/ma15217621/s1.

Author Contributions: Conceptualisation and methodology, M.W.; supervision, L.S. and M.W. investigation L.S., B.S., A.K., M.M., P.B., J.M., M.D., W.D. and D.R.; writing, M.W. funding acquisition, L.S., M.D. and J.M. All authors have read and agreed to the published version of the manuscript.

Funding: This work was co-financed by the Polish National Agency for Academic Exchange under contract number BPN/BSK/2021/1/00035/U/00001, by the Slovak Research and Development Agency under the contract number SK-PL-21-0041 and by the Scientific Grant Agency of the Slovak Republic, Grant No. 1/0727/19 and 1/0758/19, and by Wrocław University of Science and Technology subsidy.

Institutional Review Board Statement: Not applicable.

Informed Consent Statement: Not applicable.

Data Availability Statement: The data presented in this study are available in Supplementary Material.

Conflicts of Interest: The authors declare no conflict of interest.

References

1. Green, M.A.; Dunlop, E.D.; Hohl-Ebinger, J.; Yoshita, M.; Kopidakis, N.; Bothe, K.; Hinken, D.; Rauer, M.; Hao, X. Solar cell efficiency tables (Version 60). *Prog. Photovolt. Res. Appl.* **2022**, *30*, 687–701. [CrossRef]
2. Pastuszak, J.; Węgierek, P. Photovoltaic Cell Generations and Current Research Directions for Their Development. *Materials* **2022**, *15*, 5542. [CrossRef] [PubMed]
3. Barnham, K.; Ballard, I.; Barnes, J.; Connolly, J.; Griffin, P.; Kluftinger, B.; Nelson, J.; Tsui, E.; Zachariou, A. Quantum well solar cells. *Appl. Surf. Sci.* **1997**, *113*, 722–733. [CrossRef]
4. Klemmer, P.S.; Mityagin, Y.A.; Telenkov, M.P.; Nagaraja, K.K.; Elantsev, D.A.; Amiri, S. Resonant tunneling in GaAs/AlGaAs quantum well system for solar photovoltaics. *Superlattices Microstruct.* **2020**, *140*, 106472. [CrossRef]
5. Rein, S. *Lifetime Spectroscopy: A Method of Defect Characterisation in Silicon for Photovoltaic Applications*; Springer Science & Business Media: Berlin/Heidelberg, Germany, 2006.
6. Balcioglu, A.; Ahrenkiel, R.K.; Hasoon, F. Deep-level impurities in CdTe/CdS thin-film solar cells. *J. Appl. Phys.* **2000**, *88*, 7175–7178.
7. Kim, J.W.; Song, G.H.; Lee, J.W. Observation of minority-carrier traps in In Ga N/Ga N multiple-quantum-well light-emitting diodes during deep-level transient spectroscopy measurements. *Appl. Phys. Lett.* **2006**, *88*, 182103. [CrossRef]

8. Letartre, X.; Stievenard, D.; Lannoo, M.; Barbier, E. Tunnel deep level transient spectroscopy on a single quantum well. *J. Appl. Phys.* **1991**, *69*, 7336–7338. [CrossRef]
9. Letartre, X.; Stievenard, D.; Barbier, E. Accurate determination of the conduction-band offset of a single quantum well using deep level transient spectroscopy. *Appl. Phys. Lett.* **1991**, *58*, 1047–1049. [CrossRef]
10. Colakerol, L.; Veal, T.D.; Jeong, H.K.; Plucinski, L.; DeMasi, A.; Learmonth, T.; Glans, P.A.; Wang, S.; Zhang, Y.; Piper, L.F.; et al. Quantized electron accumulation states in indium nitride studied by angle-resolved photoemission spectroscopy. *Phys. Rev. Lett.* **2006**, *97*, 237601. [CrossRef]
11. Warburton, R.J.; Michels, J.G.; Nicholas, R.J.; Harris, J.J.; Foxon, C.T. Optically detected cyclotron resonance of GaAs quantum wells: Effective-mass measurements and offset effects. *Phys. Rev. B* **1992**, *46*, 13394. [CrossRef]
12. Pan, Z.; Li, L.H.; Lin, Y.W.; Sun, B.Q.; Jiang, D.S.; Ge, W.K. Conduction band offset and electron effective mass in GaInNAs/GaAs quantum-well structures with low nitrogen concentration. *Appl. Phys. Lett.* **2001**, *78*, 2217–2219. [CrossRef]
13. Knolle, J.; Cooper, N.R. Anomalous de Haas–van Alphen effect in InAs/GaSb quantum wells. *Phys. Rev. Lett.* **2017**, *118*, 176801. [CrossRef] [PubMed]
14. Hatke, A.T.; Zudov, M.A.; Pfeiffer, L.N.; West, K.W. Shubnikov–de Haas oscillations in GaAs quantum wells in tilted magnetic fields. *Phys. Rev. B.* **2012**, *85*, 241305. [CrossRef]
15. Failla, M.; Keller, J.; Scalari, G.; Maissen, C.; Faist, J.; Reichl, C.; Wegscheider, W.; Newell, O.J.; Leadley, D.R.; Myronov, M.; et al. Terahertz quantum Hall effect for spin-split heavy-hole gases in strained Ge quantum wells. *New J. Phys.* **2016**, *18*, 113036. [CrossRef]
16. Gass, M.H.; Papworth, A.J.; Joyce, T.B.; Bullough, T.J.; Chalker, P.R. Measurement of the effective electron mass in GaInNAs by energy-loss spectroscopy. *Appl. Phys. Lett.* **2004**, *84*, 1453–1455. [CrossRef]
17. Papež, N.; Dallaev, R.; Ţălu, Ş.; Kaštyl, J. Overview of the Current State of Gallium Arsenide-Based Solar Cells. *Materials* **2021**, *14*, 3075. [CrossRef]
18. Wartak, M.S.; Weetman, P. Numerical analysis of the effective masses in InGaAsN quantum-well structures with self-consistent effects. *J. Phys Cond. Matter* **2005**, *17*, 6539. [CrossRef]
19. Wartak, M.S.; Weetman, P. Characterisation of effective masses in InGaAsN quantum well structures by computer simulations. *J. Appl. Phys.* **2005**, *98*, 113705. [CrossRef]
20. Skierbiszewski, C.; Łepkowski, S.P.; Perlin, P.; Suski, T.; Jantsch, W.; Geisz, J. Effective mass and conduction band dispersion of GaAsN/GaAs quantum wells. *Phys. E* **2002**, *13*, 1078–1081. [CrossRef]
21. Shan, W.; Walukiewicz, W.; Yu, K.M.; Ager Iii, J.W.; Haller, E.E.; Geisz, J.F.; Friedman, D.J.; Olson, J.M.; Kurtz, S.R.; Xin, H.P.; et al. Band anticrossing in III–N–V alloys. *Phys. Status Solidi B* **2001**, *223*, 75–85. [CrossRef]
22. Ściana, B.; Radziewicz, D.; Pucicki, D.; Serafińczuk, J.; Dawidowski, W.; Bielak, K.; Badura, M.; Gelczuk, Ł.; Tłaczała, M.; Latkowska, M.; et al. Influence of the AP MOVPE process parameters on properties of (In, Ga)(As, N)/GaAs heterostructures for photovoltaic applications. *Electron Technol. Conf. 2013* **2013**, *8902*, 136–143.
23. Weiss, S.; Kassing, R. Deep Level Transient Fourier Spectroscopy (DLTFS)—A technique for the analysis of deep level properties. *Solid-State Electron.* **1988**, *31*, 1733–1742. [CrossRef]
24. Yakimova, R.; Paskova, T.; Hardalov, C. Behavior of an EL5-like defect in metalorganic vapor-phase epitaxial GaAs: Sb. *J. Appl. Phys.* **1993**, *74*, 6170–6173. [CrossRef]
25. Shiraki, H.; Tokuda, Y.; Sassa, K. Bistable behavior of a medium-deep center related to EL5 and EL6 in n-type bulk GaAs. *J. Appl. Phys.* **1998**, *84*, 3167–3174. [CrossRef]
26. Weiss, S.; Beckmann, R.; Kassing, R. The Electrical Properties of Zinc in Silicon. *Appl. Phys. A* **1990**, *50*, 151–156. [CrossRef]
27. Galluppi, M.; Geelhaar, L.; Riechert, H. Nitrogen and indium dependence of the band offsets in InGaAsN quantum wells. *Appl. Phys. Lett.* **2005**, *86*, 131925. [CrossRef]
28. Nag, B.R. *Physics of Quantum Well Devices*; Springer Science & Business Media: Berlin/Heidelberg, Germany, 2001.
29. Soma, T.; Satoh, J.; Matsuo, H. Thermal expansion coefficient of GaAs and InP. *Solid State Commun.* **1982**, *42*, 889–892. [CrossRef]
30. Héroux, J.B.; Yang, X.; Wang, W.I. Optical characterization of strained InGaAsN/GaAs multiple quantum wells. *Vac. Sci. Technol. B* **2002**, *20*, 1154–1157. [CrossRef]
31. Ardali, S.; Tiras, E.; Erol, A. Integer quantum Hall effect measurement analysis in Ga0. 68In0. 32N0. 017As/GaAs quantum wells with various annealing time. *Phys. B Condens. Matter* **2021**, *621*, 413305. [CrossRef]
32. Sarcan, F.; Donmez, O.; Erol, A.; Gunes, M.; Arikan, M.C.; Puustinen, J.; Guina, M. Influence of nitrogen on hole effective mass and hole mobility in p-type modulation doped GaInNAs/GaAs quantum well structures. *Appl. Phys. Lett.* **2013**, *103*, 082121. [CrossRef]
33. Baranov, P.G.; Romanov, N.G.; Preobrazhenski, V.L.; Egorova, A.Y.; Ustinov, V.M.; Sobolev, M.M. Optically-detected microwave resonance in InGaAsN/GaAs quantum wells and InAs/GaAs quantum dots emitting around 1.3 μm. *Int. J. Nanosci.* **2003**, *2*, 469–478. [CrossRef]

materials

MDPI

Article

Large-Area Mapping of Voids and Dislocations in Basal-Faceted Sapphire Ribbons by Synchrotron Radiation Imaging

Tatiana S. Argunova [1,*], Victor G. Kohn [2], Jae-Hong Lim [3], Vladimir M. Krymov [1] and Mikhail Yu. Gutkin [4,5]

1 Ioffe Institute, Russian Academy of Sciences, Polytekhnicheskaya St. 26, 194021 St. Petersburg, Russia; v.krymov@mail.ioffe.ru
2 National Research Centre 'Kurchatov Institute', Kurchatov Sqr., 1, 123182 Moscow, Russia; kohnvict@yandex.ru
3 Pohang Accelerator Laboratory, Pohang, Gyeongbuk 37673, Republic of Korea; limjh@postech.ac.kr
4 Institute for Problems in Mechanical Engineering, Russian Academy of Sciences, 199178 St. Petersburg, Russia; m.y.gutkin@gmail.com
5 Institute of Advanced Data Transfer Systems, ITMO University, 197101 St. Petersburg, Russia
* Correspondence: argunova@mail.ioffe.ru; Tel.: +7-921-387-1057

Abstract: The understanding of structural defects in basal-faceted sapphire ribbons was improved through X-ray imaging at a synchrotron source. The combination of phase contrast and X-ray diffraction makes it possible to visualize and characterize both gas voids and dislocations in the bulk of the ribbons grown by the Stepanov–LaBelle technology. Dislocations were directly related to gas voids. X-ray diffraction topography was employed to investigate the distribution, configurations, and character of the dislocations. The formation of voids of irregular shapes was detected by large-area mapping with spatial resolution in the μm range. Computer simulations of the experimental phase contrast images of microvoids were performed. The sizes of the spherical microvoids were determined. The results are discussed with reference to the available data on the emission of dislocations from the voids. The evolution of the shape, size, and arrangement of the voids during growth provides clues on the formation of block structure in basal-faceted sapphire ribbons.

Keywords: sapphire ribbons; gas voids; dislocations; synchrotron radiation imaging

Citation: Argunova, T.S.; Kohn, V.G.; Lim, J.-H.; Krymov, V.M.; Gutkin, M.Y. Large-Area Mapping of Voids and Dislocations in Basal-Faceted Sapphire Ribbons by Synchrotron Radiation Imaging. *Materials* **2023**, *16*, 6589. https://doi.org/10.3390/ma16196589

Academic Editor: Fabrizio Roccaforte

Received: 11 September 2023
Revised: 29 September 2023
Accepted: 29 September 2023
Published: 7 October 2023

1. Introduction

X-ray topography (XRT) and phase contrast imaging (PCI) methods are optimally combined with each other at the third-generation synchrotron radiation (SR) sources. Natural narrow angular collimation of SR is essential for high angular resolution of X-ray topographs. At the same time, the high spatial coherence makes it possible to observe (through the interference phenomenon) an additional phase shift variable over the cross-section of the beam that has passed through an inhomogeneous object [1,2].

PCI in combination with XRT exploits instrumental setups on special imaging beamlines, which allow the registration of both types of images using monochromatic or white SR beams. In addition, various X-ray detectors are available for recording such images. Unique facilities provide real-time experiments that shed some light on the effects of growth, solidification, or melting of crystalline materials [3–5].

Moreover, XRT and PCI can be applied sequentially with strict reference to the same area of a specimen. In this case, the spectrum and detectors change depending on the purpose of the experiment. For Bragg adjustment using the white SR beam, so that diffraction conditions are satisfied for a set of Bragg planes, a flat detector is the ideal choice. A large-area panel made with CMOS technology registers several reflections from different crystal planes at once. However, the best results are obtained with high-resolution X-ray films. As for the intensity variation caused by a variable phase shift, it is recorded using a charge-coupled device (CCD) with a micrometer pixel size. First, the X-ray intensity is converted into visible light from a crystal scintillator. Then the lens magnifies the transmitted light onto the detector. A correspondence between crystal lattice defects and inhomogeneities (pores, inclusions, and micro-cracks) is established by comparing topographs with phase contrast images [6–10].

Real crystals are not uniform across their area. To make good products, it is important to establish not only the types but also the distribution of defects. Only the PCI technique [1,2] with a fairly large field of view is suitable for this purpose. The comparison of PCI with other methods such as coherent diffraction imaging, ptychography, and microscopy shows its effectiveness for analyzing large sample areas. The experimenter can map defects and compare the XRT and PCI maps with account for scale disparities and differences in the nature of the contrasts. We emphasize that phase contrast images are interference patterns. The real size of a micro-object can be obtained only by determining the change in the phase. At the same time, the interpretation of a topograph is often possible without the assistance of the X-ray diffraction theory.

The present paper describes the investigation of gas voids and dislocations in sapphire crystals grown by the edge-defined film-fed growth (EFG) method invented by H. LaBelle and even earlier by A.V. Stepanov [11,12]. Our goal was to highlight the contribution that XRT and PCI can make (through the use of SR) for better understanding of defect formation in basal-faceted sapphire ribbons. The high demand for shaped sapphire products (see, e.g., [13]) encourages researchers to devote more attention to their study. Nevertheless, some problems arising in the growth of certain types of articles remain poorly understood. In particular, for the mass production of substrates, EFG does not look as efficient as other technologies. The main disadvantages of basal-faceted ribbons are gas voids and higher dislocation densities compared with Czochralski and Kyropoulos grown crystals [14,15]. In the meantime, a ribbon has a smooth surface and regular shape. It can be used for a substrate in its as-grown state.

In earlier works, it was observed that gas voids played an important role in the generation of dislocations in sapphire ribbons [16,17]. A model for the block structure generation in basal-faceted ribbons was proposed and confirmed by calculations of thermal stress in a growing crystal [16]. In the present paper, special attention is paid to the characteristics of gas voids and dislocations. An unambiguous relationship between them is revealed. Some specific features of the voids associated with dislocation emission from a void are discussed. As will be shown, the combination of applied imaging techniques is a particularly suitable tool for this study.

2. Materials and Methods

The growth of basal-faceted sapphire ribbons by the Stepanov/EFG method took place in a thermal zone where heat shields were used. The growth rate was 1 mm/min. Laser adjustment of the seed relative to the pulling device and the shaper allowed the achievement of a high orientation accuracy. The orientation of ribbon surfaces deviated only slightly (1.0′–3.0′) from the (0001) plane. Optical microscopy showed no blocks. The as-grown surfaces were smooth and mirror-like. Specimens were cut perpendicular to the growth direction $[10\bar{1}0]$ so that they contained the neck portion of a ribbon.

The imaging experiments were carried out at the Pohang Light Source (PLS) in Pohang, South Korea. The PLS operated with an electron energy of 3 GeV. The 6C beamline, devoted to imaging, featured a wiggler with the total generated power of 10^7–10^9 ph/s/mm^2 at 150 mA run. The angular size of the source (as seen from a point in the specimen) was in the ~10 μ radian range, leading to the transverse coherence length of several tens of microns. The beam was monochromated by a multilayer mirror with an energy resolution of $\Delta E/E \approx 2\%$. The topography and PCI experiments were carried out at photon energies of 15 and 23 keV, respectively.

A Zyla CCD (Andor, Oxford Instruments, UK) was utilized to record the phase contrast images. The CCD had 2560 × 2160 pixel resolution and 6.5 × 6.5 μm^2 pixel size. Therefore, the maximum sample area (the field of view, FOV) that the camera could image was

16.6×14.0 mm^2. The image recording was preceded by a conversion step in which the LuAG:Ce scintillator converted the radiation into the light. An optical lens served as the coupling element between the scintillator and the CCD. When a light image was magnified by a factor of 20×, the pixel-to-object size ratio decreased. The effective pixel size and the FOV were reduced down to 0.32 μm and 819×691 μm^2, respectively.

High-speed topographs were recorded on the large FOV (64×42 mm^2) detector VHR CCD (Photonic Science, Saint Leonards, UK) with a pixel size of 16 μm. High-resolution images were taken on the fine-grain film Kodak M100. The typical exposure times were short due to the high intensity available at the 6C wiggler beamline. They ranged from several tens of seconds (on the film) to several milliseconds (on the VHR CCD). The comparison of topographs and phase contrast images could not be performed within the same experimental session due to differences in the setups and viewing angles. To achieve a better match, the phase contrast images were recorded when the sample was set in the azimuth position for X-ray diffraction.

3. Results

3.1. As-Grown Dislocations

The fact that single crystals of sapphire can be grown in the shape of basal-faceted ribbons raises questions as to the configuration of dislocations in such crystals. The study of this problem is a continuation of extensive experiments carried out by chemical etching, X-ray topography, and transmission electron microscopy methods. Initially, from a theoretical study of the crystal structure, possible slip systems in sapphire were proposed such as basal, prismatic, and pyramidal. For over 70 years of research, the existence of these slip systems has been clearly confirmed. From a topographic study, basal dislocations with $\langle 11\bar{2}0 \rangle$ Burgers vectors were revealed [18–20]. The existence of the prismatic slip system $\langle 10\bar{1}0 \rangle$, $\{1\bar{2}10\}$ was well evidenced [21,22]. The reactions of dislocations were studied in detail (see, e.g., [20,22]).

Dislocation density in sapphire for application as a substrate does not exceed 10^3 cm^{-2}, which makes it possible to directly observe dislocations in single crystals grown by the heat exchange method [23] and the Czochralski [24], Verneuil [25], and Kyropoulos [26] processes. At the same time, relatively little is known about defects in profiled articles, especially ribbons. The problem of growing basal-faceted ribbons with low dislocation density has not been solved so far.

Thin ribbons with a wide basal face are characterized by high dislocation density and block structure [17]. Dislocations formed by thermal stresses set up in the solidified material during the cooling. Using a thermal field model, the authors of [16] calculated thermoelastic stresses arising in basal and prismatic slip planes. The authors arrived at the conclusion that in a thin basal-faceted ribbon, the stress acting in the basal slip system was very low (~0.1 MPa). A further conclusion was that blocks were formed from dislocations belonging to the prismatic slip system. This result has not yet been confirmed by experimental evidence. Since the yield stress τ decreases with increasing temperature and eventually becomes $\tau_{0001} < \tau_{1\bar{1}00}$ by a factor of 3–4 at T = 1000 °C [27], it is likely that basal slip is involved in stress reduction.

Here we present our results on XRT experiments with a basal-faceted ribbon of $31 \times 1 \times 265$ ($W \times H \times L$) mm^3 in size. The optimum thickness for a projection topograph depends upon the absorption of the material for the radiation used, namely, $\mu t = 1$, where μ is the absorption coefficient and t is the specimen thickness. In our case, λ = 0.827 Å so that for sapphire, μ = 18.15 cm^{-1}; therefore, the optimum thickness is 0.5 mm. The sample thickness was two times larger than the optimal value. However, $\mu t = 1.8$ does not yet fall within the interval $\mu t = 2$–5, where the dislocations can reveal both normal and reversed contrast [28].

The front view of the sample is schematically represented in Figure 1a. The arrow located in the neck portion is parallel to the direction of growth $[10\bar{1}0]$. In the sketch, the circle and box refer to the places imaged by SR XRT and PCI, respectively. Figure 1b shows a high-speed image taken on a VHR CCD detector. Features of the defect density and distribution were quickly visualized within ~2×10^{-3} s. The following characteristics were visible. First, the rows of dislocation pileups were parallel and perpendicular to the growth direction of the ribbon. Second, the pileups consisted of tangles associated with some dislocation sources. The number of the sources eventually increased with increasing distance from the seed.

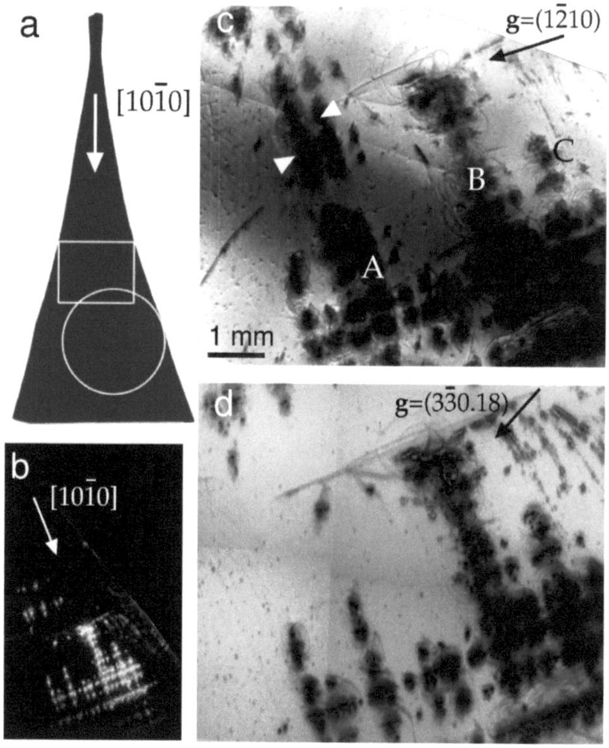

Figure 1. (**a**) Sketch of a sapphire specimen. The circle and the box show the areas imaged by XRT and PCI techniques, respectively. The arrow is the growth direction. (**b**) A high-speed topograph in back reflection taken with SR radiation (λ = 0.827 Å) on imaging detector VHR CCD, $\mathbf{g} = 3\bar{3}018$. (**c**) The topograph in transmission taken with Kodak M100 film. $\mathbf{g} = 1\bar{2}10$. (**d**) $\mathbf{g} = 3\bar{3}018$; Kodak M100.

Figure 1c,d show topographs taken with Kodak M100 fine-grain film. In high-resolution images, dense tangles were formed by curved lines that were reminiscent of dislocation semiloops. To find basal dislocations in the ribbon, a representative configuration should be revealed by a topograph taken in any $\{1\bar{2}10\}$ diffraction plane. Such a topograph of a $(1\bar{2}10)$ reflection is shown in Figure 1c. The contrast arising from pileups labeled A, B, and C stood out strongly. The rows with high dislocation density were surrounded by a relatively low dislocation density, which made it difficult to assess an average defect density.

To determine whether a particular slip pattern dominated, the contrasts in the $\mathbf{g} = 1\bar{2}10$ and $\mathbf{g} = 00018$ reflections were compared. We noticed that some contrasts visible in (c) were absent in the 00018 reflection (data not shown). To understand the variations in dislocation visibility, recall the visibility rules. Purely screw dislocations were invisible when the condition $\mathbf{g}\cdot\mathbf{b} = 0$ was satisfied, where **b** is the Burgers vector. For the case of purely edge dislocations, the visibility was least when the conditions $\mathbf{g}\cdot\mathbf{b} = 0$ and $\mathbf{g}\cdot\mathbf{n} = 0$ were satisfied simultaneously, where **n** is a normal to the slip plane.

First, consider in detail those semiloops that expand away from the big pile labeled B in Figure 1c. The images of semiloops are shown enlarged in Figure 2b. The inset shows another reflection of the same area. The white circles indicate the same location on both of the topographs. The semiloops are visible in the $\mathbf{g}_1 = 1\bar{2}10$ reflection but invisible in the $\mathbf{g}_2 = 3\bar{3}018$ reflection; therefore, $\mathbf{g}_2 \cdot \mathbf{b} = 0$, and the Burgers vector **b** of the semiloops is in the $[\bar{1}\bar{1}20]$ direction. In addition, $\mathbf{g} \cdot \mathbf{n} = 0$ because **b** lies on both the (0001) and $(3\bar{3}018)$ planes.

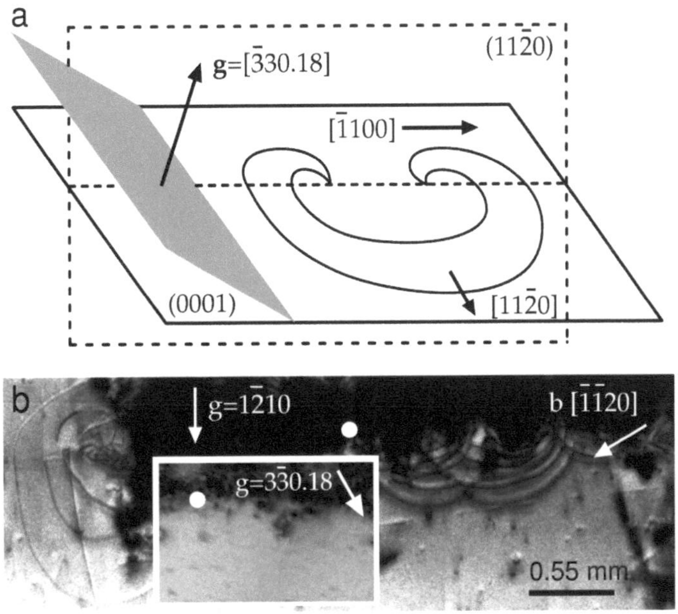

Figure 2. (**a**) Schematic relationship between the $(\bar{3}3018)$ diffracting plane and dislocation semiloops emitted by the source on the basal plane. The Burgers vectors are parallel to 1/3 $[11\bar{2}0]$. (**b**) Magnified images of semiloops in the B region of Figure 1c. The inset shows the same region where the semiloop images are extinct in the $3\bar{3}018$ reflection. The solid circles denote the same spot on both of the topographs.

A schematic representation of the semiloops along with the $(\bar{3}3018)$ reflecting plane is shown in Figure 2a. The projection of $\mathbf{g} = \bar{3}3018$ on the $[\bar{1}100]$ direction shortens the length of **g** by a factor of cos 62° (i.e., by 0.5): $\mathbf{g}_{\text{project}} = 0.74$ Å. A scalar product $\mathbf{g} \cdot \mathbf{b}$ equals 0, ∓ 3, and ± 3 for the **b** vectors $\pm 1/3$ $[11\bar{2}0]$, $\pm 1/3$ $[1\bar{2}10]$, and $\pm 1/3$ $[\bar{2}110]$, respectively. Since $\mathbf{g} \cdot \mathbf{b} = 0$ and $\mathbf{g} \cdot \mathbf{n} = 0$ are simultaneously satisfied in the $\bar{3}3018$ reflection, the dislocation semiloops in Figure 2a lie on the basal plane and belong to the (0001), $[11\bar{2}0]$ slip system.

Second, we notice an imposing group of semiloops on the left of Figure 2b. The array contains segments of different visibility. However, when the $1\bar{2}10$, $3\bar{3}018$, or 00018 topographs are examined, the semiloops are contrasted for all three reflections. It is reasonable to assume that the array lies on the prism plane $(10\bar{1}0)$ perpendicular to the growth direction.

To get an idea of dislocation sources, let us recall that gas voids are nearly always present in sapphire ribbons. Voids are usually observed in thin surface layers located at a depth of about 100 μm [29]. In the present study, the back reflection geometry was employed to reveal dislocations in a thin slice of the ribbon close to the surface. In order to evaluate the observable depth $t_{\text{obs.}}$ of dislocations, we used the relationship between $t_{\text{obs.}}$ and the absorption depth t_μ [30,31]. For low-absorbing crystals, the dislocation visibility in the back reflection was mainly determined by t_μ with the assumption that the reflected intensity of the X-rays fell down to 10% of the incident power. Considering both paths of the incident and diffracted X-rays, t_μ is calculated as follows [30]:

$$t_\mu = \frac{\ln 10 \cdot \sin\omega}{\mu} \cdot \frac{1}{1 + \sin\omega / \cos(90^\circ - 2\theta + \omega)}, \quad (1)$$

where μ = 18.147 cm^{-1}, ω = 68° is the incident angle, and θ = 40.3° is the Bragg angle. The observable depth $t_{\text{obs.}}$ = 0.23 mm is about 1/5 of the ribbon thickness.

In the back reflection geometry, the strain fields of dislocations lying deeper in the crystal did not contribute significantly to the image. However, when the upper limit of the dislocation density of about 10^6 cm^{-2} was reached, the overlap of the dislocation strain fields resulted in a darkening of the region near the dislocation sources. The intensity increase was due to the strong reflectivity of a defective crystal. Interestingly, as images of basal dislocations faded in the $3\bar{3}018$ reflection, some features became noticeable. Only features with a characteristic size of tens of microns were detected. We can suggest that when a void is large enough, it creates diffraction contrast caused by a change in thickness. As we will see later, oversized voids can merge into cavities.

Note that sources emitting dislocations on the basal plane of the sapphire were proved to operate as Frank–Read sources to emit basal dislocations [32]. Our topographic analyses revealed basal dislocations with the $[\bar{1}\bar{1}20]$ Burgers vector direction. Dislocations with the $(10\bar{1}0)$ prismatic slip plane were also observed. Those dislocations, which gave rise to pileups labeled A, B, and C, stood out strongly in all the obtained topographs. They might have had Burgers vector directions in the basal and prism planes. No experimental evidence of a predominant slip system in the specimens under consideration was found. Finally, in Figure 1 the $[10\bar{1}0]$—line-direction rows marked with arrows in (c) disappear in (d). These tangled dislocations, which are rendered invisible by the use of the back reflection, might lie deeper than the penetration depth of the X-rays.

3.2. Inhomogeneities

In the present work, micro-inhomogeneities in sapphire ribbons were revealed by the PCI method. Compared with optical microscopy, which is limited to thin slices due to the reduced focus length, the PCI technique provided high-quality images in thin and thick specimens. In the lensless scheme, a nearly parallel SR beam was used together with a pixelated detector. The spatial resolution of the image was related to the pixel size. When an optical lens magnified the image contrasts onto the CCD, the pixel-to-object size ratio was reduced. As a consequence, the FOV became smaller. An experimenter registered only a small fragment of the specimen shown by the detector. Eventually, the fragments could be assembled into a map up to several centimeters in size.

A topograph was recorded at a magnification of unity. The FOV size was shaped up due to slits, beam conditioners, and natural angular collimation of SR. The topographs were subsequently enlarged optically. In order to compare the topographs and the phase contrast images to find common features, a correct scaling needed to be determined. The scale showed a correlation between the topographic distances and the number of pixels in a digital phase contrast image. We emphasize that in the far-field conditions, the phase contrast image had a larger size compared with a small object. We further emphasize that far-field patterns were formed at rather small distances from a micro-object to the detector. Therefore, an inverse problem solution was necessary. Yet, it was always possible to measure large lengths and neglect the relative error.

Figure 3 contains the same region of the ribbon imaged by XRT (a) and PCI (b) methods. The small spots that look darker than the background in (b) can be inclusions, generally formed during the growth process. Impurities in the melt, e.g., molybdenum (Mo) from the Mo shaper, are captured by the crystallization front [33]. We observed inclusions mainly in the neck portion. To our knowledge, the impurity effects on the generation of dislocations in the sapphire ribbons were not observed, and inclusions of foreign phases were not considered here. The gas void images in Figure 3b have a simple structure consisting of a black ring around the edge and a light color in the middle. Despite the fact that the distribution of large voids can be controlled by a number of growth parameters (see, e.g., [13,14,16]), a microvoid is still an extremely interesting object of research. There are very few works on the quantitative characterization of microvoids in sapphire crystals using X-ray imaging techniques. The present article fills this gap.

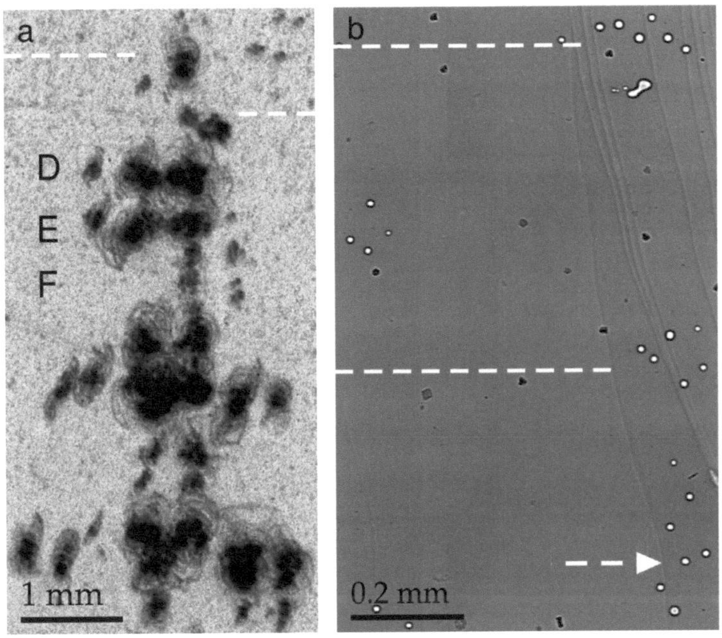

Figure 3. (**a**) General view of dislocations at the beginning of ribbon growth. The arrangement is revealed at the upper boundary of the box in Figure 1a. Lang topography; AgK_α radiation, $\bar{3}300$ reflection of sapphire. (**b**) Phase contrast image of gas voids located in the same area with dislocations. The dashed lines cross each image at the same distance from the seed.

Several features are apparent in Figure 3: the tangles of dislocation semiloops labeled D, E, and F (a); the small voids of circular shape; and the groups of voids (b). In addition, a sparse distribution of inclusions between the voids and the as-grown surface relief is shown in (b). The dashed lines are located at the same distance from the seed. A direct comparison of the voids (b) and dislocations (a) shows that the voids were very much like the dislocations in their distribution and location. One can conclude that the voids were the cause of the dislocations.

Previous XRT experiments on sapphire ribbons were carried out in defective regions when the lattice distortion was high [16,17]. The nucleation centers were not visible because of an increased background intensity occurring when a crystal was imperfect. Unfortunately, the Bragg diffraction contrast was not adequate for studying the nature of defects, when the overlap of images prevented a meaningful interpretation. On the contrary, the PCI confirmed the assignment of the nucleation centers to gas voids. The relaxation of thermal stress began with the formation of glide dislocation semiloops at the voids. The same pattern was repeated further in the direction of growth.

A small scale of the topograph in Figure 3a allows seeing almost the entire specimen area where defects were formed at the early stage of the ribbon growth. Compared with the topograph, the scale of the phase contrast images was larger. The same observable area is shown in Figures 3b and 4. Each image was taken while looking down the growth axis from the seed. The images were connected according to the levels marked with dashed arrows. The distribution of gas voids was noteworthy: they were grouped and distributed across the width of the ribbon.

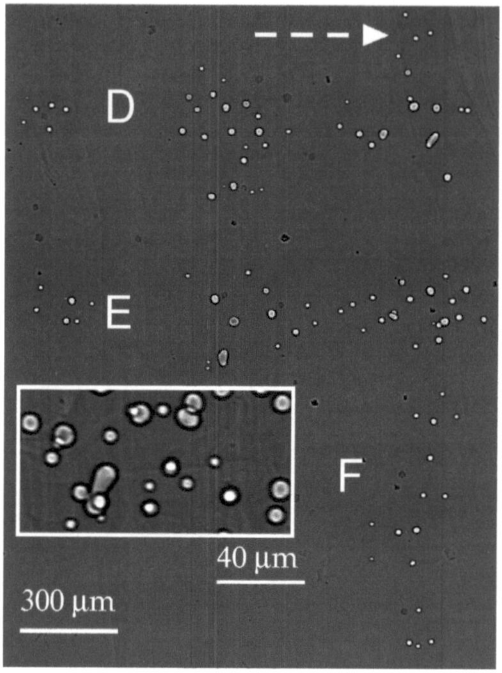

Figure 4. Gas voids in the part of the specimen later to grow as compared with the part in Figure 3b. The two parts match each other along the dashed arrow. The inset shows magnified images of voids farther away from the seed.

Consider first the sequences of the voids located between the left and the right of Figure 4. They are labeled D and E. The narrow gap between D and E is followed by a wider one. On the right, one sees some small voids (labeled F) running from the top to the bottom of the gap. A very similar motif can be found in the topograph below the dashed lines. D, E, and F denote rows of tangled dislocations (Figure 3a). Rows D and E, which lie across the ribbon width, are followed by row F running in the direction of growth. According to their size, the tangles were presumably generated by the larger (horizontal) and smaller (vertical) groups of voids.

Farther from the seed, the sequences of voids began to form in parallel to the direction of growth. The voids became closer, so that the SR beam, passing through the specimen, intersected pairs of voids. Such closely located configurations are shown in the inset in Figure 4. Under the action of thermal stress, voids can interact and merge. The coalescence can lead to the transformation of the shapes. Our observations indicated that the distortion of spherical shape proceeded gradually due to some changes in the growth conditions. The nature and evolution of non-spherical voids still remain unclear. In-depth understanding of this phenomenon requires the analysis of void sizes.

3.3. The Size of the Gas Voids

If an X-ray beam passed through a sample with inhomogeneous thickness and/or density, the wave field acquired an additional phase shift, variable in the direction perpendicular to the propagation direction. A void entailed a phase shift proportional to $s\Delta\delta$, where s is the void thickness, and δ is a decrement of the refraction index $n = 1 - \delta$ of materials. In the hard X-ray range ($E > 6$ keV), the phase shift of a coherent SR beam could be detected when the sample thickness changed from at least ~0.1 μm to several microns. In the near field, i.e., at a short distance z from the sample, the intensity oscillations were simple in shape but varied in amplitude and period. For a larger z, the sinusoidal variation in the intensity specific to Fresnel zones was formed. The sizes of the microvoids could be determined using computer simulations. To make certain that the obtained cross section diameters are correct, one has to register and fit the images of the same void at some different distances z (see, e.g., [34,35]).

As for spherical voids, their simulations were carried out in an earlier work [36]. Unlike these, we chose an experimental picture that showed the overlay of two voids located at a certain distance from each other. To our knowledge, the fitting of such configurations has not been reported. Figure 5 shows the phase contrast images recorded on the CCD at the distance z_1 = 1.5 cm (a) and z_2 = 20 cm (b). We note that the width of the void boundaries changed and became larger with the distance z, which was a consequence of the interference nature of the images. The boundaries did not overlap, so the void diameter could be roughly estimated from the image pixels. The preliminary estimates showed that the average diameter at both distances z_1 and z_2 was approximately 60 pixels. For a given pixel size of 0.325 μm, we obtained ≈20 μm. One can see that the diameters of the voids were greater than the diameter of the first Fresnel zone $2r = 2(\lambda z)^{1/2}$, which was equal to $2r_1$ = 2.2 μm and $2r_2$ = 8.1 μm at distances z_1 and z_2, respectively. The width of the edges increased with the distance z, but the average diameter did not change.

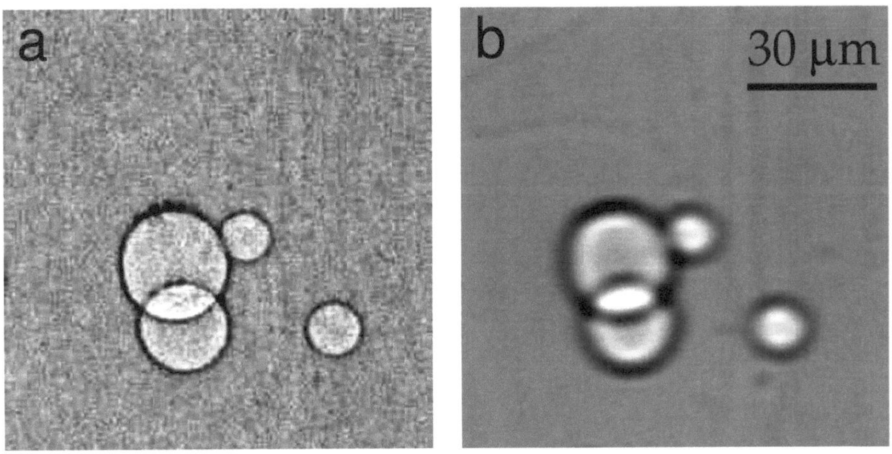

Figure 5. Experimental phase contrast images of some gas voids recoded at sample-to-detector distance z_1 = 1.5 cm (**a**) and z_2 = 20 cm (**b**).

The inline phase contrast setup did not allow an evaluation of the depth of a void in a sapphire ribbon. It was, therefore, impossible to know the distance between voids since this distance did not influence the simulated images. The sizes of voids can be different. Below we will consider a model applicable to a specific case, namely, two identical spheres with a diameter of 20 μm. Their centers were shifted diagonally by $8 \times 2^{1/2}$ = 11.3 μm. The calculation was performed by the XRWP (X-ray Wave Propagation) program [37]. The program calculated a series of two-dimensional patterns using formulas of the phase contrast theory of three-dimensional objects. The wave propagation through a substance, in which the electron density varied, was described by the transmission function of the object. Since the object containing voids was not uniform, the thickness was a variable function of the coordinate. Propagation in a free space was calculated according to the Huygens–Fresnel principle as the convolution of the wave function with the Fresnel propagator [1].

The convolution was computed through the Fourier transform method. The fast Fourier transformation (FFT) method was applied [38].

Prior to describing the results of the simulation, it is necessary to emphasize that for the imaging techniques, the recorded image resulted from a convolution of the sample response and the detector characteristics. In addition, the experimental pattern depended on the size of the SR source. A good approximation of the source is the model in which every point of its transverse size radiates independently, and the radiation intensity obeys the Gaussian law, i.e., the source may be characterized by the half-width of the Gaussian function. The structure of an image also depended on statistical and instrumental noise. The signals recorded using a CCD were superimposed on a background. When the calculated pattern was compared with the experimental one, it was necessary to normalize the latter. In order to obtain a good fit between the simulation and the experiment, the constant background must be subtracted.

At the initial stage, the simulation was carried out for fully coherent monochromatic radiation emitted by a point SR source and recorded using an ideal detector. The program calculated theoretical images, which contained sharp peaks of high intensity. The sample area containing these peaks was very small. A comparison of the image on the left in Figure 6 with the image on the right shows that the effect became more pronounced at the long distance $z_2 = 20$ cm. Prior to making comparisons with experimental images, the calculated images were averaged. We employed the convolution of the two-dimensional intensity distribution on the detector with a two-dimensional Gaussian function. Since the asymmetry of the directions along and across the SR source was not noticeable, we used the symmetric two-dimensional Gaussian. The convolution was performed for different values of the full width at half maximum (FWHM), which were the same in two directions.

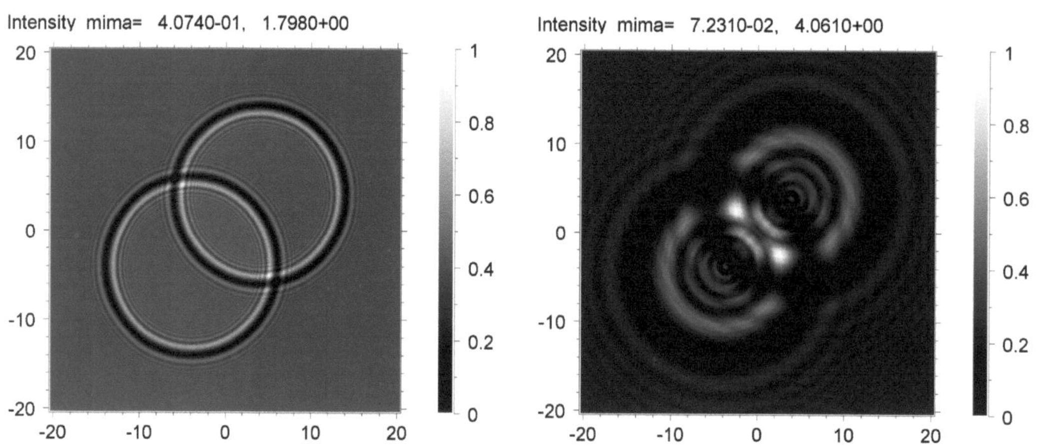

Figure 6. Theoretical phase contrast images of two spheres with 20 μm diameter calculated for the distances $z_1 = 1.5$ cm (**left**) and $z_2 = 20$ cm (**right**).

The best fits to the experimental data were selected. Figure 7 shows the averaged theoretical images of the voids calculated for the distance $z_2 = 20$ cm using the convolution with the Gaussian of the FWHM = 3 μm (left) and 4 μm (right). Let an optimal Gaussian FWHM be 3.5 μm. We emphasize that the value of FWHM must depend on the distance z. There were, therefore, good reasons for the automatic variation in two parameters during the simulation: the value of the FWHM and the sample-to-detector distance z. If we assume that the averaging occurred due to the source size, a decrease in z by a factor of 13.3 caused a decrease in FWHM by the same factor. Thus, we might expect that for the distance of 1.5 cm, an optimal FWHM was 0.26 μm.

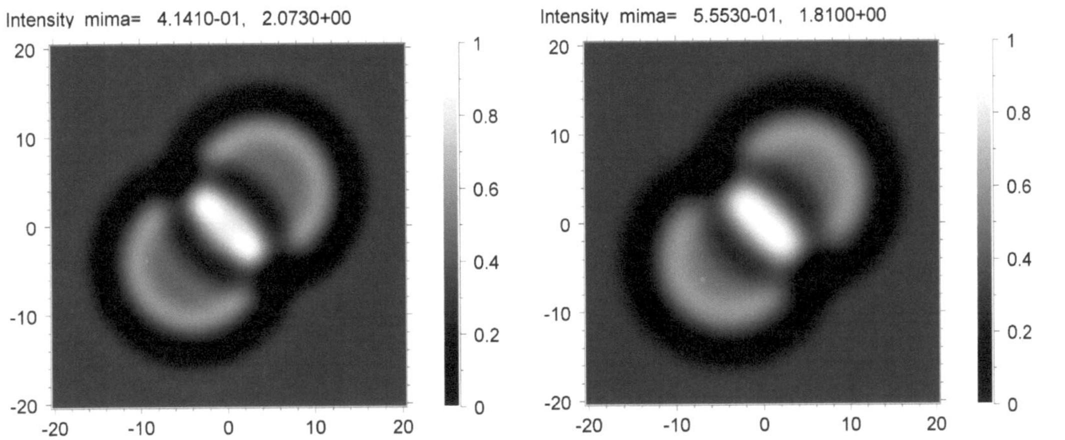

Figure 7. Phase contrast images calculated for the distance z_2 = 20 cm and averaged using convolution with Gaussian of FWHM = 3 μm (**left**) and 4 μm (**right**).

However, this effect is not observed in Figure 8, which contains the image simulations by FWHM = 0.26 μm (left) and 1 μm (right). The theoretical image on the right exhibits the distinctive effect of a large FWHM. It is more consistent with the experimental picture than the image on the left in Figure 8. To sum up, the averaging arising from the source size was observed but only to a certain extent. The given example clearly shows that imaging detectors and other aspects of the experimental procedure (including a sample vibration) should not be neglected when extracting quantitative information from an image.

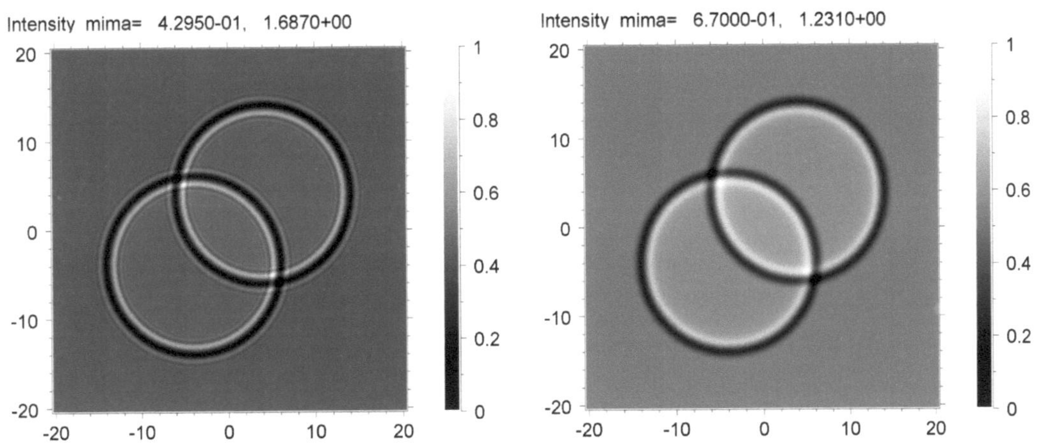

Figure 8. Phase contrast images calculated for the distance z_1 = 1.5 cm and averaged using the Gaussian of FWHM = 0.26 μm (**left**) and 1 μm (**right**).

4. Discussion

The generation of dislocations on voids in sapphire ribbons is obviously related to thermal stress relaxation at the stage of crystal cooling. For the basal slip of dislocations in the sapphire, there is an empirical formula that determines the critical shear stress τ_{cb} in the dependence of the temperature [27]:

$$\ln \tau_{cb} = \ln \tau_{0b} - 0.0052T, \tag{2}$$

where τ_{0b} = 109 GPa and T is the absolute temperature given in Kelvins. For example, at T = 2273 K, (2) results in $\tau_{cb} \approx 0.8$, which is of the order of magnitude of the thermal stresses in the growing sapphire crystals in the pre-melting state [16]. Thus, the dislocations can easy glide in this state of the growing sapphire crystals. However, the process of their nucleation needs careful examination.

Lubarda et al. [39] considered the emission of a straight edge dislocation from a cylindrical void under uniform biaxial tensile stress σ and showed that the critical stress value required to emit the dislocation was

$$\frac{\sigma_{cr}}{G} \geq \frac{b/R}{\sqrt{2}\pi(1-\nu)} \frac{\left(1+\sqrt{2}\rho b/R\right)^4+1}{\left(1+\sqrt{2}\rho b/R\right)^4-1}, \tag{3}$$

where G is the shear modulus, b is the Burgers vector magnitude, R is the void radius, ρ is the normalized cutoff radius, $\rho = w/b$, for the dislocation stress field at the dislocation core, and w is the width of the core. It is seen that the critical stress σ_{cr} decreases with an increase in the void radius R. However, when $R >> b$ (this is our case), this dependence saturated, and $\sigma_{cr}/G \to 1/[4\pi\ (1-\nu)\rho]$. For typical values ν = 0.21 and ρ = 1, this resulted in $\sigma_{cr} \approx G/10$. This characteristic value was extremely high and not available under the crystal growth conditions.

Yan et al. [40] suggested an alternative approach for estimating the critical stress needed for dislocation emission from a void. They noticed that in reality, the voids were faceted, and therefore, one can expect the dislocation emission from the edges at the void surface. In this case, "the critical length determining dislocation nucleation is not the void diameter but the length of the void edge and the stress to nucleate the dislocation scales with this length, similar to that from a Frank–Read source" [40]. As a result, the authors came to the following formula for the critical shear stress:

$$\tau_{cr} = \frac{Gb}{L}, \tag{4}$$

which is similar to that describing the activation of a Frank–Read source. Here L is the length of the void edge. Since the largest edge of a faceted void is roughly equal to the void radius, $L \approx R$, for R = 10 μm and b = 0.5 nm, (4) results in $\tau_{cr} = 5 \times 10^{-5}\ G$.

At relatively low temperatures, the shear modulus of sapphire is about $G \approx 150$ GPa, which would provide us the value of $\tau_{cr} \approx 7.5$ MPa. This value is approximately 10 times higher than the magnitude of the thermal stresses in growing sapphire crystals in the pre-melting state [16]. However, in this state, the shear modulus also can be smaller by the order of magnitude than that under low temperatures [41].

Thus, one can conclude that dislocations can be generated by the edges of faceted voids under the action of the thermal stresses in growing sapphire crystals. The inhomogeneous distribution of these stresses across the growing crystals [16] is responsible for the inhomogeneous generation of dislocations by the voids, when the voids in some regions of the crystal do emit the dislocations, while the voids in other neighboring regions do not. The role of internal gas pressure in the voids on the process of dislocation emission needs special consideration.

5. Summary

High technical standards for basal-faceted sapphire ribbons intended for windows, mirrors, and substrates make it necessary to improve single crystallinity and perfection of these products. XRT and PCI is a powerful combination to understand the relationships between the defects and inhomogeneities in crystals. The topographs and phase contrast images obtained with monochromatic SR beam allowed us to directly relate dislocations with gas voids.

By using large-area mapping of voids, we found that once nucleated, they tended to group and to spread across the width of the ribbon. When the amount of voids in the growing ribbon increased, they spread along the growth direction as well. The resulting distribution pattern was similar to a "grid" structure. Single spherical microvoids did not possess a dislocation generating activity. Moreover, the determination of void sizes using computer simulations revealed a low activity of larger spherical voids as well. In contrast, dislocations were associated with the strain fields of nonspherical cavities. The analysis of the data suggested that the voids united to form larger aggregates, which eventually developed into cavities of heterogeneous shape and slightly irregular surface. Particularly, the XRT data showed that the dislocation density increased and that the density of the dislocation sources became much higher in association with the "grid" pattern, especially in the vicinity of the dense groups of voids. At the initial stage of plastic deformation, no other mechanisms were observed.

Dislocation semiloops located near the centers of their generation were shown by XRT. A topographic $\mathbf{g}\cdot\mathbf{b}$ analysis revealed basal dislocations with $\langle 1\bar{2}10\rangle$ Burgers vectors. Prismatic semiloops belonging to the $(10\bar{1}0)$, $[1\bar{2}10]$ slip system were also present. The curved dislocation lines tended to assemble into tangles, in which no unambiguous determination of the slip system was reached because of the high density of dislocations. In addition, peculiar changes in the visibility of some tangled dislocations suggested that they were located at different depths. No evidence of the predominant slip system in the specimens under consideration was found.

Author Contributions: Conceptualization, T.S.A., V.G.K. and M.Y.G.; software, V.G.K.; investigation, T.S.A., V.G.K., J.-H.L., V.M.K. and M.Y.G.; original draft preparation, T.S.A.; writing and editing, T.S.A., V.G.K., M.Y.G., J.-H.L. and V.M.K. All authors have read and agreed to the published version of the manuscript.

Funding: The work received funding from the Ministry of Science and Higher Education of the Russian Federation. The work of T.S.A. was supported by the project 075-15-2021-1349. The work of V.G.K. was supported by the project 075-15-2021-1362. The work of M.Yu.G. was carried out at the expense of a state task at the Institute for Problems in Mechanical Engineering of the Russian Academy of Science.

Institutional Review Board Statement: Not applicable.

Data Availability Statement: Data are contained within the article.

Acknowledgments: T.S.A. would like to acknowledge S.Y. Martyushov from the Technological institute for super-hard and novel carbon materials for obtaining a X-ray topograph using a laboratory X-ray source and A.V. Myasoedov from the Ioffe Institute for consultations.

Conflicts of Interest: The authors declare no conflict of interest.

References

1. Snigirev, A.; Snigireva, I.; Kohn, V.; Kuznetsov, S.; Schelokov, I. On the possibilities of X-ray phase contrast microimaging by coherent high-energy synchrotron radiation. *Rev. Sci. Instrum.* **1995**, *66*, 5486–5492. [CrossRef]
2. Cloetens, P.; Barrett, R.; Baruchel, J.; Guigay, J.P.; Schlenker, M. Phase objects in synchrotron radiation hard X-ray imaging. *J. Phys. D Appl. Phys.* **1996**, *29*, 133–146. [CrossRef]
3. Buffet, A.; Reinhart, G.; Schenk, T.; Nguyen-Thi, H.; Gastaldi, J.; Mangelinck-Noël, N.; Jung, H.; Härtwig, J.; Baruchel, J.; Billia, B. Real-time and in situ solidification of Al-based alloys investigated by synchrotron radiation: A unique experimental set-up combining radiography and topography techniques. *Phys. Status Solidi A* **2007**, *204*, 2721–2727. [CrossRef]
4. Baruchel, J.; Di Michiel, M.; Lafford, T.; Lhuissier, P.; Meyssonnier, J.; Nguyen-Thi, H.; Philip, A.; Pernot, P.; Salvo, L.; Scheel, M. Synchrotron X-ray imaging for crystal growth studies. *Comptes Rendus Phys.* **2013**, *14*, 208–220. [CrossRef]
5. Becker, M.; Regula, G.; Reinhart, G.; Boller, E.; Valade, J.-P.; Rack, A.; Tafforeau, P.; Mangelinck-Noël, N. Simultaneous X-ray radiography and diffraction topography imaging applied to silicon for defect analysis during melting and crystallization. *J. Appl. Cryst.* **2019**, *52*, 1312–1320. [CrossRef]
6. Mancini, L.; Reinier, E.; Cloetens, P.; Gastaldi, J.; Härtwig, J.; Schlenker, M.; Baruchel, J. Investigation of structural defects and inhomogeneities in Al-Pd-Mn icosahedral quasicrystals by combined synchrotron X-ray topography and phase radiography. *Phil. Mag. A* **1998**, *78*, 1175–1194. [CrossRef]

7. Cloetens, P.; Baruchel, J.; Guigay, J.P.; Ludwig, W.; Mancini, L.; Pernot, P.; Schlenker, M. Bragg and Fresnel diffraction imaging using highly coherent X-rays. *Microsc. Microanal.* **1998**, *4*, 376–377. [CrossRef]
8. Gastaldi, J.; Mancini, L.; Reinier, E.; Cloetens, P.; Ludwig, W.; Janot, C.; Baruchel, J.; Härtwig, J.; Schlenker, M. The interest of X-ray imaging for the study of defects in real quasicrystals. *J. Phys. D Appl. Phys.* **1999**, *32*, A152–A159. [CrossRef]
9. Baruchel, J.; Cloetens, P.; Härtwig, J.; Ludwig, W.; Mancini, L.; Pernot, P.; Schlenker, M. Phase imaging using highly coherent X-rays: Radiography, tomography, diffraction topography. *J. Synchrotron Rad.* **2000**, *7*, 196–201. [CrossRef]
10. Argunova, T.S.; Gutkin, M.Y.; Je, J.H.; Kang, H.S.; Hwu, Y.; Tsai, W.-L.; Margaritondo, G. Synchrotron radiography and X-ray topography studies of hexagonal habitus SiC bulk crystals. *J. Mater. Res.* **2002**, *17*, 2705–2711. [CrossRef]
11. Stepanov, A.V. Growing of crystalline germanium plates. *Sov. Phys. Solid State* **1966**, *8*, 449.
12. LaBelle, H.E., Jr. Growth of Inorganic Filaments. U.S. Patent 3,471,266, 7 October 1969.
13. Katyba, G.M.; Zaytsev, K.I.; Dolganova, I.N.; Shikunova, I.A.; Chernomyrdin, N.V.; Yurchenko, S.O.; Komandin, G.A.; Reshetov, I.V.; Nesvizhevsky, V.V.; Kurlov, V.N. Sapphire shaped crystals for wave-guiding, sensing and exposure applications. *Prog. Cryst. Growth Charact. Mater.* **2018**, *64*, 133–151. [CrossRef]
14. Akselrod, M.S.; Bruni, F.J. Modern trends in crystal growth and new applications of sapphire. *J. Cryst. Growth* **2012**, *360*, 134–145. [CrossRef]
15. Bruni, F.J. Crystal growth of sapphire for substrates for high-brightness, light emitting diodes. *Cryst. Res. Technol.* **2015**, *50*, 133–142. [CrossRef]
16. Kuandykov, L.; Bakholdin, S.; Shulpina, I.; Antonov, P. Model of a block structure generation in basal-faceted sapphire ribbons. *J. Cryst. Growth* **2005**, *275*, e625–e631. [CrossRef]
17. Shul'pina, I.L.; Bakholdin, S.I.; Krymov, V.M.; Antonov, P.I. Study of the real structure of basal-plane-faceted sapphire ribbons. *Bull. Russ. Acad. Sci. Phys.* **2009**, *73*, 1364–1369. [CrossRef]
18. Lommel, J.M.; Kronberg, M.L. X-ray diffraction micrography of aluminum oxide single crystal. In *Direct Observation of Imperfections in Crystals*; Newkirk, J., Wernick, J., Eds.; John Wiley & Sons: New York, NY, USA, 1962; p. 543.
19. Takano, Y.; Kohn, K.; Kikuta, S.; Kohra, K. X-ray studies of dislocation structures in a sapphire crystal. *Jpn. J. Appl. Phys.* **1970**, *9*, 847–848. [CrossRef]
20. Caslavsky, J.L.; Gazzara, C.P.; Middleton, R.M. The study of basal dislocations in sapphire. *Phil. Mag.* **1972**, *25*, 35–44. [CrossRef]
21. Scheuplein, R.; Glbbs, P. Surface structure in corundum: Etching of dislocations. *J. Am. Ceram. Soc.* **1960**, *43*, 458–472. [CrossRef]
22. Cadoz, J.; Castaing, J.; Phillips, D.S.; Heuer, A.H.; Mitchell, T.E. Work hardening and recovery in sapphire (α-Al_2O_3) undergoing prism plane deformation. *Acta Metall.* **1982**, *30*, 2205–2218. [CrossRef]
23. Chen, W.M.; McNally, P.J.; Shvyd'ko, Y.V.; Tuomi, T.; Danilewsky, A.N.; Lerche, M. Dislocation analysis for heat-exchanger method grown sapphire with white beam synchrotron X-ray topography. *J. Cryst. Growth* **2003**, *252*, 113–119. [CrossRef]
24. May, C.A.; Shah, J.S. Dislocation reactions and cavitation studies in melt-grown sapphire. *J. Mater. Sci.* **1969**, *4*, 179–188. [CrossRef]
25. Castillo-Rodríguez, M.; Muñoz, A.; Castaing, J.; Veyssière, P.; Domínguez-Rodríguez, A. Basal slip latent hardening by prism plane slip dislocations in sapphire (α-Al_2O_3). *Acta Mater.* **2010**, *58*, 5610–5619. [CrossRef]
26. Sen, G.; Caliste, T.N.T.; Stelian, C.; Baruchel, J.; Barthalay, N.; Duffar, T. Synchrotron X-ray diffraction imaging studies of dislocations in Kyropoulos grown Ti doped sapphire crystal. *J. Cryst. Growth* **2017**, *468*, 477–483. [CrossRef]
27. Lagerlöf, K.P.D.; Heuer, A.H.; Castaing, J.; Rivière, J.P.; Mitchell, T.E. Slip and twinning in sapphire (α-Al_2O_3). *J. Am. Ceram. Soc.* **1994**, *77*, 385–397. [CrossRef]
28. Lang, A.R. X-ray topography: Methods and interpretation. In *Diffraction and Imaging Techniques in Material Science*, 2nd ed.; Amelinckx, S., Gevers, R., Van Landyut, J., Eds.; North-Holland Publishing Company: Amsterdam, The Netherlands, 1978; Volume 2, pp. 623–714.
29. Bakholdin, S.I.; Maslov, V.N.; Nosov, Y.G. Specific features of the morphology and distribution of gas inclusions in single-crystal sapphire ribbons grown by the Stepanov method. *Crystallogr. Rep.* **2014**, *59*, 762–767. [CrossRef]
30. Ishiji, K.; Kawado, S.; Hirai, Y.; Nagamachi, S. Determination of observable depth of dislocations in 4H-SiC by X-ray topography in back reflection. *Jpn. J. Appl. Phys.* **2017**, *56*, 106601. [CrossRef]
31. Yao, Y.; Sugawara, Y.; Ishikawa, Y. Identification of Burgers vectors of dislocations in monoclinic β-Ga_2O_3 via synchrotron X-ray topography. *J. Appl. Phys.* **2020**, *127*, 205110. [CrossRef]
32. Farber, B.Y. Dislocation Velocities and Dislocation Structure in Cubic Zirconia and Sapphire (α-Al_2O_3) Single Crystals. Doctoral Dissertation, Case Western Reserve University, Cleveland, OH, USA, 1994.
33. Novak, R.E.; Metzl, R.; Dreeben, A.; Berkman, S. The production of EFG sapphire ribbon for heteroepitaxial silicon substrates. *J. Cryst. Growth* **1980**, *50*, 143–150. [CrossRef]
34. Agliozzo, S.; Cloetens, P. Quantification of micrometer-sized porosity in quasicrystals using coherent synchrotron radiation imaging. *J. Microscopy* **2004**, *216*, 62–69. [CrossRef]
35. Kohn, V.G.; Argunova, T.S.; Je, J.H. Quantitative hard X-ray phase contrast imaging of micropipes in SiC. *AIP Adv.* **2013**, *3*, 122109. [CrossRef]
36. Argunova, T.S.; Kohn, V.G.; Krymov, V.M. Study of defects in shaped sapphire crystals by synchrotron X-ray phase contrast imaging. *St. Petersb. Polytech. Univ. J.* **2023**, *16*, 146–152.
37. Kohn, V.G. Available online: http://xray-optics.ucoz.ru/XR/xrwp.htm (accessed on 10 September 2023).

38. Cooley, J.W.; Tukey, J.W. An algorithm for the machine calculation of complex Fourier series. *Math. Comput.* **1965**, *19*, 297–301. [CrossRef]
39. Lubarda, V.A.; Schneider, M.S.; Kalantar, D.H.; Remington, B.A.; Meyers, M.A. Void growth by dislocation emission. *Acta Mater.* **2004**, *52*, 1397–1408. [CrossRef]
40. Yan, Z.; Liu, Z.; Kong, X.; Yao, B.; An, Q.; Jiang, S.; Zhang, R.; Beyerlein, I.; Zheng, S. Effect of void morphology on void facilitated plasticity in irradiated Cu/Nb metallic nanolayered composites. *J. Nucl. Mater.* **2022**, *558*, 153380. [CrossRef]
41. Nie, G.; Bao, Y.; Wan, D.; Tian, Y. Measurement of the high temperature elastic modulus of alumina ceramics by different testing methods. *Key Eng. Mat.* **2018**, *768*, 24–30. [CrossRef]

materials

Article

Control of Coherent Light through Microperiodic Director Modulation in Nematic Films under Low-Voltage DC Electric Field

Georgi B. Hadjichristov

Laboratory of Optics and Spectroscopy, Georgi Nadjakov Institute of Solid State Physics, Bulgarian Academy of Sciences, 72 Tzarigradsko Chaussee Blvd., 1784 Sofia, Bulgaria; georgibh@issp.bas.bg

Abstract: This work addresses the achievement of efficient control of laser light transmission through stationary microperiodic parallel stripe textures formed in films of nematic liquid crystals (NLCs) in planar-oriented cells upon a direct-current (DC) electric field. By varying the field intensity and, thereby, the field-induced periodic modulation of the nematic director and hence the complex transmittance function corresponding to the longitudinal domain texture induced in NLC films with initial planar alignment, the intensity of a linearly polarized laser beam passed through the films can be well controlled. In 25 μm-thick films of room-temperature NLCs pentylcyanobiphenyl (5CB), this results in a low-voltage (~4 V) sharp and deep V-shaped behavior of their electro-optically controlled transmittance. Such a reversible electro-optical effect is interesting for active control of laser beam intensity and other applications. The relevant physical mechanism is analyzed and explained.

Keywords: nematic liquid crystals; optical phase grating; coherent optical processes; light scattering; light diffraction; laser beam intensity control

Citation: Hadjichristov, G.B. Control of Coherent Light through Microperiodic Director Modulation in Nematic Films under Low-Voltage DC Electric Field. *Materials* **2023**, *16*, 6014. https://doi.org/10.3390/ma16176014

Academic Editors: Fabrizio Roccaforte and Alexander V. Baranov

Received: 24 July 2023
Revised: 12 August 2023
Accepted: 29 August 2023
Published: 1 September 2023

1. Introduction

Among the variety of liquid crystal (LC) textural formations [1–3], field-induced spatial patterns and ordered textures offer attractive possibilities for field-commanded effects and applications, such as controllable shifting, angular deflection, scattering, and diffraction of light, for use in diffractive, adaptive, and non-linear optics, along with microscopy and electro-optics [4,5]. In some cases, regular field-induced grid-like patterns based on periodic modulations and orientation patterns in LC media, in particular nematic LCs (NLCs), are suitable to use in optical devices, such as optical switches and filters for laser beams, optical phase gratings, similarly to the well-known various types of electrically-driven diffraction gratings in NLCs, e.g., [6], and electro-optically addressed NLC tunable diffraction and phase gratings, e.g., [7,8]. Such spatial, polarization, and phase modulators of light have found useful applications in photonics, optical information processing and fiber-optic communications [9–11], laser beam steering [12,13], programmable shaping of femtosecond laser pulses [14,15], reconfigurable generation of optical vortices for manipulation of laser beams and light pattern formation [16,17] and in other modern scientific research fields.

Nowadays, NLC gratings with generated spatial patterns and thus customized diffraction patterns have attracted much attention in both industry and scientific research due to their simple preparation, cost effectiveness, and high performance, including diffraction efficiency, tuneability, and polarization sensitivity [18–21]. Diffraction grating effects exhibited by homogeneously aligned NLC layers with a microperiodic distortion of their director field were thoroughly analyzed and elucidated [22–24]. The appearance and characteristics of electric field-driven texture patterns in NLC layers depend on both dielectric permittivity anisotropy and electrical conductivity anisotropy, the initial director orientation, as well as other initial conditions, system parameters, the LC cell, the characteristics of the NLC material, and possibly the additives included in it. Relevant physical mechanisms (electrostatic, electrohydrodynamic, and other standard and nonstandard models, including flexoelectricity) have been developed to describe and explain various field-induced textural formations in NLCs [3,25–32].

Of special research and practical interest are some types of electric field-induced spatially periodic and highly regular and stable stripe patterns formed in NLC materials. Such longitudinal domains (LDs) in NLC films have been comprehensively investigated both theoretically and experimentally by electrical, dielectric, and electro-optical (EO) measurements [33–37], including numerous diffraction studies in planarly aligned NLC layers, e.g., [22,24,36,38–40].

In the work presented here, scientific and applied physics interest is focused on LDs formed in planarly-aligned films of NLCs, for instance, pentylcyanobiphenyl (5CB), exposed to a direct current (DC) electric field. 5CB has a stable nematic phase at room temperature and a relatively high positive dielectric anisotropy, and planar cells with this NLC exhibit well-developed electric field-induced LDs [37]. The present experimental study of DC voltage-controlled laser beam transmission, diffraction, and scattering in planar 5CB nematic films demonstrates effective and low-voltage EO control of the intensity of the laser beam passed through them utilizing DC voltage-induced microperiodic director-modulation textures. Extending previous works [41,42], a more detailed investigation of DC voltage-commanded laser beam transmission, diffraction, and scattering by planar cells with 5CB was correlated with optical microscopy observations and with results obtained from polarization and time-domain measurements. The effect observed was also characterized with respect to the laser beam wavelength and the angular orientation of the 5CB films relative to the incident laser beam. The advantages, limitations, and applicability of the proposed approach for enhancing the optical contrast ratio of coherent light transmission through nematic films were discussed. The aim of this investigation is not to explore in detail the microperiodic director modulations in NLC layers themselves, their physics, and the various characteristics of the electrically-induced spatial patterns (domains) in the studied NLC materials, but to investigate and identify the effect of such well-known for a long time textures on the transmission of coherent light through NLCs films and to put them into action for practical use through the EO effect observed. The textural domains in the studied planar NLC films were characterized only to the extent necessary for the EO application proposed here.

2. Materials and Methods

2.1. NLCs Films

Experiments were carried out on planarly-aligned samples of the NLCs 4′-*n*-pentyl-4-cyanobiphenyl (5CB) and 4′-*n*-heptyl-4-cyanobiphenyl (7CB) supplied by Merck and used as received. These NLCs are members of the cyanobiphenyl family, which was one of the first commercially available nematic materials for use in LC displays. Due to their outstanding properties at room temperature, i.e., high chemical stability and sensitivity to applied electric fields, they are still commonly employed. At room temperature, both compounds exhibit a stable nematic LC phase. They are optically birefringent materials. For instance, 5CB is characterized by extraordinary and ordinary refractive indices n_e = 1.706 and n_0 = 1.532, respectively, at wavelength λ = 633 nm and 25 °C [43]. 5CB and 7CB have a relatively large positive dielectric anisotropy, $\Delta\varepsilon = \varepsilon_{||} - \varepsilon_{\perp} = 8.2$ and $\Delta\varepsilon = 6.7$, respectively, at 20 °C and at an electric field frequency f = 1 kHz [44]. The value of the real part of the dielectric susceptibility of these nematics along the preferred molecular direction ($\varepsilon_{||}$) is about twice higher than that in the transverse direction ($\varepsilon_{\perp}$) [44].

Planar films of NLCs with a planar alignment were prepared in commercial flat-panel LC glass cells (KSRO-25/B111N1NSS Up/Low, manufactured by E.H.C. Co. Ltd., Tokyo, Japan) with a 25 μm gap. The inner surfaces of the two glass plates of the cells were covered (by the manufacturer) with an ultrathin, transparent, electrically conductive layer of indium-tin-oxide (ITO) that served as electrodes. The ITO glasses had a polyimide overcoat with unidirectional rubbing. In 5CB films in such LC cells, stable DC-induced texture formations, such as those studied here, are absolutely reliably induced.

The NLCs in the isotropic liquid phase were injected into the cells by capillary forces. Before injecting, the cells and the NLCs were heated to a temperature above the clearing point (the nematic-isotropic phase transition temperature) in order to avoid a non-uniform alignment of the NLCs. The formation of the nematic LC phase was established with polarizing optical microscopy (POM) by observing birefringence between crossed polarizers. The temperature range for the nematic phase of 5CB was 24–35 °C and 28.5–42 °C for 7CB. The parallel-rubbed polyimide ultrathin surface layers provide NLC molecular alignment—they force the confined NLC molecules to orient themselves homogeneously parallel to the rubbing direction. The quality of the orientation of the nematic films was checked using POM. The strong planar alignment of NLCs with an overall orientation of the nematic director along the rubbing direction of the cells was confirmed.

2.2. *Electro-Optical Measurements*

The optical transmittance and diffraction from the prepared nematic films were investigated using a non-focused beam of He-Ne laser HNL050RB (Thorlabs GmbH, Munich, Germany) operating at a wavelength (λ) of 632.8 nm and having an optical power of 5 mW and optical noise of less than 0.2%. The linear polarization of the laser beam (>10^4:1) was selected by the rotatable Thorlabs LPVISB050-MP nanoparticle linear film polarizer. With this polarizer in Thorlabs' KS05RS kinematic rotation mount, one can set the polarization direction of the laser beam with an accuracy of $\pm 0.5°$.

Some tests were performed with a temperature-stabilized diode-pumped continuous wave (c.w.) solid-state laser DPGL-4007 (Photop Suwtech Inc., Shanghai, China), with Nd:YVO4 crystal, λ = 532 nm, 100 mW, linearly polarized (>500:1)). Produced by intracavity second harmonic generation, the output of this laser source did not contain IR radiation. Both green and He-Ne laser beams had TEM_{00} spatial profiles with a Gaussian intensity distribution and a divergence of 0.8 mrad and 1 mrad, respectively. The power stability of these laser sources was better than $\pm 0.5\%$. A 5 mW laser diode emitting at 405 nm (beam divergence ~1.5 mrad) was also used.

The LC cells were mounted on a micro-manipulating translation-rotation stage. This allows illumination at a desired angle of incidence of the laser beam. In most of the experiments, the incident laser beam was directed normally to the nematic film plane (or, more correctly, nearly normally, to avoid optical interference from reflections from optical elements). A part of the film about 2 mm in diameter was illuminated. The laser power incident on the nematic films was kept at ~1 mW.

A DC electric field was applied across the two ITO-coated glass plates of the LC cell (i.e., the electric field direction coincided with the laser beam direction). The electrically active area of the cells was 10 mm × 10 mm. The experiment setup for EO measurements is shown in Figure 1. The light transmitted through the 5CB cells was detected by a photodiode. For measurement of the spatially selected forward spread of scattered light, a large-aperture lens assembly was used to collect and focus the light onto a photodiode (as schematically shown in Figure 1b). In this case, proper spatial filtering of the light was performed. As for the light of diffraction peaks and other fine localized diffraction features, they were carefully separated by a small circular aperture (iris diaphragm properly open) or by a pinhole (1 mm diameter) in front of the photodiode (Figure 1a). When the laser beam incidence angle was varied, a large-aperture photodiode was used in the measurements. The X and Y coordinate axes in the XYZ reference system shown in Figure 1 are related to the film plane, with the X axis parallel to the initial orientation of the nematic director (denoted as $\mathbf{N}_0$, at zero field).

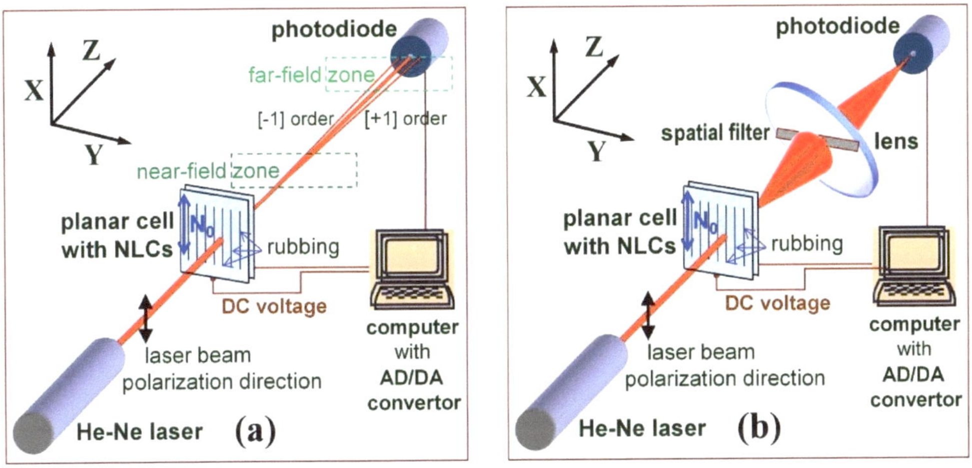

Figure 1. Schematic of the experimental setup for measurement of coherent light transmission and diffraction (**a**) and coherent light scattering (**b**) of a laser beam behind the cell with NLC film in the experiments in this work.

The light intensity was measured using a multi-channel digital-analog/analog-digital conversion interface card (Decision Group Inc., Taiwan, China) installed in a computer slot for programmable data acquisition. This high-precision data conversion card provides both the driving DC voltage and the digitization of the photodiode signal in a range of 12 bits with a conversion time of 60 μs. The computer-generated digital-to-analog pulse formation by the card is characterized by a current setting time of 0.5 μs and nonlinearity of less than 0.2%. The behaviors of DC voltage-dependent laser beam transmittance and diffraction/scattering behind the cells were recorded in voltage steps of 0.1 V. The interval between the data acquisitions was equal to 30 s, and the averaging of 10 measurements was done during 10 s at each step. The pause of 30 s was sufficient to attain the stationary state of the nematic in the cells.

In the series of measurements in which the angle of the polarization direction of the incoming laser beam was varied, the light intensity measurements were carried out under identical experimental conditions and using a reference channel to control the laser beam intensity in order to accurately determine the intensity of the detected light transmitted through the cell. In the experiments in which the EO effect with a high dynamic range was measured, a low-noise photodetector with an optical-power working range of $1–10^6$ and a measurement uncertainty of ±3% was employed—a Thorlabs PM100 power-meter equipped with a silicon photodiode power-sensor S120VC. The lower limit of this unit was 10–50 nW (in the dark).

The temperature of the studied cells was maintained by a Mettler FP82 hot stage and was controlled with an accuracy of 0.1 °C. In most experiments, the temperature was stabilized at 25 °C. Also, measurements by varying the temperature of the cells in the range of 25–32 °C were performed. A polarizing optical microscope NU-2 (Carl Zeiss Jena GmbH, Jena, Germany) was used to observe pattern formation in the studied LC cells. DC electric field-induced texture changes in the prepared NLC films were inspected in the voltage range of 0–10 V. Images of coherent light diffraction/scattering resulting from a He-Ne laser beam passed through the NLC cells were displayed on an imaging screen placed transversally to the beam behind the cells. Pictures and videos of the diffraction pattern were taken with a digital camera VG-130-D-715, 4300 × 3200 pixels (Olympus Imaging China Co. Ltd., Beijing, China) in the dark (with room lights off). The images of light patterns were processed and analyzed with Linux IMAL software (version 6.0).

3. Results and Discussion

3.1. *Longitudinal Domain Texture in 5CB Planar Cells under DC Electric Field: Polarizing Optical Microscopy Data*

At an appropriate DC voltage (V_{DC}) applied to the prepared 5CB films, their morphology displays regular parallel-striped textures well observable by optical microscopy. Two types of such textural formations were distinguished: wide-formed longitudinal domains and narrow-striped rubbing-induced longitudinal texture.

3.1.1. Wide-Formed Electrically-Induced Longitudinal Domains

When the voltage V_{DC} applied to the initially planarly oriented 5CB films was above a well-defined threshold value of V_{form} = 4.2 V, stationary stripe texture patterns in these films were clearly visible by POM (Figure 2a). The patterns were parallel to the direction of the rubbing of the cell plates—the initial orientation of the 5CB LC molecules (the direction *X*, see Figure 1). Such DC field-induced longitudinal texture formations in nematic planar cells (along the planar director orientation at the confining slides of the cells) are well known, e.g., LDs reported by Aquire and co-workers in their comprehensive study of regular structures in 5CB NLCs under the joint action of DC and alternating-current (AC) voltages [37].

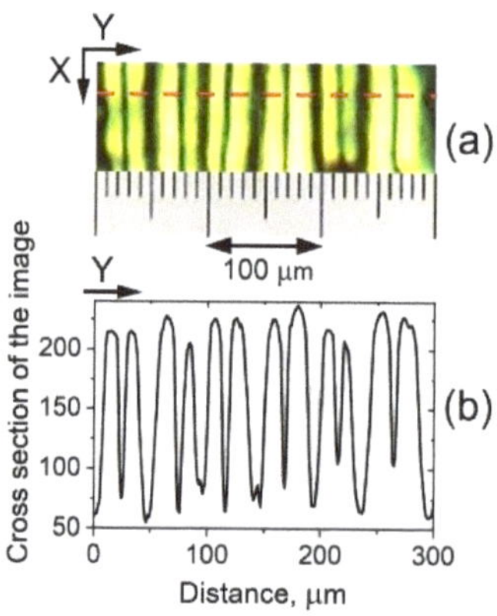

Figure 2. (**a**) POM image of the texture formed in planar 5CB film (thickness 25 μm) at V_{DC} = 4.5 V. The micrograph was taken by a slightly uncrossed polarizer and analyzer with their axes in the *X* and *Y* directions, respectively (see Figure 1). The polarization of the input light was parallel to the initial alignment of the NLCs (the rubbing of the cell plates along the *X* direction). (**b**) Cross-sectional profile of texture image (**a**) digitized at the section along a preselected line indicated by a red dashed line.

In our case, the stationary LDs can be attributed to flexoelectric domains. Similar domains are long known for NLC films under DC, very low-frequency AC, or the joint action of DC and AC electric fields. Such textures depend very strongly on the conditions (the applied voltage, the NLC film thickness, the state of the boundaries, their treatment, the molecular anchoring at the walls, and other factors) [1,45–50]. The physical mechanisms of the domain's appearance have been studied for years by numerous research groups. The observed bright LDs are divided by thin dark stripes in the middle (Figure 2a). The width of the LDs was larger than the cell thickness. For the texture shown in Figure 2a, a mean width of 46.7 μm (standard deviation ± 4.5 μm) was estimated by averaging over a lot of cross-sectional profiles of the obtained micrographs. At different locations on the film, the length and number of the LDs vary, suggesting a surface contribution to domain formation.

The stationary periodic LDs result from the static deformations of the nematic director, whose initial orientation $\mathbf{N}_0$ (in the absence of an applied electric field) is in the direction *X* (see Figure 1). As a sequence of alternating dark and bright stripes observable by POM, these domain patterns represent a periodic spatial modulation of the director. Using POM, their best contrast was achieved when the polarizer and analyzer were a few degrees off the perpendicular, suggesting that the periodic modulations of the director are exclusively in the plane of the films (the X–Y plane). Additional observations indicated that the spatial patterns shown in Figure 2a are flexo-dielectric walls (more specifically, dielectric flexoelectric walls) [1,51]. This static flexoelectric instability is different from the flexoelectric domains of the Vistin-Pikin-Bobylev type [45–49]. The latter type of patterned flexoelectric instability is a bulk flexoelectric effect occurring in nematic layers with relatively strong planar anchoring at the substrates. Such domains formed along the initial orientation of the nematic are volume flexoelectric deformations. In contrast, the flexo-dielectric walls present flexoelectric deformations in the bulk of the planar nematic film but very near the electrodes of the LC cell.

Some inhomogeneity in the LD texture pattern can be present over the 5CB films. There were regions with less or more regular formation of LDs (Figure 3). This arises from the boundary conditions at the confining plates of the LC cell and from the inhomogeneity of the anchoring of the NLC molecules. By POM, the LDs were also visible at $V_{DC} > V_{form}$ (Figure 3a), but a stable periodic array of parallel stripes was observed in a relatively short voltage range. At V_{DC} higher than 5.8–6 V, undulations and fragmentation of the LDs started to develop (Figure 3b). At $V_{DC} > V_{hd} = 6.5$ V the electrically induced hydrodynamic processes [52] in the 5CB cells were enhanced to a degree that disrupted the LDs, clearly visible under the microscope (Figure 3c). It should be noted that the voltage value V_{hd} varies slightly; e.g., the same 5CB in other but identical planar cells also shows V_{hd} of about 6 V or slightly less. On the other hand, for some of the prepared 5CB films, this periodic texture could retain up to 8 V. With decreasing V_{DC} in the same range, from 10 V to 0 V, the morphological changes were repeated in reverse at the corresponding voltage values (after the short time, less than 30 s, necessary for the stationary state of the NLC soft material in the cells, monitored by an oscilloscope during the experiments). Also, it was checked that the EO V-shaped curve of the 5CB film is fully repeatable after heating the LC cell above the temperature of the NLC-to-isotropic phase transition of 5CB (ca. 34 °C) and by subsequent cooling to room temperature. In doing so, the nematic phase and EO properties of the 5CB nematic were recovered, and the voltage values V_{th} and V_{hd} were the same as before heating.

3.1.2. Narrow-Striped Rubbing-Induced Longitudinal Texture

POM also revealed the presence of a closely spaced, narrow-formed LD texture in the 5CB planar cells under study. As with the aforementioned type of wide-formed V_{DC}-induced LDs, this microtexture was also along the X-axis and was stable over time. These spatial patterns are due to the static microperiodic orientational modulation of the director and appear as fine quasi-periodic parallel stripes along the rubbing of the LC cells. The so-formed quasi-linear grating of deformations (pattern consisting of alternating dark and bright stripes, Figure 4a) was most easily viewed by shadowgraph technique or by POM between slightly uncrossed polarizers when the applied voltage is below V_{form}. The narrow-striped director modulation was also present at the zero field. The observed fine-stripe texture follows the regular scratches –the rubbing of the cell plates, which determine the formation of the initial NLC texture. The spatial periodicity of the stripes along Y is estimated to be 6–8 μm (Figure 4b).

POM studies show that EO changes in the microscopic optical response of these textural formations under an applied DC voltage in the range of 1–10 V were relatively slow and hardly observable. The narrow-striped LDs are surface irregularities called by Hinov et al. "rubbing-induced domains", or more precisely "rubbing-induced surface texture" [53]. They are due to complex flexoelectric effects and are located in a very thin region with a thickness of ~1 μm close to the electrodes of the LC cell [53]. At $V_{DC} > V_{form}$, these fine stripes were "suppressed" by the wide-formed field-induced LDs discussed in the previous Section 3.1.1. The DC field-induced flexoelectric changes of these textural formations are related not only to bulk interactions but also to surface interactions that depend on the planar anchoring of the LC molecules at the substrates.

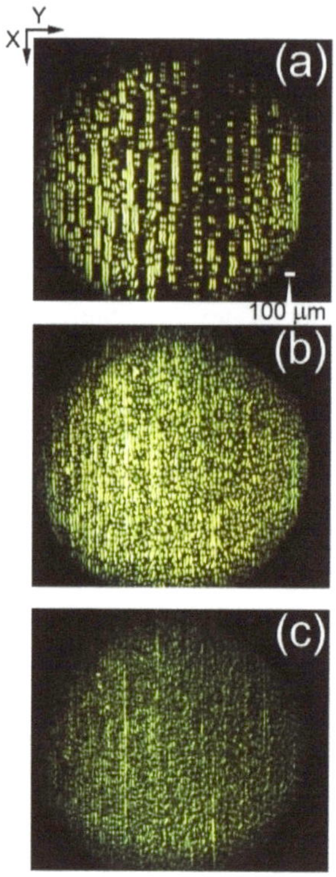

Figure 3. As in Figure 2a, photomicrographs captured for 5CB nematic film in planar cell at DC voltage: 4.5 V (**a**); 5.8 V (**b**) and 6.5 V (**c**).

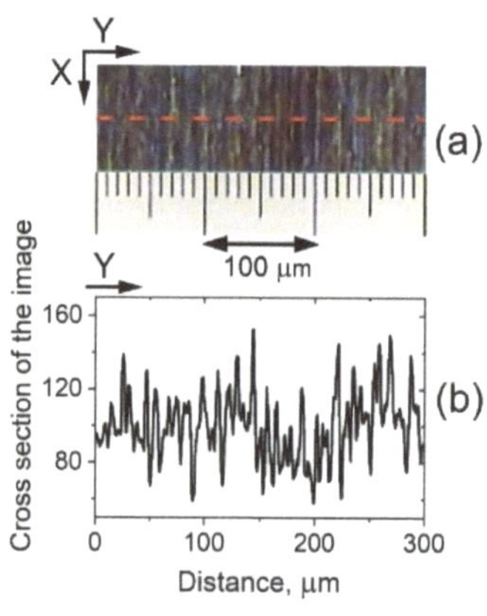

Figure 4. (**a**) POM image of the texture formed in a 25 μm-thick planar-oriented layer of nematic 5CB at V_{DC} = 3.3 V. Slightly uncrossed polarizers (set along *X* and *Y*). The input light polarization was parallel to the initial alignment of the NLCs (i.e., along *X*—in the direction of the rubbing of the cell plates); (**b**) Cross-sectional profile of texture image (**a**) digitized at the section along a preselected line indicated by a red dashed line.

3.2. Interaction of Coherent Light with LDs in 5CB Planar Films under DC Electric Field

As molecular orientation patterns, both types of DC electric field-induced spatially periodic parallel textures in planar nematic film (Section 3.1) can be considered diffraction gratings (in spectral regions where the nematic material is transparent, or at least not strongly absorbing). As such, when coherent light interacts with a planar nematic film, the LDs can give rise to characteristic diffraction patterns.

3.2.1. Optical Diffraction by Wide-Formed Electrically-Induced LDs

From an application point of view, the electrically formed spatially periodic array of LDs in 5CB planar nematic films is attractive for producing optical diffraction. For textures shown in Figures 2 and 4, the Klein–Cook parameter [54], defined as $Q = 2\pi \lambda_0 L/\Lambda^2 n_0$ (where λ_0 is the wavelength of light in vacuum, L is the interaction length (active grating thickness), Λ is the grating spacing, and n_0 is the mean refractive index), is $Q << 1$, i.e., the interaction of the laser beam with the LDs is under Raman–Nath conditions (at normal incidence of light). Since the diameter (D = 2 mm) of the spot of the incident laser beam (a plane wave) on the surface of the studied NLC film is much greater than, Λ, the Raman–Nath diffraction regime predicts that the diffracted light in the far-field zone consists of sharp and well-separated lines—the diffraction pattern contains many diffraction orders with intensities given by Bessel functions (not taking into account propagation diffraction effects inside the NLC film, which give rise to a far-field diffraction less or more smeared out).

Upon illumination with a He-Ne laser beam whose polarization is along the LDs (i.e., the direction of the rubbing of the cell plates, the X axis, Figure 1a) and when the voltage applied to the 5CB planar cell is above V_{form} (Figure 5a), two bright lateral diffraction peaks arise from these periodic textural formations. The observed diffraction pattern seems to be Fraunhofer diffraction, such as those from a thin harmonic diffraction grating (see, for example, [55]). No diffraction was present when the beam polarization direction was orthogonal to the rubbing of the cell plates. The diffracted light intensity was predominantly distributed in two side maxima (the first-order diffraction peaks, numbers +1 and −1, respectively) in addition to the central peak associated with the zero-th-order diffracted light. Behind the 5CB cell, this triplet light beam was spatially localized and had a relatively low spatial divergence (Figure 1a). The divergence of the zero-th-order diffracted light beam was a little higher than that of the incident laser beam.

Fourier transform (Figure 5c) of a typical texture image of the studied 5CB cells with well-developed LDs, i.e., at V_{DC} well above V_{form} (e.g., the image shown in Figure 5b), is close to the diffraction pattern (Figure 5a) observed when the same LD texture (keeping the corresponding value of V_{DC}) was illumined with highly coherent light (monochromatic and low-divergent). This suggests that the regular field-induced LD texture in the 5CB cells acts as an electrically driven diffraction grating when illuminated by a monochromatic plane light wave.

Figure 6 shows far-field diffraction patterns from a He-Ne laser beam propagating through a 5CB planar cell at $V_{DC} > V_{form}$. The spatial frequency of the formed LD grating texture (the grating period), calculated from the measured angular spacing of the features in the triplet diffraction pattern in Figure 6b, was Λ~45 μm, in accordance with the average width of LDs, estimated from optical micrographs (see Section 3.1.1). Note that the spatial period of the resulting texture pattern (Figure 2b) estimated by the cross-sectional profiles of the micrographs was 23–25 μm (depending on the examined location on the 5CB layer).

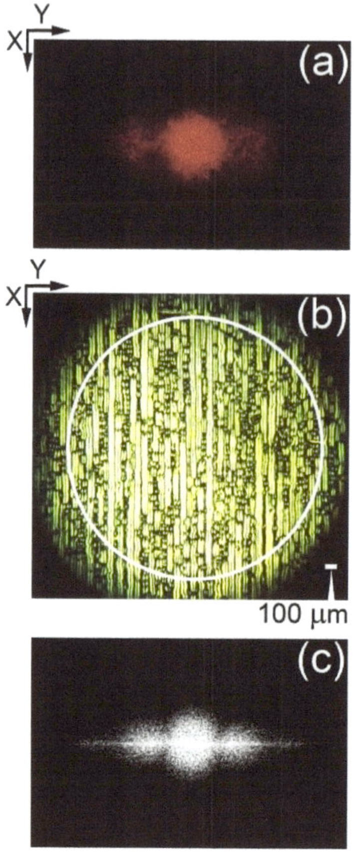

Figure 5. (**a**) Diffraction pattern obtained with a laser beam (He-Ne laser, $\lambda = 632.8$ nm) incident on a planar 5CB film (thickness 25 μm) at $V_{DC} = 5.2$ V. The diffraction was registered in transmission, in the near-field zone; the cell-to-screen distance was 15 cm. The polarization of the incident laser beam was along the direction of the rubbing of the 5CB cell; (**b**) Optical microscopy image of the texture formed in the 5CB film at $V_{DC} = 5.2$ V. A low magnification (×4) of the microscope was used in order to scale the image as large as the spot diameter of the incoming laser beam (indicated with a circle); (**c**) Fourier transform of image (**b**).

3.2.2. Coherent Light Scattering/Diffraction by Striped Textures in 5CB Planar Films under DC Electric Field

Diffraction Pattern

Besides Fraunhofer diffraction, another coherent optical process, namely coherent light scattering (CLS), was also present for planarly aligned 5CB films under a DC electric field. This process results in a distinctly localized spatial pattern of diffuse scattered light, seen as a halo or ring in Figure 6. As experimentally observed for the studied 5CB films, CLS arises at DC voltages lower than V_{form} (see Section 3.1.1), i.e., at increasing V_{DC}, CLS precedes the Fraunhofer diffraction from the wide LDs discussed in Section 3.2.1. Well visible in the near-field zone behind the 5CB cells, a strong CLS of the laser beam propagating through the 5CB planar film appears as a diffuse diffractive halo of speckle light (speckle diffraction patterns being random fluctuations in the multiply scattered intensity of coherent light). The light from CLS registered on a transversal screen as a halo (Figure 7) and was spatially spread in a light cone around the direction of the propagating laser beam (Figure 1b). The cone angle of diffuse CLS was roughly estimated to be at least 0.025 sr, as measured at a full-width-at-half-maximum intensity of the light behind the 5CB cell at V_{DC} = 4.2 V (Figure 7b). It should be noted that such an angular spread is considerably larger than that of light scattering from spontaneous fluctuations of anisotropy in NLCs without an electric field applied, e.g., [56,57].

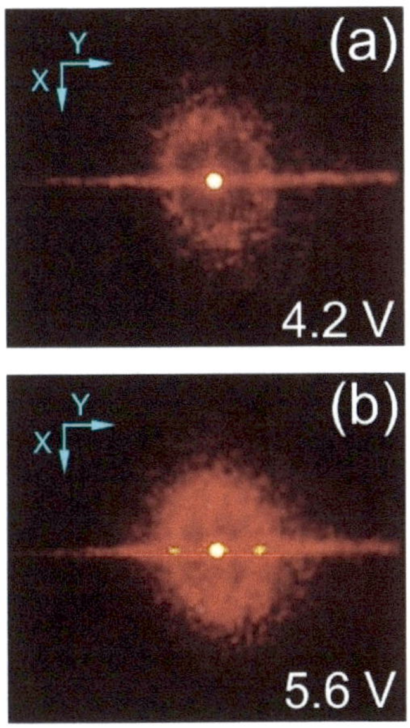

Figure 6. Far-field diffraction pattern observed on a white-paper transversal screen behind a planar cell with 5CB film with a thickness of 25 μm. The cell was illuminated with a He-Ne laser beam whose polarization was along the direction of the rubbing of the cell plates (direction *X*). The pictures were taken under identical conditions; cell-to-screen distance of 85 cm. The DC voltage applied to the cell: V_{DC} = 4.2 V (**a**) and 5.6 V (**b**).

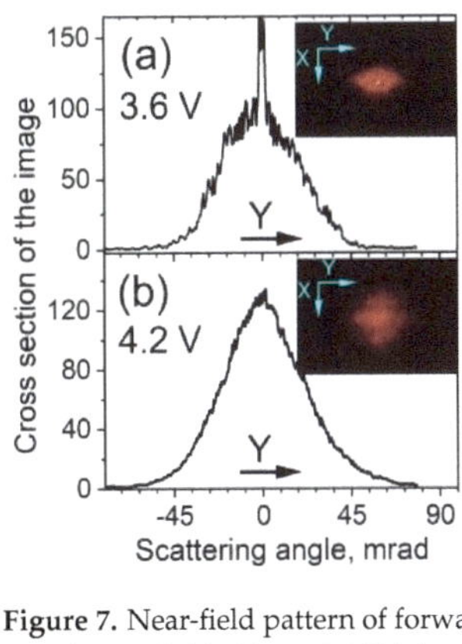

Figure 7. Near-field pattern of forward CLS behind planar cell with 5CB film (25 μm thickness) upon DC electric field: V_{DC} = 3.6 V (**a**) and V_{DC} = 4.2 V (**b**). The images were captured on a black-paper transverse screen, screen-to-cell distance of 17 cm. The polarization of the incident He-Ne laser beam was parallel to the rubbing of the cell plates (the direction *X*). The corresponding digitized horizontal/equatorial cross-sectional profiles of CLS intensity spatial distribution (angular spread of the light intensity) are plotted.

The relatively bright, narrow, widely spread horizontal/equatorial light pattern across the center of the overall scattering/diffraction pattern from the studied 5CB cells, seen more clearly in Figure 6, is also diffuse-like and can be considered CLS. This stripe light pattern extended along a direction perpendicular to the rubbing direction of the cell plates is similar to the effect of a fine quasi-linear diffraction grating, such as one with multiple vertical slits spaced very close together. In the present case, such multiplex diffraction should be the result of a large number of narrow (micrometer) stripes formed parallel to the rubbing direction of the examined cells with planar 5CB films. More precisely, the observed elongated diffuse light pattern should be called "diffuse-diffraction stripe spatial spread/spectrum" (hereafter called DDS), being complex overlaps of many diffraction orders. In the experiment, this diffraction feature, oriented orthogonally to the rubbing direction of the LC cell, can be used for very precise adjustment of the cell with respect to the polarization direction of the linearly polarized incident laser beam.

The observed CLS is due to field-induced static inhomogeneities of NLC director field distribution (in the bulk and on the cell walls). In our case, the refractive index gradients corresponding to the field-induced orientation pattern can be associated with the generation of an array of very thin cylindrical lenses [58,59]. The stationary field-induced narrow-striped LDs in the 5CB planar nematic films can be regarded as such an array for the extraordinary component of the light incident normally to the films. When the orthogonal size (the width) of the stripes of such a parallel orientation pattern is small and of the order of the wavelength of the light transmitted through the array of LDs, then light scattering from such an electrically induced spatially periodic array takes place.

In general, CLS from the examined 5CB cells is due to all micro-sized irregularities and regularities in the illuminating volume and on the cell surfaces, including the microperiodic orientational modulation of the director appearing as fine quasi-periodic parallel stripes along the rubbing of the cells, discussed in Section 3.1.2. Produced by the interference of many diffraction fields, CLS from electrically-formed quasi-periodic micro-scale director spatial modulation can be regarded as diffraction from a multi-frequency grating associated with an LD periodicity that exhibits many spatial frequencies. In fact, many irregular stripes illuminated by the laser beam within the laser beam spot area on the surface of the LC cell correspond to many grating periods. The intensity distribution of the resulting light diffraction will be an overlap of many diffraction functions; hence, the diffraction picture will be smeared out, i.e., an effect similar to that corresponding to the Raman–Nath mode of interaction and diffraction (see Section 3.2.1).

The CLS effect is usually most pronounced for optical waves whose wavelength is roughly similar to the periods of the diffracting objects. The diffuse CLS observed here is different from the random diffraction grating effect (superposition of a large number of diffraction gratings with random amplitudes and phases) but, to some extent, is similar to the diffuse transmission of coherent light. Note that in mono-domain NLCs, the diffuse transmission due to anisotropic light scattering commonly does not follow the input light polarization and is polarization independent [60,61].

Further, a well-defined diffraction pattern as a sequence of light spots (peaks) horizontally localized (along direction Y) near the central peak can also be distinguished within the overall picture of the overlapping scattering/diffraction observed by the studied 5CB films. These features can be attributed to diffraction from a grating of a quasi-periodic structure characterized by multiple spatial frequencies. In this case, the most pronounced light peaks correspond to a higher periodicity for some spatial frequencies intrinsic to the diffractive structure. At a given wavelength, such a periodicity results in sharp diffraction peaks within the diffuse scattering field. Their localization and intensity depend on the configuration of the experiment. Similar light diffraction patterns in nematic liquid crystals with a positive dielectric anisotropy are well known [62].

As found for the considered 5CB nematic films (Figure 8b–e) under the present experimental conditions, this effect (hereafter referred to as CLS diffraction peaks, CLSDPs) was most intense at V_{DC} close to 3.7 V (Figure 8c), i.e., below the voltage V_{form} = 4.2 necessary to form the wide LDs in the films (see Section 3.1.1). Also in this case, the spatial spread of CLSDPs reaches a maximum. Actually, Figure 8c presents a cross-sectional profile of a typical picture of CLS by the studied 5CB films, with the typical superposition of diffraction features as well as the transmitted central coherent beam.

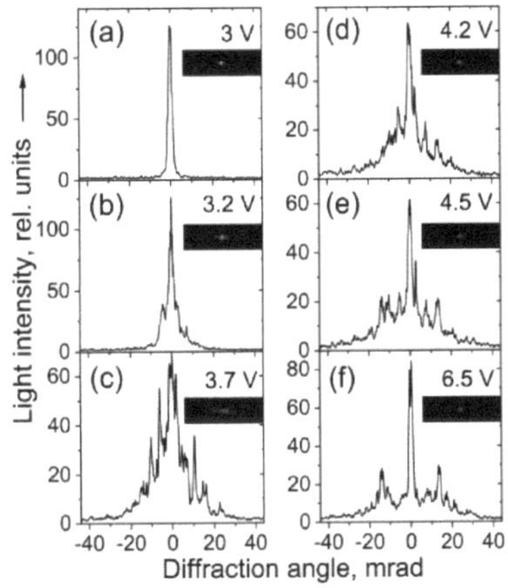

Figure 8. (**b–e**) Horizontal cross-sectional intensity profiles characterizing the CLS diffraction peaks (CLSDPs) observed by far-field imaging on a screen behind a planar cell with 25 μm-thick 5CB film upon DC electric field (V_{DC} values indicated). The polarization of the incident He-Ne laser beam was parallel to the rubbing of the cell plates. The intensity profile (**a**) can be considered as corresponding to the non-scattered transmitted laser beam. The profile (**f**) is close to that of the pattern obtained by DC voltage-driven Fraunhofer diffraction from the 5CB film.

When $V_{DC} \geq V_{form}$, for example, V_{DC} = 4.5 V (Figure 8e), the diffraction pattern transformation implies the formation of a real grating. With a further increase in V_{DC}, diffraction patterns occur (Figure 8f) that are relevant to the diffraction effect corresponding to the morphology consisting of well-developed, wide-formed LDs discussed in Section 3.2.1, i.e., in this case, a director-modulation grating takes place, which results in clearly observed first-order (+1 and −1) diffraction intensity features (recall Figure 6b). The observed change in the diffraction pattern was consistent with the change in texture of the nematic 5CB film described in Section 3.1.

The field-induced CLSDPs, diffuse CLS, and DDS were present in both forward and backward directions, being considerably stronger in the forward direction (the same applies to the observed Fraunhofer diffraction pattern). Inspection of the spatially separated forward scattered/diffracted light forming these patterns shows that the optical noise signal due to scattering from the glass cell itself (as probed by an empty cell) does not contribute to their intensity. Also, any contribution from possible stray scattering from the optical elements of the experimental setup can be excluded, taking into account the geometry of the present experiment.

Fourier Analysis

It is worth noting that Fourier transforms of images of the fine parallel stripes in the studied planar-oriented 5CB nematic films under conditions when the wide LDs are not formed, i.e., at voltage $V_{DC} < V_{form}$ applied to the cell (Figure 9a,b), resemble the observed CLS halo (Figure 7). On the other hand, the concentrated amplitudes around the central peak in the Fourier transform spectrum of a cross-sectional profile of such an image (Figure 9d) look like the intensity profile of CLSDP's pattern. In fact, Fourier analysis allows a precise inspection of the texture change in the considered 5CB planar films upon low voltages $V_{DC} < V_{form}$. In contrast, the detailed monitoring of the field-modified narrow-striped texture (Section 3.1.2) with POM is difficult.

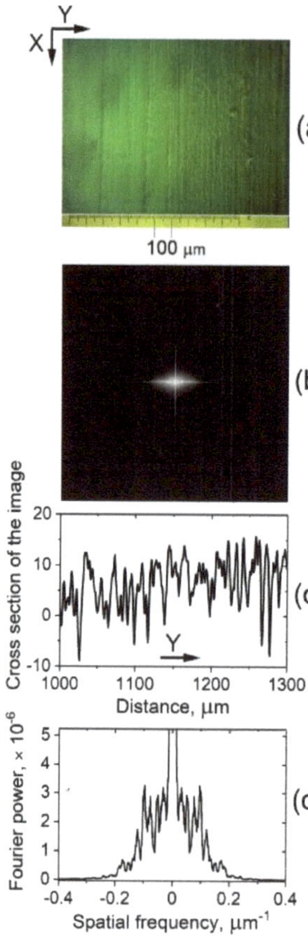

Figure 9. (**a**) Optical microscopy image of the texture formed in a 25 μm-thick planar-oriented 5CB nematic film at V_{DC} = 3.7 V; (**b**) Fourier transform of image (**a**); Panel (**c**) represents an enlarged cross-sectional profile of texture image (**a**); (**d**) is Fourier transform of the whole cross-sectional profile of image (**a**).

As an example, Figure 10a shows micrographs of the examined nematic texture as viewed at two voltages below 3.7 V. Fourier transforms of the texture images (Figure 10b) reveal that in the voltage range from 0 to 3.2 V, there is indeed no change (the micrographs taken for the texture are identical). However, at V_{DC} = 3.3 V a change in the texture is registered, as seen from the digitized images of Fourier transforms of the micrographs (Figure 10c,d). At higher voltages, e.g., V_{DC} = 3.5 V, these profiles tend to match the spatial profiles of CLSDPs shown in Figure 8c. This suggests that the CLSDPs are generated by a narrow-striped microperiodic texture that acts as a grating, as described in Section Diffraction Pattern.

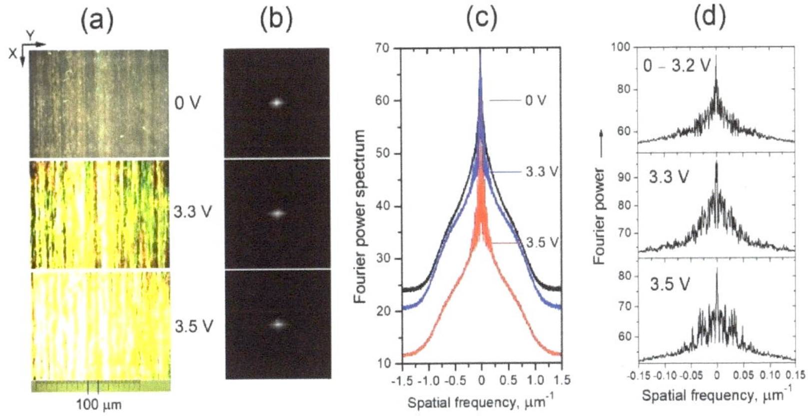

Figure 10. (**a**) Micrographs of the texture formed in a 25 μm-thick planar-oriented layer of nematic 5CB at two values of applied DC voltage. Slightly uncrossed polarizers (set along X and Y). The input light polarization was parallel to the initial alignment of the NLC (i.e., along X—the rubbing direction of the cell plates); (**b**) Fourier transforms of the images in (**a**); (**c**) Digitized images of Fourier transforms from (**b**); (**d**) Expanded view of (**c**) showing the central region of spatial frequencies.

3.3. DC Voltage-Dependent Coherent Light Transmittance of 5CB Planar Films

The spatially periodic director-field modulation in nematic planar cells upon low-voltage static electric fields (discussed in Section 3.1), combined with the strong optical anisotropy that typically characterizes any NLC phase, causes the NLC layer to act as a diffraction grating (or quasi-diffraction grating) when illuminated with a monochromatic light beam (as presented in Section 3.2). Under appropriate conditions and certain circumstances, this effect can be used for EO control of the coherent light transmittance of the NLC films.

3.3.1. Central-Beam EO Behavior

Figure 11a presents the V_{DC}-dependent intensity of the central radial part of the central beam transmitted through a planar 5CB cell. Such coherent light was separated by a pinhole in front of the measuring photodiode (recall Figure 1a). In this case, the separated coherent light was a superposition of coherent light transmission, zero-order Fraunhofer diffraction, and a small contribution of CLS. As seen in Figure 11a, at a well-defined voltage threshold V_{th} (3.2 V), the gradual increase of V_{DC} applied to planarly-aligned cells with 5CB results in a noticeable sharp decrease of the transmitted light intensity, from its maximum value (T_{max}) to its minimum (T_{min}). The latter was achieved at a V_{min} = 3.6 V. A further gradual increase in V_{DC} in the range of 3.7 V–10 V leads to a monotonical increase in the intensity of the central beam. For the two voltage ranges of the measured V-shaped transmittance curve, V_{DC} from V_{th} to V_{min} and $V_{DC} > V_{min}$ (hereafter referred to as "branch *A*" and "branch *B*"), respectively (Figure 11a), rather different EO behaviors take place. The reasons are also different.

Regarding the whole transmittance curve, the 5CB planar cells exhibited a fully reversible EO response under the conditions of the present experiment. When using the time-dividing scheme for measurements described in the Experimental Section, within the experimental uncertainty, the same V-shaped behavior was obtained by either increasing or decreasing V_{DC}. Thus, the planarly aligned 5CB cells provide a low-voltage, hysteresis-free, controllable change in the transmitted coherent light that can be useful for practical applications. Notably, branch *A* is very suitable for efficient and low-voltage modulation of laser light.

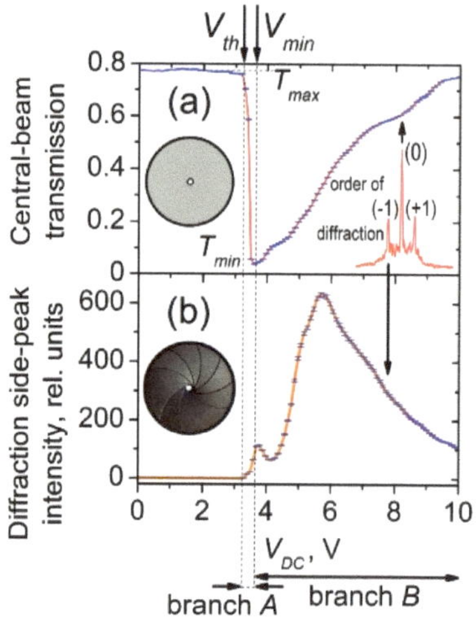

Figure 11. DC voltage-dependent intensity of (**a**) central part of the central beam, (**b**) side-diffraction peak of the diffracted-transmitted light, both measured behind a cell with 25 μm-thick planar 5CB film, in the far-field zone (85 cm distance from the cell to a photodetector with a pinhole). The inset in (**a**) illustrates a cross-sectional intensity profile of diffraction pattern, the same as the one shown in Figure 6b; the arrows indicate the measured light for the cases (**a**,**b**). The polarization of the incident He-Ne laser beam was parallel to the rubbing of the cell plates. The voltage values of the threshold (V_{th}) and for a minimum transmittance (V_{min}) are indicated on the top. At the bottom are marked the voltage ranges branch *A* and branch *B* of the V_{DC}-dependent transmittance curve (**a**), noted in the text. The error bars (in blue) correspond to the standard deviation of the measured data (after 10 averaging).

It should be noted that the considered voltage-controlled coherent light transmittance curve depends on the polarization direction of the incoming laser beam as well as on (i) the texture illuminated; (ii) temperature; (iii) the beam incidence angle; and (iv) the wavelength of the incoming monochromatic light. This is reasonable since all these factors have an influence on the refractive index of the NLCs and their change with voltage applied to the nematic film. The EO behavior of the central-beam transmittance as it depends on (i)–(iv) is presented and discussed in the Supplementary Materials. If it is not specifically stated, a zero angle of incidence is implicit in this work.

Besides the chemical structure of the NLC, the considered texture formations depend very strongly on the experimental conditions relevant to the LC cell (the NLC film thickness, the quality of the boundaries, their treatment, the alignment layers of the cell, the anchoring of the LC molecules at these layers, and other factors). Certainly, the polyimide alignment layers of the LC cells used here (as well as the way of rubbing and the rubbing's geometry) have an important role in the formation of the DC-induced stripe domains as well as in the EO effect under study (a V-shaped curve). It should be taken into account that the polyimide material has a very high resistivity, much higher than that of the active NLC film. Hence, these layers consume some portion of the energy of the DC electric field applied to the LC cell to orient the 5CB molecules and modulate the NLC director. Such an effect of highly resistive orienting layers has been well established, e.g., reported in [63,64]. Due to such a low voltage drop (actually very low), one can expect that the V-shaped curve (and the corresponding V_{th} and V_{min} values) are slightly shifted toward the higher voltage values. Further, the contact resistance at the interface between polyimide-NLC may also have some impact on the EO effect considered here. However, investigations of the effect of the material of the alignment layers as well as the rubbing parameters of the LC cells are beyond the scope of the present work.

An important factor that may affect the registration of the EO response of the studied films is the detection geometry in the measurements. The curve shown in Figure 11a is relevant to the V_{DC}-dependent central-beam transmittance of a planar 5CB cell when a small part of the transmitted light is selected for detection. This light comprises the central area of the central beam selected in our case through a pinhole with a small diameter (~1 mm). When a larger portion of the light around the center of the transmitted beam is registered, a weaker reduction of the film transmittance takes place. For instance, Figure S1b shows such V_{DC}-dependent coherent light transmittance curves measured when the light is separated by the use of an iris diaphragm (centered around the Z axis, Figure 1a) whose opening size (circular aperture) is equal to the full diameter of the transmitted laser beam (at the location of the diaphragm). When one measures all the transmitted central-beam intensity, the value of T_{min} is higher, i.e., the optical contrast ratio T_{max}:T_{min} decreases. Clearly, this is due to the contributions of the overlapping CLS processes (see Section 3.2.2) being partially included in the detection area. Being of importance for the application of the considered EO effect, light detection by measuring the beam center is most favorable. Also, the longer distance to the photodetector is profitable.

3.3.2. EO Behavior of Coherently Diffracted and Scattered Light

Voltage-Dependent Fraunhofer Diffraction

The intensity of the observed diffraction side-peaks of the triplet split of coherent light resulting from Fraunhofer diffraction by the wide-formed LDs (Section 3.2.1) also depends on V_{DC} (Figure 11b). Reasonably, this is due to the change in optical anisotropy modulation caused by the applied electric field. Since the present case is not purely diffractive, the amplitude transmittance function of the diffractive medium and its variation with V_{DC} cannot simply be deduced from the measured spatial distribution of the diffracted light intensity. This is an inverse problem that would otherwise have to be solved routinely using theory and indirect methods. Generally, the cross-sectional profile of a texture image (a micrograph taken by transmission, such as the one shown in Figure 2) represents the modulus of the complex amplitude transmittance function, which is a spatially averaging product of the transmittance coefficient and its complex conjugate.

Like the V_{DC}-dependent central-beam transmission (Section 3.3.1), the DC voltage-controlled diffracted light through the electrically-induced LDs in the considered planar nematic films was fully reversible at ascending or descending V_{DC}. It should be mentioned that the decrease in light intensity in the first diffraction orders is related to energy conversion and redistribution to the zero-th order [42].

Voltage-Dependent CLS

Regarding the V_{DC}-dependent spatial and light intensity changes of field-induced diffuse CLS, by increasing V_{DC} in the voltage range from V_{th} to V_{form}, the size of the CLS halo was gradually increased (see the photo series Pics1 in the Supplementary Materials). The same applies to the DDS stripe, but in the voltage range from V_{th} to V_{min}. The asymmetry of this pattern of CLS is like that for light scattering from NLC director anisotropy fluctuations, which follows a simple rule: the scattered intensity is highest in the direction orthogonal to the polarization direction of the incident light. This has been well established for NLCs and, more specifically, for 5CB [60]. Reasonably, the horizontal-to-vertical asymmetry of the spatially localized CLS pattern behind the examined planar cells with 5CB (CLSDPs and DDS, see Section Diffraction Pattern) may be considerable.

At the value V_{form} (4.2 V), the transversal light distribution of diffuse CLS (both circular-chaped halo and DDS stripe) reaches maximum intensity (can be seen in the photo series Pics1 in the Supplementary Materials), and the halo is transmuted into a circular ring from a diffracted light cone around the central beam of transmitted light (recall Figure 6a). When V_{DC} is above V_{form} and gradually increases, the CLS is weakened, and the intensity of the arising side-peaks of the triplet Fraunhofer diffraction pattern is enhanced (photo series Pics1 and Pics2 in the Supplementary Materials). By further increasing V_{DC}, the intensity of the CLS patterns is gradually decreasing simultaneously with the diminishing of the first-order diffracted laser beam intensity. At higher V_{DC} ($V_{DC} > 7$ V), when the electrically-induced hydrodynamics in the 5CB cells are enhanced, both Fraunhofer diffraction and CLS are greatly reduced. Still, some of the light behind the cells is diffracted and scattered by fragmented LDs (which is a superposition of field-induced periodic and aperiodic modulation of optical anisotropy).

3.3.3. Polarization Dependence

The sharp reduction of the intensity of the laser beam transmitted through 5CB planar films presented in Section 3.3.1 is polarization sensitive. Figure 12a illustrates the change in diffraction pattern behind the 5CB cell depending on the direction of polarization of the incoming laser light at a fixed value of V_{DC}. Figure 12b shows the corresponding change in the V_{DC}-dependent central-beam transmittance of the films measured under identical experimental conditions.

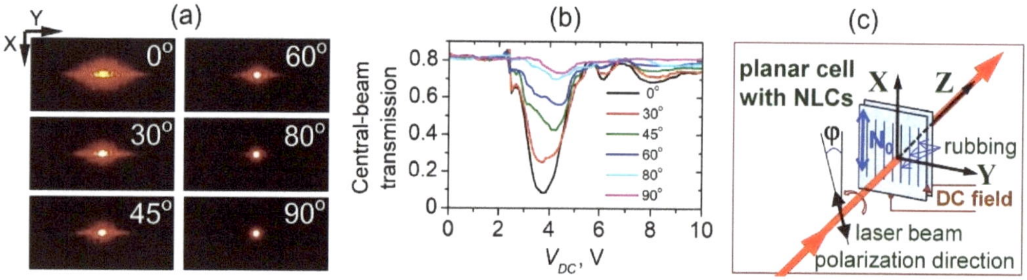

Figure 12. Changes in the diffraction pattern (pictures taken on a transversal screen) behind a 25 μm-thick planar cell with 5CB at $V_{DC} = V_{min} = 3.7$ V (**a**) and V_{DC}-dependent central-beam transmittance of the same film (**b**) for various angles φ of the He-Ne laser beam polarization toward the rubbing direction of the cell: 0°; 30°; 45°; 60°; 80° and 90°, under the same other experimental conditions. Normal incidence of the laser beam (**c**).

As seen from Figure 12b, the reduction effect is strongest when the polarization of the incident laser beam is parallel to the rubbing direction of the examined cells (direction X, see Figure 12c), i.e., to obtain maximum effect, the electric field of the incident plane-polarized light wave must be parallel to the initial (zero-field) director orientation ($\mathbf{N_0}$) and the field-induced LDs. This is consistent with the geometry of the spatially periodic orientation pattern V_{DC}-induced in the studied 5CB planar films. In the case of parallel $\mathbf{N_0}$ and laser beam polarization, the laser beam propagating through the films most effectively experiences static nematic director deformations. In turn, the diffraction grating induced in the films by the applied DC electric field does not respond to light polarized orthogonally to the direction $\mathbf{N_0}$.

Figure 13 compares the dependences of both central-beam coherent light transmission and of CLS against the angle φ between the direction of the polarization of the incident laser beam and the rubbing direction of the 5CB cell (the initial orientation of the nematic director, $\mathbf{N_0}$). The polarization-dependent central-beam transmission was measured at $V_{DC} = V_{min}$, according to the scheme shown in Figure 1a, using a pinhole as a spatial filter (Figure 13a, filter 0). It should be remembered that for the 5CB films understudy, a small portion of CLS always accompanies the central light beam and enters its measured radial zone.

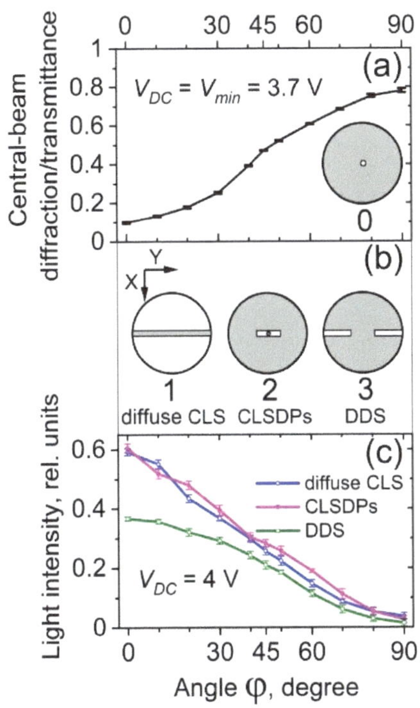

Figure 13. The intensity of central-beam diffraction/transmission (**a**) and CLS (**c**) measured for a 5CB planar cell as a function of the angle φ between the He-Ne laser beam polarization direction and the rubbing of the cell, under the same other experimental conditions. Error bars correspond to the standard deviation of the data obtained after 10^4 averaging (done during 30 s). (**b**) Illustrations of the spatial filtering performed—the blocked area on the collecting spherical lens is colored in gray; the distance cell-to-lens was 25 cm.

The polarization-dependent intensity changes of CLS patterns behind a 5CB planar cell at a fixed voltage (V_{DC} = 4 V) are given in Figure 13c. In these measurements, the different CLS patterns were separated by spatial filtering of light on a collecting lens, as shown in Figure 1b. In each of these cases, the V_{DC}-dependent intensity of the forward scattered/diffracted light was recorded while blocking the central beam of transmitted light. The diffuse CLS (the halo and its transformation into a circular diffraction ring at higher V_{DC}) was measured by using a narrow (3 mm) horizontal blocking stripe (Figure 13b, filter 1). In this way, the horizontally located diffraction/CLS patterns were rejected, i.e., CLSDPs and DDS. The accurate and complete separation of CLSDPs and DDS is a difficult task, but the applied spatial filters are acceptable solutions for their individual measurements (Figure 13b, filters 2 and 3, respectively).

It is seen from Figure 13 that, in contrast to the DC electric field-controlled central-beam transmittance of the 5CB planar cell, the light intensity of the components of field-induced CLS from the cell (diffuse CLS, CLSDPs, and DDS) was maximum when the polarization of the incident coherent plane optical wave was along the rubbing of the cell plates and was reduced when the input polarization was rotated towards the orthogonal direction.

3.4. EO Control of Coherent Light Transmission through LDs in Nematic Films—Physical Mechanism

The shift between the minimum of the light transmittance curve shown in Figure 11a and the maximum of the curve in Figure 11b implies that the sharp decrease in the intensity of the central beam of coherent light transmitted through the studied 5CB planar films has to be related to a physical process other than the diffraction splitting of the laser beam due to Fraunhofer diffraction (discussed in Section 3.2.1). A detailed inspection indicated that the observed strong reduction effect at $V_{th} < V_{DC} < V_{min}$ (i.e., branch *A* in the central-beam transmittance curve, Figure 11a) can be connected with the DC electric field-induced CLS from grating-scattering V_{DC}-induced microperiodic narrow-striped texture in these nematic films (Section 3.1.2). As mentioned in Sections 3.1.1 and 3.2.1, the wide-formed LDs in the 5CB films under study occur at a DC voltage V_{form} = 4.2 V, i.e., higher than V_{min}. Accordingly, in the voltage range $V_{th} < V_{DC} < V_{min}$ the CLS should not compete with the V_{DC}-induced Fraunhofer diffraction from the films.

When only the central beam was blocked and the total intensity of forward diffracted and scattered light from the 5CB film was measured, the recorded curve was the reciprocal of that measured for the central beam intensity (Figure 14). These opposite EO behaviors indicate the close correlation between the CLS/diffraction and the central-beam diffraction/transmission (the latter being still slightly influenced by CLS/diffraction). These optical processes are a -counter-pair. They are coupled and controlled by the electrically-driven reorientation of the 5CB molecules towards the direction of the DC electric field (which is also the direction of the incident laser beam, Z, Figure 1b) and orthogonal to the initial orientation of the director along the rubbing of the cell plates (the direction *X*).

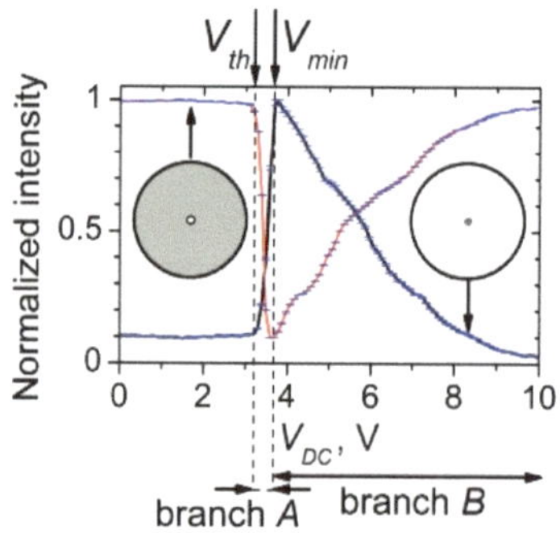

Figure 14. V_{DC}-dependent intensity of central-beam transmitted light (shown with circles) and scattered/diffracted coherent light (line), measured separately in the near-field zone at a distance of 15 cm behind a planar cell (thickness 25 μm) with 5CB. The polarization of the incident He-Ne laser beam was parallel to the rubbing of the cell plates. Both curves were obtained under identical experimental conditions, except for the spatial filtering of the light. The spatial filters employed are illustrated by inset sketches of the blocked area (colored in gray) on the collecting spherical lens. The central-beam transmitted light was measured using an iris diaphragm with an aperture diameter equal to the diameter of the input laser beam and centered on that beam. The scattered/diffracted coherent light was registered by blocking the central beam using a circular aperture with the same diameter as that beam. The error bars (in blue) correspond to the standard deviation of the measured data (after 10 averaging).

Being influenced by CLS and diffraction, the coherent light transmission in our case cannot simply be modeled as the transmittance of the nematics, considered only as birefringent media [65,66]. The propagation of an optical beam and its diffraction by a grating formed in such media due to the periodic inhomogeneity of their optical refractive index need more complex analyses and sophisticated interpretation [67]. As with other bulk NLCs, light scattering from 5CB nematic films is generally due to the anisotropy of the index of optical refraction (birefringence, Δn) of the NLC material. With a plane monochromatic optical wave of wavelength λ passed through a nematic film of thickness d, an optical phase difference (shift) is induced that, at normal incidence of the wave, is expressed as $\delta = 2\pi d\Delta n/\lambda$. The optically-induced δ can modulate the optical wave interacting with the nematic film. When an electric field is applied to the nematic film, a field-induced Δn occurs, expressed as $\Delta n = n_e(E,\psi) - n_o$, where n_e and n_o are the extraordinary and ordinary index of refraction, respectively, E is the electric field intensity, and ψ is the spatial angle between the nematic director and the direction of the incident optical beam. Thus, the corresponding field-induced δ depends on the orientation of the NLC molecules by the applied electric field.

In our case, the optical anisotropy Δn induced in NLC planar films by DC voltage V_{DC} is spatially modulated and expressed as periodic LDs—an electrically formed microperiodic array. The spatially modulated anisotropy and the V_{DC}-induced change in orientation of the nematic director result in microperiodic spatial patterns of phase shift—an electrically-induced optical phase grating whose orientation is electrically controlled. Due to the positive dielectric anisotropy of the 5CB molecules, they are forced by the applied electric field to orient themselves relative to the field direction (in our case, along the Z axis, Figure 1). Thus, the CLS and diffraction from the studied nematic films are electrically induced (via the electrically induced LDs texture in them) and electrically driven by the electrically driven reorientation of the nematic director (the local optical axis).

By measuring the integral intensity of CLS/diffraction, one cannot specify what is the origin of the sharp decrease in the intensity of the central beam, i.e., branch A of its V_{DC}-dependent transmittance curve. In order to be compared, Figure 15 presents the V_{DC}-dependent intensity curves for the transmitted light corresponding to the coherent optical processes considered above (Fraunhofer diffraction and diffuse CLS), separately measured after spatial filtering of the light behind the 5CB planar cell. Figure 15a shows the intensity of the measured light due to forward diffuse CLS as compared with the intensity of the central beam of diffraction/transmission. On the other hand, Figure 15b reports the intensity of the selected forward horizontal diffraction/CLS pattern—CLS diffraction peaks (CLSDPs) and the diffuse-diffraction spectrum (DDS) (see Section 3.2.2). Figure 15c presents the V_{DC}-dependent intensity measured for selected forward first-order diffracted light due to Fraunhofer diffraction. In this measurement in the far-wave field, the coherent light was separated by an iris diaphragm, whose circular aperture was equal to the size of the first-order diffraction pattern (peak). It should be noted that in this case, the unavoidable contribution of CLS/diffraction results in a complex curve (Figure 15c). For completeness, in Figure 15c, the V_{DC}-dependent intensity of two other bright peak-like patterns is also given, each located exactly between the central peak and the +1 or -1 order of Fraunhofer diffraction from the 5CB film. These V_{DC}-induced light peaks were measured separately in the same way in the far-field zone. Most probably, they are diffraction peaks from the CLSDPs sequence.

Comparing the data in Figure 15 for the distinguished coherent optical processes electrically driven through the V_{DC}-induced LDs in the measured 5CB planar cells, as well as the dynamic ranges of the intensity changes of their EO behaviors, one can conclude that the most active optical process that competes with the coherent transmission of the central beam in the considered small voltage range $V_{th} < V_{DC} < V_{min}$ is the CLS expressed as CLSDPs. More strictly, both CLSDPs and diffuse CLS from the narrow parallel stripes (as a fine microperiodic grating of textural LDs) are related to the sharp decrease (branch A) of the intensity of the laser beam passed through the planar 5CB films examined here.

As for the gradual decrease in CLS intensity at voltages above 4 V (Figures 14 and 15a,b), this is related to the formation of wide LDs in the studied 5CB films and the V_{DC}-induced Fraunhofer diffraction from them, respectively. Because of the positive dielectric anisotropy of the 5CB molecules, their electrically driven orientation tends to a homeotropic alignment of the initially planar 5CB nematic films. At increasing V_{DC}, this process, followed by quenching of LDs associated with a transition to a state of electrohydrodynamic instability and random motion inside the nematic films, leads to a diminishing of modulated anisotropy and the efficiency of the formed optical phase grating experienced by the incoming linearly polarized laser beam, and hence to reduced laser scattering and diffraction. This results, namely, in the decreasing wings of CLS-related curves, seen in Figures 14 and 15. At $V_{DC} > 9$ V, the beam intensity spatial profile tends to its initial shape, i.e., the scattering and diffraction disappear. Thus, the field-induced quenching of the microperiodic director-modulation textures can adequately explain branch *B* of the EO characteristic curve of the central beam of diffraction/transmission.

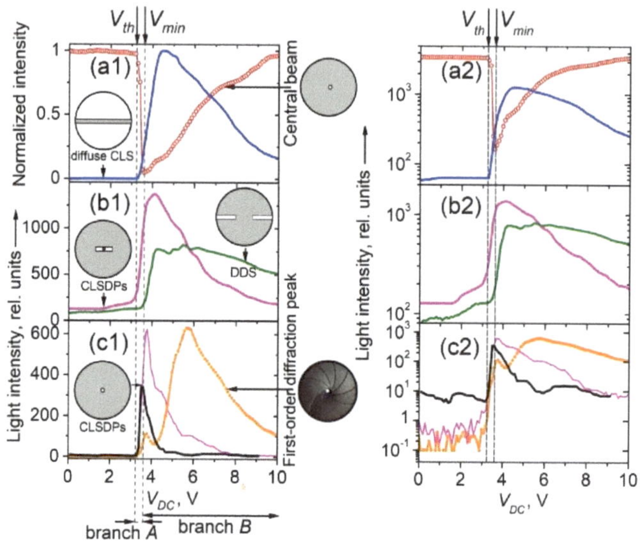

Figure 15. V_{DC}-dependent intensity of light behind a planar cell with 5CB film (thickness 25 μm): (**a**) central beam of diffraction-transmission and diffuse CLS; (**b**) CLSDPs and DDS; (**c**) the first-order diffracted light and diffracted light measured at two intermediate locations between the central peak and the first diffraction order (drawn with a thin and bold line). They, as well as the central beam of diffraction-transmission of coherent light, were measured in the far-field zone (85 cm distance to the cell, photodetector with a pinhole). In the measurement of diffuse CLS (**a**), CLSDPs and DDS (**b**), the distance cell-to-lens was 25 cm. In all cases, the polarization of the incident He-Ne laser beam was parallel to the rubbing of the cell plates. The performed spatial filtering is illustrated by inset sketches of the blocked area (shaded in gray) on the collecting spherical lens (see also Figure 1b). In the right—the same as in the left, but in logarithmic scale for the light intensity. At the bottom are marked the voltage ranges branch *A* and branch *B* of the V_{DC}-dependent transmittance curve for the central beam. The measurement error was 1.5–2.5%.

3.5. Enhancement of the Optical Contrast Ratio of Coherent Light Transmittance of Nematic Films, Electrically-Controlled by Microperiodic Director Modulation

Most likely, other NLCs can also exhibit DC low-voltage scatter spatially periodic director modulation patterns and hence suitable coherent scattering and diffraction by achieving a complex balance of various system parameters that control the formation of LDs (the periodic director modulation) and thereby controlling the complex transmittance function that determines the EO response of NLC planar films. An effective way to improve the optical contrast ratio in this case can be doping methods, for example, by including suitable nanoparticles (NPs). As practice shows, the doping of NLCs with even a small amount of NPs affects almost all important properties of the nematic materials (for example, [68,69]). In particular, metal NPs can considerably modify the texture, optical, and EO properties of such NLC-based nanocomposites compared to the host NLC material, e.g., 5CB and other cyanobiphenyls [69–72]. Moreover, there are reports showing the formation of periodic structures and stripe patterns in 5CB nematic doped with gold nanoparticles (AuNPs), which are of special interest for metamaterials and the fabrication of tunable photonic and communication devices [73,74]. Also, such nanocomposite materials can exhibit novel EO effects. The search for enhancement of the optical contrast ratio of the electro-optically controlled light transmission in nematics and related composites is a challenge.

Some efforts in this regard are focused on investigating metal NPs and hybrid metal-polymer nanostructures that can be quite effective as additives to nematics. For instance, a large reduction (~20 dB) in a short voltage range ~1.5 V) of coherent light transmission of a He-Ne laser beam has been obtained by mixtures of 5CB and 12 nm-diameter gold nanospheres at a relatively low concentration of 0.5 wt.% [41]. These AuNPs were capped with a ca. 10 nm-thick polymer layer [41]. Modifying the NLCs 5CB by adding such NPs, the nematic state holds, and the dynamic range (optical contrast ratio) of the coherent light intensity change can be markedly enhanced. Compared to the reduction effect found with identical planar cells with 5CB nematic discussed in Section 3.4, the improvement was at least one order of magnitude (Figure 16b). Such an electrically activated effect of AuNPs/5CB composite films upon low-voltage static electric field results from the larger spatial spread of light in a direction orthogonal to the polarization direction of the incident laser beam (in the present case—horizontal spread, along the Y-axis, Figure 16a).

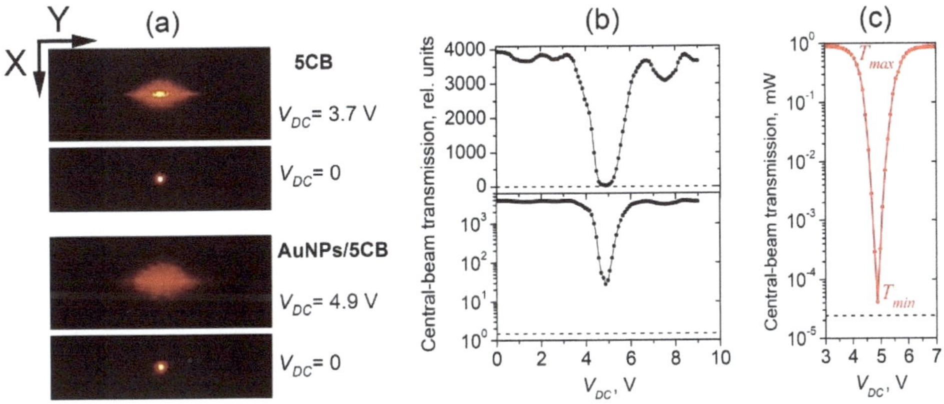

Figure 16. (**a**) Representative examples of maximum diffractive/CLS spread and laser beam intensity reduction effects resulting from He-Ne laser beam propagating through identical cells with 25 μm-thick planar nematic films of 5CB and AuNPs/5CB upon DC electric field. The pictures were taken for far-field light intensity patterns displayed on a transversal screen behind the cells. The laser beam polarization direction was parallel to the rubbing of the cell plates; the other experimental conditions were also the same. The circular beam shapes corresponding to the zero-field transmission are given for comparison; (**b**) DC voltage-dependent intensity of He-Ne laser beam passed through a 25 μm thick planar cell with a composite of 0.5 wt.% gold nanospheres in 5CB nematic. The transmitted laser beam is measured in the far-field zone (85 cm distance to the cell, photodetector with a pinhole). The curve is given in both linear and logarithmic scales. The detection limit (the dark-current signal) of the apparatus is shown with a dashed line; (**c**) As in (**b**), but measured with a higher dynamic range of photodetection.

As in the 5CB films, the complex interplay between the electrically induced light scattering and diffraction is determined by the texture that is dominant at the corresponding value of the voltage V_{DC} applied to the AuNPs/5CB film. Again the field-induced narrow-striped LDs fine-stripe microperiodic texture and the diffusive-diffractive CLS strongly decrease the intensity of the laser beam transmitted through AuNPs/5CB films, as in branch A of the V-shaped V_{DC}-dependent coherent light transmittance curve for 5CB films. There is, however, an important difference. For 5CB films, the increasing coherent light transmission at $V_{DC} > V_{min}$ is mostly due to Fraunhofer diffraction from the wide-formed LDs (Section 3.2.1). At such voltage values, these LDs become more pronounced

(Section 3.1.1). They are developed near the cell plates and over the rubbing-induced narrow-striped LDs [1,51,53]. By further increasing V_{DC}, flexo-dielectric walls in 5CB films replace the rubbing-induced surface texture. Accordingly, Fraunhofer diffraction replaces the scattering. In contrast, the EO effect observed with AuNPs/5CB planar nematic films is solely due to rubbing-induced surface texture. In this case, the strong minimization of coherent light transmission at V_{min} results from the specific angular spread of scattered-diffracted light on both sides outside the central beam at a certain V_{DC} value [41].

The contrast ratio T_{max}:T_{min} at V_{DC} = V_{min} is limited due to the same CLS. The light due to CLS appears as optical noise (coherent background) for the central beam signal, although the divergence of CLS is much larger than that of the central beam of transmission-diffraction. Due to this limitation, the contrast ratio depends on the measurement/detection scheme. Thus, the V-shaped curve for the central-beam transmission is closely related to the detection geometry and both the sensitivity and dynamic range of the photodetector. To achieve a maximum effect, the iris diaphragm in front of the photodetector should be correctly adjusted (centered around the direction of the output laser beam) and opened in the proper way in order to register only the central part of this beam. Furthermore, the polarization vector of the laser beam has to be exactly parallel to the rubbing of the cell (see Section 3.3.3), and the laser beam must be incident normally to the cell with the nematic film (Section S1.3 in the Supplementary Materials).

The dominant longitudinal texture patterns in the AuNPs/5CB planar cells viewed by POM (including observations at DC voltage in the range from 4 V to 7 V) were composed of densely spaced regular narrow (micrometer-wide) parallel stripes, oriented along the rubbing direction of the cells [41]. The dispersed AuNPs, even at the relatively low concentration of 0.5 wt.%, prevent the formation of voltage-induced wide LDs (flexo-dielectric walls) observed in the planar cells with pure 5CB. This is due to the charge-trapping effect of the polymer-capped Au metal nanospheres, which leads to ion depletion in the bulk of the 5CB nematic films [41,71] (for a thorough conceptual review of ion-trapping effects from nano-objects in LCs and related phenomena, one can refer to Garbovskiy and Glushchenko [75]).

In contrast to the 5CB films, the planar AuNPs/5CB composite films allow a full deflection of the light energy outside the beam center and, thereby, a complete EO minimization of the transmitted laser beam intensity in the far-field zone (see videos 1, 2, and 3 in the Supplementary Materials). It should be noted that the effect due to CLS/diffraction owing to the included AuNPs is much stronger than the decrease in the NLCs transmittance due to oscillations by the ordinary voltage-modulated birefringence (e.g., [66,76]). Hence, such nanocomposite nematic films are of great interest for EO applications, as discussed in the next section.

3.6. *Applicability of Coherent Light Transmission Electrically Controlled by Spatially-Periodic Director Modulation in Nematic Films under Low-Voltage Static Electric Field*

V-shaped voltage-dependent optical transmission (also termed "transmissive U-shaped EO switching") is well known for LC structures (smectic and ferroelectric) [77–83]. The specific EO control of coherent light transmittance of nematic films by scatter spatially periodic director modulation in them can be used for new modes of scatter-based EO applications exploiting the spatial patterns of optical phase shift induced in NLC films upon low-voltage DC electric fields. As presented in Section 3.5, AuNPs/5CB composites in planar-orienting cells under low-voltage static electric fields exhibit a large and sharp reduction of transmitted laser beam intensity. The registered dynamic range of this EO effect depends on the sensitivity limit and dynamic range of the photodetector and measurement unit. Practically, in this way, one can achieve an extremely high optical contract ratio T_{max}:T_{min}–higher than 10^4 (Figure 16c)—if a photodetector with such a large dynamic range is used. Such 100% efficiency can only be compared to that provided by the light deflection effects used by LC beam steering devices (for a thorough review, one can refer to, e.g., He et al. [84]). However, the LC deflectors typically require relatively high operating voltages at comparable thicknesses (25 μm) of the LC films, as well as the application of an AC electric field [84].

As noted in Section 3.5, AuNPs/5CB planar films can allow a full scattering of the light energy out of the beam center and thereby complete EO minimization of the transmitted laser beam intensity. Basically, the EO result of their operation is like that of the NLC-based polymer-dispersed liquid crystal (PDLC) devices [85–87] although the operation mechanism is different. The PDLCs operate through an electrically controllable dielectric reorientation of the LC molecules in micro- or nano-sized droplets. By applying an AC electric field of appropriate magnitude, the PDLCs are switched from the OFF-state (translucent state) to the ON-state (transparent state). The problem with PDLC devices is their relatively low optical contrast ratio, since it is practically impossible to eliminate optical scattering in the direction of the laser beam passing through the PDLC film. By using high-contrast PDLCs for high-performance EO shutters/modulators, this ratio can be as high as 30–50 [85–87], but values of 200–300 have also been reported for PDLCs with a special composite design [88–90]. Various approaches used to solve the problem of the low contrast of PDLC films require the application of a high control voltage. Alternatively, low-voltage interference effects in microscale single-layer PDLCs with large-sized NLC droplets have been proposed to increase the optical contrast ratio of coherent light transmission, making them suitable for tunable modulators of laser light [91].

However, by the reduction effect for electro-optically controlled coherent light transmission through LDs in NLC films considered in the present work, can achieve a much higher optical contrast ratio T_{max}:T_{min}. As with PDLCs, in this case, the laser light modulation is also based on electrically controlled light scattering and also depends on the optical properties of the NLC as a function of temperature and light wavelength (e.g., [43,92] for 5CB). The advantage of the EO effect of DC field-induced scatter/diffractive microperiodic textures of phase-shift regular spatial patterns in NLC films studied here is the strong reduction of light scattering in the optical path of the transmitted laser beam. Moreover, due to a splitting of the transmitted laser beam laterally in the direction perpendicular to the rubbing of the cell plates as well as due to the spatial shift of the scattered light to the periphery in the same direction (Y), the light can even be self-removed from the optical path of the transmitted laser beam, as reported for AuNPs/5CB planar nematic films under a static electric field [41].

Having a very sharp dip (fwhm less than 1.5 V), the specific low-voltage V-shaped curve of DC voltage-dependent coherent light transmittance of planar nematic films (Figure 16b,c) is certainly of interest to practice. For example, these films can be applied in the field of process control to stabilize various processes through electric feedback, which is a common application of V-shaped electrical characteristics. The state of the minimum laser beam transmission maintained by $V_{DC} = V_{min}$ applied to the cell is very sensitive to any change in this voltage level. Accordingly, the nematic film can respond to very small voltage changes, e.g., by ± 0.01 V.

Despite the complexity of the EO response of NLC media to a static electric field, a periodic switching regime based on the EO effect described here has the potential to be implemented for control functions. The effect is easily usable, but the EO switching by the studied 5CB planar films needs a time of 1 s in the configuration for measurements in the far-field zone (Figure S5a) or 20 s in the near-field zone (Figure S5f) to achieve a stable recovery of the stationary state. Hence, the maximum EO modulation frequency (repetition rate in the pulsed regime) is limited and should be less than 1 Hz or 0.05 Hz, respectively. Thus, the considered low-voltage DC electric field-driven modulation via spatially periodic director-field modulation in nematic planar cells can be used for EO control of relatively slow processes. Also, the instability and long-time dielectric relaxations inherent in nematic films, especially at DC voltages in the range of branch A of the V-shaped coherent light transmission curve (see Figure 11), may limit the applicability of this EO effect for laser intensity modulation by DC repetition pulses. Notice that this applies to the time response of planar nematic films of both 5CB and AuNPs/5CB composites [41].

The applicability of the specific EO response of planar nematic films, e.g., 5CB and AuNPs/5CB considered here, can be extended owing to the possibility that the scattering/diffractive director-field modulation pattern induced in them with DC voltage can be rapidly erased by joint AC voltage [41]. The value of the latter is also low, comparable to the DC voltage driving the scattering/diffraction effect through modulated anisotropy [41]. The erasure of the V_{DC}-induced optical phase grating (and thus the V-shaped transmittance of planar nematic films) can be relatively fast, e.g., within 0.1 s or less, depending on the strength of the externally applied AC electric field ([41]). As compared to 5CB films, AuNPs/5CB nematic films can offer faster and more stable EO modulation.

Finally, the V-shaped dependence of laser light transmittance determined by spatially periodic director modulation in NLCs can be used in low-voltage sensors and various optoelectric techniques for sensitive detection of weak dynamic electric fields. This option is useful for the detection/monitoring of events relevant to military, geo-acoustic, and biomedical applications.

4. Conclusions

Stationary longitudinal domains (LDs) formed in planar-oriented nematic films under a low-voltage DC electric field lead to coherent light scattering and diffraction. In this way, such films can enable efficient control of the transmission of coherent light through them. The effect is maximal when the polarization of the incident light wave is along the orientation direction of the field-induced LDs, i.e., parallel to the initial (zero-field) orientation of the nematic director.

At a DC voltage from zero to 10 V, two sets of electric field-induced regular LDs are observed in 25 μm-thick nematic 5CB films in planar cells: small-period (less than 10 μm) and large-period (~60 μm) LDs, both of which are of flexoelectric origin. The first type of LDs are induced by the orientation rubbing of the alignment layers of the cell; the second type of LDs are flexo-dielectric walls and take place at a voltage higher than a well-defined threshold value. As a result of each of these two kinds of field-induced periodic modulation of the nematic director, a field-induced spatial modulation of the optical phase in the plane of the nematic films arises due to the optical anisotropy modulation.

It is proven here that the microperiodic narrow-formed LDs induced by a low-voltage (~4 V) DC field can produce a sharp and large reduction of the intensity of a laser beam transmitted through the studied nematic films. By incorporating specific additives, e.g., NPs, the optical contrast ratio achievable with this EO effect can reach 10^4, which is unattainable with conventional NLC-based devices for active light control. In this way, one can switch or greatly modulate the intensity of a laser beam propagating through planar nematic films if the beam polarization is parallel to the initial orientation of the nematic director.

Furthermore, the field-induced regular wide director-modulation spatial patterns of optical phase shift in planar nematic films can also be used for DC voltage-controllable amplitude modulation of coherent light. In planar 5CB films at DC voltages from 4 V to 10 V, the intensity of the diffraction splitting of the transmitted laser beam can be almost linearly commanded by the applied DC field, and such EO behavior is also of practical significance.

The nature of the observed EO effects is elucidated. The distinct voltage regions corresponding to the involved coherent optical processes (coherent light scattering and diffraction from DC voltage-induced gratings of optical anisotropy modulation) are exactly specified. The relationship of these processes to the DC voltage-dependent coherent light transmittance of planar nematic films is defined.

Of relevance to EO applications, advanced high-performance nematic materials (e.g., NLC-based nanocomposites or hybrid materials) can be designed that have reversible low-voltage EO scattering/diffraction responses through field-induced microperiodic director modulation and field-controlled director orientation. Such electrically controlled optical phase gratings and non-absorbing light diffusers can be adopted for various micro-optic and photonic applications.

Supplementary Materials: The following supporting information can be downloaded at: https://www.mdpi.com/article/10.3390/ma16176014/s1, Figure S1: DC voltage-dependent laser beam transmission through 5CB film; Figure S2: DC voltage-dependent laser beam transmission through 5CB film at various temperatures; Figure S3: DC voltage-dependent laser beam transmission through 5CB film at various angles of incidence, Figure S4: DC voltage-dependent laser beam transmission through 5CB film at various wavelengths, Figure S5: Temporal behavior of DC voltage-dependent laser beam transmission through 5CB film, Figure S6: Optical microscopy images of texture in 7CB film and Fourier transforms Figure S7: DC voltage-dependent laser beam transmission through 7CB film Video: Pics1.mp4: Pictures of the far-field diffraction pattern behind 5CB film, Pics2.mp4: Pictures of the far-field diffraction pattern behind 5CB film (at a lower laser beam intensity), Movie 1.avi: Time evolution of the light pattern behind AuNPs/5CB film at 4.6 V, Movie 2.avi: Time evolution of the light pattern behind AuNPs/5CB film at 4.9 V, Movie 3.avi: Time evolution of the light pattern behind AuNPs/5CB film at 5 V [92].

Funding: This research was carried out within the framework of the project CoE "National Center of Mechatronics and Clean Technologies", BG05M2OP001-1.001-0008-C01 (by the European Regional Development Fund) as a part of the Operational Programme "Science and Education for Smart Growth 2014–2020".

Institutional Review Board Statement: Not applicable.

Informed Consent Statement: Not applicable.

Data Availability Statement: The data presented in this study are available on request from the corresponding author.

Acknowledgments: The author expresses his gratitude to N. Scaramuzza from the University of Calabria, Italy, for providing samples (cells) with LCs and gold nanoparticles. The author wishes to thank A. Petrov and Y. Marinov from ISSP-BAS (Sofia, Bulgaria) for fruitful discussions. The author is grateful to Y. Marinov, with whom the POM images were taken.

Conflicts of Interest: The author declares no conflict of interest. The funders had no role in the design of the study, in the collection, analysis, or interpretation of data, in the writing of the manuscript, or in the decision to publish the results.

References

1. Hinov, H.P. *Electro-Optic Effects in Thermotropic Liquid Crystals*; Nova Science: New York, NY, USA, 2015.
2. Buka, A.; Kramer, L. (Eds.) *Pattern Formation in Liquid Crystals*; Springer: New York, NY, USA, 1996.
3. Eber, N.; Buka, A.; Krishnamurthy, K.S. Electrically driven structures in bent-core nematics. *Liq. Cryst.* **2022**, *49*, 1194–1222. [CrossRef]
4. Dierking, I. *Textures of Liquid Crystals*; Wiley-VCH Verlag GmbH & Co. KGaA: Weinheim, Germany, 2003.
5. Kyeremah, C.; La, J.; Gharbi, M.A.; Yelleswarapu, C.S. Exploring different textures of a nematic liquid crystal for quantitative Fourier phase contrast microscopy. *Opt. Laser Technol.* **2022**, *147*, 107631. [CrossRef]
6. Zhang, J.; Ostroverkhov, V.; Singer, K.D.; Reshetnyak, V.; Reznikov, Y. Electrically controlled surface diffraction gratings in nematic liquid crystals. *Opt. Lett.* **2000**, *25*, 414–416. [CrossRef]
7. Huang, S.Y.; Huang, B.Y.; Kang, C.C.; Kuo, C.T. Diffraction and polarization properties of electrically-tunable nematic liquid crystal grating. *Polymers* **2020**, *12*, 1929. [CrossRef] [PubMed]
8. Algorri, J.F.; Morawiak, P.; Zografopoulos, D.C.; Bennis, N.; Spadlo, A.; Rodríguez-Cobo, L.; Jaroszewicz, L.R.; Sánchez-Pena, J.M.; Lopez-Higuera, J.M. Multifunctional light beam control device by stimuli-responsive liquid crystal micro-grating structures. *Sci. Rep.* **2020**, *10*, 13806. [CrossRef]
9. Chavel, P.; Sawchuk, A.A.; Strand, T.C.; Tanguay, A.R.; Soffer, B.H. Optical logic with variable-grating-mode liquid-crystal devices. *Opt. Lett.* **1980**, *5*, 398–400. [CrossRef]
10. Soffer, B.H.; Margerum, J.D.; Lackner, A.M.; Boswell, D.; Tanguay, A.R., Jr.; Strand, T.C.; Sawchuk, A.A.; Chavel, P. Variable grating mode liquid crystal device for optical processing computing. *Mol. Cryst. Liq. Cryst.* **1981**, *70*, 145–161. [CrossRef]
11. Tangonan, G.L. Variable-grating-mode liquid crystals for fibre-optic applications. *Electron. Lett.* **1985**, *21*, 701–702. [CrossRef]
12. McManamon, P.F.; Bos, P.J.; Escuti, M.J.; Heikenfeld, J.; Serati, S.; Xie, H.; Watson, E.A. A review of phased array steering for narrow-band electrooptical systems. *Proc. IEEE* **2009**, *97*, 1078–1096. [CrossRef]
13. Morris, R.; Jones, C.; Nagaraj, M. Liquid crystal devices for beam steering applications. *Micromachines* **2021**, *12*, 247. [CrossRef]
14. Weiner, A.M. Femtosecond pulse shaping using spatial light modulators. *Rev. Sci. Instrum.* **2000**, *71*, 1929–1960. [CrossRef]
15. Frumker, E.; Silberberg, Y. Femtosecond pulse shaping using a two-dimensional liquid-crystal spatial light modulator. *Opt. Lett.* **2007**, *32*, 1384–1386. [CrossRef] [PubMed]

16. Zhekova, M.; Maleshkov, G.; Stoyanov, L.; Stefanov, I.; Paulus, G.G.; Dreischuh, A. Formation of multi-spot focal arrays by square-shaped optical vortex lattices. *Opt. Commun.* **2019**, *449*, 110–116. [CrossRef]
17. Stoyanov, L.; Zhekova, M.; Stefanov, A.; Stefanov, I.; Paulus, G.G.; Dreischuh, A. Zeroth- and first-order long range non-diffracting Gauss–Bessel beams generated by annihilating multiple-charged optical vortices. *Sci. Rep.* **2020**, *10*, 21981. [CrossRef]
18. Chen, H.; Tan, G.; Huang, Y.; Weng, Y.; Choi, T.H.; Yoon, T.H.; Wu, S.T. A low voltage LC phase grating with switchable diffraction angles. *Sci. Rep.* **2017**, *7*, 39923. [CrossRef]
19. Katayama, K.; Kato, D.; Nagasaka, K.I.; Miyagawa, M.; Sohn, W.Y.; Lee, K.W. Origin of optical nonlinearity of photo-responsive liquid crystals revealed by transient grating imaging. *Sci. Rep.* **2019**, *9*, 5754. [CrossRef]
20. Shin, Y.; Jiang, Y.; Wang, Q.; Zhou, Z.; Qin, G.; Yang, D.K. Flexoelectric-effect-based light waveguide liquid crystal display for transparent display. *Photon. Res.* **2022**, *10*, 407–414. [CrossRef]
21. Liu, C.; Guo, Z.; Wu, J.; Yuan, D.; Zhou, G.; Tang, B.; Ye, H. Customized design of aperiodic liquid crystal grating for generation of multiple optical patterns. *Crystals* **2023**, *13*, 300. [CrossRef]
22. Zenginoglou, H.M.; Kosmopoulos, J.A. Geometrical optics approach to the nematic liquid crystal grating: Leading term formulas. *Appl. Opt.* **1989**, *28*, 3516–3519. [CrossRef]
23. Zenginoglou, H.; Kosmopoulos, J. Linearized wave-optical approach to the grating effect of a periodically distorted nematic liquid crystal layer. *J. Opt. Soc. Am. A* **1997**, *14*, 669–675. [CrossRef]
24. Zhang, J.; Xiang, Y.; Ding, X.; Hao, L.; Kaur, S.; Mohiuddin, G.; Pal, S.K.; Salamon, P.; Eber, N.; Buka, A. Electric-field-induced patterns in a hockey-stick nematic. *J. Mol. Liq.* **2022**, *366*, 120239. [CrossRef]
25. Kramer, L.; Pesch, W. Electrohydrodynamics in Nematics. In *Physical Properties of Liquid Crystals: Nematics*; EMIS Datareviews, No. 25; Dunmur, D.A., Fukuda, A., Luckhurst, G.R., Eds.; Inspec: London, UK, 2001; pp. 441–454.
26. Chillingworth, D.; Golubitsky, M. Symmetry and pattern formation for a planar layer of nematic liquid crystal. *J. Math. Phys.* **2003**, *44*, 4201–4219. [CrossRef]
27. Qiu, X.L.; Ahlers, G. Dynamics of fluctuations below a stationary bifurcation to electroconvection in the planar nematic liquid crystal N4. *Phys. Rev. Lett.* **2005**, *94*, 087802. [CrossRef]
28. Buka, A.; Eber, N.; Pesch, W.; Kramer, L. Convective patterns in liquid crystals driven by electric field. In *Self-Assembly, Pattern Formation and Growth Phenomena in Nano-Systems*; NATO Science Series II, Mathematica, Physics and Chemistry; Golovin, A.A., Nepomnyashchy, A.A., Eds.; Springer: Dordrecht, The Netherlands, 2006; Volume 218, pp. 55–82. [CrossRef]
29. Buka, A.; Eber, N.; Pesch, W.; Kramer, L. Isotropic and anisotropic electroconvection. *Phys. Rep.* **2007**, *448*, 115–132. [CrossRef]
30. Krekhov, A.; Pesch, W.; Eber, N.; Toth-Katona, T.A. Buka, Nonstandard electroconvection and flexoelectricity in nematic liquid crystals. *Phys. Rev. E Stat. Nonlinear Soft Matter Phys.* **2008**, *77*, 021705. [CrossRef]
31. Buka, A.; Toth-Katona, T.; Eber, N.; Krekhov, A.; Pesch, W. The role of flexoelectricity in pattern formation. In *Flexoelectricity in Liquid Crystals: Theory, Experiments and Applications*; Buka, A., Eber, N., Eds.; World Scientific: Singapore, 2013; Chapter 4; pp. 101–135. [CrossRef]
32. Krekhov, A.; Decker, W.; Pesch, W.; Eber, N.; Salamon, P.; Fekete, B.; Buka, A. Patterns driven by combined ac and dc electric fields in nematic liquid crystals. *Phys. Rev. E Stat. Nonlinear Soft Matter Phys.* **2014**, *89*, 052507. [CrossRef]
33. Delev, V.A.; Krekhov, A.P.; Kramer, L. Crossover between flexoelectric stripe patterns and electroconvection in hybrid aligned nematics. *Mol. Cryst. Liq. Cryst.* **2001**, *366*, 849–856. [CrossRef]
34. Kumar, P.; Krishnamurthy, K.S. Gradient flexoelectric switching response in a nematic phenyl benzoate. *Liq. Cryst.* **2007**, *34*, 257–266. [CrossRef]
35. Krishnamurthy, K.S.; Palakurthy, N.B.; Yelamaggad, C.V. Confined electroconvective and flexoelectric instabilities deep in the freedericksz state of nematic CB7CB. *J. Phys. Chem. B* **2017**, *121*, 5447–5454. [CrossRef]
36. Eber, N.; Palomares, L.O.; Salamon, P.; Krekhov, A.; Buka, A. Temporal evolution and alternation of mechanisms of electric-field-induced patterns at ultralow-frequency driving. *Phys. Rev. E Stat. Nonlinear Soft Matter Phys.* **2012**, *86*, 021702. [CrossRef]
37. Aguirre, L.E.; Anoardo, E.; Eber, N.; Buka, A. Regular structures in 5CB liquid crystals under the joint action of ac and dc voltages. *Phys. Rev. E Stat. Nonlinear Soft Matter Phys.* **2012**, *85*, 041703. [CrossRef] [PubMed]
38. Toth-Katona, T.; Eber, N.; Buka, A. Temporal response to harmonic driving in electroconvection. *Phys. Rev. E Stat. Nonlinear Soft Matter Phys.* **2011**, *83*, 061704. [CrossRef] [PubMed]
39. Pesch, W.; Krekhov, A. Optical analysis of spatially periodic patterns in nematic liquid crystals: Diffraction and shadowgraphy. *Phys. Rev. E Stat. Nonlinear Soft Matter Phys.* **2013**, *87*, 052504. [CrossRef]
40. Xu, M.Y.; Zhou, M.J.; Xiang, Y.; Salamon, P.; Eber, N.; Buka, A. Domain structures as optical gratings controlled by electric field in a bent-core nematic. *Opt. Express* **2015**, *23*, 15224–15234. [CrossRef] [PubMed]
41. Hadjichristov, G.B.; Marinov, Y.G.; Petrov, A.G.; Bruno, E.; Marino, L.; Scaramuzza, N. Electro-optically responsive composites of gold nanospheres in 5CB liquid crystal under direct current and alternating current joint action. *J. Appl. Phys.* **2014**, *115*, 083107. [CrossRef]
42. Hadjichristov, G.B.; Marinov, Y.G. Optical diffraction by using electrically-controlled spatially patterned nematic pentyl-cyanobiphenyl films under static electric field. *Mol. Cryst. Liq. Cryst.* **2016**, *632*, 9–20. [CrossRef]
43. Li, J.; Wen, C.H.; Gauza, S.; Lu, R.; Wu, S.T. Refractive indices of liquid crystals for display applications. *J. Disp. Technol.* **2005**, *1*, 51–61. [CrossRef]

44. Ratna, B.R.; Shashidhar, R. Dielectric properties of 4′-n-alkyl-4-cyanobipheniyls in their nematic phases. *Pramana* **1976**, *6*, 278–283. [CrossRef]
45. Barnik, M.I.; Blinov, L.M.; Trufanov, A.N.; Umanski, B.A. Flexo-electric domains in liquid crystals. *J. Phys.* **1978**, *39*, 417–422. [CrossRef]
46. Hinov, H.P.; Vistin, L.K. Parallel and cross-like domains due to d.c. and low frequency (<2 Hz) electric fields in nematic liquid crystal layers with negative dielectric anisotropy. *J. Phys.* **1979**, *40*, 269–292. [CrossRef]
47. Bobylev, Y.P.; Chigrinov, V.G.; Pikin, S.A. Threshold flexoelectric effect in nematic liquid crystal. *J. Phys. Colloq.* **1979**, *40*, C3-331–C3-333. [CrossRef]
48. Pikin, S.A. *Structural Transformations in Liquid Crystals*; Gordon and Breach Science Publishers: New York, USA, 1991.
49. Hinov, H.P.; Marinov, Y. Theoretical considerations and experimental illustration of the electro-optic behavior of longitudinal flexoelectric domains under the joint action of DC and AC voltages: The case of strong anchoring. *Mol. Cryst. Liq. Cryst.* **2009**, *503*, 45–68. [CrossRef]
50. Marinov, Y.G.; Hinov, H.P. On the threshold characteristics of the flexoelectric domains arising in a homogeneous electric field: The case of anisotropic elasticity. *Eur. Phys. J. E Soft Matter* **2010**, *31*, 179–189. [CrossRef] [PubMed]
51. Hinov, H.P.; Bivas, I.; Mitov, M.D.; Shoumarov, K.; Marinov, Y. A further experimental study of parallel surface-induced flexoelectric domains (PSIFED) (flexo-dielectric walls). *Liq. Cryst.* **2003**, *30*, 1293–1317. [CrossRef]
52. de Gennes, P.G.; Prost, J. *The Physics of Liquid Crystals*; Clarendon Press: Oxford, UK, 1993.
53. Hinov, H.P.; Bivas, I.; Mitov, M.D.; Shoumarov, K. Rubbing-induced surface textures in nematic MBBA layers and their behaviour under applied d. *c. or a.c. voltages. Liq. Cryst.* **2003**, *30*, 945–959. [CrossRef]
54. Klein, W.R.; Cook, B.D. Unified approach to ultrasonic light diffraction. *IEEE Trans. Sonics Ultrason* **1967**, *SU-14*, 123–134. [CrossRef]
55. Goodman, J.W. *Introduction to Fourier Optics*, 2nd ed.; Roberts & Co.: Greenwood Village, CO, USA, 2005; p. 80.
56. Marusii, T.Y.; Reznikov, Y.A.; Reshetnyak, V.Y.; Soskin, M.S.; Khizhnyak, A.I. Scattering of light by nematic liquid crystals in cells with a finite energy of the anchoring of the director to the wails. *Sov. Phys. JETP* **1986**, *64*, 502–507.
57. Val'kov, A.Y.; Romanov, V.P.; Shalaginov, A.N. Fluctuations and light scattering in liquid crystals. *Usp. Fiz. Nauk* **1994**, *164*, 149–193. [CrossRef]
58. Helfrich, W. Conduction-induced alignment of nematic liquid crystals: Basic model and stability considerations. *J. Chem. Phys.* **1969**, *51*, 4092–4105. [CrossRef]
59. Penz, P.A. Voltage-induced vorticity and optical focusing in liquid crystals. *Phys. Rev. Lett.* **1970**, *24*, 1405–1409. [CrossRef]
60. Strazielle, C.; Coles, H.J. Depolarized light scattering studies of the nematogen pentyl cyanobiphenyl. *J. Phys.* **1979**, *40*, 895–900. [CrossRef]
61. Wiersma, D.S.; Muzzi, A.; Colocci, M.; Righini, R. Time-resolved anisotropic multiple light scattering in nematic liquid crystals. *Phys. Rev. Lett.* **1999**, *83*, 4321–4324. [CrossRef]
62. Takase, A.; Sakagami, S.; Nakamizo, M. Light diffraction and light scattering in nematic liquid crystals with a positive dielectric anisotropy. *Jpn. J. Appl. Phys.* **1975**, *14*, 228–230. [CrossRef]
63. Kamanina, N.V.; Vasilenko, N.A. Influence of operating conditions and interface properties on dynamic characteristics of liquid-crystal spatial light modulators. *Opt. Quant. Electron.* **1997**, *29*, 1–9. [CrossRef]
64. Kamanina, N.V.; Berendyaev, V.I. Influence of solid–liquid crystal interface on characteristics of liquid crystal cells. *Proc. SPIE* **1998**, *3292*, 134–158. [CrossRef]
65. Gruler, H.; Scheffer, T.J.; Meier, G. Elastic constants of nematic liquid crystals. *Z. Naturforsch. A* **1972**, *27*, 966–976. [CrossRef]
66. Blinov, L.M.; Chigrinov, V.G. *Electrooptic Effects in Liquid Crystal Materials*; Springer: New York, NY, USA, 1994; p. 149.
67. Johnson, R.V.; Tanguay, A.R., Jr. Optical beam propagation method for birefringent phase grating diffraction. *Opt. Eng.* **1986**, *25*, 252235. [CrossRef]
68. Tripathi, P.; Uttam, R.; Kumar, S.; Dabrowski, R.; Dhar, R. Enhancement of the physical parameters due to the dispersion of functionalised gold nanoparticles in a room temperature nematic liquid crystal. *Liq. Cryst.* **2023**, *50*, 240–248. [CrossRef]
69. Shivakumar, U.; Mirzaei, J.; Feng, X.; Sharma, A.; Moreira, P.; Hegmann, T. Nanoparticles: Complex and multifaceted additives for liquid crystals. *Liq. Cryst.* **2011**, *38*, 1495–1514. [CrossRef]
70. Ouskova, E.; Lysenko, D.; Ksondzyk, S.; Cseh, L.; Mehl, G.H.; Reshetnyak, V.; Reznikov, Y. Strong cubic optical nonlinearity of gold nanoparticles suspension in nematic liquid crystal. *Mol. Cryst. Liq. Cryst.* **2011**, *545*, 1347–1356. [CrossRef]
71. Hadjichristov, G.B.; Marinov, Y.G.; Petrov, A.G.; Bruno, E.; Marino, L.; Scaramuzza, N. Electro-optics of nematic/gold nanoparticles composites: The effect from dopants. *Mol. Cryst. Liq. Cryst.* **2015**, *610*, 135–148. [CrossRef]
72. Brouckaert, N.; Podoliak, N.; Orlova, T.; Bankova, D.; De Fazio, A.F.; Kanaras, A.G.; Hovorka, O.; D'Alessandro, G.; Kaczmarek, M. Nanoparticle-induced property changes in nematic liquid crystals. *Nanomaterials* **2022**, *12*, 341. [CrossRef] [PubMed]
73. Qi, H.; Hegmann, T. Formation of periodic stripe patterns in nematic liquid crystals doped with functionalized gold nanoparticles. *J. Mater. Chem.* **2006**, *16*, 4197–4205. [CrossRef]
74. Lesiak, P.; Bednarska, K.; Lewandowski, W.; Wójcik, M.; Polakiewicz, S.; Bagiński, M.; Osuch, T.; Markowski, K.; Orzechowski, K.; Makowski, M.; et al. Self-organized, one-dimensional periodic structures in a gold nanoparticle-doped nematic liquid crystal composite. *ACS Nano* **2019**, *13*, 10154–10160. [CrossRef]

75. Garbovskiy, Y.; Glushchenko, I. Nano-objects and ions in liquid crystals: Ion trapping effect and related phenomena. *Crystals* **2015**, *5*, 501–533. [CrossRef]
76. Wu, S.T.; Efron, U.; Hess, L.D. Birefringence measurements of liquid crystals. *Appl. Opt.* **1984**, *23*, 3911–3915. [CrossRef]
77. Panarin, Y.P.; Panov, V.P.; Kalinovskaya, O.E.; Vij, J.K. On the V-shaped switching in antiferroelectric liquid crystals. *Ferroelectrics* **2000**, *246*, 35–42. [CrossRef]
78. Copic, M.; Maclennan, J.E.; Clark, N.A. Influence of ions on the "V-shaped" electro-optic response of ferroelectric liquid crystals. *Phys. Rev. E Stat. Nonlinear Soft Matter Phys.* **2001**, *63*, 031703. [CrossRef]
79. Strangi, G.; Versace, C.; Scaramuzza, N.; Bruno, V. Unipolar "V-shaped" switching in chiral smectic C (Sm–C*) liquid crystals bounded by an ion-store film. *J. Appl. Phys.* **2002**, *92*, 3630–3635. [CrossRef]
80. O'Callaghan, M.J. Switching dynamics and surface forces in thresholdless "V-shaped" switching ferroelectric liquid crystals. *Phys. Rev. E Stat. Nonlinear Soft Matter Phys.* **2003**, *67*, 011710. [CrossRef]
81. Blinov, L.M.; Palto, S.P.; Podgornov, F.V.; Moritake, H.; Haase, W. Hysteresis-free electro-optical switching in conductive ferroelectric liquid crystals: Experiments and modelling. *Liq. Cryst.* **2004**, *31*, 61–70. [CrossRef]
82. Engström, D.; Rudquist, P.; Bengtsson, J.; D'have, K.; Galt, S. Analog low-loss full-range phase modulation by utilizing a V-shaped switched ferroelectric liquid-crystal cell in reflective mode. *Opt. Lett.* **2006**, *31*, 2906–2908. [CrossRef]
83. Manjuladevi, V.; Panarin, Y.P.; Song, J.K.; Vij, J.K.; Sadashiva, B.K. V-shaped electro-optic response observed in a chiral ferroelectric smectic liquid crystal. *Appl. Phys. Lett.* **2008**, *93*, 093507. [CrossRef]
84. He, Z.; Gou, F.; Chen, R.; Yin, K.; Zhan, T.; Wu, S.T. Liquid crystal beam steering devices: Principles, recent advances, and future developments. *Crystals* **2019**, *9*, 292. [CrossRef]
85. Jain, A.K.; Deshmukh, R.R. An overview of polymer-dispersed liquid crystals composite films and their applications. In *Liquid Crystals and Display Technology*; Ghamsari, M.S., Carlescu, I., Eds.; IntechOpen: London, UK, 2020; Chapter 2; pp. 11–78. [CrossRef]
86. Saeed, M.H.; Zhang, S.; Cao, Y.; Zhou, L.; Hu, J.; Muhammad, I.; Xiao, J.; Zhang, L.; Yang, H. Recent advances in the polymer dispersed liquid crystal composite and its applications. *Molecules* **2020**, *25*, 5510. [CrossRef] [PubMed]
87. Zhang, H.; Miao, Z.; Shen, W. Development of polymer-dispersed liquid crystals: From mode innovation to applications. *Compos. Part A Appl. Sci. Manuf.* **2022**, *163*, 107234. [CrossRef]
88. Zhong, T.; Mandle, R.J.; Goodby, J.W.; Zhang, C.; Zhang, L. Thiol-ene reaction based polymer dispersed liquid crystal composite films with low driving voltage and high contrast ratio. *Liq. Cryst.* **2020**, *47*, 2171–2183. [CrossRef]
89. Shen, W.; Wang, L.; Chen, G.; Li, C.; Zhang, L.; Yang, Z.; Yang, H. A facile route towards controllable electric-optical performance of polymer-dispersed liquid crystal via the implantation of liquid crystalline epoxy network in conventional resin. *Polymer* **2019**, *167*, 67–77. [CrossRef]
90. He, Z.; Yu, P.; Zhang, H.; Zhao, Y.; Zhu, Y.; Guo, Z.; Ma, C.; Zhang, H.; Miao, Z.; Shen, W. Silicon nanostructure-doped polymer/nematic liquid crystal composites for low voltage-driven smart windows. *Nanotechnology* **2021**, *33*, 085205. [CrossRef] [PubMed]
91. Hadjichristov, G.B.; Marinov, Y.G.; Petrov, A.G. Gradient polymer-disposed liquid crystal single layer of large nematic droplets for modulation of laser ligh. *Appl. Opt.* **2011**, *50*, 2326–2333. [CrossRef]
92. Li, J.; Gauzia, S.; Wu, S.T. High temperature-gradient refractive index liquid crystals. *Opt. Express* **2004**, *12*, 2002–2010. [CrossRef] [PubMed]

materials

MDPI

Article

A Low Temperature Growth of Cu_2O Thin Films as Hole Transporting Material for Perovskite Solar Cells

Anna L. Pellegrino [1,*], Francesca Lo Presti [1], Emanuele Smecca [2,*], Salvatore Valastro [2], Giuseppe Greco [2], Salvatore Di Franco [2], Fabrizio Roccaforte [2], Alessandra Alberti [2,*] and Graziella Malandrino [1,*]

[1] Dipartimento di Scienze Chimiche, Università degli Studi di Catania, INSTM UdR Catania, Viale Andrea Doria 6, 95125 Catania, Italy

[2] National Research Council-Institute for Microelectronics and Microsystems (CNR-IMM), Zona Industriale Strada VIII No. 5, 95121 Catania, Italy

* Correspondence: annalucia.pellegrino@unict.it (A.L.P.); emanuele.smecca@imm.cnr.it (E.S.); alessandra.alberti@imm.cnr.it (A.A.); gmalandrino@unict.it (G.M.)

Abstract: Copper oxide thin films have been successfully synthesized through a metal–organic chemical vapor deposition (MOCVD) approach starting from the copper bis(2,2,6,6-tetramethyl-3,5-heptanedionate), $Cu(tmhd)_2$, complex. Operative conditions of fabrication strongly affect both the composition and morphologies of the copper oxide thin films. The deposition temperature has been accurately monitored in order to stabilize and to produce, selectively and reproducibly, the two phases of cuprite Cu_2O and/or tenorite CuO. The present approach has the advantages of being industrially appealing, reliable, and fast for the production of thin films over large areas with fine control of both composition and surface uniformity. Moreover, the methylammonium lead iodide (MAPI) active layer has been successfully deposited on the ITO/Cu_2O substrate by the Low Vacuum Proximity Space Effusion (LV-PSE) technique. X-ray diffraction (XRD), field emission scanning electron microscopy (FE-SEM), and atomic force microscopy (AFM) analyses have been used to characterize the deposited films. The optical band gap (E_g), ranging from 1.99 to 2.41 eV, has been determined through UV-vis analysis, while the electrical measurements allowed to establish the p-type conductivity behavior of the deposited Cu_2O thin films with resistivities from 31 to 83 Ω cm and carrier concentration in the order of $1.5–2.8 \times 10^{16}$ cm^{-3}. These results pave the way for potential applications of the present system as a hole transporting layer combined with a perovskite active layer in emergent solar cell technologies.

Keywords: HTL layer; chemical vapor deposition; hybrid perovskite

Citation: Pellegrino, A.L.; Lo Presti, F.; Smecca, E.; Valastro, S.; Greco, G.; Di Franco, S.; Roccaforte, F.; Alberti, A.; Malandrino, G. A Low Temperature Growth of Cu_2O Thin Films as Hole Transporting Material for Perovskite Solar Cells. *Materials* **2022**, *15*, 7790. https://doi.org/10.3390/ma15217790

Academic Editor: Jian-Zhang Chen

Received: 6 October 2022
Accepted: 2 November 2022
Published: 4 November 2022

1. Introduction

Recently, copper-oxide compounds represent one of the most studied classes of semiconducting materials. The main advantages of these materials are related to the exceptional possibility of tuning the optical and electronic properties within their semiconducting behavior [1,2]. Copper oxide-based materials play a significant role in many technological applications, ranging from sensing [3] to catalysis [4], from photodetector (e.g., in combination with ZnO) [5,6] to the photo-electrochemical splitting of water [7,8]. In recent years, copper-based materials in the form of thin films have become objects of interest also in solar cell devices [9–11].

Indeed, together with the other classes of p-type layers applied in solar cell technology, organic materials such as PEDOT and spiro-OMeTAD [12] or 2,4,6-triarylpyridine [13], and transition metal oxides such as MoO_3, V_2O_5, WO_3, NiO and Cu_2O have been explored as efficient hole transporting layers [14,15]. In particular, among the three most common and stable phases, Cu_2O, Cu_4O_3, and CuO, named cuprite, paramelaconite and tenorite, respectively, the cuprite phase (Cu_2O) has been intensively studied and applied as a p-type semiconducting material in solar cell devices [9–11,16,17]. Among the advantages of Cu_2O

with respect to the organic hole transporting layers are efficient hole collection, optical transparency, stability, a wide variety of synthetic strategies and low-cost production. Besides, Cu_2O is environmentally stable and its main component, Cu, is earth-abundant and non-toxic. Other interesting Cu-based hole-transport materials, such as CuI or CuSCN layers, have been applied to inverted planar perovskite solar cells [18,19].

Cu_2O as a thin film layer has been synthesized by several methods, including vapor and solution approaches. The most studied approaches are solution-phase methods [20,21], hydrothermal [22], electrodeposition [23], sputtering [24], molecular beam epitaxy [25], atomic layer deposition [26], thermal oxidation of copper [27], and chemical vapor deposition [28,29]. Among these approaches, metalorganic chemical vapor deposition (MOCVD) represents one of the most promising techniques due to its tunability processes, easy scaling up and industrial applicability. Moreover, the CVD approach offers the possibility of fine-tuning the composition of the Cu–O phases by easily altering the operating conditions and the chemical nature of the precursors [30].

Copper (I) complexes, e.g., Cu(hfa)(cod) [31], and [(cod)Cu(tfb-tfea)] [32] [hfa = 1,1,1,5,5,5-hexafluoro-2,4-pentanedionate, cod = 1,5-cyclooctadiene, and tfb-tfea = N-(4,4,4-trifluorobut-1-en-3-on)-6,6,6-trifluoroethylamine], have been applied to the deposition of Cu_2O films, but metalorganic copper (II) compounds are also widely used as CVD precursors for the deposition of copper (I) oxide thin films, due to their thermal stability and clean decomposition step during the evaporation process [33]. Among them, the most common copper adducts are copper(II) β-diketonates, i.e., $[Cu(acac)_2]$, $[Cu(tfa)_2]$, $[Cu(hfa)_2\cdot tmeda]$, and $Cu(tmhd)_2$ (acac = 2,4-pentanedionate, tfa=1,1,1-trifluoro-2,4-pentanedionate, tmeda=N,N,N′,N′- tetramethylethylenediamine, tmhd= 2,2,6,6-tetramethyl-3,5-heptanedionate) [28,29,34–37].

To the best of our knowledge, only a few works in the literature report the thermal-CVD fabrication of pure Cu_2O thin films without the use of H_2 flow as a reducing agent at relatively low temperatures. Maruyama et al. reported the CVD fabrication of CuO films at T = 300 °C [37], Lay et al. reported the CVD fabrication of copper thin film in the range of T: 275–300 °C [38], Condorelli et al. stabilized the pure Cu_2O phase at a deposition temperature of 300 °C from $Cu(acac)_2$ precursor depending on the partial pressure of the oxygen flow [34]. Gupta et al. reported the photo-assisted MOCVD of Cu_2O films starting from $[Cu(tmhd)]_2$ complexes at a deposition temperature of 750 °C [39].

In the present work, we propose an in-depth study of the metalorganic chemical vapor deposition process for the reproducible and selective fabrication of both cuprite Cu_2O and tenorite CuO copper oxide thin films starting from a β-diketonate copper (II) precursor, i.e., the bis(2,2,6,6-tetramethyl-3,5-heptanedionate) copper, $Cu(tmhd)_2$. The MOCVD process has been tested in the temperature range of 250–400 °C, allowing the selective and reproducible fabrication of Cu_2O on a large area at the lowest temperature of 250 °C, and a mixture of Cu_2O–CuO or the pure CuO at higher temperatures. A sequential deposition of the methylammonium lead iodide (MAPI) layer allowed us to test copper oxide as the substrate for the vapor deposition of $MAPbI_3$ film to realize the first part of a solar cell with a planar inverted structure. The present approach represents a simple, easily scalable, and industrially appealing process for the production of compact and homogeneous copper oxide films at relatively low temperatures. X-ray diffraction (XRD), field-emission scanning electron microscopy (FE-SEM), and atomic force microscopy (AFM) analyses allowed an accurate determination of the physicochemical properties of the deposited layers, while Hall Effect measurements enabled verifying the p-type conductivity of the deposited films. Finally, UV-vis spectra have been carried out for the determination of the optical band gap.

2. Materials and Methods

2.1. Cu_2O Synthesis

The $Cu(tmhd)_2$ compound was purchased from Sigma-Aldrich and used without further purification. The depositions were performed in a horizontal, hot-wall reactor under reduced pressure, using argon (150 sccm) as a carrier gas, and oxygen (200 sccm) as a reacting gas. The Ar and O_2 flows were controlled using MKS 1160 flow controller units and were introduced in proximity to the reaction zone. The vacuum inside the reactor was maintained through a scroll pump unit and monitored at the value of 4 Torr using MKS Baratron 122AAX. The films were deposited on Si (001) and quartz/ITO substrates in the 250–400 °C temperature range. The precursor source was kept at 130–140 °C for an efficient vaporization process. Each section was heated independently, with ±2 °C accuracy, using K-type thermocouples and computer-controlled hardware.

2.2. MAPI Deposition

The $MAPbI_3$ films were deposited by a Low Vacuum Proximity Space Effusion (LV-PSE) technique with specifically customized vacuum deposition equipment, provided by Kenosistec s.r.l. Lead iodide powders (99.99% purity) were purchased from Sigma Aldrich (St. Louis, MO, USA). Methyl ammonium iodide was purchased from Dyenamo AB. All materials were used as received without any further purification. The LV-PSE technique consists of a sequential deposition of PbI_2 and MAI via physical sublimation from powders at a base pressure of ~2×10^{-2} mbar with the crucibles taken at 350 °C and 135 °C, respectively. The substrate was posed at a medium-range distance with respect to the sources (~2 cm). During the first step, PbI_2 was deposited, then the conversion into $MAPbI_3$ occurred through an adsorption-incorporation-migration mechanism fully described in previous works [40–42].

2.3. Characterization

Structural characterization was performed using a Smartlab Rigaku diffractometer in grazing incident mode (0.5°) operating at 45 kV and 200 mA equipped with a rotating anode of Cu K_α radiation. Film morphologies were investigated using field emission scanning electron microscopy (FE-SEM) ZEISS SUPRA 55 VP. The films deposited on glass were Au-coated before FE-SEM characterization. Topographic characterization was performed through Atomic Force Microscopy (AFM) adopting an Au-coated silicon probe with a nominal 35 nm tip curvature radius and a typical force constant of 0.1 N. The AFM images were obtained in contact mode. Before and after each measurement the noise level was 0.01 nm. The UV-Visible absorption spectra were recorded using a Jasco V-650 spectrophotometer. The spectra were recorded in the wavelength range from 250 to 700 nm for Cu_2O thin films deposited on ITO-quartz substrates. Electrical characterization of the material was carried out by Hall effect measurements at room temperature using MMR H50 equipment. For this purpose, Van der Paw structures were fabricated, defining Cu/Au Ohmic contacts at the four corners of 1 cm × 1 cm Cu_2O samples grown on a SiO_2 substrate.

3. Results

An MOCVD approach has been successfully applied to the synthesis of copper oxide in form of a thin film starting from $Cu(tmhd)_2$ complex. All the depositions allow the fabrication of compact and homogeneous thin films in the area of 2 cm × 2 cm on Si and quartz/ITO substrates. The effect of deposition temperature has been accurately studied in terms of both phase composition and morphology structures.

A complete overview of the X-ray diffraction (XRD) analysis of copper oxide thin films deposited by MOCVD in the 250–400 °C temperature range is reported in Figure 1. At the lowest temperature of 250 °C, the pattern (red line) exhibits peaks at 29.64°, 36.50° and 42.43°, which can be assigned to the 110, 111 and 200 reflections of the pure, polycrystalline cuprite phase (PDF card n. 077-0199). At a deposition temperature of 300 °C, the pattern (blue line) presents an additional peak at 38.89°, which can be ascribed to 111/200 reflections of CuO traces. In fact, at higher temperatures (350 °C—orange line) together with the previous one, the additional peak at 35.58° points to the formation of a mixture of Cu_2O and CuO phases. Finally, at 400 °C, the pattern in green shows the formation of the pure tenorite phase CuO, with the presence of signals at 35.57°, 38.92° and 48.80° related to the

002, 111/200, and -202 reflections, respectively (PDF card n. 00-045-0937), while no peaks related to Cu_2O are present.

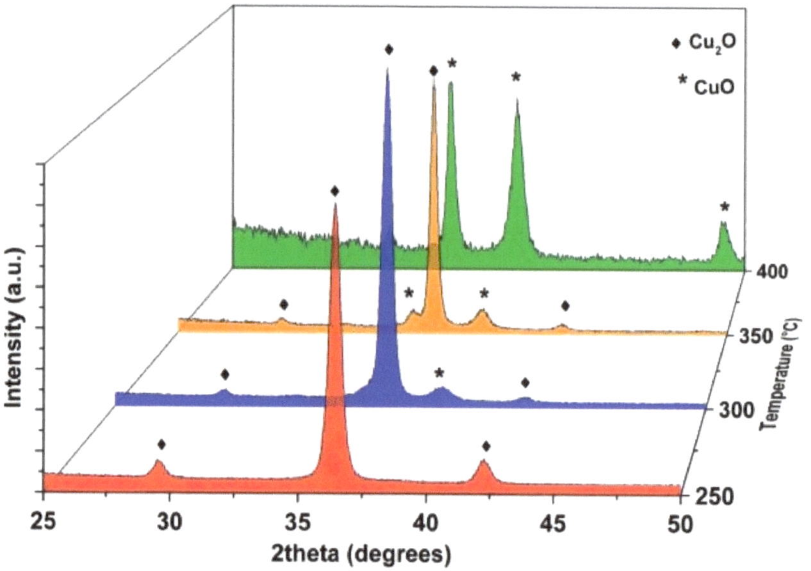

Figure 1. X-ray diffraction patterns of the Cu_2O and CuO thin films prepared by MOCVD on the Si substrate at different temperatures from 250 °C (red line), 300 °C (blue line), 350 °C (orange line) to 400 °C (green line).

The observed trend indicates a strong effect of the deposition temperature in the selective formation of Cu(I) and Cu(II) oxides, with the lowest temperature (250 °C) stabilizing the Cu_2O phase and the highest temperature (400 °C) stabilizing the CuO one. At even lower temperatures of 200 °C, only very small, isolated nuclei are found (Figure S1a), thus indicating that 250 °C is the lowest temperature to produce significant precursor decomposition. This finding is also supported by the EDX data (Figure S1b), which do not show any Cu peak; the Cu amount is lower than the detection limit of the technique.

The Cu(I) and/or Cu(II) oxide formation can be rationalized considering a balance between two aspects: (i) the decomposition mechanism of the $Cu(tmhd)_2$ precursor in the reactor, which involves the organic component of the complex acting as a reducing agent; (ii) the oxidant atmosphere under the present deposition conditions [29]. Therefore, at lower temperatures, the first aspect is prevalent and tends to stabilize the Cu(I) oxide phase due to the reducing action of the organic component; on the contrary, at higher temperatures, the second aspect prevails, resulting in the mere formation of Cu(II) oxide.

Differing from the present findings, Gupta et al. [39] reported the fabrication of Cu_2O films from the same $Cu(tmhd)_2$ precursor at a much higher deposition temperature of 750 °C, while in present experiments at a temperature of 300 °C the Cu(I) already starts to oxidize to Cu(II) and at 400 °C pure CuO forms. The reason why Gupta et al. stabilized the Cu_2O at higher temperatures is that in addition to N_2, used in their study as a carrier gas, a mixture of N_2O and oxygen was used during deposition. The N_2O was acting as a reducing agent in regard to the Cu(II), which would have stabilized for the effect of temperature. In our experiments, starting from a Cu(II) precursor, we succeeded in growing highly homogeneous Cu_2O layers at very low temperatures, with the tmhd organic ligand acting as a reducing agent. Finally, the nature of the films deposited by Gupta et al. [39] had a reported thickness of only 15 nm and TEM observed islands in the order of 100–200 nm.

It is worth noting that in the literature only a few works report the stabilization of the cuprite crystalline phase at low temperatures, with the lowest reported temperature of 300 °C for vapor deposition approaches [28,29,32,33]. In the present case, the formation of pure Cu_2O films has been achieved at even lower temperatures and actually, the thermal budget related to a 50 °C difference is a great advantage. In fact, the present optimized process, using a deposition temperature as low as 250 °C, paves the way for a wide application of the current Cu_2O synthetic approach also on a variety of polymeric substrates.

Morphology features of CuO and Cu_2O thin films have been studied by field emission scanning electron microscopy (FE-SEM). The FE-SEM image of the Cu_2O films deposited at 250 °C on Si (100) (Figure 2a) shows the formation of a very compact and homogeneous layer in which, probably due to the low fabrication temperature, a nanostructured feature is barely visible. Indeed, at higher deposition temperatures, i.e., 300 °C and 350 °C in Figure 2b,c, respectively, the films show a porous structure, probably caused by the coalescence of small grains of the order of tens of nanometers during the film growth. Finally, at 400 °C, the morphology appears much more nanostructured with the formation of plate-like grains of the order of 200–400 nm assembled into column-like structures (Figure 2d). The evident change in morphologies as a function of fabrication temperatures may be attributed on the one hand to the different contributions of nucleation and growth processes, on the other hand to the different Cu–O crystalline phase arrangements. The cross-section images display a growth trend, with an increase from 220 nm to 420 nm at 250 °C and 300 °C (Figure 2e,f), and up to 740 nm for the layer obtained at 350 °C (Figure 2g). The film deposited at 400 °C shows, indeed, a thicker and rugged profile, with a thickness of about 1250 nm (Figure 2h). Specifically, growth rates vary from 3, 7, 12 and 21 nm/min for 250, 300, 350 and 400 °C, respectively. To explain the dependence of the growth rate as a function of temperature, the Arrhenius plot (Figure 3) has been derived, even though the formed Cu–O phases differ in the investigated temperature range.

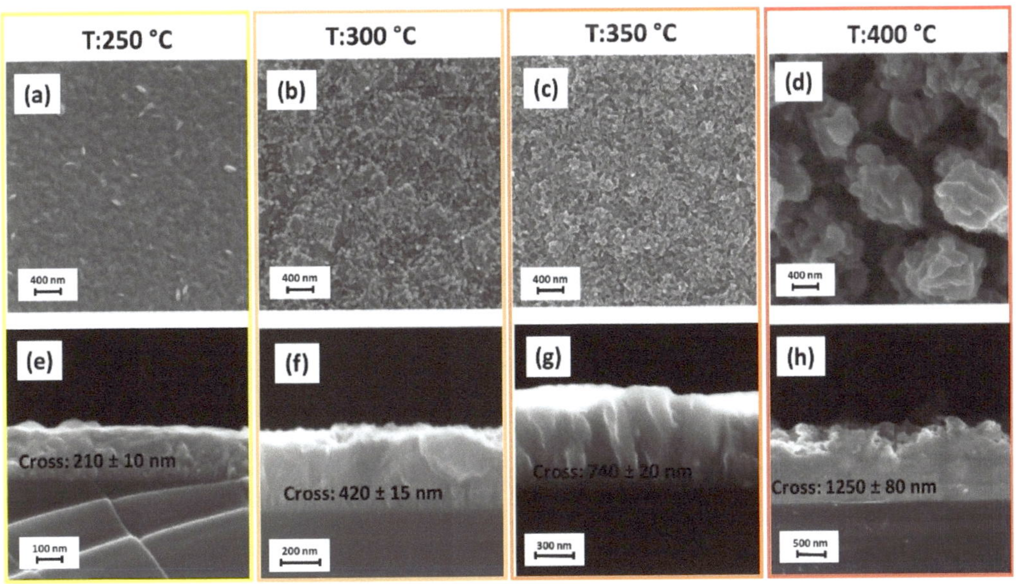

Figure 2. FE-SEM images of Cu_2O and CuO thin films prepared by MOCVD on Si substrate deposited at 250 °C: (**a**) plan view and (**e**) cross-section; at 300 °C: (**b**) plan view and (**f**) cross-section; at 350 °C: (**c**) plan view and (**g**) cross-section; at 400 °C: (**d**) plan view and (**h**) cross-section.

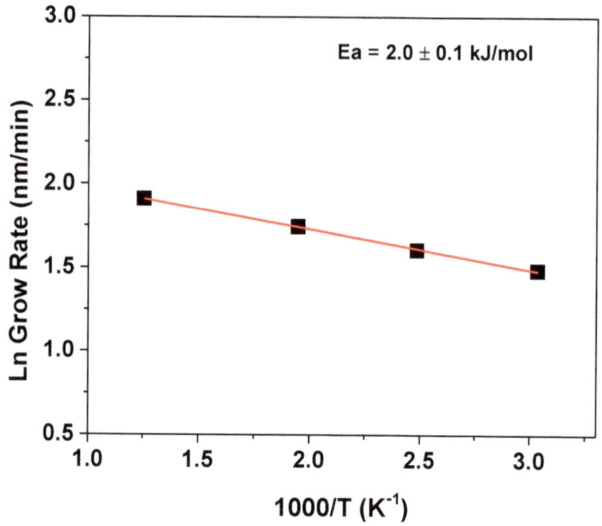

Figure 3. Arrhenius plot of the dependence of the logarithm of growth rate vs. 1000/T. Solid line is the linear best fit.

Despite the increase in thickness variation with increasing temperature, the growth rate dependence does not suggest the presence of a kinetic regime and points to a mass transport-limited regime. This evidence is supported by the derived low activation energy, obtained from the slope of the linear fitting analysis, equal to 2.0 ± 0.1 kJ/mol.

Hence, optical absorptions have been evaluated in order to determine the band gap of the semiconductor Cu_2O thin films. Particularly, UV-vis spectra have been measured for Cu_2O thin films of various thicknesses, deposited on ITO/quartz substrate.

With the aim of applications as a hole transporting layer (HTL) in solar cell devices, the thickness of the Cu_2O films has been finely tuned and optimized. Therefore, different depositions have been carried out at 250 °C and with different vaporization temperatures and deposition times, i.e., (I) T_{vap}: 140 °C, t: 30 min; (II) T_{vap}: 140 °C, t: 15 min and (III) T_{vap}: 130 °C t: 15 min, in order to have different-thickness Cu_2O layers. In particular, under these conditions, three samples have been obtained with thicknesses of 230, 140, and 90 nm, respectively. The cross-section and the morphologies are reported in Figure S2.

The absorbance spectra in the UV-vis range of the ITO/quartz substrate and the Cu_2O_230 nm, Cu_2O_140 nm and Cu_2O_90 nm films are displayed in Figure 4a. The absorbance spectra show high transparency up to 300 nm for the ITO substrate, and an absorption edge localized between 450 and 500 nm for the Cu_2O films, which corresponds to the excitonic band gap of Cu_2O films. The spectra of the Cu_2O_230 nm, Cu_2O_140 nm and Cu_2O_90 nm films display an increase in absorbance as a consequence of the thickness increase, a behavior already observed for films grown with different techniques [43]. Transmission spectra of ITO/quartz and Cu_2O thin films are reported in Figure 4b, while in Figure S3 the transmission of the ITO/quartz annealed at 250 °C is compared to the untreated ITO/quartz substrate. Regardless of thickness, all Cu_2O films absorb strongly in the UV-vis region between 250 and 450 nm, resulting in a drop in the percentage of transmission. As the thickness of the hole transporting layer is significantly reduced, the transmittance in the visible range increases. In fact, up to λ = 480 nm, the transmittance for the thickest Cu_2O_230 nm and Cu_2O_140 nm is minimal and then begins to rise, reaching only 20% for the Cu_2O_230 nm and 40% for the intermediate Cu_2O_140 nm at λ = 500 nm. Cu_2O_90 nm thin film, on the other hand, achieves acceptable transmittance levels (about 60%) at λ = 450 nm and reaches a higher 70% at λ= 650 nm. The present results agree with the literature data [44,45], and point to the potential applicability of the presently grown Cu_2O film as an HTL layer for photovoltaic applications [46].

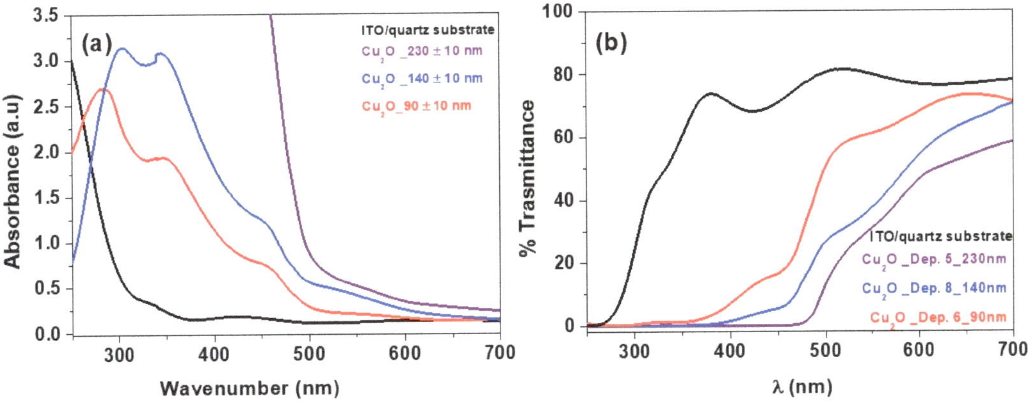

Figure 4. UV-vis absorbance (**a**) and transmittance (**b**) spectra of the ITO/quartz substrate (black line); Cu_2O on ITO/quartz with thickness of 230 nm (purple line); 140 nm (blue line) and 90 nm (red line).

Tauc's plot obtained from the relation of $(\alpha h\nu)^2$ versus $(h\nu)$ is calculated from the optical measurements reported in Figure S4 for the ITO/quartz substrate and in Figure 5 for the Cu_2O films. Using a linear extrapolation, the values of the direct optical band gap of the Cu_2O_230 nm, Cu_2O_140 nm and Cu_2O_90 nm films are evaluated to be 1.99, 2.04 and 2.41 eV, respectively. In order to evaluate the reproducibility of the process and to have statistically significant results, the band gaps from three different samples for each deposition condition of the films Cu_2O_230 nm, Cu_2O_140 nm and Cu_2O_90 nm have been extrapolated yielding average values of 1.97 ± 0.02, 2.03 ± 0.03 and 2.37 ± 0.06 eV, respectively. It is interesting to note that the band gap increases upon decreasing film thickness. This band gap spreading effect has been previously observed for Cu_2O films [47] and in similar thin film studies [48,49].

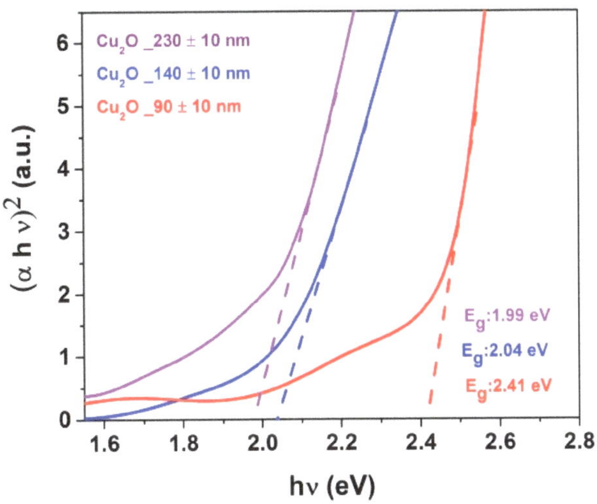

Figure 5. Tauc's plot $[(\alpha h\nu)^2$ against photon energy $(h\nu)]$ of Cu_2O films on ITO_quartz with thickness of 230 nm (purple line); 140 nm (blue line) and 90 nm (red line).

These values compare well with literature data of Cu_2O samples deposited through spatial atomic layer deposition [31,50,51].

In order to determine the type of conductivity of the deposited Cu_2O films, Van der Pauw and Hall Effect measurements have been performed at room temperature.

As an example, Figure 6 shows a typical current-voltage (I-V) measurement carried out on a Van der Pauw structure fabricated on the Cu_2O sample. The inset in Figure 6 shows schematically the geometry of the sample and the I-V measurement setup. From the linear fit of the experimental data, the sheet resistance R_{SH} of the Cu_2O films has been determined.

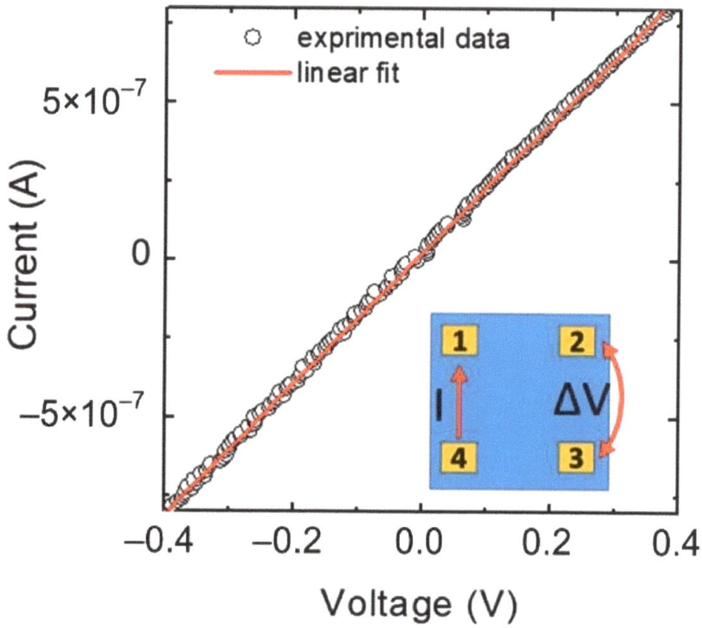

Figure 6. Typical current-voltage (I–V) measurement carried out on a Van der Pauw structure fabricated on Cu_2O sample. The inset schematically illustrates the geometry on the sample and the electrical I-V measurement setup.

Thereafter, Hall effect measurements have been performed by applying a magnetic field of 0.1 T perpendicular to the sample surface. These measurements allowed us to determine the type of carriers responsible for the conduction and their concentration [52].

Then, combining the values of the sheet resistance, determined through the Van der Pauw measurements, with the carrier density, extracted under the application of the magnetic field, the carrier mobility has been determined.

Noteworthy, from the sign of the Hall voltage it can be concluded that all the deposited films are p-type. Moreover, no significant differences in the electrical properties have been observed by varying the Cu_2O film thickness from 90 nm to 230 nm.

The hole concentration p is in the order of 1.5–2.8 $\times$ 10^{16} cm^{-3}, while the values of the Hall mobility are in the range 4–7 $cm^2V^{-1}s^{-1}$. Table 1 summarizes the main parameters extracted by the Van der Pauw and Hall effect measurements. The values reported in the table have an error of about $\pm$20% arising from the average of different measurements acquired on the Van der Pauw structure in the different orientations.

Table 1. Summary of the main parameters extracted by Van der Pauw and Hall effect measurements at room temperature.

Sample	Carrier Type	Resistivity (Ω cm)	Carrier Concentration (cm^{-3})	Mobility ($cm^2\ V^{-1}\ s^{-1}$)
230 nm	Holes	2	1.5×10^{16}	5.2
140 nm	Holes	73	2.0×10^{16}	4.3
90 nm	Holes	32	2.8×10^{16}	7.0

These differences are likely due to the high contact resistance of the metal pads at the four corners of the Van der Pauw structures, but a slight non-uniformity in the film's electrical properties cannot be excluded.

The observed values are comparable to or even better than literature data which give resistivity in the range of 20–10^3 Ω cm [53,54] and carrier concentrations of 10^{15}–10^{18} cm^{-3} [53,54].

Afterward, pure cuprite thin films are deposited on a conductive layer, i.e., ITO on a quartz substrate, starting from the optimized MOCVD parameter conditions. The Cu_2O_ITO_quartz system has been used for the sequential deposition of the MAPI active layer as part of the stack of a solar cell with a planar inverted architecture [13,40,55,56]. An accurate study has been conducted for each step of the multilayer MAPI_Cu_2O_ITO assembly. The system has been grown on a surface of 2 cm × 2 cm.

X-ray diffraction patterns of the different layers have been reported in Figure 7. The pattern in brown, related to the MAPI/Cu_2O/ITO system, displays several peaks pointing to the formation of a crystalline multilayer and associated with: (i) the pure Cu_2O phase; (ii) the MAPI reflection peaks attributed through comparison with literature data [57]; (iii) CuI impurities. No peaks associated with the ITO substrate are detected due to the thickness of the MAPI_Cu_2O layers. The detection of the CuI phase is likely a result of some interaction of the MAPI layer with the Cu_2O thin film. This phenomenon is supported by the study of the interaction between the MAPI and the Cu electrode, which has been demonstrated to produce CuI [58].

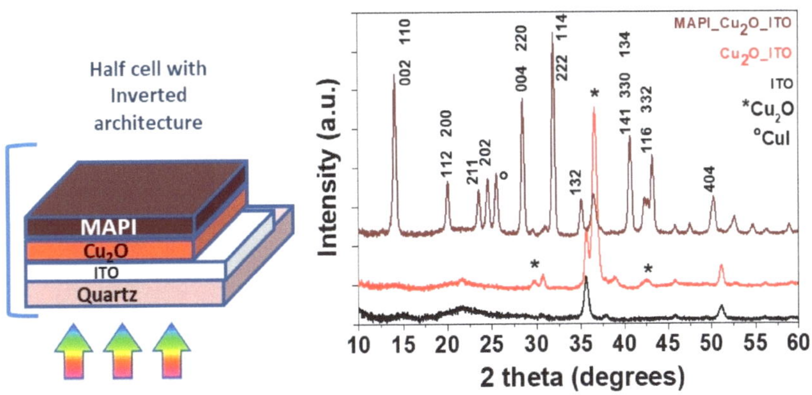

Figure 7. X-ray diffraction patterns of ITO substrate (black line), Cu_2O thin film prepared by MOCVD on the ITO substrate at 250 °C (red line) and the multilayer system MAPI/Cu_2O/ITO substrate (brown line). The * indicates the Cu_2O peaks. On the left, a cartoon is shown of the planar inverted architecture realized up to the MAPI deposition.

FE-SEM images in Figure 8 show the different morphologies of the multilayer structure from the ITO substrate (Figure 8a) and the Cu_2O film (Figure 8b) to the MAPI layer (Figure 8c). The ITO substrate is a compact homogeneous nanostructured layer with regular grains of about 50 nm and a thickness of about 305 nm. The Cu_2O thin film, deposited on top of the ITO substrate, displays a flat morphology (Figure 8b) similar to the one found for the Cu_2O film grown on Si (see Figure 2a), with a whole thickness of 525 ± 10 nm. This thickness, considering the previous estimation of 305 nm for the ITO, leads to a net Cu_2O layer of about 220 ± 10 nm.

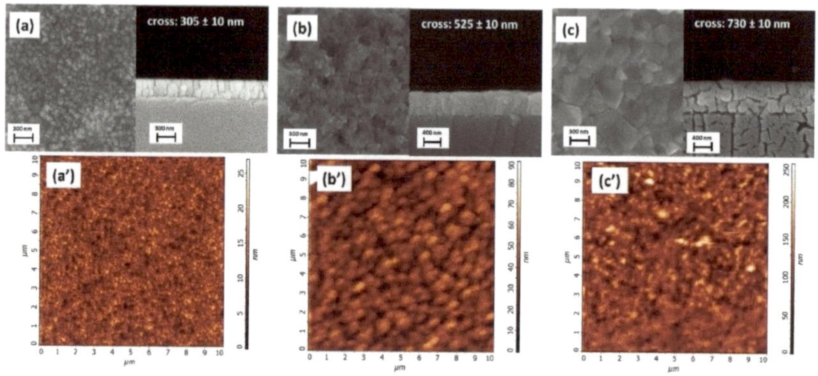

Figure 8. FE-SEM plan view and cross-section images of (**a**) ITO_quartz substrate; (**b**) Cu_2O thin film deposited at 250 °C on ITO substrate; (**c**) MAPI film deposited on Cu_2O_ITO_quartz substrate. The corresponding 10 µm × 10 µm AFM topographical images (**a'**–**c'**) are also shown.

Finally, the MAPI film in Figure 8c shows a nanostructured compact surface with smooth grains of hundreds of nanometers in size. The estimated thickness of the MAPI layer is around 275 ± 10 nm, derived from the difference in the cumulative thickness of the ITO_Cu_2O_MAPI system of 800 ± 10 nm.

Finally, the atomic force microscopy (AFM) characterization of the three layers (Figure 8a'–c') confirms the homogeneity of the films as well with fully coalesced grains on a larger area of 10 µm × 10 µm, and a root mean square (RMS) roughness of about 2.9 nm, 8.9 nm and 26.6 nm (measured on areas of 4 µm × 4 µm) for ITO, Cu_2O and MAPI films, respectively. The RMS roughness values of three different Cu_2O samples, deposited under the same conditions of the Cu_2O_230 nm, range from 7.8 to 10.1 nm. The low RMS roughness of the Cu_2O layer is comparable to literature data on films grown on Si, further confirming these films as suited for the growth of the perovskite layer.

4. Conclusions

In summary, a simple approach has been optimized to produce a stack composed of MAPI/Cu_2O/ITO/quartz. The extremely tunable MOCVD process allows the selective and reproducible fabrication of pure, uniform and highly compact Cu_2O films on various substrates. In addition, the very low deposition temperature of the Cu_2O layer makes this process appealing for deposition on temperature-sensitive substrates, such as polymers. In fact, the low operating temperature is a crucial issue and the capability to operate at 250 °C, using a process that is already industrially applied on a large scale, represents a breakthrough for the production of solar cells on plastic flexible supports. The electrical measurements and band gap values confirm the potentiality of the Cu_2O layers as p-type semiconducting materials for solar cell devices. Finally, preliminary studies have been carried out to produce the MAPI/Cu_2O/ITO stack, allowing us to scrutinize problematic issues regarding the MAPI/Cu_2O interface.

Supplementary Materials: The following supporting information can be downloaded at: https://www.mdpi.com/article/10.3390/ma15217790/s1, Figure S1: FE-SEM plan-image and EDX spectrum of the Cu_2O film deposited at 200 °C; Figure S2: FE-SEM images in plain and cross of Cu_2O thin films; Figure S3: Transmittance spectra of the ITO/quartz substrate; Figure S4: Tauc's plot of the ITO/quartz substrate.

Author Contributions: Investigation, A.L.P. and F.L.P.; methodology E.S. and S.V.; writing—original draft preparation, A.L.P.; data curation E. S., G.G., S.D.F. and F.R.; supervision G.M.; writing—review and editing, A.A. and G.M.; funding acquisition, A. A. and G.M. All authors have read and agreed to the published version of the manuscript.

Funding: This research was funded by the University of Catania for financial support within the PIACERI research program UNICT 2020-22 Linea 2 and by the Italian Ministry of University and Scientific Research (MIUR), PON ARS01_00519; CUP B88D19000160005.

Institutional Review Board Statement: Not applicable.

Informed Consent Statement: Not applicable.

Data Availability Statement: Additional data are reported in the Supplementary Materials.

Acknowledgments: The authors thank the Bionanotech Research and Innovation Tower (BRIT) laboratory of University of Catania (Grant no. PONa3_00136 financed by the Italian Ministry for Education, University and Research, MIUR) for the diffractometer facility. A.L.P. thanks the Ministero dell'Università e della Ricerca within the PON "Ricerca e Innovazione" 2014-2020 Azioni IV.4 program.

Conflicts of Interest: The authors declare no conflict of interest.

References

1. Al-Jawhari, H.A. A review of recent advances in transparent p-type Cu_2O-based thin film transistors. *Mater. Sci. Semicond. Process.* **2015**, *40*, 241–252. [CrossRef]
2. Meyer, B.K.; Polity, A.; Reppin, D.; Becker, M.; Hering, P.; Klar, P.J.; Sander, T.; Reindl, C.; Benz, J.; Eickhoff, M.; et al. Binary copper oxide semiconductors: From materials towards devices. *Phys. Status Solidi B* **2012**, *249*, 1487–1509. [CrossRef]
3. Lupan, O.; Cretu, V.; Postica, V.; Ababii, N.; Polonskyi, O.; Kaidas, V.; Schütt, F.; Mishra, S.K.; Monaico, E.; Tiginyanu, I.; et al. Enhanced ethanol vapour sensing performances of copper oxide. *Sens. Actuators B* **2016**, *224*, 434–448. [CrossRef]
4. Mohammed, A.M.; Mohtar, S.S.; Aziz, F.; Mhamad, S.A.; Aziz, M. Review of various strategies to boost the photocatalytic activity of the cuprous oxide-based photocatalyst. *J. Environ. Chem. Eng.* **2021**, *9*, 105138. [CrossRef]
5. Elfadill, N.G.; Hashim, M.R.; Saron, K.M.A.; Chahrour, K.M.; Qaeed, M.A.; Bououdina, M. Ultraviolet Visible photo-response of p-Cu_2O/n-ZnO heterojunction prepared on flexible (PET) substrate. *Mater. Chem. Phys.* **2015**, *156*, 54–60. [CrossRef]
6. Ghamgosar, P.; Rigoni, F.; Shujie You, S.; Dobryden, I.; Kohan, M.G.; Pellegrino, A.L.; Concina, I.; Almqvist, N.; Malandrino, G.; Vomiero, A. ZnO-Cu_2O core-shell nanowires as stable and fast response photodetectors. *Nano Energy* **2018**, *51*, 308–316. [CrossRef]
7. Jun Seo, Y.; Arunachalam, M.; Ahn, K.-S.; Hyung Kang, S. Integrating heteromixtured Cu_2O/CuO photocathode interface through a hydrogen treatment for photoelectrochemical hydrogen evolution reaction. *Appl. Surf. Sci.* **2020**, *551*, 149375. [CrossRef]
8. Li, C.; Fang, T.; Hu, H.; Wang, Y.; Liu, X.; Zhou, S.; Fu, J.; Wang, W. Synthesis and enhanced bias-free photoelectrochemical water-splitting activity of ferroelectric $BaTiO_3$/Cu_2O heterostructures under solar light irradiation. *Ceram. Int.* **2021**, *47*, 11379–11386. [CrossRef]
9. Masudy-Panah, S.; Zhuka, S.; Tana, H.R.; Gong, X.; Dalapati, G.K. Palladium nanostructure incorporated cupric oxide thin film with strong optical absorption, compatible charge collection and low recombination loss for low cost solar cell applications. *Nano Energy* **2018**, *46*, 158–167. [CrossRef]
10. Nguyen, V.S.; Sekkat, A.; Bellet, D.; Chichignoud, G.; Kaminski-Cachopo, A.; Muñoz-Rojas, D.; Favre, W. Open-air, low-temperature deposition of phase pure Cu_2O thin films as efficient hole-transporting layers for silicon heterojunction solar cells. *J. Mater. Chem. A* **2021**, *9*, 15968–15974. [CrossRef]
11. Kim, S.; Jung, Y.S.; Hong, J.S.; Kim, K.H. Fabrication of copper oxide-based heterojunction thin film solar cells using sputtering. *Mol. Cryst. Liq. Cryst.* **2018**, *677*, 10–18. [CrossRef]
12. Calil, L.; Kazim, S.; Gratzel, M.; Kurias, S.A. Hole-Transport Materials for Perovskite Solar Cells. *Angew. Chem. Int. Ed.* **2016**, *55*, 14522–14545. [CrossRef] [PubMed]
13. Duan, L.; Chen, Y.; Jia, J.; Zong, X.; Sun, Z.; Wu, Q.; Xue, S. Dopant-Free Hole-Transport Materials Based on 2,4,6-Triarylpyridine for Inverted Planar Perovskite Solar Cells. *ACS Appl. Energy Mater.* **2020**, *3*, 1672–1683. [CrossRef]
14. Messmer, C.; Bivour, M.; Schön, J.; Hermle, M. Requirements for Efficient Hole Extraction in Transition Metal Oxide-Based Silicon Heterojunction Solar Cells. *J. Appl. Phys.* **2018**, *124*, 085702. [CrossRef]
15. Chen, W.; Wu, Y.; Yue, Y.; Liu, J.; Zhang, W.; Yang, X.; Chen, H.; Bi, E.; Ashraful, I.; Grätzel, M.; et al. Efficient and stable large-area perovskite solar cells with inorganic charge extraction layers. *Science* **2015**, *350*, 944–948. [CrossRef]
16. Markose, K.; Shaji, M.; Bhatia, S.; Nair, P.R.; Saji, K.J.; Antony, A.; Jayaraj, M.K. Novel boron doped p-type Cu_2O thin film as hole selective contact in c-Si solar cell. *ACS Appl. Mater. Interfaces* **2020**, *12*, 12972–12981. [CrossRef]
17. Jayathilaka, C.; Kumara, L.S.R.; Ohara, K.; Song, C.; Kohara, S.; Sakata, O.; Siripala, W.; Jayanetti, S. Enhancement of Solar Cell Performance of Electrodeposited Ti/n-Cu_2O/p-Cu_2O/Au Homojunction Solar Cells by Interface and Surface Modification. *Crystals* **2020**, *10*, 609. [CrossRef]

18. Khadka, D.B.; Shirai, Y.; Yanagida, M.; Miyano, K. Ammoniated aqueous precursor ink processed copper iodide as hole transport layer for inverted planar perovskite solar cells. *Sol. Energy Mater. Sol. Cells* **2020**, *210*, 110486. [CrossRef]
19. Ye, S.; Sun, W.; Li, Y.; Yan, W.; Peng, H.; Bian, Z.; Liu, Z.; Huang, C. CuSCN-Based Inverted Planar Perovskite Solar Cell with an Average PCE of 15.6%. *Nano Lett.* **2015**, *15*, 3723–3728. [CrossRef]
20. Nitta, R.; Kubota, Y.; Kishi, T.; Yano, T.; Matsushita, N. One-step direct fabrication of phase-pure Cu_2O films via the spin-spray technique using a mixed alkaline solution. *Mater. Chem. Phys.* **2020**, *243*, 122442. [CrossRef]
21. Karle, S.; Rogalla, D.; Ludwig, A.; Becker, H.-W.; Wieck, A.D.; Grafen, M.; Devi, A. Synthesis and evaluation of new copper ketoiminate precursors for a facile and additive-free solution-based approach to nanoscale copper oxide thin films. *Dalton Trans.* **2017**, *46*, 2670–2679. [CrossRef] [PubMed]
22. Ibupoto, Z.H.; Khun, K.; Lu, J.; Willander, M. The synthesis of CuO nanoleaves, structural characterization, and their glucose sensing application. *Appl. Phys. Lett.* **2013**, *102*, 103701. [CrossRef]
23. Borkar, R.; Dahake, R.; Rayalu, S.; Bansiwal, A. Copper Oxide Nanograss for Efficient and Stable Photoelectrochemical Hydrogen Production by Water Splitting. *J. Electron. Mater.* **2018**, *47*, 1824–1831. [CrossRef]
24. Lakshmanan, A.; Alex, Z.C.; Meher, S.R. Cu_2O thin films grown by magnetron sputtering as solar cell absorber layers. *Mater. Sci. Semicond. Process.* **2022**, *148*, 106818. [CrossRef]
25. Huo, W.; Shi, J.; Mei, Z.; Liu, L.; Li, J.; Gu, L.; Du, Z.; Xue, Q. High-index Cu_2O (113) film on faceted MgO (110) by molecular beam epitaxy. *J. Cryst. Growth* **2015**, *420*, 32–36. [CrossRef]
26. Hu, X.; Schuster, J.; Schulz, S.E.; Gessner, T. Surface chemistry of copper metal and copper oxide atomic layer deposition from copper(II) acetylacetonate: A combined first-principles and reactive molecular dynamics study. *Phys. Chem. Chem. Phys.* **2015**, *17*, 26892–26902. [CrossRef]
27. Karapetyan, A.; Reymers, A.; Giorgio, S.; Fauquet, C.; Sajti, L.; Nitsche, S.; Nersesyan, M.; Gevorgyan, V.; Marine, W. Cuprous oxide thin films prepared by thermal oxidation of copper layer. Morphological and optical properties. *J. Lumin.* **2015**, *159*, 325–332. [CrossRef]
28. Kobayashi, H.; Nakamura, T.; Takahashi, N. Preparation of Cu_2O films on MgO (1 1 0) substrate by means of halide chemical vapor deposition under atmospheric pressure. *Mater. Chem. Phys.* **2007**, *106*, 292–295. [CrossRef]
29. Condorelli, G.G.; Malandrino, G.; Fragalà, I. Metal-Organic Chemical Vapor Deposition of Copper-Containing Phases: Kinetics and Reaction Mechanisms. *Chem. Mater.* **1994**, *6*, 1861–1866. [CrossRef]
30. Arana-Chavez, D.; Toumayan, E.; Lora, F.; McCaslin, C.; Adomaitis, R.A. Modeling the Transport and Reaction Mechanisms of Copper Oxide CVD. *Chem. Vap. Depos.* **2010**, *16*, 336–345. [CrossRef]
31. Sekkat, A.; Nguyen, V.H.; Masse de La Huerta, C.A.; Rapenne, L.; Bellet, D.; Kaminski-Cachopo, A.; Chichignoud, G.; Muñoz-Rojas, D. Open-air printing of Cu_2O thin films with high hole mobility for semitransparent solar harvesters. *Commun. Mater.* **2021**, *2*, 78. [CrossRef]
32. Jürgensen, L.; Höll, D.; Frank, M.; Ludwig, T.; Graf, D.; Schmidt-Verma, A.K.; Raauf, A.; Gessner, I.; Mathur, S. Controlled growth of Cu and CuO_x thin films from subvalent copper precursors. *Dalton Trans.* **2020**, *49*, 13317–13325. [CrossRef] [PubMed]
33. Pousaneh, E.; Korb, M.; Dzhagan, V.; Weber, M.; Noll, J.; Mehring, M.; Zahn, D.R.T.; Schulz, S.E.; Lang, H. β-Ketoiminato-based copper(II) complexes as CVD precursors for copper and copper oxide layer formation. *Dalton Trans.* **2018**, *47*, 10002–10016. [CrossRef] [PubMed]
34. Condorelli, G.G.; Malandrino, G.; Fragalà, I. Kinetic Study of MOCVD Fabrication of Copper(I) and Copper(II) Oxide Films. *Chem. Vap. Depos.* **1999**, *5*, 21–27. [CrossRef]
35. Liu, H.; Nguyen, V.H.; Roussel, H.; Gélard, I.; Rapenne, L.; Deschanvres, J.-L.; Jiménez, C.; Muñoz-Rojas, D. The Role of Humidity in Tuning the Texture and Electrical Properties of Cu_2O Thin Films Deposited via Aerosol-Assisted CVD. *Adv. Mater. Interfaces* **2019**, *6*, 1801364. [CrossRef]
36. Barreca, D.; Gasparotto, A.; Maccato, C.; Tondello, E.; Lebedev, O.I.; Van Tendeloo, G. CVD of Copper Oxides from a β-Diketonate Diamine Precursor: Tailoring the Nano-Organization. *Cryst. Growth Des.* **2009**, *9*, 2471–2480. [CrossRef]
37. Maruyama, T. Copper oxide thin films prepared by chemical vapor deposition from copper dipivaloylmethanate. *Sol. Energy Mater. Sol. Cells* **1998**, *56*, 85–92. [CrossRef]
38. Lay, E.; Song, Y.-H.; Chiu, Y.-C.; Lin, Y.-M.; Chi, Y. New CVD Precursors Capable of Depositing Copper Metal under Mixed O2/Ar Atmosphere. *Inorg. Chem.* **2005**, *44*, 7226–7233. [CrossRef]
39. Gupta, N.; Singh, R.; Wu, F.; Narayan, J.; McMillen, C.; Alapatt, G.F.; Poole, K.F.; Hwu, S.; Sulejmanovic, D.; Young, M.; et al. Deposition and characterization of nanostructured Cu_2O thin-film for potential photovoltaic applications. *J. Mater. Res.* **2013**, *28*, 1740–1746. [CrossRef]
40. Smecca, E.; Valenzano, V.; Valastro, S.; Deretzis, I.; Mannino, G.; Malandrino, G.; Accorsi, G.; Colella, S.; Rizzo, A.; La Magna, A.; et al. Two-step $MAPbI_3$ deposition by Low-Vacuum Proximity Space-Effusion for high-efficiency inverted semitransparent perovskite solar cells. *J. Mater. Chem. A* **2021**, *9*, 16456–16469. [CrossRef]
41. Smecca, E.; Jena, A.K.; Deretzis, I.; Valastro, S.; Sanzaro, S.; Mannino, G.; Bongiorno, C.; La Magna, A.; Miyasaka, T.; Alberti, A. $MAPbI_3$ Deposition by LV-PSE on TiO_2 for Photovoltaic Application. *Front. Electron.* **2021**, *2*, 726171. [CrossRef]
42. Alberti, A.; Smecca, E.; Valastro, S.; Deretzis, J.; Mannino, G.; Bongiorno, C.; Fisicaro, G.; La Magna, A. Perovskite Solar Cells from the viewpoint of innovation and sustainability. *Phys. Chem. Chem. Phys.* **2022**, *24*, 21549–21566. [CrossRef] [PubMed]

43. Aithssi, A.; Atourki, L.; Labchir, N.; Ouafi, M.; Abouabassi, K.; Elfanaoui, A.; Ihlal, A.; Bouabid, K. Optical and dielectric properties of electrochemically deposited p-Cu_2O films. *Mater. Res. Express* **2020**, *7*, 016424. [CrossRef]
44. Xue, J.; Shen, Q.; Liang, W.; Liu, X.; Bian, L.; Xu, B. Preparation and formation mechanism of smooth and uniform Cu_2O thin films by electrodeposition method. *Surf. Coat. Technol.* **2013**, *216*, 166–171. [CrossRef]
45. Chen, A.; Long, H.; Li, X.; Li, Y.; Yang, G.; Lu, P. Controlled growth and characteristics of single-phase Cu_2O and CuO films by pulsed laser deposition. *Vacuum* **2009**, *83*, 927–930. [CrossRef]
46. Sawicka-Chudy, P.; Sibinski, M.; Pawelek, R.; Wisz, G.; Cieniek, B.; Potera, P.; Szczepan, P.; Adamiak, S.; Cholewa, M.; Glowa, L. Characteristics of TiO_2, Cu_2O, and TiO_2/Cu_2O thin films for application in PV devices. *AIP Adv.* **2019**, *9*, 055206. [CrossRef]
47. Chua, D.; Kim, S.B.; Li, K.; Gordon, R. Low Temperature Chemical Vapor Deposition of Cuprous Oxide Thin Films Using a Copper(I) Amidinate Precursor. *ACS Appl. Energy Mater.* **2019**, *2*, 7750–7756. [CrossRef]
48. Ben Rabeha, M.; Khedmia, N.; Fodhaa, M.A.; Kanzaria, M. The Effect of Thickness on Optical Band Gap and N-type Conductivity of $CuInS_2$ Thin Films Annealed in Air Atmosphere. *Energy Procedia* **2014**, *44*, 52–60. [CrossRef]
49. Sönmezoğlu, S.; Arslan, A.; Serin, T.; Serin, N. The effects of film thickness on the optical properties of TiO_2–SnO_2 compound thin films. *Phys. Scr.* **2011**, *84*, 065602. [CrossRef]
50. Zheng, W.; Chen, Y.; Peng, X.; Zhong, K.; Lin, Y.; Huang, Z. The Phase Evolution and Physical Properties of Binary Copper Oxide Thin Films Prepared by Reactive Magnetron Sputtering. *Materials* **2018**, *11*, 1253. [CrossRef]
51. Murali, D.S.; Kumar, S.; Choudhary, R.J.; Wadikar, A.D.; Jain, M.K.; Subrahmanyam, A. Synthesis of Cu_2O from CuO thin films: Optical and electrical properties. *AIP Adv.* **2015**, *5*, 047143. [CrossRef]
52. Schroder, D.K. *Semiconductor Material and Device Characterization*, 3rd ed.; John Wiley & Sons, Inc.: Hoboken, NJ, USA, 2006.
53. Resende, J.; Nguyen, V.-S.; Fleischmann, C.; Bottiglieri, L.; Brochen, S.; Vandervorst, W.; Favre, W.; Jiménez, C.; Deschanvres, J.-L.; Nguyen, N.D. Grain-boundary segregation of magnesium in doped cuprous oxide and impact on electrical transport properties. *Sci. Rep.* **2021**, *11*, 7788. [CrossRef] [PubMed]
54. Guo, Y.; Lei, H.; Xiong, L.; Li, B.; Chen, Z.; Wen, J.; Yang, G.; Li, G.; Fang, G. Single phase, high hole mobility Cu_2O films as an efficient and robust hole transporting layer for organic solar cells. *J. Mater. Chem. A* **2017**, *5*, 11055–11062. [CrossRef]
55. Yuan, J.; Chen, Y.; Liu, X.; Xue, S. Dopant-free Hole-transporting Materials for $CH_3NH_3PbI_3$ Inverted Perovskite Solar Cells with an Approximate Efficiency of 20%. *ACS Appl. Energy Mater.* **2021**, *4*, 5756–5766. [CrossRef]
56. Huang, J.; Ge, C.; Qin, F.; Zou, Y.; Zhou, Y.; Li, W.-S.; Gao, X. Achieve Better Performance of Inverted Perovskite Solar Cells by Using the Fluorinated Polymer as the Electron Transporting Layer. *ACS Appl. Energy Mater.* **2022**, *5*, 2522–2530. [CrossRef]
57. Baikie, T.; Fang, Y.; Kadro, J.M.; Schreyer, M.; Wei, F.; Mhaisalkar, S.G.; Graetzel, M.; White, T.J. Synthesis and crystal chemistry of the hybrid perovskite $(CH_3NH_3)PbI_3$ for solid-state sensitized solar cell applications. *J. Mater. Chem. A* **2013**, *1*, 5628–5641. [CrossRef]
58. Udalova, N.N.; Nemygina, E.M.; Zharenova, E.A.; Tutantsev, A.S.; Sudakov, A.A.; Grishko, A.Y.; Belich, N.A.; Goodilin, E.A.; Tarasov, A.B. New Aspects of Copper Electrode Metamorphosis in Perovskite Solar Cells. *J. Phys. Chem. C* **2020**, *124*, 24601–24607. [CrossRef]

materials

MDPI

Article

Conductometric H_2S Sensors Based on TiO_2 Nanoparticles

Yassine Alaya [1], Malek Madani [1], Noureddine Bouguila [1], Lassaad El Mir [1], Enza Fazio [2], Carmelo Corsaro [2] and Giovanni Neri [3,*]

1 Laboratory of Physics of Materials and Nanomaterials Applied at Environment (LaPhyMNE), Faculty of Sciences in Gabes, Gabes University, Gabes 6072, Tunisia; yassine1alaya@gmail.com (Y.A.); malek.madani@univgb.rnu.tn (M.M.); bouguila.nour@gmail.com (N.B.); lassaad.elmir@fsg.rnu.tn (L.E.M.)

2 Department of Mathematical and Computer Sciences, Physical Sciences and Earth Sciences, University of Messina, Viale F. Stagno D'Alcontres 31, 98166 Messina, Italy; enza.fazio@unime.it (E.F.); carmelo.corsaro@unime.it (C.C.)

3 Department of Engineering, University of Messina, C.da Di Dio, 98166 Messina, Italy

* Correspondence: gneri@unime.it

Abstract: High-performance hydrogen sulfide (H_2S) sensors are mandatory for many industrial applications. However, the development of H_2S sensors still remains a challenge for researchers. In this work, we report the study of a TiO_2-based conductometric sensor for H_2S monitoring at low concentrations. TiO_2 samples were first synthesized using the sol-gel route, annealed at different temperatures (400 and 600 °C), and thoroughly characterized to evaluate their morphological and microstructural properties. Scanning electronic microscopy, Raman scattering, X-ray diffraction, and FTIR spectroscopy have demonstrated the formation of clusters of pure anatase in the TiO_2 phase. Increasing the calcination temperature to 600 °C enhanced TiO_2 crystallinity and particle size (from 11 nm to 51 nm), accompanied by the transition to the rutile phase and a slight decrease in band gap (3.31 eV for 400 °C to 3.26 eV for 600 °C). Sensing tests demonstrate that TiO_2 annealed at 400 °C displays good performances (sensor response Ra/Rg of ~3.3 at 2.5 ppm and fast response/recovery of 8 and 23 s, respectively) for the detection of H_2S at low concentrations in air.

Keywords: sol-gel; TiO_2 nanopowder; thermal treatment; H_2S sensor

Citation: Alaya, Y.; Madani, M.; Bouguila, N.; El Mir, L.; Fazio, E.; Corsaro, C.; Neri, G. Conductometric H_2S Sensors Based on TiO_2 Nanoparticles. *Materials* **2024**, *17*, 3283. https://doi.org/10.3390/ma17133283

Academic Editor: Abderrahim Yassar

Received: 3 June 2024
Revised: 28 June 2024
Accepted: 30 June 2024
Published: 3 July 2024

1. Introduction

With the swift evolution of the global industry and the desire to improve air quality, hydrogen sulfide (H_2S) has been recognized as one of the highly concerned pollution gases, commonly emitted by industries operating in the fields of pulp and paper manufacturing, natural gas, biological decomposition of organic waste material, and crude petroleum [1–3]. H_2S is a hazardous chemical, colorless, and extremely flammable [4,5]. At low concentrations, it has an odor of rotten egg, which may cause coughing and sore throat and eyes, while people exposed to high concentrations (300–500 ppm) may experience the human olfactory nerve system and the collapse of the cardiovascular system [6]. Therefore, it is mandatory to develop H_2S sensors with good performance.

Among the variety of sensors used for gas sensing, conductometric sensors have proven to be pretty attractive for detecting a variety of gases, since they are easy to fabricate, low cost, and simple to operate [7–9]. In the literature, there are many reports on H_2S sensors based on metal oxide semiconductors (MOS), such as Fe_2O_3 [10], CuO [11], ZnO [12], WO_3 [13], and NiO [14]. However, TiO_2-based H_2S sensor development is still scarce, and their gas-sensing performance needs to be improved [15]. TiO_2 has intriguing physical and chemical features, making it a promising choice for gas sensor applications due to its distinct allotropic phases (anatase, rutile, and brookite) [6]. This involves the microstructural, morphological, and defect characteristics, which can play a crucial role in enhancing sensor response. Meanwhile, selecting synthesis techniques for the TiO_2 nanoparticles is a vital step to achieving a larger surface area with higher roughness. Various physical and chemical routes are commonly used for the synthesis of TiO_2 nanoparticles, such as Pulsed Laser Deposition (PLD) [16], sol-gel [17], thermal evaporation [18], sputtering [15], spray pyrolysis [19], and Atomic Layer Deposition (ALD) [20]. TiO_2 has a high surface area, enhancing its interaction with gas molecules and improving sensitivity [21–23]. It is chemically stable and corrosion-resistant, ensuring long-term durability and reliability. TiO_2 exhibits excellent photocatalytic activity [24], significantly changing its conductivity when exposed to light and gas molecules, which enhances sensitivity and response time. The material can be synthesized in various nanostructured forms, providing greater surface area and more active sites for gas adsorption. Additionally, TiO_2 is cost-effective and abundantly available, making it an economical choice for gas sensor development. Therefore, these features provide the TiO_2-based sensor with great sensitivity and selectivity for hydrogen sulfide.

In this study, we have synthesized TiO_2 nanoparticles (NPs) with the modified sol-gel method using ethyl alcohol under supercritical conditions, which requires lower energy consumption and allows the synthesis of materials with high purity and homogeneity. We investigated their structural, morphological, and optical properties and their performances in gas-sensing for detecting low hydrogen sulfide concentrations in the range from 0.5 to 4 ppm. The developed sensor exhibited enhanced sensitivity, selectivity, and fast response/recovery times to H_2S.

2. Experimental Section

2.1. Synthesis of TiO_2 Nanopowder

TiO_2 nanopowder was prepared using the protocol of El Mir et al. [25,26] based on the following steps. First, 15 mL of Titanium (IV) isopropoxide $Ti(OC_3H_7)_4$ (97%, from Sigma–Aldrich, Saint Louis, MO, USA) was dissolved in 45 mL of methanol blended with 2 mL of acetic acid (CH_3COOH). The mixture was kept under magnetic stirring until the precursors were completely dissolved. The resulting solution was then poured into the autoclave to achieve drying in supercritical conditions of 250 mL of ethanol (Tc = 243 °C; Pc = 63.6 bars), with a heating rate of 45° C/h. Afterward, the as-obtained nanopowder was calcined for 2 h in air at different temperatures, (T = 400 °C) and (T = 600 °C). For the preparation of the TiO_2 conductometric sensor, a quantity of 1 mg of TiO_2 powder was sonicated for 30 min with 1 mL of deionized water. The gas sensor was manufactured in the temperature range between 20 °C and 25 °C. A scheme of the synthesis procedure is illustrated in Figure 1.

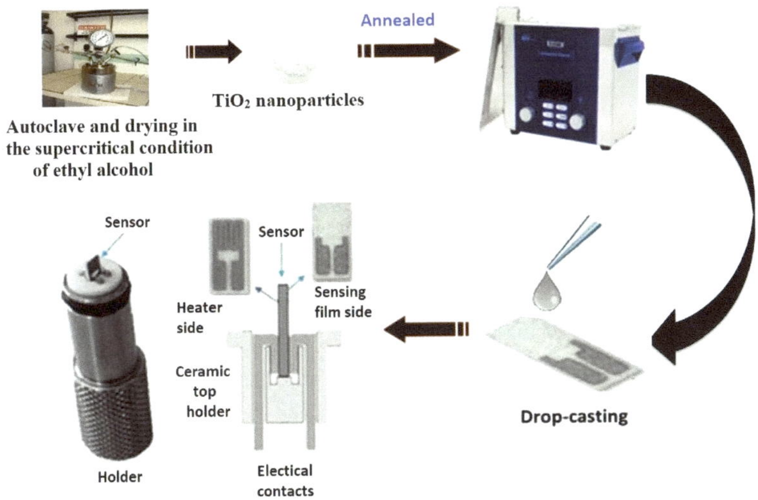

Figure 1. Fabrication process of the TiO_2 conductometric sensor.

2.2. Characterizations

Microstructural analysis was determined with a D2 phaser Bruker X-ray diffractometer (Bruker, Billerica, MA, USA) using the Cu Kα line (0.159 nm) in the 10–80° 2θ range. FT-IR spectra were recorded utilizing a PerkinElmer spectrometer (PerkinElmer, Waltham, MA, USA) equipped with a universal attenuated total reflectance (ATR) sampling accessory. The UV–visible diffuse reflectance spectra (UV–visible DRS) were measured using a Shimadzu (Kyoto City, Japan) 2600–2700 spectrometer with $BaSO_4$ as a reference. Raman spectra of the samples were recorded using the XploRa Raman spectrometer (Horiba Scientific, Piscataway, NJ, USA) equipped with an Olympus BX-40 microscope (Olympus, Tokyo, Japan) (objective ×50 focal length), Peltier cooled CCD detector, 532 nm diode laser, and 600 L/mm grating. The laser power was 5 mW, and the acquisition time was 30 s. Two to ten spectra were registered for each sample at different positions to verify sample homogeneity. The reference spectrum of Si (peak position of 521 cm^{-1}) was measured to avoid temperature drift. Scanning Electron Microscope (SEM) images were taken using a Zeiss (Oberkochen, Germany) (Gemini II) microscope at the acceleration voltage of 5 kV.

2.3. Gas Sensing Tests

The gas sensing tests were carried out with sensors fabricated by printing TiO_2 on the sensor device with a heating element and Pt-interdigitated electrodes. For the tests, the sensor devices were introduced into the test chamber. An Agilent E3632A instrument (Agilent, Santa Clara, CA, USA) was employed for setting the operating temperatures, whereas the resistance of the TiO_2 sensing layer was measured with an Agilent 34970A multimeter (Agilent, Santa Clara, CA, USA). H_2S sensing tests were carried out under a flow of dry synthetic air of 100 cc/min, operating at temperatures from 100 to 400 °C, with H_2S gas concentrations of 0 to 4 ppm. The gas response, S, is defined as the ratio R_a/R_g for n-type behavior, where R_a is the baseline resistance in dry synthetic air and R_g is the electrical resistance at different gas concentrations. The response time, τ_{res}, and recovery time, τ_{rec}, were defined as follows. Response time, τ_{res}, i.e., the time required for the sensor to reach 90% of the saturation resistance after injection of the target gas, and recovery time, τ_{rec}, i.e., the time required for the sensor to reach 90% of the resistance baseline value in air. These were also evaluated.

3. Results and Discussion

3.1. Sample Characterizations

TiO_2 samples synthesized using the sol-gel route and annealed at different temperatures (400 and 600 °C) were first thoroughly analyzed by different characterization techniques. The SEM images of the TiO_2 sample annealed at 400 °C are reported in Figure 2.

The sample annealed at 400 °C is not completely homogeneous, showing regions with different microscopic features characterized by randomly distributed and non-uniform clusters of TiO_2 (Figure 2a,b). However, all regions show a high porosity on a nanometric scale with grain size in the range of 10–20 nm (Figure 2c,d).

Figure 3 reports characteristic SEM images of the TiO_2 sample annealed at 600 °C. The morphology of this sample is more homogeneous, with a bigger grain size in the 30–50 nm range and a fractal-like structure induced by calcination. In addition, it can be noted that, at the highest annealing temperature, the collapse of the mesostructure occurs, which could be caused by the crystallization of the amorphous titania into nanosized anatase particles and/or with the transition from the anatase to the rutile phase.

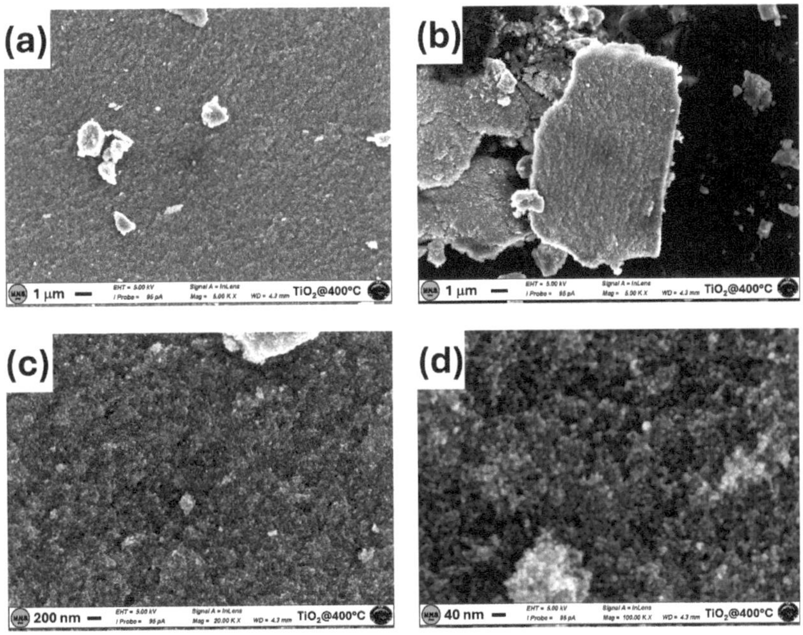

Figure 2. SEM images of the TiO_2 sample annealed at 400 °C acquired at 5k magnifications in two different regions. (**a**,**b**) At 20k magnifications (**c**) and 100k magnifications (**d**).

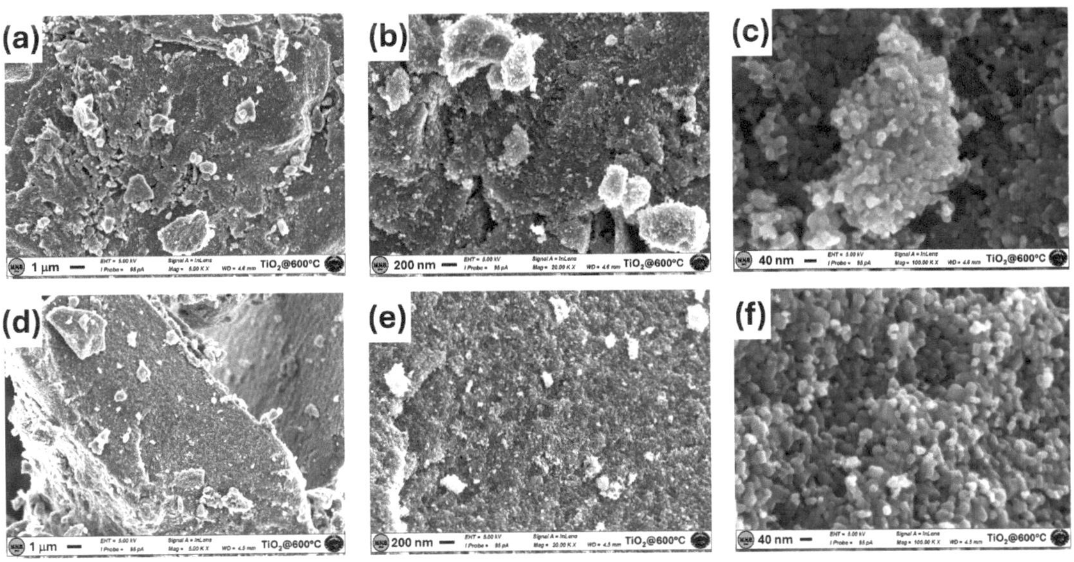

Figure 3. SEM images of the TiO_2 sample annealed at 600 °C were acquired in two different regions at 5k magnifications (**a**,**d**), 20k magnifications (**b**,**e**), and 100k magnifications (**c**,**f**).

To investigate this transition phase process, further characterizations have been carried out. The vibrational properties were investigated by Raman measurements, the profiles of which are reported in Figure 4. The three Raman peaks centered at about 386, 509.3, and 630.3 cm^{-1} (inset in Figure 4) are assigned to the Raman active modes of the anatase

TiO_2 crystalline structure, while the peak at about 472.5 cm^{-1} is associated with the Raman active modes of the rutile crystalline phase [27]. The bands in the region higher than 1500 cm^{-1} are due to C-C and C-H/C-O contributions, which is in good agreement with FTIR data. Ultimately, Raman evidence indicates that the two TiO_2 phases characterize the investigated samples. However, the relative intensity of the peaks of the anatase phase compared to the peak associated with the rutile phase is different in the two samples. This indicates that the phase transition from the anatase phase to the rutile phase occurs at the highest temperature.

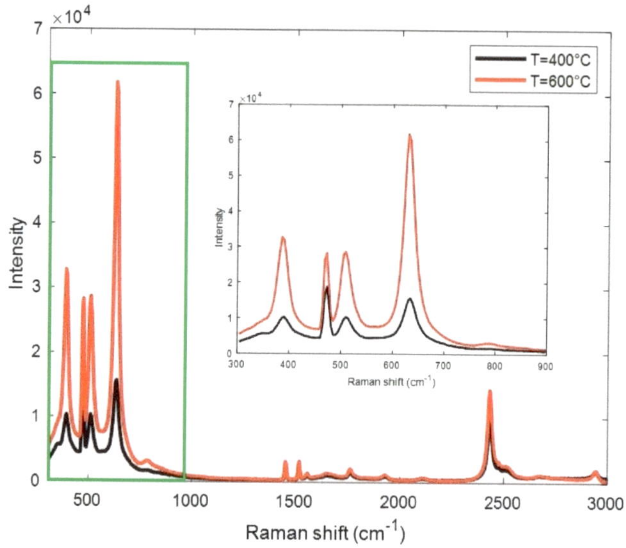

Figure 4. Raman spectra of the samples annealed at 400 °C and 600 °C. The inset shows an enlargement of the spectral region between 300 and 900 cm^{-1}.

In Figure 5, the XRD patterns of both TiO_2 samples are shown. In Figure 5a, which shows the XRD pattern of TiO_2 powder annealed at 400 °C, we identified the (101), (103), (004), (112), (200), (105), (211), (204), (116), (220), and (215) diffraction peaks ascribed to the TiO_2 tetragonal structure in the anatase phase (JCPDS 21-1272) [28,29]. As shown in Figure 5b, upon increasing the annealing temperature to 600 °C, the diffraction peaks are narrower and slightly more intense. We can also discern the orthorhombic structure of TiO_2 as discerned by its characteristic (011) peak centered at 31.7° (JCPDS 80-5176) [30].

The Rietveld refinements of the crystal structures of the as-prepared TiO_2 samples were carried out using the FullProf software (https://www.ill.eu/sites/fullprof/, accessed on 6 May 2024). The method employs a least-squares procedure to compare Bragg intensities and those calculated from a possible structural model. In the first step of refinement, the global parameters, such as background and scale factors, were refined. In the next step, structural property parameters such as lattice parameters, profile shape and width parameters, preferred orientation, asymmetry, isothermal parameters, atomic coordinates, and site occupancies were refined in sequence.

The average crystallite size and the lattice strain were calculated according to the Williamson–Hall method using the following equation [31]:

$$\beta \cos\theta = \frac{K\lambda}{D} + 4\varepsilon \sin\theta \quad (1)$$

where β is the peak full width at half maximum (FWHM), θ is the Bragg angle, K is the shape factor (0.9), λ is the incident wavelength (λ = 1.5406 Å), and ε is the film strain. The trend of $\beta\cos\theta$ as a function of $4\sin\theta$ for the investigated samples is shown in Figure 5c,d.

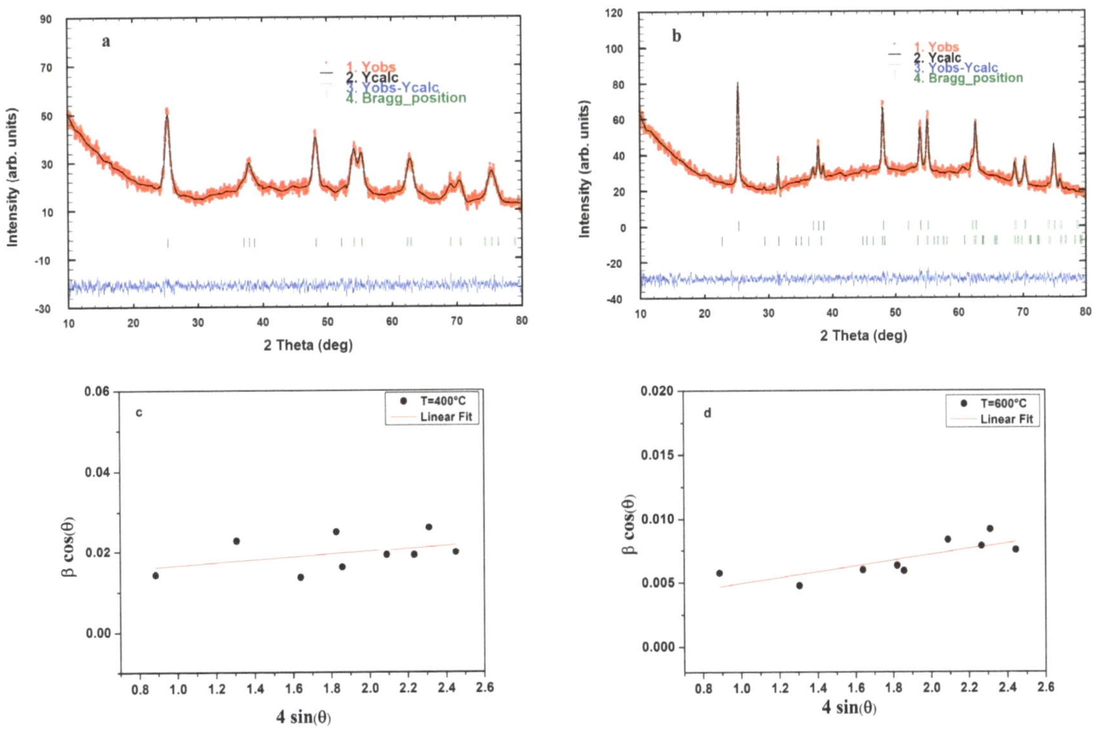

Figure 5. (**a**,**b**) Rietveld refinement of the X-ray diffraction profile and (**c**,**d**) Williamson–Hall plots of TiO_2 nanopowders calcined at 400 °C and 600 °C.

The fitting quality of the experimental data is assessed by computing parameters such as the 'goodness of fit' χ^2, the Bragg R-factor, and the Rf-factors (Profile R-factor (Rp), Weighted Profile R-factor (Rwp), and Expected R-factor (Rexp)). The values of these structural parameters are reported in Table 1.

Table 1. Fitting parameters of the Rietveld refinement on DRX profiles of TiO_2 annealed at 400 °C and 600 °C.

Sample	χ^2	Bragg R-Factor	R Factors	Rp	Rwp	Rexp
TiO_2 (400 °C)	1.86	11.0	11.2	5.34	6.47	4.75
TiO_2 (600 °C)	2.57	2.83	1.55	4.37	5.37	3.35

The results deduced from the Rietveld refinements of the XRD profiles are reported in Table 2, giving information about the significant variation of the phase composition and the crystallite size. It emerges that the crystallite size of the TiO_2 sample increases upon increasing the annealing temperature.

Table 2. Crystallographic properties of TiO_2 annealed at 400 °C and 600 °C.

Sample	Crystalline Phase	Space Group	Lattice Parameters (Å)	Crystallite Size (nm)	Strain (×10^{-5})	Phase Composition (%)
TiO_2 (400 °C)	Tetragonal (anatase)	$I4_1/amd$	a = b = 3.7761 c = 9.4950	11	0.00359	100
TiO_2 (600 °C)	Tetragonal (anatase)	$I4_1/amd$	a = b = 3.7868 c = 9.5174	51	0.00224	76.95
	Orthorhombic (rutile)	Pnma	a = 5.1124 b = 3.2054 c = 6.0871	-	-	23.05

Figure 6 shows the FTIR spectra of TiO_2 samples in the 400–4000 cm^{-1} range. The spectrum of the sample annealed at 400 °C shows two broad bands, centered at about 500 and 860 cm^{-1}, which are assigned to the Ti–O bending and Ti−O−Ti stretching vibrations [32], respectively. Furthermore, the barely visible contributions at about 1240 and 1340 cm^{-1} are ascribed to the C-H twisting and bending vibrational modes [33,34], whereas the two bands at around 1630 and 3310 cm^{-1} correspond to the presence of related hydroxyl groups (Ti-OH) and those of water molecules [35,36]. As expected, when calcination temperature increases, peaks relative to the hydroxyl groups and adsorbed C-H disappear. At the same time, we observed a slight change in the diffraction profile of TiO_2, indicating the rearrangement of the Ti-O network to facilitate the crystallization of TiO_2 [32].

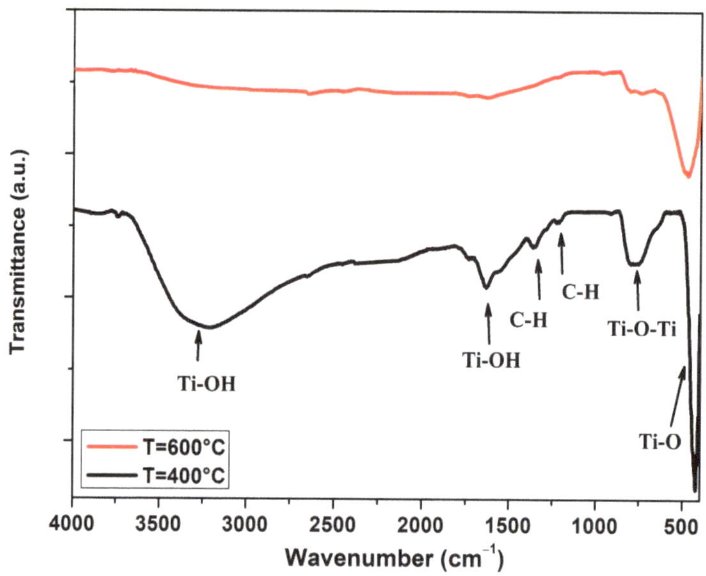

Figure 6. FTIR spectrum of TiO_2 nanopowder annealed at 400 °C and 600 °C.

This agrees with what is known from the literature [37], which reports that although the oxygen content remains constant up to annealing temperatures of 900 °C, when there is an increase in temperature, there is also an increase in "O^- species" due to the hydroxyl groups and carbon impurities desorbing from the surface. This process, due to the localized charge transfer between anionic and cationic frameworks during thermally induced reduction, favors the sensing mechanism of H_2S, which reacts with the adsorbed oxygen species to form SO_2 and H_2O (see Section 3.3).

UV–visible absorption measurements were carried out to investigate the changes in optical transitions of TiO_2 nanostructures caused by annealing. The Kubelka–Munk equation was used to calculate the absorption spectra of the samples from the diffuse reflectance spectra [26]. Figure 7a shows absorbance spectra with wavelengths ranging from 250 to 800 nm for both samples synthesized using the sol-gel method. The absorbance of the nanostructures is around 90% in the UV range and decreases dramatically beginning in the visible range. Seemingly, annealing has no significant impact on the absorbance of the ceramic in the UV range. The bandgap energy E_g of TiO_2 nanostructures was estimated according to the Tauc method, following Equation (2) [29,38,39]:

$$(\alpha h\nu)^2 = A(h\nu - \text{Eg}) \quad (2)$$

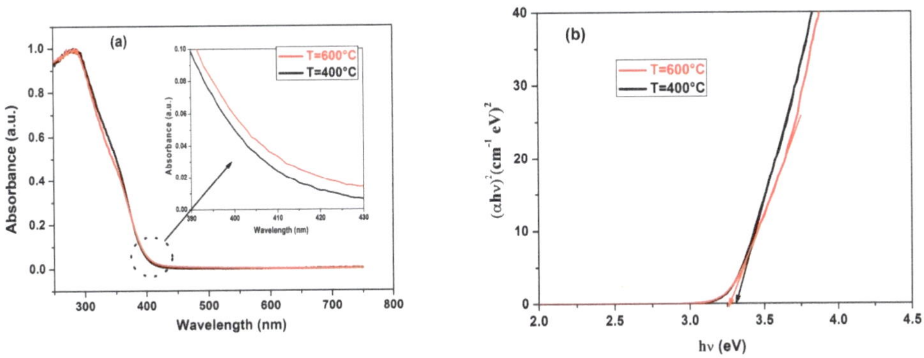

Figure 7. Absorbance spectra (inset shows the observed shift) (**a**) and Tauc plots (**b**) of TiO_2 nanopowders prepared using the sol-gel technique and annealed at different temperatures.

In this equation, α is the absorption coefficient, A is a constant, and $h\nu$ is the photon energy. Extrapolating the linear part of the curve to the $h\nu$-axis yielded the optical band gap, as illustrated in Figure 7b. The estimated band gap energy values were 3.31 eV and 3.26 eV for samples annealed at 400 °C and 600 °C, respectively. Similar behavior was reported in the literature for TiO_2 nanostructures synthesized using the sol-gel method [29].

3.2. Gas Sensing Tests

Before investigating the gas sensing properties, the baseline resistance of the TiO_2 layer, denoted as R_a, versus operating temperature has been investigated (see Figure 8). The sensor baseline displays a higher resistance at low temperatures. As the temperature increases, the resistance baseline decreases because of the thermal excitation of electrons into the conduction band, indicating the semiconductor behavior of TiO_2. The data have further shown that TiO_2 (600 °C) is more resistive, due to the presence of the rutile phase, compared to TiO_2 (400 °C).

Operating temperature is also an important parameter to take into account for the gas sensing response. Indeed, temperature influences the adsorption/desorption processes of gases occurring on the sensing surface, as well as their reaction rate with adsorbed oxygen on the TiO_2 surface, and consequently the sensor response. As the above Figure 8 demonstrates, at temperatures lower than 300 °C, the baseline resistance is very high. To evaluate the optimal operating temperature for the sensing tests, the sensor was exposed to 1.5 ppm H_2S gas at temperatures ranging from 300 to 400 °C (Figure 9). Based on the results obtained, 350 °C appears to be the best operating temperature for this sensor, which displayed a high response to H_2S and short response/recovery times.

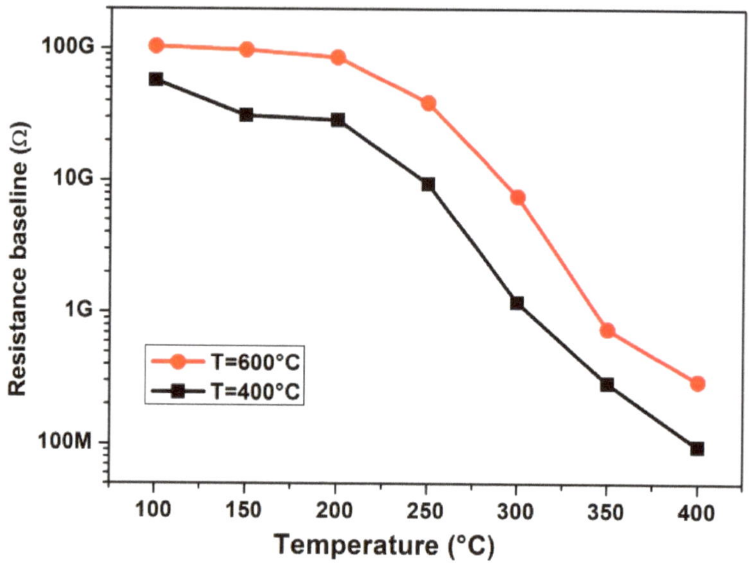

Figure 8. Baseline resistance of TiO_2-NP sensors vs. operating temperature.

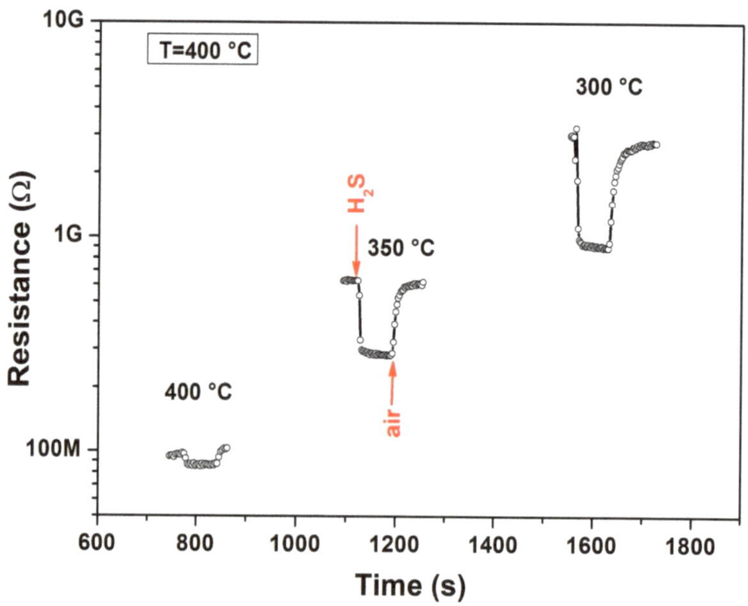

Figure 9. Resistance vs. time of TiO_2 NPs (400 °C)-based sensor for different operating temperatures (@1.5 ppm H_2S concentration).

In these operating conditions (Figure 10a), the sensor response of TiO_2 (400 °C) was registered to be 3.26 for 2.5 ppm of H_2S, higher than that reported for the TiO_2 (600 °C) sensor. In addition, we confirmed this result in Figure 10b, which depicts the evolution of the sensor response as a function of the H_2S concentration for both sensors. This finding is consistent with the results of XRD and SEM analyses. Indeed, it is well-known that when the grain size of the sensing material is small enough, it substantially impacts the gas sensing properties [37,38]. In addition, the sensor annealed at 400 °C has a larger surface-to-volume ratio due to the smaller grain size, thus further justifying the larger response compared to the TiO_2 (600 °C) sensor. The lower response for the sensor annealed at 600 °C could therefore be related to the improvement in the crystallinity of TiO_2 nanoparticles. The rearrangement of the atoms is a process that reduces the gas adsorption on the surface [39].

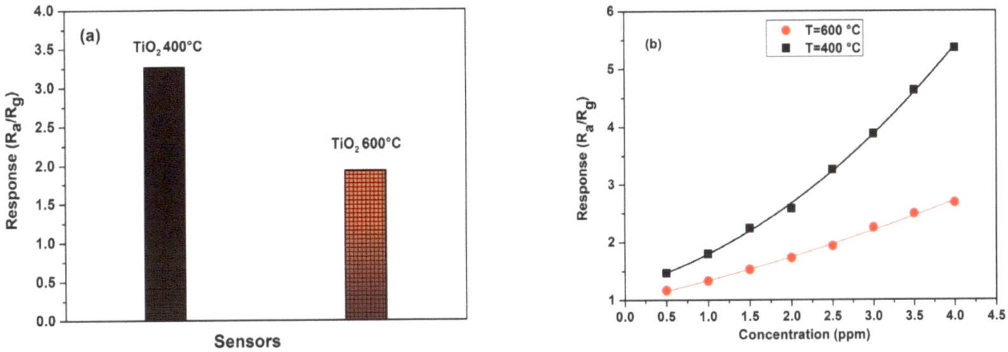

Figure 10. Sensor response of TiO_2 (400 °C) and TiO_2 (600 °C) (**a**) at a H_2S concentration of 2.5 ppm; (**b**) response versus H_2S concentration at a temperature of 350 °C.

Apart the above structural considerations, the effect of the different phases (anatase and rutile) on the sensing response cannot be excluded. The advantages of using anatase or rutile in gas sensing have been discussed for a long time and depend on many variables such as the target gas and operating temperature. For example, Zakrzewska and Radecka discovered that rutile-dominated TiO_2 nanomaterials exhibited higher sensitivity towards hydrogen than those with the prevailing anatase [40]. This phenomenon could be accounted for by band alignment and electron transfer from rutile to anatase to facilitate oxygen pre-adsorption. On the contrary, by using density functional theory (DFT) to study the adsorption and reaction of H_2S on TiO_2 anatase (101) and rutile (110) surfaces, it has been demonstrated that the adsorption and dissociation of hydrogen sulfide at the TiO_2 anatase surface require a lower energy barrier compared to at the anatase surface [41]. This latter finding indicates that the presence of anatase at a high concentration (100%) is a factor to take into account when considering the sensor response enhancement of H_2S.

The sensing performance of the TiO_2 (400 °C) sensor was further investigated by exposing the fabricated sensors to different concentrations of H_2S gas. Figure 11a shows the plotted gas response to H_2S gas sensed by the TiO_2 (400 °C) sensor at an operating temperature of 350 °C. The response amplitude of the sensor increases with H_2S concentration in the range of 0.5 to 4 ppm. Moreover, in Figure 11b, it can be observed that the response increases almost linearly with the concentration. The sensor is also sufficiently sensitive at the lowest concentration (0.5 ppm) of H_2S tested. This result suggests that it can be promising for the sensing of hydrogen sulfide in practical applications.

The response and recovery times are two very important characteristics of gas sensors in practical applications. The response and recovery times of the TiO_2 (400 °C) sensor as a function of various H_2S concentrations at the operating temperature of 350 °C are presented in Figure 12. The measured response and recovery times are short. Indeed, in the H_2S concentration range of 0.5 to 4 ppm, the response time is slower than 10 s and the recovery time is slower than 31 s.

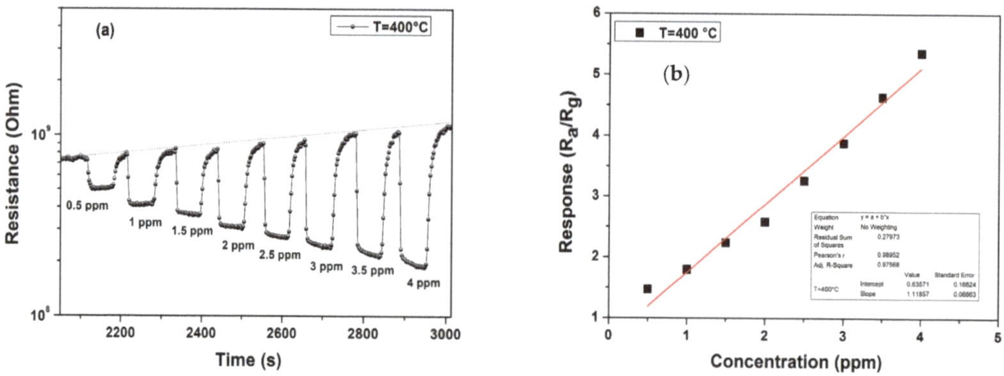

Figure 11. (**a**) Resistance vs. time for different concentrations and (**b**) response vs. concentration of the TiO_2-NP sensor at an operating temperature of 350 °C.

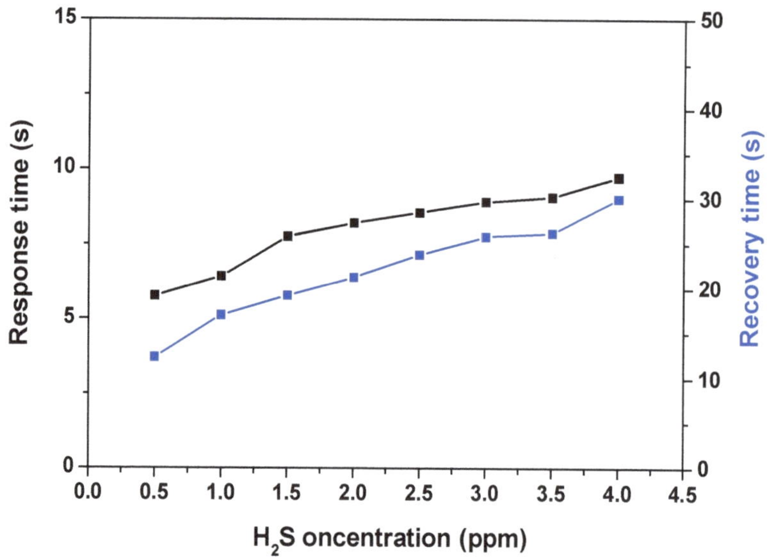

Figure 12. Response and recovery time vs. H_2S concentrations of TiO_2-NPs at an operating temperature of 350 °C.

The gas sensing selectivity of the TiO_2 (400 °C) sensor against different gases, i.e., nitrogen dioxide, carbon monoxide, and hydrogen, was also studied (Figure 13). The selectivity patterns indicate that, for all the interfering gases, it presents low responses, and therefore exhibits excellent selectivity to H_2S.

Repeatability is an important indicator for measuring the reliability of the sensor response and the stability of the sensor. Figure 14 shows the reproducibility of the sensor when exposed to three consecutive pulses of 4 ppm of H_2S gas at the working temperature of 350 °C. It is observed that the response and recovery characteristics are almost reproducible.

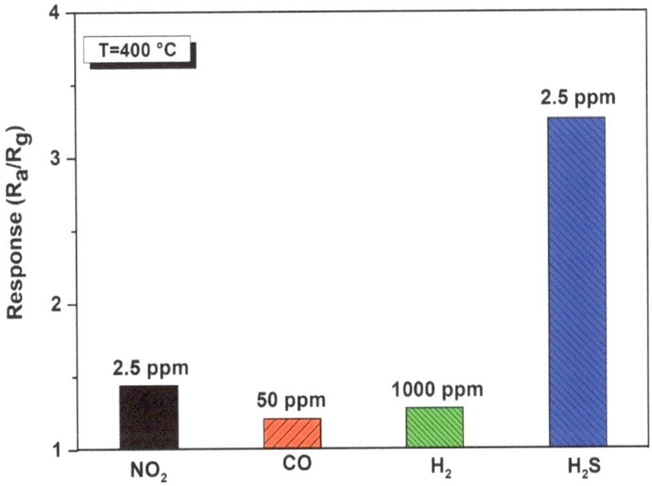

Figure 13. Selectivity pattern of the TiO_2-NPs sensor at an operating temperature of 350 °C.

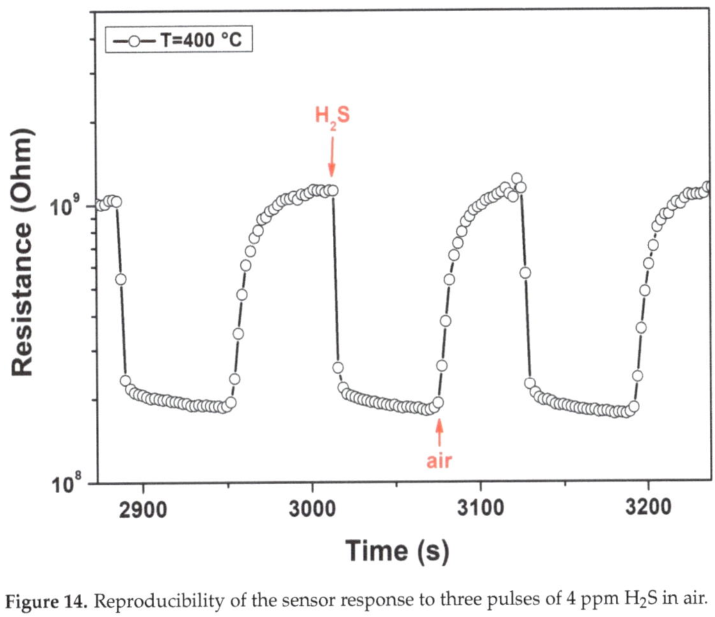

Figure 14. Reproducibility of the sensor response to three pulses of 4 ppm H_2S in air.

3.3. Gas Sensing Mechanism

The gas sensing mechanism of the developed sensor is explained by the change in the conductance of the semiconducting TiO_2 sensing layer. Herein, the conductivity of the sensor is modified by the phenomenon of target gas adsorption-desorption, which causes variations in the electrical conductivity of the sensing layer. The kinetics of gas adsorption and desorption are critical to the performance of gas sensors, influencing their sensitivity, response time, and recovery time. The adsorption process is enhanced by a high surface area, optimal pore size, and high surface energy. Materials such as nanostructured titanium dioxide (TiO_2) are ideal due to their large surface area and chemical stability, providing numerous active sites for the adsorption of gas molecules. Desorption depends on factors such as binding energy and temperature. Strong interactions between gas molecules and the sensor surface can slow down desorption, resulting in longer recovery times. Increasing the temperature can facilitate faster desorption by providing the necessary energy to overcome binding forces. Sensor design must balance these kinetics to achieve rapid detection and quick recovery. Enhancing selectivity involves modifying the sensor material, such as doping TiO_2 with elements such as silver or platinum, to tailor the interaction strength with specific gases. These modifications optimize both adsorption and desorption rates, ensuring that the sensor performs reliably and efficiently. Understanding and optimizing these kinetic processes are then essential for developing high-performance gas sensors capable of detecting hazardous gases accurately and swiftly.

When the sensor is exposed to air, oxygen molecules are adsorbed on the surface and extract electrons from the conduction band [39,40]. Oxygen molecules are adsorbed on the active sites of the rough grain surface as (O_2^-, O^-, and O^{2-}) by trapping electrons from the conduction band, which results in an electron depletion region [42–44].

When the TiO_2-based sensor was exposed to H_2S, it reacted with adsorbed oxygen species and released the trapped electrons back to the TiO_2 (see Figure 15). Hence, the high sensitivity to hydrogen sulfide can be attributed to its low dissociation energy compared to other gases on TiO_2 anatase, enabling it to readily react with the adsorbed oxygen [45–48]) to form SO_2 and H_2O, as seen in Equation (3). This causes the bulk release of a large concentration of free electrons, which results in the narrowing of the electron depletion region.

$$2H_2S\ (g) + 3O_2^-(ads) \rightarrow 2H_2O\ (g) + 2SO_2(g) + 3e^- \quad (3)$$

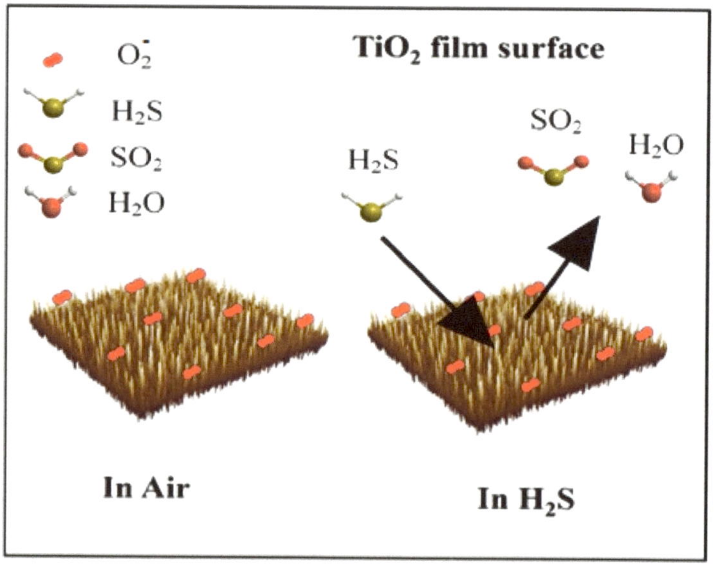

Figure 15. Gas sensing mechanism of TiO_2 (400 °C) in the presence of H_2S gas.

The good performances of our sensor compared to those reported in the literature are reported in Table 3. Remarkably, the sensor response is very high, considering the low H_2S concentration tested in our case, as well as being faster.

Table 3. Comparison of the sensing performances of the TiO_2-based sensor with other sensors reported in the literature.

Material	H_2S (ppm)	Response (Ra/Rg)	Temperature (°C)	Response Time (s)	Recovery Time (s)	Reference
TiO_2 nanoplates (Anatase)	10	4.8	300	10	-	[46]
TiO_2 nanotube (Anatase)	50	26	300	22	6	[6]
TiO_2 nanowires (Rutile)	80	11	140	-	-	[47]
TiO_2-Al_2O_3 (Rutile)	1000	38.7	650	390	480	[49]
Ag-doped TiO_2 nanofiber	100	8.5	350	-	-	[44]
CuO doped TiO_2 nanoparticle (Anatase)	50	1.78	Room temperature	14	22	[50]
TiO_2 nanoparticles (Anatase)	2.5	3.3	350	8	23	This work

Finally, we planned new tests for the evaluation of further characteristics regarding our sensor. Among these, humidity is well-known to influence the response of resistive sensors. However, the exact behavior is not predictable and various findings have been reported, depending on the metal oxide, the target gas, the operating temperature, and the humidity value [51]. Therefore, tests carried out in different humidity conditions have been planned for the near future.

4. Conclusions

In summary, titanium dioxide nanoparticles were prepared by a modified sol-gel technique and then annealed at different air temperatures (400 °C and 600 °C). The synthesized samples exhibit a polycrystalline structure, characterized by grains of pure anatase on TiO_2 (400 °C) and with about 25% of rutile in TiO_2 (600 °C). Subsequently, the samples were used to fabricate gas sensors for H_2S. The TiO_2 (400 °C)-based gas sensor was found to display the best performances in terms of high response, fast response/recovery, and good selectivity when operating at 350 °C. The obtained results confirm that the TiO_2 sample treated at 400 °C can be considered very promising for the detection of low H_2S concentrations and suitable for a variety of environmental applications.

Author Contributions: Investigation, Y.A., M.M., E.F. and C.C.; Writing—original draft, Y.A.; Writing—review & editing, G.N.; Supervision, N.B., L.E.M. and G.N. All authors have read and agreed to the published version of the manuscript.

Funding: This work has been partially funded by European Union (NextGeneration EU), through the MUR-PNRR project SAMOTHRACE (ECS00000022).

Institutional Review Board Statement: Not applicable.

Informed Consent Statement: Not applicable.

Data Availability Statement: The original contributions presented in the study are included in the article, further inquiries can be directed to the corresponding author.

Conflicts of Interest: The authors declare no conflict of interest.

References

1. Tong, X.; Shen, W.; Zhang, X.; Corriou, J.-P.; Xi, H. Synthesis and Density Functional Theory Study of Free-Standing Fe-Doped TiO_2 Nanotube Array Film for H2S Gas Sensing Properties at Low Temperature. *J. Alloys Compd.* **2020**, *832*, 155015. [CrossRef]
2. Deng, Q.; Yin, J.; Wu, X.; Zhang, T.; Wang, H.; Liu, M. Research Advances of Prevention and Control of Hydrogen Sulfide in Coal Mines. *Sci. World J.* **2019**, *2019*, e8719260. [CrossRef]
3. Jameh, A.A.; Mohammadi, T.; Bakhtiari, O.; Mahdyarfar, M. Synthesis and Modification of Zeolitic Imidazolate Framework (ZIF-8) Nanoparticles as Highly Efficient Adsorbent for H2S and CO_2 Removal from Natural Gas. *J. Environ. Chem. Eng.* **2019**, *7*, 103058. [CrossRef]
4. Lei, Y.; Wang, K.-P.; Chen, S.; Zhang, Q.; Hu, Z.-Q. A Fluorescent Probe Based on Tetrahydro[5]Helicene for Highly Selective Recognition of Hydrogen Sulfide. *Spectrochim. Acta A Mol. Biomol. Spectrosc.* **2018**, *204*, 295–300. [CrossRef]
5. Panthi, S.; Manandhar, S.; Gautam, K. Hydrogen Sulfide, Nitric Oxide, and Neurodegenerative Disorders. *Transl. Neurodegener.* **2018**, *7*, 3. [CrossRef] [PubMed]
6. Tong, X.; Shen, W.; Chen, X.; Corriou, J.-P. A Fast Response and Recovery H2S Gas Sensor Based on Free-Standing TiO_2 Nanotube Array Films Prepared by One-Step Anodization Method. *Ceram. Int.* **2017**, *43*, 14200–14209. [CrossRef]
7. Jaballah, S.; Benamara, M.; Dahman, H.; Ly, A.; Lahem, D.; Debliquy, M.; Mir, L.E. Effect of Mg-Doping ZnO Nanoparticles on Detection of Low Ethanol Concentrations. *Mater. Chem. Phys.* **2020**, *255*, 123643. [CrossRef]
8. Neri, G. First Fifty Years of Chemoresistive Gas Sensors. *Chemosensors* **2015**, *3*, 1–20. [CrossRef]
9. Jaballah, S.; Benamara, M.; Dahman, H.; Lahem, D.; Debliquy, M.; El Mir, L. Formaldehyde Sensing Characteristics of Calcium-Doped Zinc Oxide Nanoparticles-Based Gas Sensor. *J. Mater. Sci. Mater. Electron.* **2020**, *31*, 8230–8239. [CrossRef]
10. Kheel, H.; Sun, G.-J.; Lee, J.K.; Lee, S.; Dwivedi, R.P.; Lee, C. Enhanced H2S Sensing Performance of TiO_2-Decorated α-Fe2O3 Nanorod Sensors. *Ceram. Int.* **2016**, *42*, 18597–18604. [CrossRef]
11. Wu, W.; Long, J.; Guo, Y.; Zu, X.; Li, S.; Xiang, X. P-CuO/n-TiO_2 Heterojunction Nanostructure-Based Surface Acoustic Wave Sensor with Strong Electric Loading Effect for Highly Sensitive H2S Gas Sensing. *Sens. Actuators B Chem.* **2023**, *394*, 134380. [CrossRef]
12. Hosseini, Z.S.; Mortezaali, A.; Iraji Zad, A.; Fardindoost, S. Sensitive and Selective Room Temperature H2S Gas Sensor Based on Au Sensitized Vertical ZnO Nanorods with Flower-like Structures. *J. Alloys Compd.* **2015**, *628*, 222–229. [CrossRef]
13. Lee, I.; Choi, S.-J.; Park, K.-M.; Lee, S.S.; Choi, S.; Kim, I.-D.; Park, C.O. The Stability, Sensitivity and Response Transients of ZnO, SnO2 and WO3 Sensors under Acetone, Toluene and H2S Environments. *Sens. Actuators B Chem.* **2014**, *197*, 300–307. [CrossRef]
14. Dalavi, D.S.; Suryavanshi, M.J.; Patil, D.S.; Mali, S.S.; Moholkar, A.V.; Kalagi, S.S.; Vanalkar, S.A.; Kang, S.R.; Kim, J.H.; Patil, P.S. Nanoporous Nickel Oxide Thin Films and Its Improved Electrochromic Performance: Effect of Thickness. *Appl. Surf. Sci.* **2011**, *257*, 2647–2656. [CrossRef]
15. Nagmani; Pravarthana, D.; Tyagi, A.; Jagadale, T.C.; Prellier, W.; Aswal, D.K. Highly Sensitive and Selective H2S Gas Sensor Based on TiO_2 Thin Films. *Appl. Surf. Sci.* **2021**, *549*, 149281. [CrossRef]
16. Jagadale, T.C.; Nagmani; Ramgir, N.S.; Prajapat, C.L.; Debnath, A.K.; Aswal, D.K.; Gupta, S.K. Attempts to Improve the H2S Sensitivity of TiO_2 Films. *AIP Conf. Proc.* **2016**, *1731*, 080021. [CrossRef]
17. Hazra, A.; Das, S.; Kanungo, J.; Sarkar, C.K.; Basu, S. Studies on a Resistive Gas Sensor Based on Sol–Gel Grown Nanocrystalline p-TiO_2 Thin Film for Fast Hydrogen Detection. *Sens. Actuators B Chem.* **2013**, *183*, 87–95. [CrossRef]
18. Jagadale, T.; Nagmani; Aswal, D.; Gupta, S. Exclusive H2S Detection with Greater Response Using Thermally Evaporated Titanium Oxide Film. *Asian J. Mater. Chem.* **2016**, *1*, 39–42. [CrossRef]
19. Alaya, Y.; Souissi, R.; Toumi, M.; Madani, M.; Mir, L.E.; Bouguila, N.; Alaya, S. Annealing Effect on the Physical Properties of TiO2 Thin Films Deposited by Spray Pyrolysis. *RSC Adv.* **2023**, *13*, 21852–21860. [CrossRef]
20. Wilson, R.L.; Simion, C.E.; Blackman, C.S.; Carmalt, C.J.; Stanoiu, A.; Di Maggio, F.; Covington, J.A. The Effect of Film Thickness on the Gas Sensing Properties of Ultra-Thin TiO_2 Films Deposited by Atomic Layer Deposition. *Sensors* **2018**, *18*, 735. [CrossRef]
21. Wang, X.; Cui, F.; Lin, J.; Ding, B.; Yu, J.; Al-Deyab, S.S. Functionalized Nanoporous TiO_2 Fibers on Quartz Crystal Microbalance Platform for Formaldehyde Sensor. *Sens. Actuators B Chem.* **2012**, *171–172*, 658–665. [CrossRef]
22. Shin, J.; Lee, G.; Choi, M.; Jang, H.; Lim, Y.; Kim, G.-S.; Nam, S.-H.; Baek, S.-H.; Song, H.-C.; Kim, J.; et al. Atomically Mixed Catalysts on a 3D Thin-Shell TiO_2 for Dual-Modal Chemical Detection and Neutralization. *J. Mater. Chem. A* **2023**, *11*, 18195–18206. [CrossRef]
23. Cho, D.; Suh, J.M.; Nam, S.-H.; Park, S.Y.; Park, M.; Lee, T.H.; Choi, K.S.; Lee, J.; Ahn, C.; Jang, H.W.; et al. Optically Activated 3D Thin-Shell TiO_2 for Super-Sensitive Chemoresistive Responses: Toward Visible Light Activation. *Adv. Sci.* **2021**, *8*, 2001883. [CrossRef] [PubMed]
24. Alaya, Y.; Chouchene, B.; Medjahdi, G.; Balan, L.; Bouguila, N.; Schneider, R. Heterostructured S-TiO_2/g-C3N4 Photocatalysts with High Visible Light Photocatalytic Activity. *Catalysts* **2024**, *14*, 226. [CrossRef]
25. El Mir, L.; Amlouk, A.; Barthou, C.; Alaya, S. Luminescence of Composites Based on Oxide Aerogels Incorporated in Silica Glass Host Matrix. *Mater. Sci. Eng. C* **2008**, *28*, 771–776. [CrossRef]
26. El Mir, L.; Amlouk, A.; Elaloui, E.; Saadoun, M.; Pierre, A.C. Preparation and Optical Characterization of Transparent, Microporous TiO_2 Xerogel Monoliths. *Mater. Sci. Eng. B* **2008**, *146*, 69–73. [CrossRef]

27. Secundino-Sánchez, O.; Diaz-Reyes, J.; Sánchez-Ramírez, J.F.; Jiménez-Pérez, J.L.; Secundino-Sánchez, O.; Diaz-Reyes, J.; Sánchez-Ramírez, J.F.; Jiménez-Pérez, J.L. Structural and Optical Characterization of the Crystalline Phase Transformation of Electrospinning TiO_2 Nanofibres by High Temperatures Annealing. *Rev. Mex. Fís.* **2019**, *65*, 459–467. [CrossRef]
28. Ma, S.; Ye, T.; Wu, T.; Wang, Z.; Wang, Z.; Ramakrishna, S.; Vijila, C.; Wei, L. Hollow Rice Grain-Shaped TiO_2 Nanostructures for High-Efficiency and Large-Area Perovskite Solar Cells. *Sol. Energy Mater. Sol. Cells* **2019**, *191*, 389–398. [CrossRef]
29. Madani, M.; Omri, K.; Echabaane, M.; Gouadria, S.; Alharbi, F. Elaboration and Influence of Annealing Process on Physico-Chemical Properties of Silica–Titania Nanocomposites. *Appl. Phys. A* **2021**, *127*, 435. [CrossRef]
30. Chaurasiya, N.; Singh, A.; Kumar, K.; Chandra Yadav, B.; Kumar Yadawa, P.; Kumar Singh, S.; Kumar Dey, K. A Highly Sensitive Room Temperature Liquefied Petroleum Gas (LPG) Sensor with Fast Response Based on a Titanium Dioxide (TiO_2)–Reduced Graphene Oxide (r-GO) Composite. *Sens. Diagn.* **2023**, *2*, 1215–1227. [CrossRef]
31. Sen, S.K.; Paul, T.C.; Dutta, S.; Matin, M.A.; Islam, M.F.; Hakim, M.A. Effect of Gamma (γ-) Irradiation on the Structural, Morphological, Optical and Electrical Properties of Spray Pyrolysis-Deposited h-MoO_3 Thin Films. *Surf. Interfaces* **2019**, *17*, 100377. [CrossRef]
32. Rab, N.; Chong, F.K.; Mohamed, H.I.; Lim, W.H. Preparation of TiO_2 Nanoparticles by Hydrolysis of $TiCl_4$ Using Water and Glycerol Solvent System. *J. Phys. Conf. Ser.* **2018**, *1123*, 012065. [CrossRef]
33. Devanand Venkatasubbu, G.; Ramasamy, S.; Ramakrishnan, V.; Kumar, J. Folate Targeted PEGylated Titanium Dioxide Nanoparticles as a Nanocarrier for Targeted Paclitaxel Drug Delivery. *Adv. Powder Technol.* **2013**, *24*, 947–954. [CrossRef]
34. Naghibi, S.; Madaah Hosseini, H.R.; Faghihi Sani, M.A.; Shokrgozar, M.A.; Mehrjoo, M. Mortality Response of Folate Receptor-Activated, PEG–Functionalized TiO_2 Nanoparticles for Doxorubicin Loading with and without Ultraviolet Irradiation. *Ceram. Int.* **2014**, *40*, 5481–5488. [CrossRef]
35. Sánchez-Rodríguez, D.; Méndez Medrano, M.G.; Remita, H.; Escobar-Barrios, V. Photocatalytic Properties of BiOCl-TiO_2 Composites for Phenol Photodegradation. *J. Environ. Chem. Eng.* **2018**, *6*, 1601–1612. [CrossRef]
36. Hoseini, S.N.; Pirzaman, A.K.; Aroon, M.A.; Pirbazari, A.E. Photocatalytic Degradation of 2,4-Dichlorophenol by Co-Doped TiO_2 (Co/TiO_2) Nanoparticles and Co/TiO_2 Containing Mixed Matrix Membranes. *J. Water Process Eng.* **2017**, *17*, 124–134. [CrossRef]
37. Hannula, M.; Ali-Löytty, K.; Lahtonen, H.; Sarlin, E.; Saari, J.; Valden, M. Improved Stability of Atomic Layer Deposited Amorphous TiO_2 Photoelectrode Coatings by Thermally Induced Oxygen Defects. *Chem. Mater.* **2018**, *30*, 1199–1208. [CrossRef] [PubMed]
38. Pentyala, N.; Guduru, R.K.; Shnerpunas, E.M.; Mohanty, P.S. Synthesis of Ultrafine Single Crystals and Nanostructured Coatings of Indium Oxide from Solution Precursor. *Appl. Surf. Sci.* **2011**, *257*, 6850–6857. [CrossRef]
39. Al-Jumaili, B.E.; Rzaij, J.M.; Ibraheam, A.S. Nanoparticles of CuO Thin Films for Room Temperature NO_2 Gas Detection: Annealing Time Effect. *Mater. Today Proc.* **2021**, *42*, 2603–2608. [CrossRef]
40. Zakrzewska, K.; Radecka, M. TiO_2-Based Nanomaterials for Gas Sensing—Influence of Anatase and Rutile Contributions. *Nanoscale Res. Lett.* **2017**, *12*, 89. [CrossRef]
41. Huang, W.-F.; Chen, H.-T.; Lin, M.C. Density Functional Theory Study of the Adsorption and Reaction of H2S on TiO_2 Rutile (110) and Anatase (101) Surfaces. *J. Phys. Chem. C* **2009**, *113*, 20411–20420. [CrossRef]
42. Ji, H.; Zeng, W.; Li, Y. Gas Sensing Mechanisms of Metal Oxide Semiconductors: A Focus Review. *Nanoscale* **2019**, *11*, 22664–22684. [CrossRef]
43. Xiang, C.; She, Z.; Zou, Y.; Cheng, J.; Chu, H.; Qiu, S.; Zhang, H.; Sun, L.; Xu, F. A Room-Temperature Hydrogen Sensor Based on Pd Nanoparticles Doped TiO_2 Nanotubes. *Ceram. Int.* **2014**, *40*, 16343–16348. [CrossRef]
44. Franke, M.E.; Koplin, T.J.; Simon, U. Metal and Metal Oxide Nanoparticles in Chemiresistors: Does the Nanoscale Matter? *Small* **2006**, *2*, 36–50. [CrossRef]
45. Bai, J.; Zhou, B. Titanium Dioxide Nanomaterials for Sensor Applications. *Chem. Rev.* **2014**, *114*, 10131–10176. [CrossRef]
46. Tong, X.; Shen, W.; Chen, X. Enhanced H2S Sensing Performance of Cobalt Doped Free-Standing TiO_2 Nanotube Array Film and Theoretical Simulation Based on Density Functional Theory. *Appl. Surf. Sci.* **2019**, *469*, 414–422. [CrossRef]
47. Ma, S.; Jia, J.; Tian, Y.; Cao, L.; Shi, S.; Li, X.; Wang, X. Improved H2S Sensing Properties of Ag/TiO_2 Nanofibers. *Ceram. Int.* **2016**, *42*, 2041–2044. [CrossRef]
48. Lee, A.P.; Reedy, B.J. Temperature Modulation in Semiconductor Gas Sensing. *Sens. Actuators B Chem.* **1999**, *60*, 35–42. [CrossRef]
49. Arafat, M.M.; Haseeb, A.S.M.A.; Akbar, S.A.; Quadir, M.Z. In-Situ Fabricated Gas Sensors Based on One Dimensional Core-Shell TiO_2-Al_2O_3 Nanostructures. *Sens. Actuators B Chem.* **2017**, *238*, 972–984. [CrossRef]
50. Guo, W.; Feng, Q.; Tao, Y.; Zheng, L.; Han, Z.; Ma, J. Systematic Investigation on the Gas-Sensing Performance of TiO_2 Nanoplate Sensors for Enhanced Detection on Toxic Gases. *Mater. Res. Bull.* **2016**, *73*, 302–307. [CrossRef]
51. Shooshtari, M.; Salehi, A.; Vollebregt, S. Effect of temperature and humidity on the sensing performance of TiO2 nanowire-based ethanol vapor sensors. *Nanotechnology* **2021**, *32*, 325501. [CrossRef]

MDPI AG
Grosspeteranlage 5
4052 Basel
Switzerland
Tel.: +41 61 683 77 34

Materials Editorial Office
E-mail: materials@mdpi.com
www.mdpi.com/journal/materials

www.ingramcontent.com/pod-product-compliance
Lightning Source LLC
Chambersburg PA
CBHW041425210926
51850CB00013B/3357

9783725827985